Ecology and natural history of tropical bees

CAMBRIDGE TROPICAL BIOLOGY SERIES

EDITORS:
Peter S. Ashton *Arnold Arboretum, Harvard University*
Stephen P. Hubbell *Princeton University*
Daniel H. Janzen *University of Pennsylvania*
Peter H. Raven *Missouri Botanical Garden*
P. B. Tomlinson *Harvard Forest, Harvard University*

Ecology and natural history of tropical bees

DAVID W. ROUBIK
Smithsonian Tropical Research Institute

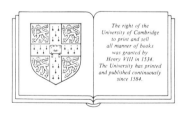

CAMBRIDGE UNIVERSITY PRESS
Cambridge
New York   Port Chester   Melbourne   Sydney

Published by the Press Syndicate of the University of Cambridge
The Pitt Building, Trumpington Street, Cambridge CB2 1RP
32 East 57th Street, New York, NY 10022, USA
10 Stamford Road, Oakleigh, Melbourne 3166, Australia

© Cambridge University Press 1989

First published 1989

Printed in the United States of America

Library of Congress Cataloging-in-Publication Data
Roubik, David. W. (David Ward), 1951–
Ecology and natural history of tropical bees
(Cambridge Tropical Biology series)
Bibliography: p.
Includes index.
1. Bees – Tropics.   2. Insects – Tropics.
I. Title.   II. Series.
QL567.96.R68   1989   595.79′9′0452623   87–23837

British Library Cataloguing-in-Publication Data
Roubik, David. W.
Ecology and natural history of tropical bees –
(Cambridge Tropical Biology series)
1. Bees – Tropics – Ecology
I. Title
595.79′9045   QL568.A6

ISBN 0-521-26236-4

Contents

	Acknowledgments	*page* ix
1	**Introduction**	1
1.1	Approaches to tropical bee biology	1
1.2	Diversity of tropical bees	4
	1.2.1 Natural groups and current distribution	4
	1.2.2 Biogeography and ages of bee groups	9
	1.2.3 Social and solitary life cycles	15
2	**Foraging and pollination**	25
2.1	Resources gathered by bees	25
	2.1.1 Sap	26
	2.1.2 Resin	27
	2.1.3 Floral lipids	29
	2.1.4 Fragrances	31
	2.1.5 Nectar	32
	2.1.6 Pollen	37
	2.1.7 Waxes	41
	2.1.8 Other resources	41
2.2	Mechanisms of resource collection	43
	2.2.1 Pollen capture	43
	2.2.2 Pollen cleaning, manipulation, and storage	51
	2.2.3 Floral lipid collection	53
	2.2.4 Retention or loss of traits	56
	2.2.5 Evolution of gathering and unloading	61
	2.2.6 Nectar harvest	65
	2.2.7 Feeding rate and efficiency	68
2.3	Foraging and flight activity	71
	2.3.1 General considerations	71
	2.3.2 Navigation and chemical ecology	73
	2.3.3 Flight range	82
	2.3.4 Thermoregulation, metabolism, and response to weather	91
	2.3.5 Social foraging	99

	2.3.6	Foraging styles and syndromes	107
	2.3.7	Foraging patterns and optimality	113
	2.3.8	Predation and mimicry	123
	2.3.9	Activity patterns of social and solitary bees	131
2.4	Pollination ecology		138
	2.4.1	General considerations	138
	2.4.2	What makes a pollinator effective?	143
	2.4.3	Specialization and coevolution	149
	2.4.4	Bees as persistent plant enemies	154

3 Nesting and reproductive biology — 161

3.1	Bee nests		161
	3.1.1	General considerations	162
	3.1.2	Physical conditions, bee cells, and life cycles	176
	3.1.3	Nest sites	185
	3.1.4	Nest construction	199
3.2	Natural enemies, associates, and defense		206
	3.2.1	General considerations	206
	3.2.2	Guard bees and kin recognition	208
	3.2.3	Colony defense, pheromones, and predation	214
	3.2.4	Parasitism, mutualists, and commensals	225
	3.2.5	Pathogens and related symbionts	229
	3.2.6	Insect natural enemies of bees	233
	3.2.7	Mites and nematodes	244
	3.2.8	Parasitic bees	249
3.3	Mating and brood production		267
	3.3.1	Selecting a mate	267
	3.3.2	Larval development and nutrition	280
	3.3.3	Reproduction of solitary bees	284
	3.3.4	Reproduction by queens and colonies	288
	3.3.5	Genetic variability of bees	302
	3.3.6	Chromosomal variation, sex determination, and sex ratios	306

4 Community ecology — 313

4.1	Bee seasonality, abundance, and flower preference		315
	4.1.1	Community studies of floral choice	317
	4.1.2	Long-term monitoring studies	327
	4.1.3	How stable are bee populations?	336
	4.1.4	In-depth studies of euglossine bees	341
4.2	Composition of bee assemblages		347
	4.2.1	The neotropics	348

	4.2.2	The paleotropics	349
	4.2.3	Are bees less diverse in the tropics?	350
4.3	Roles of bees in communities		351
	4.3.1	General considerations	351
	4.3.2	Community structure	353
	4.3.3	Ecology of Africanized honeybees in tropical America	357
	4.3.4	Tropical beekeeping	366
	4.3.5	Why are so many highly eusocial bees tropical?	380

Appendixes

A	Extant families, subfamilies, tribes, genera, and subgenera of the Apoidea: a partial checklist	390
B	Illustrated tropical bee genera, arranged by family and geographic area	398
	References	421
	Subject index	487
	Taxonomic index	504

Acknowledgments

For allowing me to draw on their support both in the forest and out, I must acknowledge with some profound feelings the three branches of my family in the United States, Panama, and Costa Rica. Parts of the manuscript benefited from the extensive comments, notes, literature, or timely corrections provided by João Camargo, Thomas Seeley, James Cane, Warwick Kerr, George Eickwort, Philip Torchio, Stephen Buchmann, Shôichi Sakagami, Brian Smith, Allison Snow, Jerome Rozen, Jr., Wu Yanru, Radclyffe Roberts, Edward Southwick, William Wcislo, Keith Waddington, Frank Parker, Friedrich Ruttner, James Ackerman, David Inouye, Fred Dyer, Gerhard Gottsberger, Benjamin Underwood, Marinus Sommeijer, Gordon Frankie, Tamiji Inoue, Paulo Nogueira-Neto, Roger Darchen, Justin Schmidt, Thomas Rinderer, Anita Collins, Ronald McGinley, Alvaro Wille, Christopher Starr, E. Gorton Linsley, and Suzanne Batra. Their contributions are deeply appreciated, and decidedly improved the development of Chapters 2 and 3. For contributing their time and guidance or for providing less direct types of inspiration, I would like to thank especially Charles D. Michener and Daniel H. Janzen, both of whom responded ably to early drafts of some chapters.

For their impeccable artistic renderings I am most grateful to Sally Bensusen, João Camargo, Arlee Montalvo, Dieter Wittmann, Shôichi Sakagami, and Charles D. Michener, and, for their excellent photographic works, to Sean Morris (Oxford Scientific Films), Stephen Buchmann, Gustl Anzenberger, Carl Hansen, George Eickwort, James Marden, Adam Messer, Scott Armbruster, and Victor Kranz.

The contingent of biologists and collaborators to whom I wish to dedicate this book include those who have contributed much to the field work on tropical bees in their natural environment. I extend a debt of gratitude to the above colleagues and to Carlos Vergara, Thomas Kursar, Martin Aluja, Robert Carlson, Francisco Peralta, Norris Williams, Francisco Reyes, Richard B. Selander, David Carlson, Mercedes Delfinado-Baker, Edward Baker, Rollin Coville, Arlee Montalvo, Juan Barria, Hindrik Wolda, Nicolas Koeniger, Bonifacio de León, Lyson Phiri, Isaac Kigatiira, Kim Steiner, Tamiji Inoue, Dieter Wittmann, Siti Salmah, Vichian Hengsawad, Enrique Moreno, N. Michele Holbrook, Pongtep Akratanakul, German Parra, Guiomar Nates, Michael Holbrook, Ester Jaén, Robert Brooks, Douglas Yanega, Curtis Gentry, Ricardo Ayala, Stephen Bullock, Wilson Devia, Christine Copenhaver, and Susanne Renner. Onward tropical biology!

Acknowledgments

The museum material needed to complete the appendix illustrations of bee genera was made available through collections of the Smithsonian Institution or generously loaned to me by Charles D. Michener from the Snow Entomological Museum of the University of Kansas, Roy Snelling from the Natural History Museum of Los Angeles County, and Vichian Hengsawad from the Entomological Museum of the Department of Entomology, Chiang Mai University, Thailand. I thank David Wahl, Arnold Menke, Eric Grissell, Robert Carlson, and Michael Schauff for their help with wasp taxonomy.

1 Introduction

1.1 Approaches to tropical bee biology

Information gathered from the remaining intact tropical habitats might seem to deserve special treatment, for such data can be thought to represent precise, finely tuned biological interrelationships. Their evolutionary reason for existence might be revealed simply in their current ecology. Tropical data in general are often deemed more significant than comparable data taken in other habitats. Like the first-mentioned bias, this belief can encourage creative speculation, but sometimes it diminishes the accuracy with which the actual setting of a field study is examined or described. For example, tropical secondary growth that returns after the forest is cut away can be impressively green and complex. Yet the largest tree may be only 60 years old, whereas the original forest most likely contained trees exceeding 300 years in age and held several times as many species of plants and animals (Richards 1980; Myers 1984; Whitmore 1984). On a geological time scale the importance of historical factors is further magnified. Most of the Amazon Basin was the bottom of a lake some three million years ago, and some of its vegetation has on several occasions since that time been radically different from what is found there now (Leyden 1984; Absy 1985; Sanford et al. 1985; Graham 1986). The last author shows that although past climatic changes were not severe in lowland tropical areas, a vast portion of these areas became drier during glaciations. These periods constitute about 90% of the past three million years. We are still relatively ignorant of what effect such large-scale changes have on species richness and biological diversity (Mares 1986).

Despite such potential complexities, tropical studies readily uncover valuable first glimpses of biological phenomena as well as some of the interactions that produce them. Whether or not a study is replicable or the results adequately understood, if the original observations were taken correctly, they should be recorded for posterity. Tropical biology is full of time lags or gaps between original observations, subsequent research, and satisfactory explanation or conceptualization. Perhaps more often than in the realms of temperate-zone biology, the tropics offer scant opportunities to follow up initial studies. One might say that the chances of seeing something interesting are quite good; a tremendous number of species continuously interact. However, the chances of seeing a particular event may be quite remote. Furthermore, what is seen in nature seldom is easily taken into the laboratory or growing house. As a result, the published sample of tropical biology is often descriptive, or the context in which the data were gathered receives minimal attention. Although this may not diminish the value of studies primarily concerned with physiology, genetics, or certain aspects of behavior, their eventual integration within a broader framework of natural history, ecology, and evolution is hampered. The melding of tropical natural history and evolutionary biology presents philosophical as well as technical challenges (Lloyd 1984; Thornhill 1984). Good field data from natural tropical habitats are often expensive and difficult to gather. After the usual sacrifices required by such studies and the exhiliration of working in one of the most exciting biological worlds, it may seem inappropriate to offer conclusions far less impressive than the study site from which they came.

The inevitable limitations of pioneering efforts to introduce ecology and evolutionary insight into studies of tropical bees stand out in early attempts to interpret foraging behavior. Setting the stage for these first steps was the extraordinary research of Nobel laureate Karl von Frisch and his European students concerning behavioral mechanisms used by honeybees for communicating resource location and quality to nest mates. This led directly to the comparison of honeybees to another group of colonial, honey-making bees, the stingless bees (Lindauer 1961). Studies in Brazil had demonstrated the odor-droplet-following behavior of stingless bees as they were recruited by a nest mate. However, honeybees failed to direct nest mates to dishes of sugar water placed high above the ground. Since the abstract dance language of the honeybee emphasizes just two spatial variables, the distance and direction of resources from the nest, a simple theory was advanced to explain the evolution of resource exploitation by honeybees and stingless bees. The notion was that the stingless bees perform with less navigational error by following traillike chemical deposits through the complex labyrinth of the tropical forest. In contrast, the temperate-zone habitat of honeybees does not require such precision in three dimensions, and their communication behavior functions adequately using only the two dimensions of foraging distance and direction. Such ad hoc evolutionary ecology did not consider that both bee groups arose and predominate in tropical forests, that most of their foraging occurs in tree canopies (which have strong visual and olfactory characteristics lacking in dishes of sugar water), and that few stingless bees appear to forage primarily by using odor trails. Gould and Towne (1987) advance a discussion of some spatial ecological variables that could influence the evolution of honeybee foraging behavior. Again, the portion of their argument corresponding to the distribution and abundance patterns of available resources remains for future corroboration.

A strictly adaptationist interpretation of biology has met with strong and sometimes justifiable criticism (Gould and Lewontin 1979; Clutton-Brock and Harvey

1984). Such criticism does serve to sharpen scientific rigor, and it is perhaps most valuable to tropical field biology, whose studies seldom take place within a tidy laboratory. Fieldwork can lead to inference, and a carefully chosen experiment might help to test an idea. However, the appeal of well-designed experiments has at times led to their substitution for more basic work, and their artificial nature clearly illustrates the limits of attempting to explain the selective basis of an observed trait. Studies of foraging bees at dishes of sugar syrup, to which honeybees arrive in large numbers, led to numerous experiments and the discussion of colonial territoriality mediated by aggressive competition. A striking behavior of bees from different colonies in this setting is intense fighting (Kalmus 1954). At flowers, however, even moderately aggressive behavior is very rare or completely absent. To judge from other aspects of the bee's natural history, such fights fit patterns of nest robbing, but they have little or no relevance to most foraging behavior (Chapter 2).

The accumulating details of bee biology, both tropical and temperate, and of tropical ecology in general (summaries in Janzen 1983, 1986; Deshmukh 1986) are permitting steady growth in the amount of experiment and theory applied to research in the tropics. This will augment the fund of natural history information that may eventually provide answers for questions yet to be asked, and it will also direct future studies to resolve questions for which there are no hard facts. Growing raw knowledge, sophistication in approach, and collaboration are beginning to transform the type of natural history studies produced in the tropics. Even though it is not possible to gain detailed insight into the lives of more than a small portion of tropical species, a select few with outstanding research potential, such as the honeybees and stingless bees, may provide broad insights that will have many uses. The biology of tropical bees in general has the potential to advance beyond the pitfalls of storytelling or lengthy descriptions with little or no analysis. It is a field well on its way to developing into a mature and interactive science – one that can not only offer unique insights but that also lends itself to formulating and answering meaningful questions.

Why should tropical bees receive attention as subjects ripe for ecological and evolutionary treatment? One reason is that, from the purely historical perspective, all major bee lineages and the flowering plants arose, to a large extent concurrently, in the tropical habitats of the early Cretaceous (Raven and Axelrod 1974; Crepet 1983; Michener and Grimaldi 1988). Another reason is that bees at flowers and sometimes at nests are not difficult to locate and can be studied with minimal equipment. Their accessibility and the potential for direct study of ecological processes invite "explanatory" biology, an appropriate synonym and goal for evolutionary ecology. When we step into a tropical forest and begin to take note of the interactions between bees and the flowering plants, it can be likened to looking through a window on events occurring many millions of years ago. The original cast of characters has long disappeared, but some of the major interactions they participated in still take place. In terms of selection pressures, the evolution of symbioses, and coadaptations of plant and pollinator, we again witness the rise of the angiosperms.

Understanding the interactions of bees with plants, especially the vital ties that determine reproductive success and survival, is certainly one of the areas in which tropical bee ecology has much to offer. The results of bee interactions with natural enemies, perhaps including our own tropical ancestors who were keenly interested in certain bees (Chapter 4), are abundantly clear from tropical examples of bee chemical ecology, nest architecture, and behavior. Moreover, virtually all elements of tropical bee ecology relate to groups and species currently displaying widespread economic importance in temperate areas. Such bees are increasingly well known (the common honeybee, bumblebees, carpenter bees, and leaf-cutter bees), but they were probably tropical in origin (Michener 1974a; Iwata 1976; Itô 1985). There are very few studies on any such bees in the tropics and even fewer in normal habitat (Sakagami 1976; Plowright and Laverty 1984; Seeley 1985). In short, the study of tropical bees provides essential insight into the ecology and evolution of major groups of organisms in both temperate and tropical regions. Some examples emphasized in this book might seem out of place because they refer to temperate-zone studies that lack tropical counterparts. A task of future workers will be to discover the degree to which temperate and tropical bee ecologies are alike.

1.2 Diversity of tropical bees

1.2.1 Natural groups and current distribution

No one knows exactly how many species of bees now exist, particularly in the tropics. An estimated 20,000 valid names perhaps apply to bee specimens residing in the world's museums, but the proportion of new species found in groups currently under study suggest that twice this number may be more realistic. The former estimate was offered by Friese (quoted by Malyshev 1935) when about half of the formally described species were from palearctic areas. Revisional systematic work continuously reveals a multiplicity of names for single taxa, and fieldwork continues to proffer new species and genera. Presiding over the process are the taxonomists who keep careful records of these efforts and whose work entails discovery of phylogenetic relationships among Apoidea. Fossil bees are few (Section 1.2.2), so that much of what is known of bee taxonomic diversity comes from the study of the extant fauna (summaries by Michener 1974a; Cane 1983b; Duffield, Wheeler, and Eickwort 1984; Ruttner 1988).

Groups of bees sharing close common ancestors can be assigned to supraspecific categories – basically the units most easily grasped and applied in the field. A partial list has been compiled by R. B. Roberts and other bee specialists (Appendix A). Unfortunately, ideal systems for classification do not exist, due to the variety of ways that data representing the organisms themselves can be gathered and analyzed. An operational rule, increasingly followed in order to propose classifications, is that natural groups are recognized by unique features inherited from a recent common ancestor (Wiley 1981; Gould 1985). The key word is *unique,* which means independently developed during the evolution of the lineage and subsequently found among its members from that point onward. Alternative classifications are often based upon particular combinations of characteristics that allow recognition of a group, although some of the traits within this set are found among members of other groups. Groups united by so-called primitive features are sometimes discounted, but a primitive trait cannot always be distinguished from one that is derived. Systematics is a complex practice and rarely closed to modification. Once a lineage is recognized, its groups of species are further delimited within their families, subfamilies, and tribes. Through the force of tradition and its apparent utility, the next lowest category, the genus, usually receives further formal qualification to the level of subgenus. There are about 1,000 such groups. The informal "species groups" (those with no distinctive taxonomic label, such as a genus name) are the lowest echelon of this scheme. To a limited extent, all supraspecific categories may, as a result of continued study (e.g., McGinley and Rozen 1987; Sakagami and Khoo 1987), be reshuffled, split, merged with a higher category, or made into informal groups.

Geographic races or subspecies of a bee routinely prove to be different species. For example, Moure and Kerr (1950) compiled extensive distributional information on the genus *Melipona* in South America, particularly where bees thought to be racial variations of single species were found. Their work and subsequent study (Camargo, Moure, and Roubik 1988) has revealed over 40 species where previously only 12 were recognized (Schwarz 1932). Similarly, Kimsey and Dressler (1986) list 166 species now known in the tropical apid tribe Euglossini, indicating an increase of 100% in euglossine taxa during recent decades. Even the genus of the honeybee *Apis* contains more species than the four now commonly recognized. In an extensive study by Maa (1953), 28 "species" of the genus *Apis* were proposed. Slowly this analysis has made an impact; there are probably at least seven distinct species of honeybees (Ruttner 1988; Tingek et al. 1988). If such colonial bees, which are present in relatively large numbers throughout the year, still reveal new species, the chances seem very good that even higher proportions of new taxa will be found among bees that are seasonal and scarce.

Eleven families of bees are generally recognized, only some of which are identified by derived traits setting them apart from other bee families. The number of species within families varies tremendously (Rozen 1967, 1977a; Michener 1974a, 1981, 1986; McGinley 1980; Michener and Greenberg 1980; Michener and Brooks 1984). Four small families are seen by some researchers as closely related to larger families. Thus the families of bees can be grouped as Melittidae – Ctenoplectridae; Colletidae – Stenotritidae; Andrenidae – Oxaeidae; Megachilidae – Fideliidae; Halictidae; Anthophoridae; and Apidae. At the base of these phylogenetic clusters are the bees with the longest evolutionary history. They have the fewest derived traits to separate them from sphecoid wasplike ancestors from which all bees originated (Michener 1974a; Brothers 1975). The oldest large families are the Colletidae, Andrenidae, and by some indications also the Melittidae. Although the Halictidae are by anatomical signs an old and relatively primitive group, similar to the Melittidae, their ethology is particularly varied. Some are strictly solitary, and some are primitively eusocial (Section 1.2.3). Thus ecology and behavior cannot necessarily be surmised from the morphology and apparent phylogenetic history of a bee group, even though this may often be the case (Ruttner 1988). Bees that more recently emerged as distinct families display varied biology and relatively long mouthparts for imbibing nectar, and these large families are the Megachilidae, Anthophoridae, and Apidae. It should be noted, however, that four families of short-tongued bees – Colletidae, Halictidae, Andrenidae, and Stenotritidae – have a few members with very long proboscides that match exceedingly long corollas of local flowers (Hurd and Linsley 1963; Houston 1984; Laroca and Almeida 1985). Bees of the family Ctenoplectridae, and perhaps of the families Melittidae, Andrenidae, Stenotritidae, and Oxaeidae, lie somewhere in the evolutionary middle. The order in which these groups diverged from ancestral stock and their closest taxonomic affinities are areas of active research.

Bee genera can often be recognized at a glance, particularly when the bees are alive and their behavior and other features are evident. Combined with subgenera, often quite distinctive, approximately 600 generic groups exist in the African, American, and Oriental tropics (Michener 1979). About half of the approximately 300 bee genera occurring in the tropics are depicted in Appendix B. According to Michener and McGinley (in press), genera are preferably large groups that show obvious distinctions. As one becomes familiar with these groups, subgenera can also be recognized. Some subgeneric names, such as those of stingless bees and carpenter bees, will be used often in this book without reference to the genus in which they are placed, and they can be checked through Appendix A.

The genera and subgenera depicted in Appendix B give representative samples of the forms displaying the ecology and behavior that will be described. Photographs provide a "gestalt" key but in no way allow quick species identification. Written keys, even those complete enough to function for most of the bee species known within a defined area, usually mirror the same problem. Surveys and descriptions of local fauna are often incomplete, and comprehensive identification manuals are seldom available.

Global distributions of bee groups are cogently summarized by Michener (1974a, 1979) and further condensed in Duffield et al. (1984). Michener (1979) concludes that most groups have ranges explained by dispersal across the land. For the most part these dispersal events occurred after the continental separations produced by drifting of tectonic plates from some 200 to 15 million years ago. Deviations from the distributions expected through population movement across land are analyzed in the following section. Shifting bee populations, particularly during relatively cool, dry periods of glaciation, help to explain some patterns whereas others indicate the antiquity of bee lineages and dispersal routes.

The Andrenidae are scarcely found in the tropical portion of the Old World and are poorly represented in equatorial America. Their nearest relatives are the oxaeids, found in tropical and subtropical regions of North and South America. Halictidae, Anthophoridae, Apidae, and Megachilidae are the most uniformly abundant tropical bee groups. Colletids are pantropical, and a closely related family, Stenotritidae, occurs only in Australia. The Fideliidae are closely allied with the Megachilidae and have a range restricted to Chile and Africa. Melittids are primarily bees of the Old World, having fewer tropical than temperate representatives and none in Australia. The Ctenoplectridae are also an Old World group but are tropical and subtropical.

In the paleotropics there is largely a single bee fauna, which is richest in Africa; *alpha diversity* (species number) lessens through the Oriental to the Australasian regions (Michener 1979). Australia is very rich in bees of the family Colletidae. Northern and xeric areas of India are also somewhat exceptional because palearctic bees rather than Oriental species occupy these regions. Between Southeast Asia and Papuasia the bee fauna is largely Oriental, but relatively insular bee faunas of low diversity occur in the Philippines, Madagascar, and Australia. Smaller islands having some tropical habitat – New Zealand, New Caldeonia, the Solomon Islands, and Micronesia – have a depauperate bee fauna. African bees are most diverse within subtropical to temperate areas in the southern hemisphere. This pattern also applies to South America. One possible explanation is that during glacial periods, bees of warm temperate habitats displayed wide distribution, covering many former tropical areas, and that cooler, drier temperatures predominated during those times. When the glaciers receded, some temperate bee species may have remained in the highlands of at least the drier subtropical regions, whereas tropical species spread along with the moister and warmer habitats. The combination of the two fauna probably augmented the richness of bee species in some warm temperate and subtropical continental habitats.

To what extent are tropical bees unlike the bees of other areas? There are about as many known genera and subgenera in the strictly temperate regions as in the biogeographic regions that include the tropics (Michener 1979, p. 328). As shown by Michener the tropics generally lack unique tribes and genera. The biological characteristics of species and genera seem to vary widely within each area, as will be emphasized throughout this book. Primarily tropical bees are in large part social and presumably evolved as an integral part of the tropical biota. Representation of social bees is high within the tropical fauna as a whole.

Adaptive radiation by the ancient bee fauna in both temperate and tropical zones is reflected among the higher taxonomic categories of bees. From a tabular summary of the world distribution of bee tribes and subfamilies (table 3 in Michener 1979), of which 36 exist in regions that are largely tropical (e.g., Africa, the neotropics, northern Australia, and New Guinea), 31 also are found in nearby temperate areas. Another four of the tropical groups are barely represented in the temperate zone. Largely tropical are certain tribes within the Nomadinae (Halictidae), the Ctenoplectridae, Exomalopsini (Anthophoridae), Augochlorini (Halictidae), Xylocopini, Centridini (Anthophoridae), Meliponinae, Apinae, and Euglossini (Apidae), and some Megachilidae. Of the world's major biogeographic regions, the largest share of genera and subgenera is neotropical, where disproportionately rich groups are the Apidae and Exomalopsini. The Centridini (Anthophoridae) and Euglossini are confined to the American tropics and subtropical areas (Kerr and Maule 1964; Sakagami, Laroca, and Moure 1967).

Bee species richness in the tropics is similar to or even less than that in many temperate regions, as indicated by comparing the number of species inhabiting warm and dry temperate areas with the number of species in the few tropical moist forests that have been studied (Michener 1979). In Chapter 4, the types of bees involved will be reviewed in some detail. Their peculiar characteristics may in large part contribute to conservative rates of bee species formation in the tropics. The contrast between the bees and many other groups of organisms in this regard, and particularly the relative richness of species seen in tropical plants compared with temperate flora, underline the importance of such details.

A broad, qualitative difference among tropical and temperate bee groups has been indicated by Michener (1979). Many of the groups mentioned above as largely tropical are bees that seldom nest in the soil. Their aerial nesting habits are described in Chapter 3. Although tropical and temperate bees make waterproof nests and often nest in the ground (Cane 1981; Eickwort, Matthews, and Carpenter 1981; Cane and Carlson 1984; Shimron, Hefetz and Tengö 1985), the bees that

1. Introduction

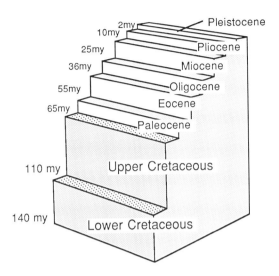

Figure 1.1. A geological time scale, drawn to indicate the millions of years, in major geological periods, that have passed since the early evolution of bees and flowering plants.

have had broad success in moist tropical areas utilize other nesting substrates or unique nesting materials.

Tropical forest bee assemblages, like those of many arid tropical areas, are largely made up of anthophorids, megachilids, and apids. The halictids frequently display similar high representation (Halictinae in the New World and Nomiinae in the Old World), as do possibly the colletids in northern Australia (Michener 1965b, 1979; Sakagami et al. 1967; Raw 1985; Eickwort 1988a). The total species known within varied tropical and subtropical regions, including a range of elevations, precipitation regimes, and vegetation types, suggest that the richness of bee families might be ranked as follows: Anthophoridae > Halictidae > Megachilidae > Apidae > Colletidae > Oxaeidae > Andrenidae > Melittidae > Fideliidae > Ctenoplectridae > Stenotritidae (Chapter 4). Ecological diversity within a bee family might correspond broadly to the number of its species, yet in terms of biomass and ecological roles, this proposed ranking of families does not necessarily hold true (Chapter 4). Considering the American bee fauna north of Mexico (Hurd 1979), the Andrenidae, Anthophoridae, Megachilidae, and Halictidae predominate.

1.2.2 Biogeography and ages of bee groups

Geological and biogeographical evidence attest to an origin of the bees by the middle Cretaceous, at least 120 million years ago (Fig. 1.1). Recently, this conclusion was reached because a North American Cretaceous fossil of a meliponine bee having an age of 80 million years was discovered in amber from New Jersey

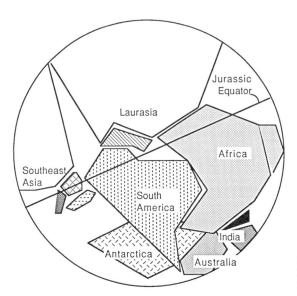

Figure 1.2. A reconstruction of the position of tectonic plates 180 million years ago (after Nelson and Platnick 1981). Classical continental drift theory places both India and Australia near the southern extremity of Africa, and diverse interpretations of seafloor spreading and plate tectonics agree that Australia and South America were in contact with Antarctica. Madagascar is shown here lying next to southeastern Africa; the major islands of Malesia are shown near Southeast Asia.

(Michener and Grimaldi 1988). Such fossil evidence is supplemented by early fossil flowers that show traits uniquely associated with pollination by bees in the early Tertiary (Crepet and Taylor 1985; Taylor and Crepet 1987). The fossil record indicates all major bee families existed in the late Eocene, 40 million years ago (Fig. 1.1). Further information on antiquity must be gleaned from the geological record, supplemented with fossil evidence when possible (Zeuner and Manning 1976; Michener 1979). As stated earlier, dispersal instead of evolution in isolation accounts for much of the current geographic distribution of bees. Particular genera and tribes separated by formidable barriers point toward remote origins, prior to sweeping geological and climatic changes. These disjunct distributions not only include many Colletidae and Fideliidae but also involve an apid group that constitutes the only pantropical bees at present found nowhere outside tropical and subtropical regions, the stingless bees (Meliponinae). The ancient distribution of this group included what is now temperate habitat because specimens of *Trigona eocenica* have been found in Baltic amber and *Trigona prisca* has been found in North American amber (Zeuner and Manning 1976; Sakagami 1978; Michener and Grimaldi 1988).

Continental drift implies that land masses forming most of the present-day tropics, except some of Southeast Asia and associated islands, were unified until 180 million years ago or later (Audley-Charles et al. 1981; see also Fig. 1.2). Although the original positions of all the present continents and the oceans surrounding them are still debated (Nelson and Platnick 1981), it is clear that Antarctica was in contact with Australia and the southern tips of South America and Africa. Most of the present tropics were contained within the great southern continent *Gondwana*. Continental fragments that later became India and Madagascar were at one time close to Africa, Antarctica, and Australia. The first three began to separate from Gondwana before central Africa detached from South America. Antarctica lost direct contact with a northward-displacing Australia at about 55 million years ago, at least 15 million years after South America had separated from it.

Stingless bees (Meliponinae) provide a clear example of a bee group that had the opportunity to disperse with drifting continents. All evidence indicates their widespread presence prior to the late Cretaceous, when close continental connections still existed between Africa, South America, and Australia. This is revealed by the current disjunct distribution of some stingless bees, likely the result of continental drift. Stingless bees of Australia, New Guinea, and South America include bee groups that are nearly identical and can be placed in the meliponine subgenus *Plebeia* (Michener 1965b), thus providing another compelling reason to accept that their origin occurred in remote times. Stingless bees similar to the *Plebeia* are found in Africa, but there are none in continental Southeast Asia, originally part of *Laurasia*, the northern paleocontinent. I interpret this as an indication that the ancestors of these bees were Gondwanan. Less certain are the affiliations of bees within a composite stingless bee group called *Tetragona* (Wille 1979b). It includes some species in Australia and New Guinea, as well as similar bees found in all of the neotropics and Southeast Asia (Sakagami 1982; Wille 1983a). The stingless bees, including species groups that still exist, must have been present in the mid-Cretaceous, and Michener (1979, p. 332) concluded the following: The fact that meliponine distribution is best explained by assuming that these bees were in Gondwanaland before its breakup should be viewed with the understanding that it was the tropical parts of South America and Africa that separated last and were separated by narrow and probably island-filled seas. It would be more accurate to say that the Meliponinae were in Gondwanaland before its fragments became so isolated that the bees were unable to cross occasionally by rafting or other means.

Fossil stingless bees preserved in amber of the middle Miocene include an extinct species of the primarily South American *Nogueirapis;* Oligocene fossils are known of the extinct subgenus *Proplebeia* (Wille and Chandler 1964; Wille 1977, 1983b; Michener 1982a). These fossils from Mexico and the Greater Antilles, and insular species of modern stingless bees found in the Lesser Antilles and northern Central America, are good evidence for bee movement between North and South America across a Caribbean arc during the Eocene or earlier. This preceded formation of the Panama landbridge by at least 35 million years (Camargo et al. 1988).

Bees believed to be the most primitive forms of the meliponines are restricted to Africa (Wille 1979b). Wille found a nearly complete sting apparatus in the African meliponine group *Axestotrigona* and suggested that the large contrasts among African meliponines indicate that they first evolved there. Apart from fossil evidence, the existence of the meliponines might be dated from the time the connection between southern South America and Antarctica existed under warm temperate conditions. Although Antarctica was severed from South America considerably later, its past warm temperate climate probably disappeared before the upper Cretaceous, as early as 100 million years ago (Michener 1979, p. 329).

The honeybees originated and spread through the Old World well after the breakup of Gondwana. They are a Laurasian group; all but four species are strictly tropical and subtropical, and only one occurs west of Oman (Michener 1974a; Ruttner 1988; Tingek et al. 1988). Diversification of honeybees is intimately related to glacier formation and retreat within the past few million years, although the most reasonable approximation of the group's origin is one dating to the Oligocene (Zeuner and Manning 1976; summaries by Culliney 1983; Ruttner 1988).

As late as 1983 (Ruttner and Maul), the existence of more than three truly reproductively isolated species of *Apis* was still in some doubt because two very similar species, *A. cerana* and *A. mellifera,* did not exist in sympatry (Ruttner 1988). Reproductive isolation among the sympatric species of *Apis* is often preserved by staggered timing of mating activity and marked differences in the structure of the male endothallus because sex pheromones are similar or identical between species (Koeniger and Wijayagunasekera 1976; Koeniger, Tingek, Mardan, and Rinderer 1988; Ruttner 1988; Tingek et al. 1988; see also Chapter 3). Cryptic species exist among superficially similar honeybees. When sympatric, the male genitalia and other features are divergent; but when no overlap exists, some traits retain similarity (Michener 1974a; Wu and Kuang 1986; McEvoy and Underwood 1988). Wu and Kuang demonstrated differences in cell size and placement on branches of the wax combs made by the two species of little honeybees in southern China. Similar differences in nesting have been noted among bees of the giant honeybee group. For instance, *Apis dorsata* in the Philippines nests close to the ground; its nests are more concealed than those of mainland *A. dorsata* and also occur individually rather than in aggregations (Seeley, Seeley, and Akratanakul 1982; Starr, Schmidt, and Schmidt 1987), and *A. andreniformis* builds unconcealed nests and is more aggressive than the other little honeybee found in the same habitat, *A. florea* (Wu and Kuang 1986). Even as preliminary data, such behavioral and bionomic differences invite further study to ascertain whether speciation has occurred (Ruttner 1988).

Five kinds of giant honeybees are geographically separated. The widespread *A. dorsata* is found from western India and Pakistan to the bay of Tonkin and Hainan

1. Introduction   13

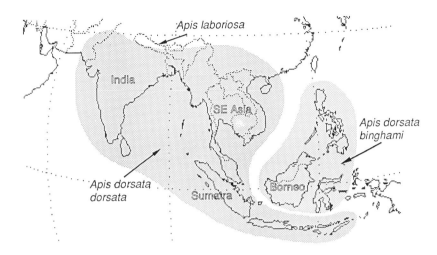

Figure 1.3. The current distribution of named members of the *Apis dorsata* group. These giant honeybees have repeatedly colonized the islands of the Sunda Shelf and probably Wallecea during glaciations; in these periods they also formed an isolated population in the Himalayas. The ranges of the most widespread populations are shown in the diagram.

Island in southern China, and south to Palawan, Sulawesi, Philippines, and most of Malesia (Fig. 1.3). It is represented by the race by *A. dorsata binghami* in Sulawesi, Philippines, and Borneo and its associated islands (Carlson, Roubik, and Milstrey in press); *A. laboriosa* is Himalayan (Maa 1953; Sakagami, Matasumura, and Itô 1980; Roubik, Sakagami, and Kudo 1985; McEvoy and Underwood 1988). The bees restricted to the Philippines, Borneo, Sulawesi, and the Himalayas are dark in color (Appendix Fig. B.21).

Spread of the giant honeybees from continental Southeast Asia onto Sumatra, Borneo, Java, and possibly Palawan occurred several times since the late Pliocene, three million years ago, because during glacial periods islands of the Sunda shelf were connected to mainland Southeast Asia due to reduction in the sea level (Audley-Charles 1981). The lowest sea levels, some 160 m lower than present, brought the best chances of interisland dispersal approximately 150,000 years ago (Heaney 1986). The giant honeybees have been able to traverse larger water barriers as swarms. Like *A. florea* [which exists on Borneo and Palawan but not on the islands of Sulawesi and Philippines (Ruttner 1988)], their nests are not built within cavities, and there is no opportunity to disperse by floating in hollow logs or via human transport (giant honeybees sting fiercely). During glaciations the giant honeybees colonized not only the just-mentioned islands but also the Lesser Sundas and possibly the Moluccas, across oceanic gaps that even in times of reduced sea level were still as large as 20 km. Giant honeybees reached Timor and

Wetar islands, but there are no honeybees recorded on Sumba (Maa 1953 and Leiftinck, quoted by Maa 1953), separated by slightly more than 30 km of deep ocean from Flores Island, where giant *Apis* exists. It thus seems reasonable to designate a dispersal limit of 20–30 km for giant honeybees, which, during glaciations and probably before tectonic activity in the area had ceased, permitted them to colonize, from Borneo, both the western Philippines via Palawan and Sulawesi via the Makassar Strait.

The last glacial period peaked some 18,000 years ago and began 100,000 years earlier (Edwards, Chen, Ku, and Wasserburg 1987). As maintained by Heaney (1986) this was a period during which the lowest sea levels were attained, and during this interval there was again the chance for interisland dispersal of giant honeybees as far as Palawan and the Philippines. However, the giant honeybees of Borneo, Palawan, Philippines, and Sulawesi have diverged to a degree implying earlier separation from the mainland populations and its maintenance despite contact during the last glacial period (Carlson et al. in press). Their cuticular hydrocarbons are distinct, whereas those of the bees inhabiting Sumatra and Sri Lanka, each connected to the mainland during the last glaciation, are no different than bees from India or Thailand. Therefore, some islands are inhabited by racially distinctive honeybees (Fig. 1.3), or perhaps even distinct species of the giant honeybee. The giant honeybee of the Himalayas has drone genitalia that are identical to those of the lowland *A. dorsata* (McEvoy and Underwood 1988). There is, however, compelling biochemical, historical, and bionomic evidence that the Himalayan honeybee *Apis laboriosa* is a distinct species. We know that *A. laboriosa* and *A. dorsata* overlap slightly in distribution in the eastern Himalayas and in Yunan province of south China, and no hybrids have been found. Further, *A. laboriosa* populations have been separated from the lowland populations for most of the past three million years, because during cool glacial periods the lowland *A. dorsata dorsata* likely retreated far from the Himalayas. Some of their parasites are distinct, although this is not the case for parasitic mites, which often transfer between *Apis* species (McEvoy and Underwood 1988; Ruttner 1988; see also Chapter 3). Tests of species status among giant honeybees and other disparate populations of the *Apis* will become complete as further biochemical, morphological, and behavioral research is carried out.

Isolated populations of honeybees and stingless bees provide an idea of the length of time in which distinctive races and species evolve. Successful colonization by such bees requires dispersal of numerous intact colonies rather than the chance dispersal of a few females by wind or other methods. On the island of Madagascar, which separated from Africa 165 million years ago, there are few known stingless bees and the smallest type of the common honeybee (Jolly, Oberlé, and Albignac 1984; Ruttner 1986; Michener pers. commun.). Because there were no bees at the time Madagascar rifted from southeast Africa, and there are no intervening islands or shallow seas, bees dispersed there across open ocean. For *Apis mellifera unicolor,* however, this was brought about almost certainly by humans who transferred log hives from East Africa when they colonized the island. No more than 1,500 years have passed since that time (Jolly et al. 1984). On the other hand, the stingless bees found on Madagascar are very tiny species of the *Hypotrigona* group that are not cultivated for honey (Portugal-Araújo 1955; Michener and Grimaldi 1988). These bees are found only on Madagascar and thus probably have a history on the island much longer than that of *A. mellifera unicolor.* Their nests must have drifted there in substantial numbers within floating logs.

It has been suggested that similar long-distance dispersal occurred in numerous wood-nesting bees (Michener 1979), among them *Xylocopa* and one of its natural enemies now established in the Galapagos of Ecuador (Linsley 1966). Furthermore, endemic allodapine bees including parasites exist on Madagascar (Michener 1977a), likely the result of dispersal over water in twigs or branches. Although the Madagascar honeybees are a small variety of African *Apis mellifera,* they are singularly unaggressive, yet seem to have been derived from a highly aggressive lowland African variety, *Apis mellifera litorea* (Chandler 1975; Ruttner 1988). Such changes occurring in a relatively short time indicate the speed with which the existence of relatively small colonizing populations and release from past selective pressures might modify bee traits. Although the original honeybees brought by human immigrants to the island were not necessarily highly aggressive, the complete absence of such behavior in the naturalized honeybees now living there suggests that there have been behavioral adjustments as well as a reduction in body size.

1.2.3 *Social and solitary life cycles*

Bee life cycles are obligately or facultatively social, or they are completely solitary. Nest aggregations of solitary bees arise in any type of habitat where bees occur and are likely related to the evolution of colony formation (Fig. 1.4). Colonies might arise through *multifoundress associations* (Eickwort 1986) when many bees are concentrated in a single area, but colonies seem more likely to form through *matrifilial associations.* The process is facilitated when the parental nest is reused by some of the female progeny, which is thought to be relatively common among solitary bees in the temperate zone (Malyshev 1935). However, the most completely social bees are almost entirely confined to tropical and subtropical areas of the world. Thus not only are the tropics where certain bees proliferated once they no longer nested in the ground, but they are now the primary and likely also

16   Ecology and natural history of tropical bees

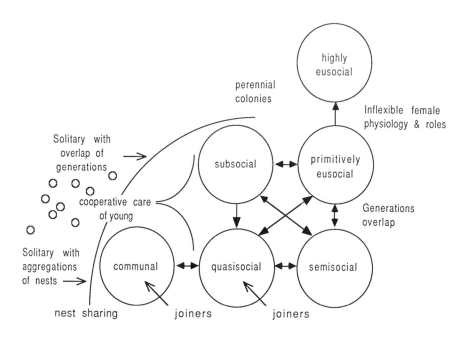

Figure 1.4. Possible evolutionary sequences from solitary life to social bee colonies. The small circles indicate nests, and large circles represent nesting colonies (modified from Sakagami and Zucchi 1978; Itô 1980; Michener 1985). Transitions between the various stages of social evolution and some of the principal associated traits are shown.

the original habitat of the highly eusocial bees. Two diverse lineages, the stingless bees and honeybees, have evolved *highly eusocial* life cycles, in which there is no solitary phase of the life cycle (Michener 1974a; Sakagami 1982). Their several hundred species and races now constitute the numerically dominant bees of most tropical forests (Chapter 4).

Social behavior and its development in bees are scrutinized by Wilson (1971), Michener (1974a, 1985), Eickwort (1975, 1981, 1986); Itô (1980), Knerer (1980), Sakagami (1982), Seeley (1985), Velthuis (1987), and also by other authors cited in Chapter 4. Somewhat broader reviews concerned with the development of sterility in social insects are provided by Hamilton (1972), Lin and Michener (1972), Starr (1979, 1984), Gadagkar (1985), Joshi and Gadagkar (1985), and Stubblefield and Charnov (1986). Sociality implies coexistence of adults in a single bee nest. When an association is matrifilial (daughters remain for some time with their mother), it can be called *eusocial,* at least until daughters prepare larval provisions and lay eggs of their own (Michener 1974a). If neither mother nor daughter can survive or produce males and queens alone, then the bee is obligately eusocial. Its social roles are inflexible.

There is some unavoidable error in applying this terminology unless the complete behavioral repertoires and the state of reproductive physiology are known for coexisting females. Even the permanently eusocial colonies of stingless bees and honeybees are occasionally composed only of sisters for a short period following replacement of the queen. Bees are usually referred to as *parasocial* if the extent and type of their social interactions are not fully understood (Michener 1974a). Brief nest cohabitation may be difficult to characterize. For example, extensive X-ray studies of xylocopine bees within their tunnels were necessary to show that they participated in cooperative brood production, mutual feeding, and nest guarding (Gerling, Hurd, and Hefetz 1983). These authors proposed a term, *metasocial,* to describe a colony that displayed all the components of social behavior but only occasional nest cohabitation by mother and daughters. However, the types of eusocial colonies initiated by some euglossines are so similar, with the exception of mutual feeding that is rare or absent in these groups (Michener 1972b; Garófalo 1985; Sakagami and Michener 1987), that the added terminology seems unnecessary.

Certain types of field data can prove to be incomplete for describing the sociality of bees. For instance, when I checked the numbers of female bees and the numbers of open brood cells being prepared in a nest of *Euglossa imperialis* at dawn, I found that they were usually equal. After a few hours there were more females than open cells, creating the impression that females were cooperatively rearing offspring (Roberts and Dodson 1967; Sakagami and Zucchi 1978). The presence of more females than open brood cells, coupled with the pollen-collecting activity of all females, is often preliminary evidence for cooperative brood rearing (Michener 1974a). However, apparent cooperative behavior is at times reversed; a female can suddenly open a provisioned cell and consume the egg, replacing it with her own. Moreover, unrelated females may be seen, at least for a short period of time, in single nesting cavities or sharing a single burrow entrance – even with different species (Michener 1974a; Itô 1980, p. 273). This may occur due to drifting of individuals between nests or during usurpation of the nest of one species by another, and it indicates considerable tolerance on the part of both resident bees and new arrivals. Recently, a subtropical nesting population of *Halictus ligatus* was found to include a small portion of females that regularly moved between nests (Packer 1986). Such females were egg layers and showed no mandibular wear (the normal correlate of nest excavation in the soil). Because they entered nests but did not carry pollen, they were ostensibly parasitic. Indeed, parasitism of daughters by mothers, or even by sisters, is potentially a force driving the formation of eusocial colonies (Stubblefield and Charnov 1986). It is known to occur in at least one temperate halictid bee (Eickwort 1986, p. 747).

Both female xylocopines and euglossines can coexist for some period in the same nest with their mother, and she may feed them. Whether or not she continues reproductive activity and they contribute to colony maintenance and thus are temporary workers are slight differences that nonetheless affect the way in which their sociality is understood. Some of the most informative types of interactions between cohabitating females are revealed only by intense study. For example, Sakagami and Maeta (1986) found frequent egg consumption by workerlike females in nests of *Ceratina japonica* and *C. flavipes* that contained two females. Late at night, the worker opened cells and replaced eggs of the queen with those of her own. These species seldom live in colonies. Female offspring that temporarily use the maternal nest before initiating nests of their own are probably a common phenomenon (Michener 1985). This could conceivably be permitted if neither they nor their mother lays eggs during this period. In effect, they would be reproductively neutral; thus conflicts and reproductive competition between reproductive females in one nest might be obviated. Kurihara, Maeta, Chiba, and Sakagami (1981) found that maternal care in small colonies of *Ceratina* inhibits the development of ovaries in the female offspring. Eickwort (1986) found that halictid colonies result from failure of daughter bees to find nesting sites shortly after emergence. Subsequently, they are dominated by their mother (see Smith 1986) or a sister in a multifoundress nest.

How did eusociality evolve in bees? An attempt to answer the related question of why permanent eusociality had the best chance of evolving in the tropics is deferred to Chapter 4, and ideas on the relative contributions of kin selection and parental manipulation are explored by many authors cited at the beginning of this section (see especially Michener 1974a, pp. 233–253). It is the comparative stages of sociality that are discussed here, and what is known of their development. The variety of social behaviors found in living bees can be imagined to form sequences, some temporary en route to eusociality for a particular species and some of a rudimentary social nature that do not culminate with cohabitation of a nest by mothers and daughters (Fig. 1.4). Allodapine bees and some *Ceratina* (Anthophoridae) are *subsocial,* denoted by parental care of immature bees (feeding of immatures and removal of larval feces) until the first brood of adults emerges (Iwata 1938; Michener 1971). A few of the females may remain as helpers in the nest made by their mother (Maeta, Sakagami, and Michener 1985), making their colony a eusocial one. The end result of social evolution in bees is a colony that displays:

1. cooperation among sisters in nest building and the feeding and care of larvae,
2. females that cannot directly produce other females, and
3. the persistent association of mother and daughters.

In its extreme, displayed only by the honeybees and stingless bees, eusociality involves the loss of foraging and nesting ability of a reproductive female. This can be interpreted as an evolutionary dead end.

Lines of social development that fall short of permanent sociality have been classified as either:
  (a) *communal* – a common nest entrance is shared by otherwise independent females;
  (b) *subsocial* – a colony comprised of an adult female that progressively feeds her immature or adult offspring;
  (c) *quasisocial* – a colony comprised of adult females of approximately the same age that cooperatively produce offspring (all females can lay fertilized eggs); or
  (d) *semisocial* – a colony of adults of the same generation, usually sisters, some of which are incapable of laying fertilized eggs (Michener 1974a; see also Fig. 1.4).

Collectively, semisocial, quasisocial, and communal colonies are referred to as *parasocial*.

Only eusocial and semisocial bees have predominantly "helper" castes that nonetheless possess functional reproductive systems, but the ovaries of a semisocial or *primitively eusocial* female can develop or regress (Michener 1974a); thus physiological changes bring about reversals in social role. Primitively eusocial colonies are most similar to those of highly eusocial bees because they may store large amounts of food and consist of a queen and her daughter workers. However, the life cycle of the colony terminates when new queens and males are produced; and because queens and workers are not ordinarily produced simultaneously, reproduction by swarming cannot occur (Section 3.3.4). Semisocial colonies could give rise to eusocial colonies if mother and reproductive progeny coexist (Michener 1974a; see also Fig. 1.4). Starr (1979) and Itô (1980) doubt that this route to eusociality was likely, but the types of colonies that seem necessary to make this transition are known in nature. A temporarily eusocial halictid bee studied by Eickwort (1986) has two generations during the temperate-zone summer, but the queen is short-lived and is replaced by one of her daughters. Semisocial and primitively eusocial phases thus occur in colonies. The frequent nesting associations of females of a single generation and the evolution of eusociality at least five times in the Halictidae (Eickwort 1986) certainly admit the possibility of semisocial precursors to eusociality. In the absence of such observations, considering some eusocial colonies as they now exist can produce a contrary impression. Indeed, tropical eusocial colonies of halictid bees typically produce a final generation of future queens and males just prior to demise of the colony (Wille and Orozco 1970; Michener 1974a; Brooks and Roubik 1983), and when several queens are present in the nest of *Bombus* or most *Apis* (Ruttner 1988), they kill each other. The reproductive female that remains in the nest becomes the colony's queen (Michener 1974a; Zucchi, quoted in Sakagami 1982). Fixed reproductive traits seem likely to produce conflict between a primary reproductive and her reproductive offspring, but this need not be so if the queen suppresses ovarian development among offspring. Moreover, the existence of parental care, notably the feeding of adult offspring, is thought to be one of the primary reasons for the evolution of highly eusocial Hymenoptera (Stubblefield and Charnov 1986). Such feeding is absent in semisocial bee colonies and in most primitively eusocial colonies (Halictidae, but see Kukuk and Schwarz 1987). Semisocial colonies could evolve into facultatively eusocial colonies, if and only if the mother queen could avoid mortal combat or extensive mutual consumption of eggs with reproductive offspring. This may be assured as a result of a univoltine life cycle in which a single generation is produced in one year, common in the temperate zone, and the hibernation (diapause) of adult bees (Sakagami and Maeta 1984; Eickwort 1986).

When pushed continuously by an ecology that favors colonies and primed by traits permitting mutual tolerance in nests, a bee population can shift between the adaptive modes represented by different types of colonial organization (Sakagami and Zucchi 1978; Eickwort 1981; Michener 1985; see also Fig. 1.4). Populations can run the entire gamut of solitary, subsocial, parasocial, and primitively eusocial associations. Among halictine and allodapine colonies, semisocial colonies can arise if the original queen of a eusocial colony dies (Eickwort 1981, p. 261). Workers of both primitively eusocial and highly eusocial colonies produce males if the queen disappears, but production of females by workers is lacking in all highly eusocial species except the South African honeybee *A. mellifera capensis*. Remnants of facultative resumption of a female reproductive role by workers may still exist in this highly eusocial bee. Most workers in colonies of many stingless bees possess developed ovaries (Sakagami, Beig, Zucchi, and Akahira 1963; summary by Bego 1982). These eggs are sometimes trophic (anucleate). The ovaries of worker *Apis* will develop in absence of the queen, and tropical races of *A. mellifera* do so most rapidly (Ruttner and Hesse 1981).

Despite certain advantages of social flexibility conferred upon facultative colonies of other tropical bees, discussed at greater length in Chapter 4, all meliponines and apines exist exclusively in colonies. Evidently, the fitness penalty for deviations from permanently eusocial colonial life is so severe that selection maintains all populations in permanent sociality. The queen and the worker are each selective agents for their partner and are involved in an obligate mutualism (Lin and Michener 1972). Gradual evolutionary shifts, both progressive and regressive, among the other social and nesting arrangements probably occur (Michener 1974a, p. 47; also Fig. 1.4), since none of these require loss of foraging and nesting ability and associated structures in queens.

Prereproductive offspring are often tolerated by "queen" Xylocopinae (Bonelli 1976; Michener 1985). Bonelli (1974) recorded nearly 30 bees of both sexes and more than one generation in a single nest of *Xylocopa somalica* in Ethiopia. A eusocial colony is formed when broods of an actively reproductive female emerge and remain within the nest, and these new individuals contribute to the production of their mother's brood. This happens for various reasons; they may fail to mate, or fail to find a nesting site, and thus come under the direct behavioral or chemical manipulation of their parent or sister (Eickwort 1986). However, some can later leave the nest and establish one of their own, or they can replace their mother in the same nest. The former primary reproductive of a xylocopine continues to care for her brood by feeding them as young adults. They later lay eggs, as will their mother, but their "personal" reproduction can occur apparently only when the primary reproductive is in a late stage of senescence or dies. The mother would presumably eat any eggs she might encounter that are not her own. Such *delayed eusociality* (Sakagami and Maeta 1977) can be defined as social behavior that is compatible with a basically solitary life cycle (Sakagami pers. commun.); it is associated with parental care that does not require absolute cessation of reproductive activity. One explanation for delayed eusociality is that it might represent to the founding female a tradeoff between promoting the survival of progeny and continuing egg production at a declining rate. In addition, the progeny that have been fed by the mother can in turn feed siblings as they emerge if the mother should die. This has been observed in *Xylocopa* in a subtropical habitat (Scholz and Wittmann in press). Without gradual senescence and declining physiological capability of egg production, delayed eusociality perhaps would not occur.

It is a statistical rather than fixed population characteristic that describes nesting associations, and hence sociality, of most nonsolitary bees. For instance, Sakagami and Maeta (1986) found that 3 of 191 newly founded nests of *Ceratina japonica* contained two or more females, as did 63 of 177 nests of the same species that were occupied since the previous year. Another *Ceratina* species displayed multifemale associations in only 3 of 2,300 natural nests. At these particular times and places, the sampled bee populations could be described as slightly social and almost solitary.

The communal and quasisocial colonies are different from other types because they have no workers or queens. Such colonies are known for bees of the Colletidae, Anthophoridae, Megachilidae, Halictidae, Apidae, and Andrenidae; but only some of the Apidae have attained permanent eusociality (Fig. 1.5) whereas the queen and worker of semisocial and subsocial colonies are scarcely differentiated in their anatomical features. They are often known only by the contrasts found among members of a single colony, which in many social bees consists only of two adult females. A queen *Ceratina, Manuelia, Pithitis,* or allodapine, for example, will tend to be larger, possess large ovaries, be inseminated, forage very little, receive food in the nest from a worker, and probably destroy the eggs laid by a

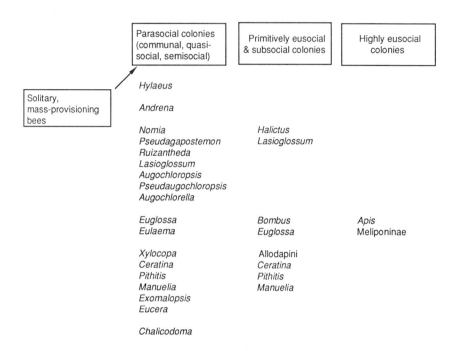

Figure 1.5. Six bee families that contain at least some social species, and the taxonomic groups showing the indicated types of colonies. The first column groups genera of the six families; from top to bottom they are Colletidae, Andrenidae, Halictidae, Apidae, Anthophoridae, and Megachilidae.

worker (Sakagami and Maeta 1977; Michener 1985). Many if not all of the same features distinguish the queen from workers in colonies of primitively eusocial halictids, *Bombus, Apis,* or stingless bees. Workers in these three groups are very unlikely, under normal circumstances, to destroy or replace the eggs of the queen. Further, the anatomical and physiological distinctions between queen and worker are unmistakable. Behavioral dominance and consumption of eggs laid by other females also seem characteristic of small euglossine colonies consisting of a mother and daughters (Garófalo 1985). The colonies of these bees are primitively eusocial, essentially similar to those of some halictids. Lack of feeding behavior among adults is probably phylogenetically determined for the halictids, and feeding may have evolved independently in Apidae or been inherited from xylocopine ancestors of apids (Sakagami and Michener 1987).

Broad ecological constraints upon female reproductive success are likely to stem from predators and parasitoids, as well as from the availability of nest sites and food. Due to such limitations, the rate of reproduction, chances for survival, and overall work efficiency of females are likely to increase in colonies (Michener

1974a, p. 245; Oster and Wilson 1978; Eickwort 1981; Houston, Schmid-Hempel, and Kacelnik 1988). Furthermore, expected fitness benefits theoretically exceed those possible through personal reproductive effort, although there must be a considerable margin of error because a queen of *Apis mellifera* is replaced only after her potential fitness has decreased by half (Seeley 1985, p. 59; see also Section 3.4).

Both predators and competitors can influence selective advantages associated with female size. When females of a colony are all of approximately the same size, chemical or behavioral interactions among them probably determine which ones produce personal offspring and which do not. Michener (1974a) reported that removal of the reproductive female immediately led to ovarian development in 65% of the halictid females remaining in the nest. The physical basis is queen harassment of workers (summary by Smith 1986). Chemical control of reproductive activity thus is very unlikely. For example, in the social halictid bee *Lasioglossum zephyrum,* egg-laying females direct several activities of foraging nest mates (Breed and Gamboa 1977; Buckle 1984; Smith 1986). Smith and Weller (in press) report that *Lasioglossum* queens use size and possibly the amount of lactone pheromone produced (larger bees produce more Dufour's gland lactone) as a cue to limit the amount of fighting between rival queens. However, the timing and amount of pheromone released is not correlated with bee size. They propose that the difference in size of queen and worker *L. malachuurum* is due to two major selective pressures. One is predators, which are more effectively repelled by a larger amount of defensive secretion from the Dufour's gland (Chapter 3), and the other is the advantage obtained in intraspecific female fighting by larger bees. Quantitative differences in odor-production capacity may have acquired definite selective advantages in the early evolution of eusociality (see also Velthuis 1987).

Michener (1985) discusses direct evolution (especially in anthophorid bees) of eusocial colonies from any of the intermediate stages (Fig. 1.4). Several stages might even exist at one time in a bee population. Evident traits that would favor eusociality include the following:

1. a high degree of relatedness among individuals in a nesting population and scant dispersal from this area;
2. a high degree of intraspecific tolerance (see Batra 1968; Eickwort 1975);
3. extended female longevity (or rapid larval development), permitting an overlap of generations;
4. extended use of a nest site, either by aggregating females or through several matrifilial generations (Michener 1974a; Eickwort 1981; Maeta et al. 1984); and
5. persistent polytypic behavior (or physiology, morphology, biochemistry, etc.) existing within a population, permitted by assortative mating or frequency-dependent mating advantage.

The last point corresponds to general maintenance of polymorphism in outcrossing populations (Maynard Smith 1982). For several genetically determined traits to exist in a balance, selection must favor different sets of traits under different circumstances or a heterozygote advantage. Fixation of certain social traits in bees seems more likely when mortality due to natural enemies is relatively common (Chapter 4).

A discussion of these and other considerations bearing on social evolution has been given by Michener (1974a, p. 253) and will not be repeated here, but a significant qualification of some field data has been confirmed since his comments on the subject (op. cit., p. 246). Studying colony size and reproductive efficiency, Michener (1964a) noted that the number of reproductive females (queens) produced per female in a eusocial colony often seemed no greater for colonies than for lone females. Small colonies appeared to reproduce more efficiently than slightly larger colonies of a few to several females (Michener 1974a, p. 245). A paradox was implied, because the first eusocial colonies were undoubtedly small, and it was not clear what *reproductive* advantage would have been conferred on females staying with their mother rather than leaving the nest. The explanation seems to lie in the *survival* advantages of colonies. Semisocial colonies of *Ceratina* and *Braunsapis* displayed a reproductivity per female similar to that of lone females, but eusocial colonies with a single queen and daughter workers more than doubled productivity (Michener 1971; Maeta et al. 1984, 1985; Sakagami and Maeta 1984). Apparent productivity of lone females and small colonies is, however, higher if only the surviving nests are considered. If selection already has eliminated many of the lone females and small colonies, then the real productivity of this group is not known and must appear greater than it is. Protected laboratory colonies, on the other hand, reveal the productivity of all females and groups. These include the bees that would have been eliminated by natural enemies or by poor choice of nesting substrate in natural conditions. Before selection has occurred, colonies of all sizes may show greater reproductive efficiency than individual females. When the successful lone females, in nature, are compared with colonies, productivity differences are hidden. Such comparisons strongly suggest that the formation of small colonies is firmly rooted in the initial success of individual females.

Continued success as a colony seems to require the division of tasks among nest mates (in addition to social roles involving reproduction), selection of a superior nest site, and added protective mechanisms against microbes, predators, and parasites. The natural history elements affecting some of these variables are reviewed in Chapters 2 and 3, and their relation to social evolution and ecology of highly eusocial bees in particular are briefly analyzed in Chapter 4.

2 Foraging and pollination

2.1 Resources gathered by bees

An understanding of bee ecology requires knowledge of the variety of resources they use to construct and defend their nests, maintain metabolism, and reproduce. These activities, crucial to reproductive fitness and survival, most certainly influenced the evolution of morphological, physiological, and behavioral traits. With few exceptions, both female bees and males gather food from flowers. In addition, except for those that usurp the nests of others, most bees gather nesting material and provisions for their larvae. The resource menu is a large one. Foraging bees normally seek plant products such as gums, resins, rotten wood, bark, fruit juices, seeds, leaves, plant hairs or trichomes, fragrances, pollen, nectar, oils, spores and rusts, sap, and the honeydew excreted by plant-feeding homopteran bugs and fungi, other natural products such as wax, animal feces, carrion, urine, and hairs, and combined plant and animal products such as cerumen (a mixture of resin and wax made by meliponine bees) and propolis (a similar mixture containing relatively little wax made by honeybees). Other substances used by bees include mud, loose soil, gravel, various salt solutions, and water. An overview of such materia prima gathered by bees is given here, followed by discussion of mechanisms of its collection.

Because so many different materials are employed, we scarcely know the degree to which the structural, biochemical, and behavioral traits of bees are suited to each. Considering the many substances transported by bees, there is a large gap in our knowledge of why some are selected. Mechanisms used for choosing among materials are largely unknown. It is here that comparative studies provide a means of predicting the materials that a bee might use, based on knowledge of its anatomy and of its phylogenetic relationship and ecological similarity to the bees that have been studied.

Individual flowers in the tropical lowlands differ somewhat from those of the temperate zone because most last only a single day (Bawa 1983; Primack 1985). Still, the functional life of some tropical flowers, both in lowlands and highlands, often reaches three to five days (Appanah 1982; Renner 1984) and for orchids can reach a few weeks (Dressler 1981; Janzen 1981a). Primack suggests that the selective advantage of a short-lived flower may include minimizing transpiration and herbivory or reducing investment in a flower that lasts long enough to be visited (a correlate of favorable weather and pollinator activity that may be expected for perhaps most tropical flowers). Nonetheless, even within a single day, a tropical flower is a varying source of food, both in quality and quantity. As will be detailed below, other plant resources display similar variation.

2.1.1 Sap

Nonfloral resources, such as the sap of woody vegetation and fruit, constitute seasonally important and perhaps even essential resources for some bees. Temperate-zone *Bombus* and *Apis* feed upon sap (Proctor and Yeo 1973; Olesen 1985), and tropical social bees at least of the subgenus *Trigona* often harvest sap from diverse sources. Plant sap is distinct from nectar and latex as well as from the gum and resin that sap can carry. Sap is aqueous, tends to have a high pH, and is carried in plant xylem or phloem. The nutritional value of sap for bees derives principally from its photosynthetic products, sucrose and other sugars, which may be taken either as honeydew or from openings in the plant. Constituents of sap as found in honeydews also include free amino acids, vitamins, minerals, and occasionally proteins (Kennedy and Fosbrooke 1972). The honeydew from Homoptera tends to be similar in composition to the phloem sap of the host plant, but fungal plant pathogens may produce honeydew exudates with different sugars from those of the host (Mower and Hancock 1975). One study has documented similar sugars in what may be called nectar-mimicking exudates of a fungal pathogen and in the nectar of its host plant, both of which are collected by bees (Batra and Batra 1985). Xylem sap tends to have lower sugar concentration than that of phloem (Durkee 1983). The sugar content of phloem sap was primarily sucrose and, from a sample of 500 tree species, ranged from 1% to 30% in concentration (Zimmerman and Ziegler 1975). In addition to carbohydrates, phloem sap contains some amino acids, proteins, vitamins, salts and elemental metals (Chino et al. 1982; Richardson, Baker, and Ho 1982; Stemmer et al. 1982; Simpson and Neff 1983). Saps collected by bees can carry fungi or bacteria; as Kevan, Saint Helens, and Baker (1983) show, the honeybee *Apis mellifera* may regularly collect sap from plant lesions infected with bacteria. Recently, Norden and Batra (1985) showed that the pungent-smelling sap of a temperate umbel is collected by the male of an anthophorid bee in order to mark its mating territory (Chapter 3).

Honeybees and other apid bees have been observed in many different regions collecting the honeydew from homopteran scale insects (Coccidae), membracids (Membracidae), and aphids; other Homoptera producing honeydew include Fulgoridae, Cicadellidae, Psyllidae, and Pseudococcidae (Way 1963; Kunkel and Kloft 1977; Crane and Walker 1985; Wagner and Cameron 1985). The availability of honeydew is influenced by formicine and myrmecine ants that protect homopterans from predators (Wilson 1971; Beattie 1985; Crane and Walker 1985) and also by competing wasps (Wagner and Cameron 1985). In the tropics, stingless bees of just two subgenera, *Oxytrigona* and *Trigona*, have been observed taking honeydew, which they gather from nymphal Coccidae and Membracidae (Salt 1929; Castro 1975; Laroca and Sakakibara 1976). Such bees deter ants from this resource, probably in part by chemical means (Wilson 1971; Castro 1975). I have watched bees tend individual immature Homoptera on leaves, day after day, even when the nymphs are considerably smaller than the bees and producing negligible, if any, resources. It seems remarkable that the bees locate and defend such minute resources, although it is conceivable that a larger aggregation of the bugs had been dispersed by rain. The same bee species that attend immature Homoptera also frequent the surfaces of leaves where honeydew secretions accumulate and are infested with fungal growth.

2.1.2 Resin

Resin is used by female bees primarily during nest construction, often serving both as protection and a building material, as well as a biologically active compound. Resins are insoluble mixtures of terpenoids and contain some volatile components that bees may use to locate them. Although all conifers produce resin, at least two-thirds of the plant species that ooze resin from a wound are tropical angiosperms (Langenheim 1969, 1973). Resin chemistry is still poorly known, but its functions include reducing infestation by microbes (Ghisalberti 1979). Resins are secreted as liquids, then harden. They are collected by bees while soft or still in liquid form (Armbruster and Webster 1979; Ghisalberti 1979). A small fraction of propolis, the term applied to combined resin, wax, and sticky substances used by European *Apis mellifera*, is soluble in organic solvents and amenable to ordinary chemical analysis. Ghisalberti (1979) states that the largest group of compounds identified in this fraction consists of flavonoid pigments. These have antifungal, antibacterial, and antiviral properties. Acids, alcohols, and aldehydes also occur in both propolis and resin. The propolis studied was from temperate areas, but it did not acquire substantial new chemicals from the bees or their wax; rather, it retained its original composition. According to the above authors, main constituents of resins are long-chain isoprene polymers derived from acetate and include mono-, di-, and sesquiterpenes, tetraterpenes (carotenoids), and triterpenes (steroids). Some monoterpenes are known repellents of ants (Eisner, Deyrup, and Meinwald 1986) and are potentially used in the exclusion of flies and ants from carrion by silphid beetles, which hints a similar role for resins carried by necrophagic bees (Section 2.1.8).

Resins are often secreted following injury or disease of plant branches, buds, or fruit, but resin canals and secretion exist normally and do not require external injury in common tropical resin-producing trees such as the caesalpiniaceous legume *Hymenaea* (Langenheim et al. 1982). The composition of resin may be species-specific for tropical plants. However, its chemical content may differ greatly among the leaves, fruits, wood, and bark of a single plant or among the individuals in a population (Langenheim 1973; Langenheim, Lincoln, and Foster 1978). Resins function as attractants for insects that oviposit or feed at wounded plants and as repellents for defoliators such as lepidopteran larvae (Langenheim and Stubblebine 1983). In the tropics, I have seen resins from broken branches or felled trees attract certain apid bees within a few minutes.

The major plant families in the tropics that produce copious resin are the legumes (Fabaceae), particularly Caesalpinioideae and Papilionoideae, and the Dipterocarpaceae (Langenheim 1969). Also noted for resin or latex (an aqueous mixture that sometimes contains resins or gums) are tropical angiosperms such as Anacardiaceae, Clusiaceae (Guttiferae), Burseraceae, Styracaceae, Hamamelidaceae, Rubiaceae, Rutaceae, Moraceae, Myrsinaceae, Euphorbiaceae, Arecaceae (Palmae), Liliaceae, Apiaceae (Umbelliferae), Zygophyllaceae, Convolvulaceae, and Asteraceae (Compositae). Pinaceae and Araucariaceae include primarily highland species and are relatively scarce in the tropics (Whitmore 1984); they are particularly attractive as a resin source for stingless bees when planted in the lowlands.

Resins from fruits of *Vismia* (Guttiferae) are extensively collected by *Melipona* in the neotropics, evident from the seeds that are embedded in the resin gathered by the bees (Fig. 2.1). Various *Melipona* collect the resin and seeds from the fruit of *Vismia* (Kerr 1978). The bees shown in the figure, so far as is known, will use only this material to build much of their nest (Roubik 1983a). It is continuously gathered and deposited on the nest entrance tube of *Melipona fuliginosa* that I have observed in the Amazon and Panama, and the fresh resin is bright orange. Recently, it has been shown that the resin of this plant genus is an insect-feeding deterrent (Simmonds et al. 1985); it might also repel some natural enemies of bees (Chapter 3).

Floral resins are largely triterpenes and are collected in lieu of nectar by several genera of neotropical bees (Armbruster and Webster 1979; Armbruster 1984; see also Fig. 2.2). Except in one instance (Armbruster and Mziray 1987), this phenomenon has not been studied carefully elsewhere in the tropics. Unlike resin secreted from woody plant parts, floral resins scarcely harden for several weeks and can thus be gathered by bees, as well as reused within the nest, over a relatively long period of time. Most species of *Clusia* (Clusiaceae) and the euphorb *Dalechampia* produce floral resins (Armbruster, Hammel pers. commun.). Many

2. Foraging and pollination 29

Figure 2.1. Workers of *Melipona fuliginosa* (Apidae) building their nest entrance with resin and seeds from the fruit of *Vismia* (Guttiferae). Original photograph by Carl Hansen.

Clusia are epiphytes, and both they and *Dalechampia* include shrubs and small trees. The resins issue from floral bractlets near stamens or, for *Clusia,* in the center of the androecium of the male flower or gynoecium of the female flower.

Latex and plant gums seem to be utilized little by bees. Latex from *Ficus elastica* is mined from the leaves of the plant by some bees, such as the meliponine subgenus *Scaptotrigona*. Gums, unlike resins, are water-soluble (Langenheim 1969) and therefore unlikely to be extensively employed in bee nest construction, except in combination with resin. Their specific roles as building materials are unclear.

2.1.3 Floral lipids

Floral lipids are usually in the form of oils and are secreted from structures called elaiophores. These small glands are visited, so far as is known, exclu-

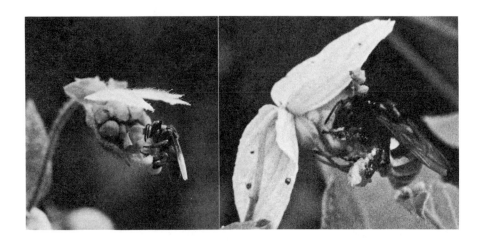

Figure 2.2. Female *Tetragona perangulata* (left) and *Eulaema meriana* collecting liquid resin from the flowers of *Dalechampia* (Euphorbiaceae). Original photographs provided by W. Scott Armbruster (from Armbruster and Herzig 1984).

sively by bees (Vogel 1969, 1974; Buchmann 1987). Other floral oils commmonly form pollenkit – lipid that covers surfaces of pollen grains. Tangible liquid masses of floral oils are only found in association with the elaiophores although similar oils are modified to produce most plant lipids, and a similarity has been noted between elaiophore products and the lipid portion of nectar (Baker and Baker 1983b). Epithelial elaiophores fill until forming a lipid blister beneath thin plant cuticle; trichome elaiophores are less common and are aggregated on the plant, producing lipids not protected by a cuticle but occasionally concealed within floral structures (Buchmann 1987).

Known floral oils are saturated fatty acids, paraffins, oils, and esters and are presented as diglycerides as well; elaiophore exudates may also contain amino acids, glucose, carotenoids, phenolics, glycosides, nonvolatile isoprenoids, and saponins (Buchmann and Buchmann 1981; Simpson and Neff 1981; Cane et al. 1983; Buchmann 1987). Elaiophore products appear to contain virtually no sugars. Their energy value is higher than that of nectar and pollen, ranging from 32 kJ/g to 43 kJ/g (Table 2.1). Elaiophores have been found on the epithelium and trichomes of floral calyces, bracts, and anthers. Recently, Steiner (1985) found that floral oil produced on glandular trichomes of gesneriads [as opposed to "glue" made in similar trichomes of some cucurbits (Vogel 1981)], may be an accessory pollenkit. The oil ostensibly aids dispersal of the small and dry pollen; Steiner observed that bees did not collect it but foraged solely on floral nectar. Most known examples of

Table 2.1. *Energy content of materials gathered by bees, determined by bomb calorimetry*

Material	Energy (kJ/g)	No. species sampled
Sap and nectar	1.4–14.2	>500
Pollen	20.5–28.5	47
Elaiophore lipids	31.6–35.6	5
Pollenkit lipids	38.5–42.7	10
Apid wax[a]	40.6–43.9	4
Vertebrate flesh	8.5–16.3[b]	3

[a]Waxes secreted by *Bombus, Apis,* and *Trigonisca.*
[b]Dry weight protein content of pork, chicken, and turkey ranges from 54% to 87% [USDA, "Nutritive Value of Foods," *Home and Garden Bulletin* 72 (1981)].
Sources: Colin & Jones 1980; Baker & Baker 1983b; Simpson & Neff 1983; Blomquist, Roubik, & Buchmann 1985; and S. L. Buchmann pers. commun.

oil flowers are neotropical, but plant species elsewhere in the tropics also produce oils (Neff and Simpson 1981; Vogel 1984; Gottsberger 1985; Buchmann 1987). Buchmann and Buchmann (1981) and Buchmann (1987) report that about 80 genera and 2,300 angiosperm species have elaiophore structures, primarily in the Malpighiaceae, Iridaceae, Scrophulariaceae, Krameriaceae, and Primulaceae. Additional groups containing tropical representatives are Melastomataceae, Orchidaceae, Gesneriaceae, and Solanaceae. However, the proportion of these that normally produce lipids or serve as food, nesting material, or accessory pollenkit has not been determined. One species of Primulaceae, *Lysimachia*, is known to produce floral oil gathered by bees as both a construction material and larval provision supplement (Cane et al. 1983). Adult bees are not known to ingest the oils.

2.1.4 *Fragrances*

Fragrances associated with nectar and the opening of flowers provide one of the primary cues that attract foraging bees. A fraction of the volatile compounds constituting the entire floral odor may be sufficient to allow bees to learn and discriminate among varieties of a single flower species (Masson 1982; Pham-Delegue et al. 1986). Fragrances are produced in specialized secretory tissues, the osmophores, located on the surfaces of flowers (Vogel 1963; Fahn 1979). Odors may also be released through other parts of the floral cuticle (Williams 1982). Metabolism of starch stored below the epidermal cells is likely required to produce the volatile fragrances of flowers (Vogel 1963; Williams and Whitten 1983). Floral fragrances may contain up to a few hundred compounds including terpenoids, aromatics, aminoid compounds, or hydrocarbons; often they are monoterpenes (Williams and Whitten 1983; Pham-Delegue et al. 1986). As shown in following sections concerning glandular chemical odors secreted by bees (see Table 2.4), many bee odors resemble floral fragrances, and it is probable that flowers attract bees by mimicking some of their odors.

Tropical orchids have received much attention due to the diversity of floral fragrances they produce and the variety of visitors they attract. Typical orchid monoterpenes include geraniol, myrcene, nerol, cineole, menthol, limonene, camphor, and geranial. Aromatics include methyl salicylate, eugenol, methyl cinnamate, and vanillin. From 2 to 18 aromatic compounds have been identified in single orchid fragrances. Many neotropical plants produce volatile oils such as those gathered from some orchids and aroids by male euglossine bees (Dodson 1962; Williams 1982). There are approximately 625 species and 55 genera of orchids in this region that have no nectar and whose pollen is not consumed by bees (Dressler 1982). Floral fragrances are the primary attractants produced by most of these species, although they are possibly of little importance to some bees that pollinate them (Section 3.3). The aromatics and terpenoids constituting such volatile floral oils also occur at aroid flowers. Nonfloral sources also exist, produced by fungi associated with rotting wood and on living tree trunks and roots, presumably derived from sap of the host plant (Dressler 1982; Ackerman 1985). A number of orchid species are deception flowers, having no chemicals or food gathered by the bees, and some species produce nectar instead of fragrances (Ackerman 1983b, 1985).

2.1.5 Nectar

A remarkable feature of nectar is its variable composition and quantity per flower, yet the sugar content of temperate and tropical nectars is nearly identical. Southwick et al. (1981) found nectars ranging from 18% to 68% sugar among spring flowers of nine plant families visited by bees in the state of New York. Roubik and Buchmann (1984) measured sugar concentrations ranging from 19% to 72%, and Roubik et al. (1986) found 10–65% sugar in nectar harvested by bees in the lowlands of Panama. The principal ingredients of nectar are mono- and disaccharides, amino acids, enzymes, lipids, proteins, nonprotein amino acids, glycosides, alkaloids, phenolics, saponins, various inorganics, organic acids, and antioxidants such as vitamin C (Baker and Baker 1982b, 1983a,b). Some of these components are potentially toxic, and nectar itself has been shown responsible for the death of foraging European honeybees (summary by Barker 1978). Toxic honeydews collected by bees are produced by Homoptera feeding on a large variety of plants, and toxic nectars are known from species among many plant families, including Solanaceae (*Datura*), Ericaceae (*Rhododendron*), Sapindaceae, Tiliaceae, and Polygonaceae. After pollen is collected by honeybees, it sometimes contains sugars that are toxic to adult bees and probably to their larvae, unless diluted by water or honey, but it is not established whether the sugars come from nectar added by bees or from pollen (Barker 1977).

Both sugar and nonsugar constituents of nectars tend to be consistent within a plant species (Baker et al. 1978; Southwick et al. 1981; Baker and Baker 1983b). Floral nectars range from about 5% to 80% sugar (Baker and Baker 1983a), an amount influenced also by evaporation of water. Nectars are secreted at concentrations perhaps no higher than 60% sugar and often at concentrations of less than 35% sugar (Corbet et al. 1979; Roubik and Buchmann 1984). The latter authors demonstrated that nectar viscosity increases markedly at concentrations higher than 55% sugar, with straightforward consequences to foragers (Sections 2.2.6 and 2.2.7). Nectars of some flowers pollinated by birds are unaffected by evaporation due to the effect of a lipid layer on the nectar surface (Corbet and Wilmer 1981; Baker et al. 1983). However, evaporation may be important in releasing floral odors into the air, which may then induce bees to probe a flower after approaching it. Furthermore, bees can in some instances visually detect nectar rewards in flowers by perceiving the accumulated nectar (Marden 1984; reviewed by Kevan and Baker 1983 and Corbet et al. 1984).

The various constituents of floral nectars change over time and in accordance to sexual expression of the plant. Sugar composition in nectar, specifically the relative amounts of simple sugars and sucrose, is often associated with visitor type – for example, bird, bee, wasp, bat, butterfly, or moth (Baker and Baker 1982b, 1983b). Furthermore, the dioecious *Triplaris* (Polygonaceae) showed this kind of difference between staminate and pistillate flowers (Baker 1978). In a detailed study of an Asian androdioecious species, Appanah (1982) not only found amino acid concentrations to be 5 to 12 times higher in nectar of pistillate flowers of *Xerospermum* (Sapindaceae) but also found that the simple hexose sugars predominated in staminate flowers. Sucrose was at least twice as concentrated in the pistillate flowers although both flower types secreted 14 amino acids in their nectars.

Some correlations are thought to exist between bee family and nectar composition. Bees having long proboscides may prefer nectar that is predominantly sucrose (disaccharide), and smaller and shorter-tongued bees may prefer nectar rich in fructose and glucose or may take nectars that are sucrose-poor, balanced, or sucrose-rich (Percival 1961; Southwick et al. 1981; Baker and Baker 1983b). However, Appanah (1982) found that the staminate flowers of *Xerospermum* were visited by only the relatively large bees, *Apis, Xylocopa,* and *Nomia,* but not by the small Trigonini. Trigonini apparently avoided nectar that was rich in fructose and glucose but readily visited the sucrose-rich pistillate flowers.

Nectar composition often changes during the life of a flower (Southwick and Southwick 1983), and some of the change may be due to fermentation caused by microorganisms (Wickerman and Burton 1952; Batra, Batra, and Bohart 1973; Baker and Baker 1983a). For example, two common species of *Ixora* (Rubiaceae) in the Philippines contain maltose in their nectar, an unexpected component of fresh nectar, and it was postulated that glucophilic enzymes introduced by foragers could account for this sugar (Rowley 1976). Gottsberger, Schrauwen, and Linskens (1984) show that nectar composition is altered by amino acids leached from pollen that falls into nectar. According to their study, the amount and type of amino acids in uncontaminated floral nectar have no correlation with sugar concentration or the taxonomic group (e.g., bee or bird) of the principal visitor. It seems that the normal situation is for pollen to contaminate nectar after visitors arrive at flowers, and an initially higher content of amino acids in the nectars of female flowers might be offset by amino acids released from pollen at male flowers.

Nectars are secreted at various times of the night or day, usually just prior to flower opening. This trait is generally uniform within flower populations, but secretion rates vary according to weather conditions. Some species resorb nectar into the nectary and thence into the sap if it is not removed, sometimes within a time span of a few hours (Appanah 1982; Cruden and Hermann 1983; Durkee 1983; Opler 1983). Other flowers secrete additional nectar following a lapse from initial secretion (Appanah 1982; Frankie and Haber 1983), but perhaps most secrete over a period of a few hours and then stop (Cruden and Hermann 1983). Nectar secretion usually begins some time before pollinators arrive at flowers, probably because nectar is secreted slowly and flower visitors do not normally visit flowers to collect relatively minute amounts of a resource.

The rate and amount of nectar secretion is variable within and among different species. For example, Opler (1983) reports that in lowland Costa Rica, from a sample of 587 species, total daily nectar secretion per flower ranged from 9.4 ml for balsa, *Ochroma pyramidale,* to 0.03 µl in a relative of the cashew, *Anacardium excelsum.* This approaches a volumetric difference of a half-millionfold. Some Asian and African plants also produce up to several milliliters of nectar which, like that of *Ochroma,* is generally taken by bats (Baker et al. 1983). The amount of nectar secreted by flowers presumably pollinated by small bees and wasps was 0.63 µl maximum in Opler's study. In the Malay peninsula, 3.5 µl of nectar is secreted by the small flowers of *Xerospermum,* visited primarily by small bees (Appanah 1982). Flowers visited by medium-size and large bees (≥100 mg, or the size of the familiar European honeybee) secrete several microliters of nectar (Real 1981; Ackerman et al. 1982; Frankie et al. 1983; Roubik and Buchmann 1984). However, Steiner (1985) has found that the amount of nectar in flowers of *Drymonia* (Gesneriaceae) in Panama, pollinated by *Epicharis* (Anthophoridae), far exceeds this amount. The mean nectar production was 264 µl over two days, during which flowers of *D. serrulata* secreted nectar of 20% sugar. Total nectar produced by flowers corresponds in general to the morphology and biomass of the flower (Opler 1983). Long tubular flowers, typically those with nectar collected by visitors with long tongues, tend to have more dilute nectars (10–35% sugar) and larger nectar volume, compared with open flowers that secrete small amounts of nectar. If there is considerable foraging activity, however, the standing nectar crop in flowers remains under a few microliters (Southwick et al. 1981) after the first visitors remove accumulated nectar.

As a cautionary note to researchers, sugar concentration in nectar is usually measured with a hand-held refractometer corrected for ambient temperature, and errors in sugar measurement may occur due to the influence of nonsugar nectar constituents (Inouye et al. 1980). A maximum error of 9–13% nectar sugar content was calculated by Inouye et al. for artificial nectars similar to those harvested by bees and birds. Refractometer readings can in some instances overestimate the sucrose composition of nectar.

Both floral and extrafloral nectaries actively change the chemical composition of xylem or phloem sap, concentrating and altering chemical composition in the secreted nectar (Durkee 1983; Simpson and Neff 1983). Extrafloral nectaries are visited by bees, ants, wasps, and other insects and occur on floral bracts, calyces, sepals, petioles, leaf bases, corollas, and developing fruit (Baker, Opler, and Baker 1978; Fahn 1979; O'Dowd 1979; Elias 1983; Opler 1983; Anderson and Symon 1985). In a single tropical locality, Keeler (1978) recorded 77 different genera of insects, most Hymenoptera, feeding from extrafloral nectar at a species of *Ipomoea*. Hagen (1986) surveys the use of extrafloral nectaries by entomophagous insects, many of which must compete with bees. In the neotropics, I have seen stingless bees of the subgenus *Trigona* harvest extrafloral nectar from malvaceous and caesalpiniaceous plants. These were common plants and relatively easy to observe. Certainly the phenomenon is not rare.

The composition of extrafloral nectar has been reviewed by Bentley (1983) and Beattie (1985) and is broadly similar to floral nectar. The composition of nectar from either type of nectary has seldom been compared to sap from the same plant, but analysis of a few species shows that sugars and those amino acids present are more concentrated in nectar than in the sap (Baker et al. 1978; Pickett and Clark 1979; Inouye and Inouye 1980). On the other hand, amino acids are generally removed from sap before it is secreted as nectar, and some sap sucrose is converted to simple sugars – fructose and glucose (Simpson and Neff 1983). Extrafloral nectars differ slightly from floral nectar in chemical composition and more often contain cysteines, but information on tropical plants is far from complete (Baker et al. 1978). Deuth (1980) compared the floral and extrafloral nectars of *Aphelandra* (Acanthaceae) in Costa Rica: Extrafloral nectar was richer in amino acids but slightly more dilute than the 31% sugar of floral nectar, which was imbibed by hummingbirds and euglossine bees. The foliar nectaries of balsa (*Ochroma pyramidalae,* Bombacaceae) were found to secrete nectar poor in amino acids but reaching 64% sugar; ants usually dominated the nectaries, but meliponine bees (Trigonini) visited them when the ants were not present (O'Dowd 1979).

Honey is the final product of collected nectars that are stored by Apidae – the stingless bees, bumblebees, and honeybees. Its chemical composition is influenced by the vessels in which bees store the honey, which are containers made with resin/wax, wax/pollen, or pure wax, respectively. Honey is collected in its ripened and viscous state only by bees raiding the nests of other bees, an interaction found in all honey-making apids (Nogueira-Neto 1970a; Dyer and Seeley 1987). Extensive reviews of honey chemistry and composition in some drier tropical and many temperate localities have been prepared by Crane (1975) and Crane, Walker, and Day (1984). The latter reference also compares these features in the nectar and honey of single plant species. Honey stored by *Apis mellifera* is usually near 80% sugar and rich in proline and other amino acids. Such amino acids often are often leached from pollen rather than being part of the nectar or added by bees (Gottsberger et al. 1984). Percival (1965) suggests that nectar of sugar concentration lower than 18% cannot be profitably converted to honey by European honeybees. Other components of honey are the simple sugars made by bee enzymes that break down sucrose to glucose and fructose, the hydrogen peroxide and gluconic acid produced by glucose metabolism, and phytocides that prevent spoilage by microbes (L. R. Batra et al. 1973; Stanley and Linskens 1974). The honey of stingless bees has many added chemicals, most undescribed, originating from the resin of honey storage containers (Gonnet, Laevie, and Nogueira-Neto 1964). It is also less concentrated, averaging 69% sugar in a survey of 27 species (Roubik 1983a). However, the honey of stingless bees, honeybees, bumblebees, and honey-making wasps contains the hydrogen peroxide formed by the action of glucose oxidase added from the hypopharyngeal glands, and this substance prevents spoilage of the nectar until it is sufficiently hygroscopic to retard microbial growth (Burgett 1974).

Several other natural sources of nectarlike substances are known. Stigmatic exudates contain sugar and may be collected by insects; gymnosperms such as *Ephedra* secrete sugary solutions from the perianth and integument (Bino and Meeuse 1981; Simpson and Neff 1983). Secretions of up to 8% sugar are found on stigmas of some aroids and palms (Croat 1980; Simpson and Neff 1981). Such exudates may contain lipids, amino acids, phenolics, alkaloids, and antioxidants

(Baker, Baker, and Opler 1973; Simpson and Neff 1983). Sap or juice of ripe or rotting fruit is occasionally gathered by eusocial bees such as *Trigona* s. str., and the subgenus *Partamona,* and also by *Apis*. It has a relatively low sugar content of 8–15% (Banziger 1981; pers. obs.). The extraction of sap through perforations made by stingless bees in woody plants often damages crops (Freire and Gara 1970) and occurs both at cultivated and native tropical species.

2.1.6 Pollen

Pollen is often the only resource collected by bees from some flowers. It is gathered extensively from flowers that produce no other resource (Section 3.4). Composition, quality, and ease of collection vary markedly among pollen species used by bees. For example, the pollen of corn may contain up to 50% water by weight; it is intensely harvested by some bees in the humid tropics. In contrast, the pollen of most angiosperm trees is about 20% water (Stanley and Linskens 1974; Roubik, Schamlzel, and Moreno 1984). Although pollen of some plants is less than 15% protein, that of others is over 60% the dry weight content.

In a manner analogous to nectar production, pollen release is variable in initiation, peak occurrence, and duration. This is due to genetic control and also the influence of the weather. Stanley and Linskens (1974) point out that anther dehiscence frequently occurs over a longer time period than the receptive period of stigmas on the same plant. Most plants dehisce in the early morning or at two peaks during the day, but dehiscence is nocturnal in others. A few flowers are nondehiscent and rely on insects to rupture the anther to release pollen. Anther sacs normally release pollen due to the breakdown of endothecial cells, caused by the action of intracellular enzymes, ambient temperature, and humidity (Stanley and Linskens 1974). Pollen is released in three basic manners: on the surfaces of anthers, in a longitudinal slit, or through a single pore. Dehiscence occurs gradually over several hours or even days in species having pores through which pollen is ejected. These flowers comprise the buzz-pollinated angiosperms (Buchmann 1983). Some of them, the common tropical legume *Cassia* (=*Senna* in part), for example, have anthers that shed pollen both through slits and pores in different regions of the flower (Stanley and Linskens 1974). In other plants, the most common type of dehiscence involves mass presentation of pollen on the surface of anthers through slitlike openings (Proctor and Yeo 1973; Stanley and Linskens 1974; Faegri and van der Pijl 1979; Crepet 1983). Explosive pollen disperal occurs in wind-pollinated taxa, whereby pollen is shed through slits (Stebbins 1974). The Asclepiadaceae and Orchidaceae disperse pollen in small sacs (pollinia) of hundreds to thousands of grains. Pollinia are transported but not consumed by bees (Fig. 2.3).

Figure 2.3. Pollinaria carried on the heads of bees. Above is a female of *Xylocopa aureipennis* (Anthophoridae) from southern India with pollinia of Asclepiadaceae attached to the face. Below is a male *Eulaema cingulata* (Apidae) with pollinaria of *Notylia* (Orchidaceae) attached to the labrum. Photo by author.

Nectarless plants vary widely in the quality of their pollens. Within the poricidally dehiscent group (in which anthers release their pollen through pores), over 70 plant families are represented, most of them predominantly tropical families (Buchmann 1983). Buchmann estimates that 5% of the angiosperms display poricidal dehiscence. Pollen grains produced by these flowers tend to be small (10–30 µm in diameter), smooth, dry, and lacking in pollenkit, a polysaccharide substance that makes grains stick to each other and the stigma (Thorp 1979; Buchmann 1983; Simpson and Neff 1983; see also Table 2.1). However, these pollens may have the highest crude protein content of any plant, from 55% to 65% (Buchmann 1983, 1986). Primarily wind-pollinated plants tend to have relatively small, smooth pollen, 20–40 µm in diameter (Whitehead 1983). Their protein content is often low, 7–15% (Stanley and Linskens 1974). Although neither of these flower types usually provides nectar, and both usually have relatively small pollen, exceptions are known. Some poricidal Melastomataceae and many Ericaceae produce nectar (Mori and Pipoli 1984; Cane et al. 1985; Renner in press); the species of Ericaceae also shed pollen tetrads. Another peculiarity of poricidally dehiscent pollens is the occasional presence of pollen that lacks protoplasm (Simpson and Neff 1983). Size dimorphism in grains of a species also occurs, resulting in differences of more than an order of magnitude in grain volume (Buchmann 1983). A positive correlation has been noted between plants displaying differing grain size and those with variable sexual expression (variability in the relative proportions of pollen and stigmas of an individual) or variation in the size and shape of flowers (Stanley and Linskens 1974; Cruden and Lyon 1985).

Most pollen collected by bees in temperate areas falls within the range of 10 to 100 μm in diameter and has an average size of 34 μm (Roberts and Vallespir 1978); this seems likely to be representative for tropical bees as well. Taken as a whole, pollen grains range from 5 μm to >300 μm (*Manihot esculenta*, a euphorbiaceous tuber crop, has the largest pollen grains I have seen) and have volume differences of up to five orders of magnitude. The tiniest grains are a million times smaller than the largest. The surface texture of pollen collected by bees varies considerably and includes smooth, reticulate, and spiny grains. A number of plant groups visited by bees present pollen grains in groups of four or more (polyads) and include Mimosoideae, Cucurbitaceae, Rubiaceae, Ericaceae, Annonaceae, Hippocrateaceae, Ochnaceae, Gentianaceae, Winteraceae, and Araceae. Onagraceae also produce pollen in groups held together by viscin threads, and several bees that collect the pollen have specialized behavior and structure to obtain and manipulate the masses of pollen so gathered (Linsley, MacSwain, and Raven 1963; Roberts and Vallespir 1978).

Pollen of the nonporicidal taxa frequently have a 20–35% protein content, occasionally reaching 40% protein, such as among some Agavaceae (Stanley and Linskens 1974). Caloric values of pollen did not seem to vary significantly in studies of 42 species of wind- and insect-pollinated monocots and dicots (Colin and Jones 1980). Colin and Jones found that dicots ranged from 23,240 J/g of pollen to 28,215 J/g, and wind-pollinated monocots were between 20,523 J/g and 25,414 J/g. However, these values were obtained by analyses that included energy values of the pollen exines, which are largely untouched by bee digestion and may include a large portion of the pollen biomass (Stanley and Linskens 1974; Klungness and Peng 1984). Several angiosperm families produce pollen with high starch content – Asteraceae and Onagraceae, for example (Baker and Baker 1979). Many wind-pollinated and self-pollinating (autogamous) taxa produce starchy pollen, yet starch in pollen grains is also common among flowers normally pollinated by bees and other animals (Baker and Baker 1982a). Starch in pollen is toxic to some bees (Schmidt et al. in press)

The European honeybee has been observed collecting fungal spores, hyphae, conidia, and asci. In Panama, I have seen the stingless bees *Trigona fulviventris* and *T. nigerrima* collect liquid exudates and possibly spores from the stinkhorn fungus, *Dictyophora*. Collection of a fungal exudate by *T. fuscipennis* takes place on tree trunks, both in Central and in South America. Other substances honeybees have been seen gathering are the plant rust *Mycelia,* macroconidia of *Neurospora*, and flour from seeds and tubers such as wheat and manioc (Maurizio 1953; Shaw and Robertson 1980). All have low nutritional value and are apparently rarely collected, even though conidia strongly reflect ultraviolet light, which is generally attractive to bees (Shaw and Robertson 1980). Dispersal of ultraviolet-reflectant conidia that has been noted in solitary bee species (Batra and Batra 1985) reveals a relationship analogous to pollination of angiosperms dependent on a similar mechanism for attracting bees (Section 2.4). When collected, such unusual resources are gathered in large quantities when other, superior resources are unavailable.

In addition to crude protein content ranging from 7% to 65%, pollen may contain from a few to 20 amino acids (Beiberdorf, Gross, and Weichlen 1961; Kauffeld 1980) and various lipids, vitamins, minerals, carbohydrates, toxins, and potential feeding deterrents (Stanley and Linskens 1974; Southwick and Pimentel 1981). Barker (1977) reported that pollen contains chemicals toxic to honeybees, such as raffinose, lactose, stachyose, xylose, arabinose, galactose, galacturonic acid, gluconic acid, and pectin. Experiments with the stingless bee *Scaptotrigona postica* indicate that some of the same carbohydrates, such as raffinose, lactose, and arabinose, cannot by themselves support metabolism of worker bees and are undigestable, but not toxic, at concentrations of 7.5% (Zucoloto 1979). However, while sorbitol is ingested by this stingless bee, it is not imbibed and apparently not tasted by temperate area *Apis mellifera*. Mannose was found to be toxic to both bee species (Zucoloto and Penedo 1977). Galactose is a toxin to honeybees and has been found in stigmatic exudate (Barker and Lehner 1977).

Many tropical plants, notably aroids, palms, bromeliads, and diverse monocot and dicot families, produce crystals of calcium oxalate; these "raphides" and "druses" are mixed with pollen at dehiscence (Pohl 1941; Kugler 1942; Maurizio 1953; Stanley and Linskens 1974). These substances may deter feeding by certain animals (they certainly produce a prickling sensation on the human tongue), but they are not thought to damage the alimentary canal of bees (Stanley and Linskens 1974). They do, however, dissociate in water to produce oxalic acid, a common plant defense against herbivores (Rosenthal and Janzen 1979). Toxins such as these are thought to be eliminated by some mutualistic microbes associated with herbivores (Jones 1985). Discovery of similar actions within the provisions or gut of bees would shed light on many aspects of bee resource use. Preliminary work with a stingless bee species to which I fed a mixture of raphide-containing pollen of *Impatiens* and normal larval provisions suggested that larval mortality was unaffected by the raphides. The presence of raphide crystals and druses often goes undetected by palynologists and mellitologists interested in bee diet composition; crystals break down upon heating and dilution during preparation of slides for microscopic analysis.

2.1.7 Waxes

Wax is eaten, although apparently not digested, by at least the larvae of some bumblebees (Michener 1974a), and waxes are collected by some *Apis* and meliponine bees from abandoned or even occupied nests in order to construct new ones (Chapter 3). The energetic value of wax is higher than 40 kJ/g, making it a valuable resource (Table 2.1). Honeybee wax differs from the wax of meliponines and bombines in having a slightly higher melting point and complex esters; pure wax of all three apid subfamilies contains hydrocarbon monoester, primary alcohols, and free fatty acids (Blomquist, Chu, and Renaley 1980; Tulloch 1980; Blomquist, Roubik, and Buchmann 1985; Hepburn 1986). Saturated hydrocarbons constitute over 90% of the wax produced by primitive stingless bees of the *Trigonisca* group but make up less than 75% of bumblebee wax and less than 50% of European honeybee wax.

2.1.8 Other resources

The flesh of dead animals is used in place of other protein sources by certain neotropical bees (Roubik 1982a). Such necrophages are both obligate and facultative. Three of the former are known, apparently all of a single, closely related group of *Trigona* (Camargo and Roubik pers. obs.). The obligate necrophage appears to deposit digestive enzymes on the animal carcass, because it carries the partly digested flesh in a liquid slurry that is held in the crop and the flesh is imbibed with the mouthparts as though it were nectar. Small resources, such as dead lizards and frogs, are dominated by the bees and consumed rapidly, but a wide range of reptiles, amphibians, birds, fishes, and larger animals is harvested by the necrophages. Their quality as a source of amino acids, lipids, protein, carbohydrate, vitamins, and minerals must vary considerably with taxon and stage of decomposition. These carrion feeders shun animals that have been dead for more than a short time or are infested with fly larvae. However, they rapidly and efficiently harvest new carrion, which is converted into a greenish gray glandular material and then stored and used as though it were pollen (Roubik 1982a). The nutritional value of flesh gathered by necrophagous bees, despite having lower energy content than pollen, may be superior due to a substantially higher protein content (Table 2.1).

Unusual resources of bees indicate harvest of inorganic salts from many sources, and information on their role in nutrition (particularly synthesis) or other functions is incomplete. A correlate of the following discussion is that bees at times accumulate an excess of metabolic water (Bertsch 1984), which might lead to excretion rates accelerating the loss of ions that must be replenished. Sodium salts,

42 Ecology and natural history of tropical bees

Figure 2.4. Three female *Centris longimana* (Anthophoridae) (see arrows) imbibing salts at a seepage area along a river bank in the Colombian Amazon. Photo by author.

potassium, and phosphates are likely candidates as the primary resources bees seek by visiting urine, feces, blood, recently washed clothing, and areas where washing is done (Fig. 2.4), and campfire ashes. Barrows (1974) found that halictid bees chose NaCl when given the opportunity of selecting several nonfood substances. The collection of "moisture" or "sweat" from animals and carcasses has been noted in stingless bees, honeybees, and, not surprisingly, the "sweat bees," Halictidae (Schwarz 1948; Michener 1974a). Wet ashes are visited by a variety of apid bees in Sumatra (Salmah pers. commun.). Facultative necrophagy involving mastication of flesh by foraging bees and even utilization of dried or fresh blood (Chance 1983; Crewe 1985; Baumgartner and Roubik in press) suggests that some bees supplement the normal pollen and nectar diet with animal products. A correlation with local soil type is likely because the elemental constitution of soils is readily expressed in pollen and nectar (Bromenshenk et al. 1985). Low phosphate content could account especially for bees' propensity to forage at washing sites and ashes. I have seen large aggregations of bees at these sites in the Colombian Amazon, where *Trigona, Melipona, Apis* (Apidae), *Xylocopa, Centris, Ptilotopus, Peponapis* (Anthophoridae), *Megachile, Coelioxys* (Megachilidae), *Augochlora,* and *Augochloropsis* (Halictidae), both males and females, forage throughout the day on soil substrates by regurgitating water and then reimbibing it. These are the common bee families and genera in this area and in other neotropical regions, and more extensive observation might therefore reveal that most local bees gather inorganic salts in this manner. An individual bee repeats the process of regurgitation and imbibement several times; when forced to expel its crop contents, these appear to be muddy water, and their sugar content is negligible. For all but the obligate necrophages, the neotropical *Trigona hypogea* group, animal products are probably ingested to collect ions or salts. Many details of facultative necrophagy, mineral collection, and related behavior await investigation.

Material collected by the tropical apids (*Eulaema, Eufriesea,* and some meliponines) includes animal feces. Such material is either incorporated directly in construction of the nest or is stored in separate containers by some *Eulaema* and *Melipona* (Chapter 3). A number of meliponine bees forage excrement and carry away bits of fecal material on the hindlegs. There is no proof that this is used only as nesting material, and it may conceivably provide a source of nitrogen or phosphates. Feces may be used by some tropical bees as the starting material for production of a germicide (Chapter 3).

2.2 Mechanisms of resource collection

Bees have many traits functional in procuring and transporting plant products. Transport behavior is relatively uniform although the collecting behavior and structures are extremely varied. Bee behavior and floral structures are likely to have coevolved when the specializations of each are extreme (Sections 2.4.3 and 2.4.4.), but the opportunism of foraging bees is still apparent in their activity and seasonality (Chapter 4).

2.2.1 Pollen capture

Females transport pollen externally on *scopae,* the brushy areas usually on hindlegs, *corbiculae,* the scooped-out areas on hindtibiae, or internally mixed with honey or nectar. Some traits are apparently functional adaptations and are unique to particular bee taxa. These involve specialized mechanisms, for example, for manipulating very small pollen or oil (Section 2.2.3). On the other hand, many traits are widely shared among bees, likely demonstrating the many uses of ancestral behavior and anatomy. Roberts and Vallespir (1978) review sites on which female bees carry their pollen (Table 2.2 and Fig. 2.5). Among the melittids, one

Table 2.2. Modes of pollen transport by bees

Bee family	Pollen-carrying structures							Pollen-load additives		
	Hindleg				Propodeum	Metasoma (sterna)	Internal (crop)	Dry	Nectar	Oil
	Trochanter	Femur	Tibia	Basitarsus						
Stenotritidae	rarely	freq.	freq.	rarely				✓		
Colletidae (Hylaeinae, Euryglossinae)	occas.	freq.	freq.			freq.	~exclus.	✓	✓	
Oxaeidae	occas.	freq.	freq.					✓		
Halictidae	rarely	freq.	freq.	rarely		rarely		✓		
Andrenidae (Panurginae)	occas.	freq.	freq.		rarely	rarely		✓		
Melittidae			freq.	freq.				✓	✓	✓
Ctenoplectridae			freq.	freq.						✓
Fideliidae						~exclus.		✓		
Megachilidae						~exclus.		✓		
Anthophoridae		rarely	freq.	freq.		rarely		✓	✓	✓
Apidae			freq. ~exclus.				rarely			✓

Abbreviations: occas., occasionally; freq., frequently; ~ exclus., almost exclusively.
Sources: Modified from Thorp (1979) and Roberts & Vallespir (1978).

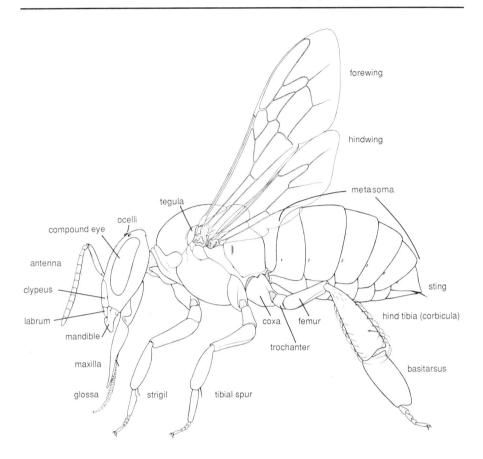

Figure 2.5. General external anatomy of a worker apid bee (*Apis mellifera*). Original drawing provided by J. M. F. Camargo.

of the more primitive of living bee families (Michener and Brooks 1984), there are species that carry pollen of quite variable consistency. Different taxa carry their pollen either dry, mixed with nectar, or mixed with oil. Regarding the transport sites, apids, melittids, ctenoplectrids, and megachilids all show little variation whereas the other bee families display wide variation. A distinctive feature of the Apidae, however, is that almost all of its more than 1000 species place pollen solely on the hindtibiae. Only the *cleptobiotic* (robbing) apids carry all pollen in their crop, mixed with nectar (Section 3.2.8). Further, apids do not transport pollen without moistening it with regurgitated nectar. Many other bees are able to transport unmoistened pollen. This ability is largely due to dense and frequently plumose hairs that are often relatively long or dense on the portion of the body receiving pollen (Fig. 2.6). Evolution of more elaborate or dense body hairs may

Figure 2.6. Hindlegs of parasitic and normal anthophorid bees. On the lower left is the leg of the cleptoparasite *Triepeolus*, which is never used to carry pollen, contrasted with that of *Svastra*, showing the brushlike scopae used to carry pollen to the nest. Original drawing by Denis J. Brothers, provided by C. D. Michener.

have been associated with the pollen-collecting habit of bees as they diverged from sphecoid wasp ancestors (Brothers 1975).

Internal transportation of pollen, mixed with nectar or honey and then passed into the crop, occurs to varying degree among all nonparasitic bees (Michener 1974a; Roberts and Vallespir 1978; Thorp 1979; Houston 1981). Females of the colletid subfamilies Euryglossinae and Hylaeinae consume all pollen they gather while collecting it at flowers. It is then carried to their nests mixed with nectar in the crop. In addition, workers of *Lestrimelitta* and *Cleptotrigona,* both obligate robbers of other stingless bees (Meliponinae, Section 3.2.8), carry pollen mixed with honey in order to transfer it from a host's nest to their own. Pollen transport sites and structures for manipulating pollen have been completely lost in these bees although they use the hindlegs to carry cerumen or resin pillaged from other bee nests (Portugal-Araújo 1958; Nogueira-Neto 1970a; Sakagami, Roubik, and Zucchi in press). A number of nonparasitic bees, for example *Ceratina* and *Xylocopa* (Anthophoridae), appear to collect pollen in both manners and display a moderate reduction of scopae. Explanations for this behavior are lacking.

Electrostatic charge may contribute to the pollen-holding capacity of the bees, particularly those with dense hair on body parts contacting flowers (Erickson 1975). In honeybees, which have a moderately hairy thorax and base of the head, foragers have a slight positive charge, which is augmented during flight. The pollen in anthers tends to have a negative charge (Stanley and Linskens 1974) and thus readily clings to the bodies of bees.

Ordinarily, bees can collect pollen simply by contacting the surfaces of anthers and then grooming pollen adhering to hairs and, especially for relatively sticky pollen, elsewhere on their body (Section 2.2.2). Bees usually collect pollen by combined working of the forelegs and mouthparts, even shredding or destroying anthers in the process, and by movement of the body over the anthers and flower parts. Pollen on stigmas is apparently not harvested by bees, presumably because it has already germinated and lost nutritional value, or possibly due to toxic stigmatic exudates mentioned previously. Subtle behavioral or anatomical features may come into play even during relatively passive pollen capture. Many bees press strongly downward on a flower while collecting pollen as they grasp its edges with their claws. Their bodies deflect petals and stamens with what appear to be specialized and rapid movements. I have noticed the weak buzzing and rocking motions of *Melipona* as they visit flowers with fully exposed and accessible pollen (*Mimosa* and *Rinorea,* for example). Other bees are known to display similar combinations of gathering behavior, as they use methods particularly suited to harvesting poricidally dehiscent pollen while visiting anthers of other types of flowers (Buchmann 1985).

Buzz-collecting is an outstanding example of bee behavior modified to cope with a protective mode of pollen presentation found in many tropical plants. Such buzz-foragers have specializations useful for collecting the small, dry pollen in tubular anthers. Bees cling to the anthers while shivering their flight muscles. During this process, pollen is sprayed from the anther tips and is caught in hairs over the surface of the bee's body (Thorp 1979; Buchmann 1983). Three basic means are employed for pollen collection from plants with poricidal anthers:

1. buzzing while clinging to anthers;
2. "milking" the anthers with the mandibles while also buzzing, which is a variation of the first behavior and only works well when anthers are somewhat soft; and
3. gleaning pollen fallen onto other parts of the plant [Michener, Winston, and Jander (1978); see the review of older literature by Thorp and Estes (1975)].

None of these methods appears to be completely restricted to poricidally dehiscent flowers (Buchmann 1985). Pollen thieves remove pollen from tubular anthers by inserting the proboscis, and destructive flower visitors extend this behavior by biting off anther segments (Wille 1963; see also Section 2.4.4). Although few exact measurements have been made, the vibration frequency of a bee buzzing anthers for pollen is on the order of 4–5 kHz, and its duration varies from roughly 0.1 sec to 10.0 sec. Shorter visits seem to occur when more pollen is available (Buchmann 1983). However, some bees may detect such pollen rewards and, like bees foraging nectar, respond by remaining longer on a flower to harvest its pollen. Cane (pers. commun.) has found that foragers of *Habropoda* (Anthophoridae) spend more time buzzing the individual flowers of *Vaccinium* that offer more pollen. Pollen is completely concealed in the anthers, and it offers no external visual cue. However, bees such as *Centris* groom themselves rapidly while visiting the flowers, possibly detecting whether they have acquired pollen, because the grooming leg is passed by the mouthparts after scraping the head and thoracic notum (Snow and Roubik 1987).

It is probably significant that the most numerous and conspicuous tropical bees, Apinae and most Meliponinae (tribe Trigonini), lack appropriate behavior for legitimately harvesting pollen from poricidal anthers. Moreover, other apids that are less abundant or widely distributed in the tropics, *Melipona* and *Bombus*, do in fact employ buzz-collecting. The honeybees do nothing more than glean such pollen, removing some that has been released during visitation by another bee, whereas a small minority of the Trigonini destroy anthers to remove it (Section 2.4.4). Thus the highly eusocial bees are largely unable to use this pollen source. In addition, few or no Ctenoplectridae, Megachilidae (Neff and Simpson 1988), or Fideliidae appear to collect such pollen, either as gleaners, destructive flower visitors, or legitimate visitors.

Most bee families have many genera and species that vibrate anthers, including those of some flowers displaying longitudinal dehiscence. Part of the specific behavior necessary to extract pollen at a poricidal plant species is probably learned by the forager – for example, by finding where to grasp anthers with legs and/or mandibles. Some species may more readily learn the appropriate behavior or produce vibrations of the correct intensity to extract pollen from anthers that, in different species, vary much in size, pollen content, and thickness. Common buzz-collecting tropical bees include all nonparasitic euglossini; many anthophorids, notably *Xylocopa, Centris, Epicharis, Gaesischia, Exomalopsis, Amegilla, Anthophora, Habropoda,* and *Thygater,* the apids *Melipona* and *Bombus* ; most Colletidae and Oxaeidae; a few Andrenidae, Melittidae, and Stenotritidae; and various medium-size and large Halictidae (Houston and Thorp 1984; Buchmann 1985; Michener pers. commun.; Roubik pers. obs.). The small anthophorids such as *Exomalopsis,* the Halictidae, and various *Melipona,* some species of which are only 7 mm in length and weigh less than 30 mg, include the smallest bees that buzz anthers. The other bees are considerably larger, ranging from 13 mm to 30 mm in length and weighing up to more than a gram. Like other types of pollen collection, buzz-collection can result in pollen deposition on the top of the head or thorax – usually the lower clypeus, labium, and labrum of the face or the anterior thorax (nototribic pollen collection); deposition on the venter, usually between the coxae and femora of mid- and hindlegs and the metasoma (sternotribic collection); or placement on the side of the bee (pleurotribic collection). The loud buzzing sound of tropical bees is often the only indication of their presence as they forage at poricidally dehiscent flowers in the canopy or otherwise hidden from view. Experience in locating bees using their loud bursts of sound led Linsley (1962a) and Michener (1962) to draw attention to this novel method of pollen collection, extending and clarifying some notes of earlier observers (Sprengel 1793).

The Apidae have a unique and highly derived structural modification of the hindtibia, the corbicula or pollen basket, used to carry pollen, resin, wax, and other substances (Fig. 2.7). The face of the corbicula is usually widened, concave, and smooth with a few long, thick hairs or setae. Its anterior edge is rimmed with long setae curving over the area that a full load would occupy; posterior setae follow the contour of the corbicula and extend its effective area. Each of the setae may have the potential of indirectly registering load sizes by the degree of its displacement, in the manner of setal mechanoreception found in many insects (Ford et al. 1981). Loading of pollen in the apid corbicula, among all but the Meliponinae, occurs through the motion of the basitarsus and its distinctive basal pollen press (Michener et al. 1978; Wille 1979a). The trigonine subgenus *Scaura* is unique in that all its three species possess a greatly enlarged hindbasitarsus that is dragged across flower parts on which pollen has fallen, after which it is passed to the middle tarsus and then onto the corbicula (Laroca and Lauer 1973; Michener et al.

Figure 2.7. Hindleg of a worker apid bee, showing the corbicula of *Scaptotrigona* (Apidae) on which the pollen load is carried, fringed with long setae. Original photograph by Stephen L. Buchmann.

1978). This gleaning technique is also employed by Trigonini with no special leg morphology, such as *Plebeia* and *Trigonisca,* and even by honeybees. Some anthophorid and megachilid bees have long hairs on the maxillae, used to gather pollen from narrow corollas (Stephen, Bohart, and Torchio 1969). Pollen robbing and its associated behavior involve biting through anthers and then removing pollen with the labial palpi, glossa, and specialized hairs; this will be discussed further in the context of bees as plant enemies (2.4.4).

The area used for transporting the pollen load is called a scopa for other, non-apid bees and is formed by specialized bristles alone with no accessory pollen press. Its location is on the hindlegs, metasomal sterna, or propodeum. The metasomal scopae of megachilids and fideliids have rows of long, stiff, backward-slanting setae across the sternites (Pasteels and Pasteels 1975). Their pollen loads, unlike those of other bees, virtually cover the entire metasomal sternum but are present nowhere else on the body. At least one anthophorid bee, *Eremapis,* possesses keeled setae used to carry pollen on the underside of the abdomen, and some colletids appear to capture leguminous pollen with these "scooping setae" (Neff 1984; Simpson, Neff, and Moldenke quoted by Neff 1984).

2.2.2   Pollen cleaning, manipulation, and storage

The general subject of grooming by bees has a robust significance in their ecology. Several areas on their body surface are not reached by grooming, and external travelers such as mites, phroetic parasites, and pollen are found in these places (Kimsey 1984b, see also Section 3.2.7). Kimsey reported that in euglossine bees such "free zones" are ordinarily located on the underside of the head and neck region, the anterior face, the propodeal region and part of the upper surface of the abdomen. The first metasomal tergum and perhaps most crevices on the body seem likely to constitute free zones in other types of bees.

Grooming and often the moistening of pollen permit female foragers to position it in specific transport sites on the body and also to initiate the process of converting pollen to suitable nest provisions for larvae (Section 3.3.2). Bees may be airborne while executing these movements or may perform them while hanging from the vegetation by their mandibles, or while resting there. Pollen gathered on the head and thorax is passed by most bees from legs on the same side of the body (ipsilateral) from fore- to mid- and thence to the hindlegs. Grooming of the head employs the forelegs although grooming of the folded tongue in long-tongued species may employ mid- and hindlegs (Kimsey 1984b). The thorax is groomed with the midtarsal segments, whereas the wings and portions of the abdomen are groomed using the hindlegs. Pollen is passed to the corbiculae by synchronous movements that are bilateral, that is, duplicated by legs on both sides of the body (Kimsey 1984b). An exception to this route occurs among *Trigona* that damage anthers to extract pollen with the mouthparts; these species transfer pollen from the forelegs to the thoracic venter before further manipulation (Nogueira-Neto 1957; Michener et al. 1978; Roubik in press a). Rapidly rubbing together the hindlegs while hovering or in flight, the contralateral hindbasitarsus pushes pollen upward on the opposite basitarsus. The apid bees *Apis*, *Bombus*, and most meliponines and euglossines add nectar to pollen while it is being groomed and, in the final stage of pollen packing, transfer pollen from the middle basitarsi, one at a time, to contralateral hindbasitarsi (Ribbands 1953; Michener et al. 1978; Kimsey 1984b). In the meliponines and euglossines, pollen is most often transferred between ipsilateral midtarsi to the hindbasitarsi. All apids use the ipsilateral midlegs to smooth and pat down a corbicular load.

Grooming and pollen manipulation movements originate with the process of bodily cleaning; grooming by bees includes a number of typical behavioral sequences (Farish 1972; Jander 1976; Jander and Jander 1978; Michener et al. 1978; Thorp 1979; Kimsey 1984b) and can be related to apoid phylogeny and the use of pollen and other plant products. In a review of traits associated with pollen collection, Thorp (1979) summarizes pollen manipulation, stating that it requires integrated cleaning behavior with a successive series of grooming structures to move pollen to rear transport sites. Loading of the pollen transport apparatus involves the substitution of scraping movements that transfer pollen to this site, rather than cleaning motions to merely discard debris removed during grooming. For the most part, grooming behaviors have persisted from the aculeate ancestors of apoids. Structural traits for capturing and retaining pollen or oil were added.

Discussing wing-grooming patterns of bees and sphecid wasps, Jander and Jander (1978) describe seven basic patterns among apoids and the lack of marked differences between Sphecidae and Apoidea. A sequence of cleaning behavior usually occurs: Cleaning of wing surfaces and bases is followed by cleaning of the hindleg with the midleg and cleaning of the abdomen. The middle leg scrapes the outer surface of the ipsilateral hindleg and is then passed between appressed hindbasitarsi, which then are rubbed against each other. Antennal cleaning is also homologous among Hymenoptera; virtually all species pass the antenna through the cleft between the tibial spur and basitarsus of the ipsilateral foreleg, the location of the special antennal cleaning gap or strigil (Fig. 2.5). Similarly, head and thoracic cleaning of the Sphecidae and Apoidea take place with forward movements of the fore- or midlegs. In contrast to most other Hymenoptera, bees were found by Jander (1976) to clean the forelegs with the midlegs rather than the mouthparts. Midlegs of bees have two types of cleaning structures. Combs are formed by rows of close, stiff, bristlelike setae that are slanted diagonally and have blunt ends. Brushes are dense pads of relatively short hairs, either oriented perpendicular to the cuticular surface or aligned at an angle to it. Both combs and brushes tend to be smaller in male and parasitic bees, or they have been lost during evolutionary change (Section 2.2.4).

A multitude of distinctive modified bristles or setae exist for manipulating pollen; they parallel but are less spectacular than the oil-handling apparatus depicted in the following section. Virtually all pollen-collecting bees have grooming brushes or combs on the forelegs and on the hind- and midbasitarsi and tibiae (Thorp 1979; Neff 1984 and the illustrations therein). Five additional locations of grooming structures tend to vary among bee families. Two of these are upon appendages of the mouthparts, the galea and stipes. All families but the Oxaeidae and Fideliidae have these grooming combs (Thorp 1979; Winston 1979a). Almost all female bees seem to eat pollen while visiting flowers, and the grooming of pollen from mouthparts takes place during pollen consumption by female foragers (Jander 1976). The remaining pollen grooming structures are on the midlegs. These are combs on the inner surface of the tibia or the ventral base of the femur and brushes on the ventral base of the femur or the ventral surface of the trochanter (Fig. 2.5). In the Apidae, the midlegs have brushes only on the trochanters (Kimsey 1984a,b), whereas all other families have genera with at least two different types of midleg brushes or combs. The Andrenidae and Colletidae have all four types represented among their taxa. These data are of interest because they suggest that the oldest bee lineages have responded to the variety of their food sources through structural evolution, to a greater degree than have more recent bee groups. Whether behavioral versatility in the more derived bees takes the place of morphological adaptations is still uncertain.

2.2.3 Floral lipid collection

The basitarsal setae have become further modified to combs and setal pads on the fore- and midtarsi of most bees that collect oil or other nonnectar liquids from flowers (Fig. 2.8); inner hindtibial spurs have also been modified to include flattened, pectinate teeth for manipulating oil by squeezing it from the brushes on which it is collected (Vogel 1969, 1974, 1981, 1984; Buchmann and Buchmann 1981; Neff and Simpson 1981; Cane et al. 1983; Buchmann 1987). Scraping and perforating fixtures are found on the appendages of the oil-collecting bees. Blade-like setae on fore- and midtarsi are used to rupture the lipid blister elaiophores of the Malpighiaceae. Typically, the plant oils are scraped or adsorbed (mopped) from flowers and passed directly by the mid- or forelegs to the scopae of ipsilateral hindlegs. The oil-transferring leg is drawn across the juncture of the hindtibia and basitarsus, and the oil is deposited in scopae that are likely to be modified to hold the liquid. A mat of short, dissected hairs holds the oils by capillarity, and long, stiff unbranched hairs may support the mass on the scopae (Linsley and Cazier 1963; Roberts and Vallespir 1978). Oil and pollen are often mixed together (Table 2.2), and the oil forms a sticky mass on the long scopal hairs. The specialized pectinate hindtibial spurs may be needed to manipulate this load, but similar structures occur in bees that do not collect floral lipids (Neff and Simpson 1981; Michener pers. commun.). There is at present no description of how oil–pollen or pure oil loads are removed by bees in their nests. My observations in Panama of *Centris analis*, a bee that deposits oil on a finished brood cell cap, indicate that the hindlegs are cleaned as the ipsilateral midlegs are rubbed against them and the side of the abdomen.

Floral lipids used to feed larvae or construct part of the nest are gathered by temperate and tropical bees, but most oil-collecting species are tropical. Bees that gather oil belong to the anthophorid tribes Centridini (*Centris, Ptilotopus*, and *Epicharis*), Exomalopsini (*Paratetrapedia, Tapinotaspis*, and *Chalepogenus*) and

Figure 2.8. Oil-gathering combs and brushes on the legs of anthophorid bees. Upper left and middle: *Tapinotaspis*, subgenus *Tapinorhina*, ventral forebasitarsus and enlargement of same area. Upper right and lower left: *Tapinotaspis chalybaea*, anterior midbasitarsus and enlargement of same area. Lower and middle right: *Tetrapedia maura*, anterior foretarsus and enlargement of same area. Abbreviations on figures signify the following: cc = combs of modified setae; sp = specialized decussate setae; stc = strigular comb (appearing in the strigulus gap normally used for antennal cleaning; see Fig. 2.5). Original photographs provided by John L. Neff (Neff and Simpson 1981).

Tetrapediini (*Tetrapedia*). All of these genera are found in the neotropics as well as in the American subtropics (Neff and Simpson 1981). These last authors speculate that some other anthophorines may occasionally gather oils, including *Anthophora*, *Lanthanomelissa*, and *Exomalopsis*. The structures required for oil collection are, however, inconspicuous in these bees.

Observations in the neotropics indicate the oils are collected to some extent by stingless bees having no specialized oil-collecting structures. Ways in which they use the plant lipids are uncertain. At least some species of neotropical subgenera of *Trigona*, such as *Tetragona dorsalis* and *Trigona spinipes*, *T. pallens*, and *T. cilipes*, collect oils (summary by Steiner 1985; Roubik pers. obs.). These oils are probably not ingested but instead used for nest construction. *Melipona* reportedly collects oils from the melastome *Mouriri* (Renner 1984), but it now appears unlikely that this plant produces the lipids (Renner in press).

2. Foraging and pollination   55

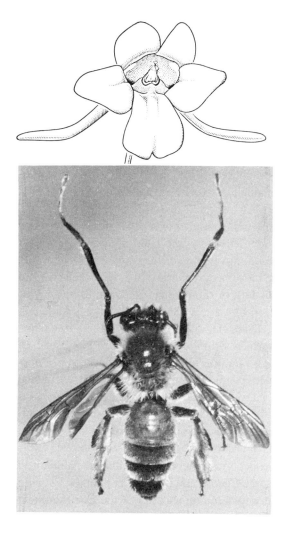

Figure 2.9. An oil collecting female of *Rediviva emdeorum* (Melittidae) and one of the flowers that provides oil (*Lysimachia,* Primulaceae). Bees insert the elongate forelegs in the long lateral floral spurs to adsorb oil. Original photograph and drawing provided by Stephen Vogel (Vogel and Michener 1985).

Both temperate and tropical Melittidae (including the holarctic *Macropis*) collect oils (Michener 1981; Cane et al. 1983; Vogel and Michener 1985). The last authors describe the association existing between a bee of southern Africa, *Rediviva emdeorum,* and an endemic plant genus of the Scrophulariaceae (Fig. 2.9). The female bee has immensely elongate forelegs that fit inside the pair of elongate spurs where the floral oil is presented. Here is one of the most striking known examples of bee–flower coadaptation and probable coevolution (Section 2.4). Several other bees of this genus collect oil but gather it from several plant species (Manning and Brothers 1986).

From field studies by Michener (Michener and Greenberg 1980) and Vogel (1981) on *Ctenoplectra* in Kenya and Borneo, we know that an Old World tropical bee family has evolved unique behavior and collecting equipment related to oil-producing flowers, apparently just the Cucurbitaceae. Female *Ctenoplectra* have a hindtibial spur that is a crescent-shaped comb, used to remove the oil adsorbed onto brush setae located in patches on the underside of the abdomen. Ctenoplectrids swing the abdomen from side to side while walking over trichome elaiophores, mopping up the oils and grooming them from the venter, between swings, with their crescentic combs.

The functional morphology of oil-collecting structures has many known behavioral correlates to flower and elaiophore type. Neff and Simpson (1981) describe oil-collection patterns that are "two-legged" (exomalopsine bees) or "four-legged" (centridine bees). These refer to the number of sepals, buds, or trichomes from which oil can be collected simultaneously; the latter style requires tarsal modifications on four legs. Collection movements of the forelegs differ for different plant families, with clockwise movement required of the right foretarsus and counterclockwise movements in the left, or the reverse. Neff and Simpson emphasize the functional sophistication of traits such as the enlarged, flattened, apically curved overlapping setae that are flexible but maintained in a row when scraped across an irregular plant surface. Strong ridges, in which giant setae of tarsal combs of *Centris* and *Tetrapedia* are inserted, and the forebasitarsi of *Paratetrapedia* that undergo extreme expansion and flattening are also structures unique to oil-gathering bees. The evolutionary sequence in which these modifications came about is unclear, but Neff and Simpson suggest that putative ancestral oil-collecting bees may have had unspecialized setae on the abdomen and legs that allowed collection of trichome lipids or nectars. These later became modified to the current specializations of bees that collect from trichome or epithelial elaiophores, and have secondarily been lost in some *Centris* and other anthophorids. Because no known oil-collecting bee is a recent, common ancestor of the Centridini and Exomalopsini, even within the Anthophorinae there is no evidence as to whether oil collecting evolved independently or exists as a homologous trait that arose only once in the ancestors of this group.

2.2.4 Retention or loss of traits

When the gathering or transport behavior of some bees differs fundamentally from most other bees, it is usually obvious in the bees' general appearance. However, the retention of some apparently useless traits in haploid male bees and sometimes in workers of social species (Michener 1974a, p. 245; see also Section 3.3.5) is as much a part of bee biology as is their loss among females.

Obligate robbers (Section 3.2.8), parasitic and male bees, and the queens of obligately social species have a smaller number of branched hairs and a similar reduction in most other structures associated with pollen collection (see, for example, Michener 1978; see also Fig. 2.6). The reduction principle often has application to predicting the parasitic lifestyle (Section 3.2.8) of a bee. For example, the absence of oil-manipulation apparatus in the African bee *Ctenoplectrina,* despite lack of behavioral data, implies a parasitic lifestyle and adaptive divergence of females from a closely related host, the oil- and pollen-gathering *Ctenoplectra* (Michener and Greenberg 1980). The traits of oil collection are secondarily lost in these and other bees (Neff and Simpson 1981; Sakagami and Itô 1981). Similarly, a parasitic lifestyle was implied for Hawaiian hylaeines that, compared with their probable hosts, lacked hairs on their labial palpi used to rake pollen grains. Host and proposed parasite were assigned to the same genus *Neoprosopis* because the differences between them were cryptic (Perkins and Forel 1899).

A few male bees have specialized foraging structures that are absent in their females. Only the males of euglossine bees possess modified foretarsal pads of dense brushes, midbasitarsal combs, and dense overlapping hairs ridging the sides of a large hindtibial slit to which substances are passed via the basitarsal combs. An expanded hindtibia containing spongy tissue is used to receive these collected substances (Cruz-Landim et al. 1965; Dressler 1982; Kimsey 1984a,b). Anatomical differences from females have been found in the alimentary canal in males of the neotropical *Oxaea flavescens,* which ostensibly only perforate flowers when foraging and thereby imbibe nectar unusually rich in amino acids (Camargo, Gottsberger, and Gottsberger 1984).

In most other bees, both the males and females eat some pollen during their adult lives (probably the result of imbibing pollen in nectar for males) and have alimentary structures that remove pollen from the nectar in the crop. Pollen is removed from the crop (honey stomach) and passed into the midgut by a filter of hairs within the contractile mouth of the proventriculus. Three pairs of rectal pads are sites of high metabolic activity that resorbs amino acids, ions, and water from the excreta (Blum 1986). Both of these structures are greatly reduced in male *Oxaea* mentioned above. Camargo et al. (1984) suggest these modifications are due to the copious release of floral amino acids into nectar, resulting from flower perforation by male *Oxaea.* The male bees do not come into contact with anthers and thus cannot possibly ingest pollen even during grooming of the face and head.

Secondary loss of the corbicula, pollen-capturing hairs on the mouthparts, and associated structures for pollen harvest has occurred in the cleptobiotic apids *Lestrimelitta* and *Cleptotrigona,* which never visit flowers (Nogueira-Neto 1970a; Michener 1974a). Species that never use pollen in any form, such as the *T. hypogea* group of the neotropics, lack a corbicula and also lack hairs on the

Figure 2.10. The corbiculae of *Trigona pallens* (left) and of a worker of the *Trigona hypogea* group (Apidae). The former is a facultative necrophage and collects carrion with the mouthparts in addition to carrying pollen on the corbicula. The second species is an obligate necrophage and never transports pollen on its hindlegs. Original photographs provided by Stephen L. Buchmann.

mouthparts that apparently allow congeneric species to pull pollen from within tubular anthers of poricidally dehiscent plant species (Wille 1963; Roubik 1982; Renner 1983). This bee, however, does collect resin on the hindtibia (Roubik 1982a). In Figure 2.10 a worker of the *Trigona hypogea* group is compared with a worker of closely related *T. pallens*. The latter also collects carrion of dead animals, which is the only source of protein for the *T. hypogea* group. The facultative carrion collectors of the subgenus *Trigona* have long, hooked hairs on the labial palpus (Fig. 2.11). These are used to extract pollen from anthers that they have chewed open (Nogueira-Neto 1957; Renner 1983). In contrast, workers of the *T. hypogea* group have short, straight hairs on the palps, indicating that they have lost these pollen collecting structures. The worker of the *obligate* necrophagic bees has no depressed, wide area on the hindtibia for transport of pollen, whereas the species that sometimes use flesh but also collect pollen possess a normal corbicula (Fig. 2.10). This group of stingless bees, a single subgenus, often robs

2. Foraging and pollination 59

Figure 2.11. The specialized sinuate, curved hairs on the labial palpus of *Trigona pallens* (Apidae) used for removing the pollen from anthers that it perforates. Photo by author.

flowers for both pollen and nectar and has toothed mandibles that apparently enable the worker bees to efficiently remove floral tissue (Roubik 1982a,b).

Mandibular teeth used in combat and resource collection seem to be maladaptive for those Trigonini in which they are not used. The subgenerically distinctive mandibular teeth are much reduced or absent in males of the *Trigona* species (Fig. 2.12) and most other stingless bees. In contrast, mandibular teeth of megachilids are conspicuous in both sexes. The explanation seems to be that both sexes of *Megachile* must emerge from cocoons and usually leaf cells that may be tough and require considerable effort to break apart upon emergence (Section 3.1.4). Mandibular teeth in males might enhance their emergence from cells; the females could also use them to perforate leaves and build nests.

Although there is apparently some cost associated with retention of pollen-collecting traits that are no longer employed – for example, as a parasitic life style is evolved – males of various bees retain some features that would seem beneficial only to females. Such traits may have selective neutrality and are merely not suppressed, which is another way of suggesting that if there were strong selection

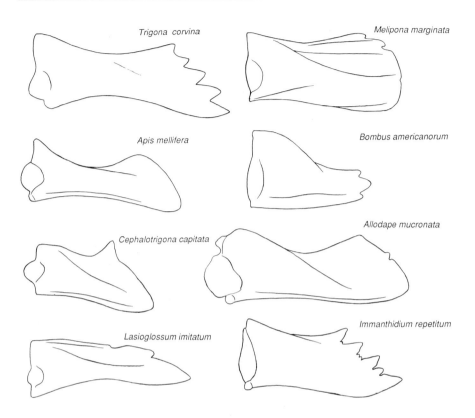

Figure 2.12. Diversity of bee mandibles. The taxa shown are females or workers belonging to the familes Apidae, Halictidae, Megachilidae, and Anthophoridae. Original drawings provided by C. D. Michener (Michener 1974a).

pressures against them, either the traits or their bearer would have disappeared. Males of oil-collecting anthophorids have retained to a degree the combs and brushes used in oil collection while apparently completely lacking this behavior (Sakagami and Itô 1981). Male colletid bees have a glossa like that of the females (McGinley 1980), the dorsal surface of which is used to apply a glandular secretion to the lining of the cell (Section 3.1). Application of the Dufour's gland contents to cell walls is the only known use for this peculiar bifurcated glossal structure, and it is unlikely to be useful in nectar fogaging by males (Fig. 2.13). Males of certain bumblebees even have very weak corbiculae, for which there is no conceivable function (Sakagami and Itô 1981). Selection has promoted the genetic dominance of rather specialized female traits having great importance to reproductive success, so that they are even expressed in the hemizygous males (see Section 3.3.5; see also Fig. 2.16).

2. Foraging and pollination   61

Figure 2.13. Glossal specializations of the male colletid bee, *Colletes*, showing the bifid glossa that produces four extensions of the distal tongue. Original photograph by Carl Hansen.

2.2.5   Evolution of gathering and unloading

The processes by which the corbicula is packed by Apidae and the general evolution of gathering and unloading behavior in bees have been considered by Michener et al. (1978) and Kimsey (1984a). The envisioned primitive-to-derived sequence involves initial carrying of pollen in the crop, as do some modern Colletidae (Section 2.2.1) and then evolution of transport in a brushy scopa on hindlegs, as in most nonapid bees. It is suggested that the latter method is more efficient, because larger loads are carried during a single foraging trip. However, the pollen on the abdomen is still largely lost because it cannot be groomed and then placed in scopae, or it can only be brushed off into a cell in the nest. In itself, brushing into a cell could be sufficient, but apids have evolved grooming movements that include scraping of the abdomen with the inner surfaces of the hindbasitarsi. The rastellum, a comb on the distal end of the hindtibia (Fig. 2.14), is further used by apids to scrape pollen from the contralateral basitarsus and is associated with a unique series of "pumping" movements displayed by hovering

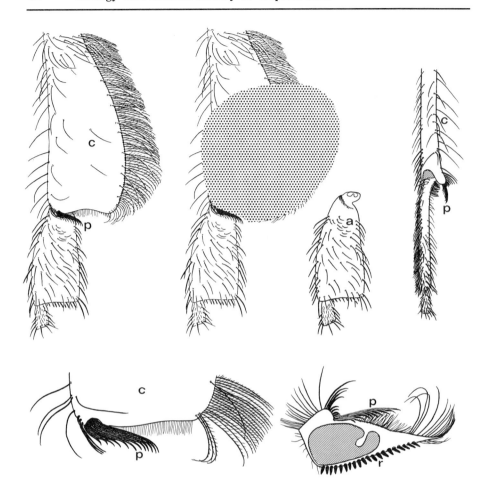

Figure 2.14. Hindleg specializations used for pollen manipulation by the apid bee *Trigona pallens*. Upper left: lateral view of left hindtibia and hindbasitarsus and full pollen load. Upper right: hindbasitarsus showing auricle, and oblique view of right hindtibia and basitarsus. Lower left: outer apex of left hindtibia. Lower right: oblique lower view of the distal left hindtibia with the hindbasitarsus removed. Abbreviations signify the following: c = corbicula or "pollen basket"; p = penicillum, r = rastellum or "pollen comb"; a = auricle or "pollen press." Original drawing provided by C. D. Michener (Michener et al. 1978).

apids while foraging. A rastellum is lacking in some meliponine groups (Michener et al. 1978; Camargo and Moure 1983). In all apids, pollen is loaded from the tibial-basitarsal joint and pushed upward onto the outer surface of the corbicula. A further derived method is that of the megachilids and presumably the fideliids, in which the hindlegs, rather than the middle legs, are used to transfer pollen to the scopa, which is on the metasoma of the abdomen instead of the hindlegs. Because their hindlegs also can brush the abdomen, these bees should retain much of the pollen captured on this part of the body. Moreover, evolution of the apid corbicula, which is largely an open, smooth area, allows bees to save time when dislodging the pollen load. This is accomplished with a single movement of the midleg by apids, yet requires several strokes in other bees.

An interpretation that ended here would overlook an outstanding feature of the apid corbicula. It is more than a pollen transport device that is occasionally used for unusual materials such as feces, mud, cerumen, or wax. It has an equally important function in transporting resin and gummy substances used as building materials. These substances are much stickier and less easy to remove from hairs or combs than are pollen and oils. Apart from corbicular structure, an indication that natural selection affects resin manipulation is that the process of passing resin back to the hind corbicula is the only strictly asynchronous (i.e., one-sided) grooming pattern known in euglossine bees (Kimsey 1984b). If apid corbiculae have undergone modifications in order to be readily cleaned, it may also be necessary for apids to make a sticky mass of their pollen by moistening it with nectar. Similarly, melittid and andrenid bees that mix nectar with pollen can be contrasted to related species by their small, sparse and only moderately branched scopal hairs (Thorp 1979). In all these groups a strong correlation appears between scopal or corbicular structures that do not tenaciously hold pollen and the behavior of moistening pollen loads with nectar. An apt generalization is provided by the local word for the many stingless bee species of Sumatra, western Indonesia: *galo-galo*, or bees that clean sticky substances from the legs (Inoue pers. commun.). Many worker stingless bees, however, normally have hardened resin on their hindlegs at all times, implying that it can no longer be removed. The possibility that strong-smelling resins serve some function in forager communication has been mentioned, but no experimental data are available (Roubik 1980a). Nest mates help to remove resin from the legs of returning honeybee foragers (Seeley 1985, p. 131), and this behavior possibly occurs in nests of other apids.

The Megachilidae also use resin in large quantities for nest construction but carry it as a bolus with the mandibles (Pasteels 1977b; Armbruster 1984; Messer 1984). Some indication has been given that anthophorids carry resin in the scopae (Michener and Lange 1958a; Batra and Schuster 1977; Roubik and Michener 1980), but this is probably erroneous. Resinlike scopal loads of *Epicharis* were found to be elaiophore lipids (Buchmann and Roubik pers. obs.), presumably collected for feeding larvae and making brood cells (Fig. 2.15). Among the apids there are no other known special structures or behaviors for carrying and manipulating resin, although the mandibles of stingless bees are used extensively to mine resin from trees having thin bark or from leaves and branches (Silva et al. 1968; Freire and Gara 1970).

Figure 2.15. A complete cell of *Epicharis zonata* (Anthophoridae) taken from sandy soil (photo from Roubik and Michener 1980). The shining inner cell lining is partly made from lipid waxes collected from flowers (Buchmann and Roubik, unpublished).

Figure 2.16. Heads of a female and male *Chalicodoma pluto* (Megachilidae) showing the greatly expanded mandibles and labrum used by the female when transporting a bolus of resin for nest construction. Males also possess a large labrum. Original photographs provided by Adam Messer (Messer 1984).

The large megachilid bee *Chalicodoma pluto* of the Moluccas, Indonesia, does have a structural modification that aids carrying of resin (Messer 1984; see also Fig. 2.16). The greatly enlarged labrum of the female supports a ball of resin measuring 10 mm in diameter, which is transported for nest construction (Fig. 2.16). Pasteels (1977a) summarizes the work of earlier authors in separating the megachilid group Anthidiinae into two tribes based on ethological and morphological features; the suggested tribal classifications separated bees that manipulated different nesting materials.

Methods of resin and pollen collection influence the foraging behavior of bees.

No apid can collect pollen and resin during the same foraging bout. Some megachilids and possibly fideliids, due to placement of pollen on metasomal scopae, forage pollen and resin in the same trip and at the same flower (Armbruster 1984). Armbruster adds that the collection of resin from flowers by apids prevents them from grooming the pollen removed during the same flower visit and possibly enhances their value as pollen vectors.

2.2.6   Nectar harvest

Differences in feeding structures correspond to major ecological and phylogenetic divisions among bees. After examining mouthparts of 290 species selected from most bee tribes, Michener and Brooks (1984) concluded that the traditional division of bees into long-tongued and short-tongued groups is supported not by tongue length but by other morphological characteristics (also see Michener 1984). The feeding canal of adult bees is formed by the modified labium (glossa) sheathed in long-tongued bees by curved labial palpi and galeae. When the mouthparts are retracted, they are folded against the head in a flattened Z (Figs. 2.17 and 2.18). The distal segment, composed of the glossa, labial palpae, and galeae, projects backward. Harder (1983) shows that the rate of sucrose solution imbibement increases exponentially with the body mass of long-tongued bees but only arithmetically for the short-tongued bee species. The former group includes Fideliidae, Megachilidae, Anthophoridae, and Apidae. Designation of a bee as a member of a particular category has some drawbacks and inadequacies. Intermediates do exist, such as most panurgine Andrenidae and the Ctenoplectridae (Michener and Greenberg 1980; Michener 1984). Different parts of the proboscis have lengthened or shortened independently, and thus the Halictinae have some species with a long glossa or prementum but are nonetheless lumped together with the short-tongued bees. Euglossine bees, notably some *Euglossa, Eulaema*, and *Eufriesea*, have extremely long tongues, longer than the body (Appendix Figs. B.207 and B.240), and these perhaps constitute yet another category.

Beyond the anatomical tip of the glossa, most long-tongued bees have an additional expanded and flattened structure called the *flabellum* (Fig. 2.19). The flabellum may have specialized chemotactile sensitivity, similar to that reported for the setae of the labial palpus and galea of the honeybee (Whitehead 1978). Sensillae like those on the labial palpi that respond to sugars have been noted on the glossa, foretarsus, and antenna, but have yet to receive experimental attention. Frisch (1967, p. 512) was the first to note honeybee sensitivity to different types of sugars and concluded that their detection took place at the mouthparts. Michener and Brooks (1984) state that no bee having a truly short tongue possesses a distal specialization of the glossa. Some of the long-tongued bees have no flabellum

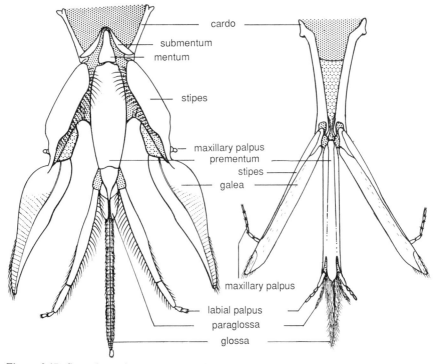

Figure 2.17. General mouthpart structures of long-tongued (right) and short-tongued bees. The first, at the left, is *Apis mellifera* (Apidae) and the second *Pseudaugochloropsis* (Halictidae). Original drawing by Barry Siler, provided by C. D. Michener (Michener 1974a).

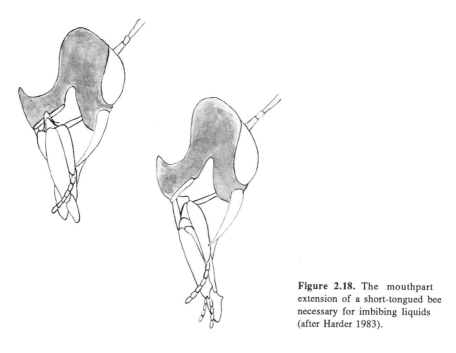

Figure 2.18. The mouthpart extension of a short-tongued bee necessary for imbibing liquids (after Harder 1983).

2. Foraging and pollination   67

Figure 2.19. Diversity in structure of the distal glossa of bees. From top to bottom: *Xylocopa* (Anthophoridae), showing the circular flabellum; *Thrinchostoma* (Halictidae), showing brushlike distal glossa; *Perdita* (Andrenidae), showing flabellum at tip. Original photographs by Robert Brooks, provided by C. D. Michener (Michener and Brooks 1984).

whereas bees having a relatively short tongue length often possess the flabellum.

Variations in glossal complexity sometimes represent modification associated with nest construction or provisioning; they are not solely related to nectar feeding. Some structures of the glossa have been lost in parasitic species, implying that these are needed for nest construction or provisioning rather than feeding (Section 2.2.4). The proboscidal traits of female colletids that are also shared by males

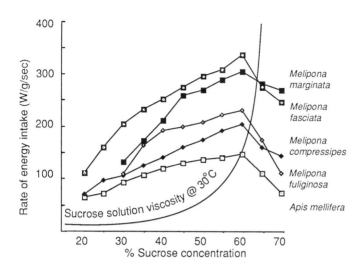

Figure 2.20. Sucrose solution viscosity and its correlation with the relative imbibing rates of five apid bees. Worker bees were given sucrose solutions at 30° C (after Roubik and Buchmann 1984).

(Batra 1980a; McGinley 1980) are associated with nest preparation, but no field observations have been made of possible advantages of the bifid glossa in nectar feeding by the females.

2.2.7 Feeding rate and efficiency

Considering the varied sizes and shapes of the bee glossa and mouthparts, there should be a priori differences in the rate at which bees can harvest nectar of differing viscosity and from different floral types (Harder 1983; Heynemann 1983; Roubik and Buchmann 1984). The morphological differences among species are stable and may lead to resource partitioning of nectar sources. Aside from the often critical differences that floral structure enforces in regulating visitor type, resource partitioning should reflect parameters of foraging success that are dependent upon physical properties of nectar and the bees themselves (Roubik and Buchmann 1984; see also Fig. 2.20). Comparative ingestion rates of bees imbibing 27–30% sucrose solutions are given in Table 2.3 and include species that are not among the equivocal bee-tongue-length categories. The comparisons of four sympatric *Melipona* and eight Trigonini show that larger bees tend to harvest sucrose solutions more rapidly. In a comparison of several temperate-zone bees of similar sizes but differing proboscidal length, Harder (1982, 1983) showed that small bees of either

Table 2.3. *Bee imbibing rates measured for 27–30% sucrose solutions near 30° C*

Bee taxa	Live bee weight (mg)	Imbibing rate (µl/sec)
Short-tongued bees		
Andrenidae		
Andrena geranii	25	0.13
Andrena carlini	81	0.23
Halictidae		
Augochlorella striata	15	0.08
Halictus ligatus	33	0.07
Agapostemon virescens	42	0.16
Long-tongued bees		
Anthophoridae		
Ceratina duplex	15	0.10
Melissodes apicata	69	0.34
Apidae		
Trigona pallens	10	0.30
Scaptotrigona luteipennis	22	0.74
Cephalotrigona capitata	31	1.23
Melipona fasciata	60	2.20
Melipona fuliginosa	125	2.60
Bombus ternaris[a]	62–102	0.5–0.9

[a]Females of differing sizes ("castes").
Sources: Harder 1982; Roubik & Buchmann 1984; Roubik & Yanega unpublished.

category imbibed nectar at a similar rate, but larger long-tongued bees fed more rapidly than short-tongued bees of the same size. An important distinction in feeding behavior in the two groups is that the short-tongued bees extend and retract the labium (Fig. 2.18), whereas this structure remains stationary in long-tongued bees. Harder also notes that the galeae and the sheathlike basal segments of the labial palpi compress the sides of the glossa while the long-tongued bee feeds, thus assisting the movement of liquid through the glossa toward the mouth; the labial palpi of short-tongued bees cannot assist fluid movement in this manner. Lapping movements of the glossa are used by both short- and long-tongued bees to initiate fluid transport, but in short-tongued bees this is accomplished in a distinctive way, by the rocking back and forth of the prementum. The glossa is retracted behind the tips of the galeae for such bees, but no further than their tips in the long-tongued bees.

A cautionary note applies to field comparisons of bee feeding ability. The functional length of the proboscis, or "tongue length," has been used to determine the effective reach and imbibing efficiency of bees collecting nectar from flowers. Efficiency implies cost/benefit ratio (Section 2.3.1), but it is often equated with the extent to which bee feeding structures correspond to the physical properties of floral resources. An estimate of feeding potential efficiency is commonly derived by measuring the distance from the base of the prementum to the tip of the extended glossa (Brian 1957; Heinrich 1976a; Morse 1977; summary by Harder 1982). As an index of potential flower species used by coexisting bees, based on functional corolla tube length of nectar flowers, this method may be acceptable. Of course, some bees have an extended face or slender head, which can also be inserted in the flower, and small bees crawl inside larger flowers. Furthermore, the functional proboscis length should be measured differently for short-tongued and long-tongued bees (Harder 1983). For the former, it is the length of the labium, and for the latter the glossal length. These different measurements provide an equitable means of gauging the potential imbibing efficiency of different bees on a given flower species.

Harder (1983) suggests that short-tongued bees of any size are at a competitive disadvantage relative to long-tongued bees, despite some general advantages if they happen to have a large body size and corresponding larger musculature of the glossa and cibarial muscles drawing nectar to the mouth. However, the expectations for bee sizes on plants having many relatively open flowers might not follow this prediction, and floral morphology could intervene strongly in determining the feeding efficiency of bees. For example, among sympatric bumblebees, the longer-tongued species avoid flowers with shorter corollas, presumably due to the disadvantage of feeding less efficiently from them than do bees having shorter proboscides (Pyke 1982). Moreover, if there are no adequate footholds, a bee having a long tongue cannot profitably remain in hovering flight to imbibe the nectar from a short-corolla flower. In the tropical highlands of the Americas, large colletids with short tongues, such as *Ptiloglossa* and *Crawfordapis,* coexist through the year with several *Bombus* of similar size (Plowright and Laverty 1984; Roubik and Michener 1984). This may suggest that they utilize different species for nectar. Among the extremely long-tongued bees, additional foraging expenditure is required after a bee reaches a flower with a correspondingly long corolla. One of the most common euglossine bees in Panama is the very-long-tongued *Euglossa imperialis* (Roubik and Ackerman 1987). I have watched this bee alight on the tip of a flower, rear back and extend its glossa to full length, insert the tip of the glossa in the lip of the corolla, and then move forward to insert the proboscis and head in the flower. These movements are similar to drawing a sword and then inserting it into a sheath, and they are quite time-consuming.

Finally, tongue length for certain types of social bees is sometimes variable, and no simple index is likely to reflect ecological range in feeding habits. *Bombus* and other less notably polymorphic bees such as some halictids and *Ceratina* display differences in body and proboscidal dimensions and thus are comparable to several monomorphic bee species in the variety of nectar sources to which they have access. Although there are no data for tropical bees, an example from a radically different type of environment serves to emphasize the adaptive significance of tongue length variation in bees. In Denmark and Fennoscandia, where bumblebees are largely dependent on a single type of flower during a season (Pamilo, Pekkarinen, and Varvio-Aho 1981), worker size variation is slight, and proboscidal dimensions concur with corolla tube lengths. A wider variety of significant nectar sources may therefore relax selection against size variation and promote polymorphism (see also Waddington, Herbst, and Roubik 1986).

2.3   Foraging and flight activity

The behavioral flexibility of bees is best characterized by their mobility and flight activity, which determine success in foraging, mating, and reproduction. Further, the ecological relationships between bees and flowering plants are dependent upon many physical parameters of the foraging environment and the ways that bees perceive and respond to them.

2.3.1   General considerations

The most widely accepted basis for understanding forager behavior involves appraisal of foraging costs and benefits. Costs are measured in terms of resources expended on a foraging task, but they may also include seemingly peripheral matters such as predation at flowers, the chances of mating, overheating while in flight, generation of metabolic water, nest usurpation and parasitism, and any number of risks associated with foraging decisions. The benefits can also be considered in a variety of ways. Foraging success of a bee can be measured by varied criteria. These include its efficiency (benefit/cost), the net gain (gain – cost), and gross gain (Cheverton, Kacelnic, and Krebs 1985; Waddington 1985). These can be further qualified as to net rate of intake (gain – cost)/time, and net harvest efficiency (gain – cost)/cost (Seeley 1986). All success may ideally be quantified in terms of biological fitness, which is ultimately the genetic contribution to future generations. Fitness is generally augmented through either maximization of rate or efficiency of energetic gain, nutrient gain or material acquisition, and the minimization of time and material expenditure derived from making a correct foraging choice.

A formidable task is implicit in applying this approach, since costs and benefits should ideally be known within the large foraging arena to which a bee has access. Furthermore, costs and benefits must be measured over an interval of time sufficient to include some of the important natural selection agents, which are often completely unknown. One might postulate that fitness of a bee frequently at risk from predators is determined principally by its harvesting rate, and that the fitness of a long-lived or relatively secure bee (or bee colony) is better served by maximizing harvesting efficiency. There is no guarantee that either of these particular strategies applies unless certain details of fitness are understood. A recent treatment of foraging theory emphasizes the fundamental importance of recognizing the right model for a foraging situation and type of organism (Stephens and Krebs 1986).

The problem of understanding the scale and dimension of the foraging region is not trivial. Fieldwork and conceptualization of bee foraging behavior have been developed primarily in artificial feeding enclosures, agricultural areas, forests with few plant species, and alpine meadows. A mature tropical lowland forest, with vegetation extending some 60 m aboveground (Richards 1980; Whitmore 1984), is simply not comparable. It presents hundreds of significant resources as well as stress sources (Jander 1975), temperatures, wind currents that differ between strata, and a broad array of vegetation forms and plant growth habits. Besides its sheer physical complexity, the olfactory dimensions in this setting, including floral fragrances and the foraging odors of bees, are undoubtedly augmented far beyond many temperate zone habitats.

Forager behavior should be closely tied to foraging energetics (Heinrich and Raven 1972; Heinrich 1975) and the assessment of foraging profit (Krebs and McCleery 1984; Pyke 1984; Stephens and Krebs 1986; Houston et al. 1988). To the degree that these factors impinge on the survival and reproductive success of the bee, and given a genetic basis for bee behavior, a sound working hypothesis is that bees should respond in a predictable fashion (Oster and Wilson 1978; Glasser 1984). Much if not most of foraging behavior has sizable components contributed by learning. It depends on such variable features of the environment as the distribution and abundance of different resource types, in addition to specific floral characteristics. The complexity of the resources themselves can limit the number of resource types sampled (and successfully manipulated) by a forager (Lewis 1986). Therefore, the foraging process is subject to trial-and-error learning but also depends on less flexibile behavior patterns that provide a limited set of responses to foraging experience. Application of foraging theory to behavior patterns of foraging bees is still a subject undergoing rapid development and modification (Waddington and Holden 1979; Stephens et al. 1986). It will continue to reveal deficiencies that cause empiricists to ponder the advantages of applying incomplete "optimality" theory or of rising to the task of increasing its complexity (or simplicity!) to include behavior unaccounted for by the basic theory (Glasser 1984).

The behavior of a bee that is reasonably adapted to its environment will be determined by the perceived resource landscape to which it has access, which includes shifting elements of resource quality, dispersion, quantity, and competition. Bees or colonies must decide whether to continue foraging, basing this decision on the degree of local resource depletion and general resource availability in the habitat (Stephens and Krebs 1986). Colony behavior ostensibly involves regulating the proportion of scouts and recruits (Johnson, Hubbell and Feener 1987). For such social species, it will also depend on various attributes of the foragers and their colonies. Specifically, foraging choices will be made to some extent by the colony, whereby the choice of an individual bee can be altered and even nullified (Sections 2.3.5 and 2.3.7). Other factors are relevant too, such as sunlight, wind, temperature, humidity, potential mates, and natural enemies (Frisch 1967). Each of these elements relates to bee foraging patterns and may assume central importance at some time.

2.3.2 Navigation and chemical ecology

Bees navigate by using varied types of information gathered by different senses (Frisch 1967; Michener 1974a; Martin and Lindauer 1977; Dyer and Gould 1983; Wehner 1984; Fent and Wehner 1985; Wehner and Rossel 1985; Gould 1986, 1987; Free 1987). A central orientation cue is the horizontal component of direction toward the sun. The changes in sun position are recorded by bees to determine an axis for global orientation. This constitutes a relative compass or "sun compass." Further cues are taken from landmarks, planes of polarized light, and airborne odors. In addition, geomagnetic cues are detected in magnetically sensitive tissue containing iron in the head and prothorax, potentially giving bees an absolute compass. Recently, Gould (1986) demonstrated that honeybees do not require learned sequential combinations of cues to move to and from their nest, but carry a cognitive map of their environment that allows them to navigate novel routes.

This extensive knowledge has been compiled based primarily on studies of European *Apis mellifera* but probably includes the range of mechanisms found among bees. As Dyer and Gould (1983) make clear, there is general redundancy in the information provided by the various orientation cues, resulting in the assurance that navigation to and from resources can take place under varied conditions. For tropical *Apis*, study of the nocturnal flight of *A. dorsata* shows that it forages all night long when the moon is half-full or larger and that its sensitivity to light is far greater than that of temperate *A. mellifera* (Dyer 1985). The same phenomenon has been documented for several solitary, nocturnal bees (Kerfoot 1967a,b). These have large ocelli and will fly only if the moon is visible near the onset of night.

Although illumination from the moon was necessary for flight in all of these species, the moon itself was definitely not an orientation cue. Bees likely remembered the aziumuthal course of the sun to obtain nocturnal compass bearings; they used moonlight, and probably odors or other general cues to detect landmarks, which provided the primary means of navigation.

Although the sun allows bees to judge position in space, the time of day is known to them primarily due to an internal clock (Saunders 1982). This ability is vital as a means of arriving at a known resource at the appropriate time as well as of compensating for the passage of time during foraging. Bees must learn the timing of flower opening, as they also learn and anticipate the placement of feeders for experimental studies. After such studies are terminated, honeybees or stingless bees previously trained to a spot may continue to arrive for several days or even weeks (Frisch 1967).

During cloudy weather or when the sun is not directly visible, the sun compass is set by perception of polarized light. Recent work has shown that specialized ommatidia of the bee compound eye serve as the detectors of the polarized light, the planes of which radiate at an angle perpendicular to the sun's path (Wehner and Rossel 1985). The ocelli may also detect polarized light (Fent and Wehner 1985; Rossel and Wehner 1986). Wehner and Rossel provide convincing evidence that bees do not instantaneously know the position of the sun from reading the ultraviolet e-vector polarization patterns in a clear patch of sky. Rather, they scan the sky while in flight to determine the sun's azimuth. Thus bees construct and maintain a simple model of the external world, with the main orientation axis defined by the sun's path, by utilizing polarized light. But bees seemingly fail to extract more precise information. The honeybee e-vector compass is designed as simply as possible, in a manner that provides correct orientation information most of the time but also must produce orientation errors averaging 22.5° (Wehner and Rossel 1985). When certain social bees provide information to nest mates on resource location (Section 2.3.5), the recruits that respond to it often scatter around the target. These errors can be interpreted as perceptual or adaptive in nature (Gould and Towne 1987). Certain perceptual errors are inherent to the apoid mechanism of measuring distance, which relys upon the energetic cost of flight. Upwind flight from the nest, or flight uphill, cause bees to overestimate the distance flown from the nest (Frisch 1967), but this perceptual error is only significant to the degree that the flight path or meteorological conditions vary among multiple flights to the same point. Other anomalies exist in the communication of foraging distance when the flight path is not horizontal (Gould and Towne 1987). Single bee orientation mechanisms, however extraordinary, are fallible.

Orientation behavior of bees at a feeding site or nest site is readily observed. Bees leaving the nest often turn and hover a few centimeters in front of the nest entrance (Malyshev 1935). When departing from a nest or foraging site, bees may make repeated and increasingly large looping flights before flying off in a straight path. Small bees approaching or leaving flowers or nest sites display rapid zigzag movements also, but this is probably to improve resolution and depth of field perception of an object, because their eyes are close together. Foraging bees of all sizes may fly in a meandering pattern when locating the source of a floral odor to which they are attracted (Kullenberg, Borg-Karlson, and Kullenberg 1984). As postulated for other insects (Cardé 1984; Janzen 1984), bees may fly in a path perpendicular to the wind until an attractive odor is encountered and then fly upwind.

Chemical ecology of odors that affect bees' behavior so thoroughly pervades their biology that it is impossible to confine their discussion to a small section of the present book. An introduction to the topic is provided here. Bees have a remarkable sensitivity to some chemicals commonly used in orientation or navigation outside of the nest and near the nest entrance. Odor perception by *Apis* is from 10 to 100 times more sensitive than that of humans to substances such as floral odors and bee pheromones (summary by Michener 1974a), and thus small amounts of an odor should influence behavior. Free (1987) emphasizes that the acute sensitivity of bees to pheromones produces immediate behavioral responses that can be observed and quantified. On the other hand, most pheromones have many components and several functions, so that the exact correspondence between specific chemicals and behavior is often unknown. Moreover, visual and olfactory stimuli often combine to evoke a particular behavior. Free's work also calls attention to other probably general aspects of bee odors used outside of the nest. Odors of *Apis mellifera* seem to have components specific to each colony, but the degree of odor specificity to other bee species is unknown. Further, it is not clear whether a directional component of odors – for example, those deposited by a bee visiting several flowers in a row – is apparent to other bees.

Some bees recognize their individual nests both by chemical markers (Steinmann 1976; Shimron, Hefetz, and Tengö 1985; Anzenberger 1986; Free 1987; Hefetz 1987) and by visual cues. Although often cryptically shaped and located, the nest entrance has features that may be characteristic for each species (Chapter 3). Among tropical halictid bees, for example, *Halictus* and *Lasioglossum* sometimes have nest entrance "turrets" lined with a white material of glandular origin (Brooks and Roubik 1983; Brooks and Cane 1984). The *Dufour's gland* (Chapter 3) products lining the entrance of another halictid, *Evylaeus,* are recognizable at the individual level and may indicate bee kin groups and ownership of nests (Hefetz, Bergström, and Tengö 1986; Hefetz 1987). Nests of stingless bees *Partamona pearsoni* in the Amazon have white entrances made of highly reflectant sand grains (Camargo pers. commun.). The nest entrances of many *Melipona* species, including the Mexican and Central American *Melipona beecheii* and the South American *M. compressipes*, have highly ultraviolet-reflectant whitish glandular secretions applied to the entrance structure (Buchmann and Roubik pers. obs.). Females in aggregations of nesting anthophorids such as African *Xylocopa* (Anzenberger 1986) and *Eucera* (Shimron et al. 1985) place individually distinctive odors at the nest entrances, allowing them to recognize their own nests. Host nest odors are recognized by parasites of Hymenoptera (Tengö et al. 1982; Cederberg et al. 1983; Section 3.2), and similar species-specific odor cues in nests of tropical bees are likely but have not been documented. Close-range orientation pheromones at nest entrances of *Apis mellifera* are at least strongly implicated by the behavior of bees, although they and many such postulated pheromones for which bioassays have been made are undescribed (Free 1987). Mandibular gland attractants are common in solitary bees; odors that are highly attractive to bees are apparently mimicked by flowers that require bee pollination (Borg-Karlson and Tengö 1986). However, the degree to which the same chemicals guide bees to resources is scarcely known for most bees.

Chemicals followed from the nest to a resource appear to be used only by some bees, all of them meliponine bees of tropical forests. Mandibular gland odor droplets deposited on vegetation are followed like a trail by meliponine groups such as *Trigona, Oxytrigona, Scaptotrigona, Meliponula, Cephalotrigona, Lestrimelitta*, and *Nannotrigona*; *Partamona* are believed to guide recruits by releasing odors as they fly through the air (summaries by Kerr 1969 and Michener 1974a). The chemical composition of mandibular gland substances in the stingless bees is known for few species, and these same substances, in higher concentration, function to some degree in eliciting alarm behavior and nest defense (Section 3.2). The volatile compounds of stingless bee mandibular glands have been studied in relatively few meliponines, so far including primarily neotropical species, but the list is rapidly growing (Table 2.4). Mating aggregations of stingless bees draw both males and virgin queens to a site (Nogueira-Neto 1970b; Engels and Engels 1984); these appear to form due to attractiveness of the cephalic secretions from mandibular glands. I have noticed stingless bee odors at a distance of a few meters, but the distances at which conspecific odors or those of other bee species are perceived are not well known. However, there should be formidable restrictions to the foraging and reproductive activity range of trail-laying stingless bees, due to the small total amount of odor that a bee can deposit. So far as is known, the stingless bees place small odor droplets every few meters. Male bees such as *Andrena, Bombus, Centris,* and *Xylocopa* also mark mating or feeding territories with pheromones from glands in the hindlegs, head, thorax, or abdomen (Frankie and Vinson 1977, 1984; Frankie, Vinson, and Lewis 1979; Coville et al. 1986; Vinson, Frankie, and Williams 1986; Williams, Vinson, and Frankie 1987; An-

Table 2.4. *Volatile secretions of bees, associated with foraging, mating, and defense*

Bee family and group[a]	Volatile glandular substances[b]	Function	References
Colletidae			
Colletes	citral, geranial, linalool (MG) lactones, acids (DG)	male attractants germicides?	Cane & Tengo 1981; Hefetz et al. 1979a Duffield et al. 1984
Hylaeus	citral, neral (MG)	unknown	Blum & Bohart 1972; Bergstrom & Tengo 1973
Prosopis	lactones, ethyl esters (DG)	germicides?	Duffield et al. 1980
Ptiloglossa, Crawfordapis	lactones, ethyl esters (DG)	germicides?	Cane 1983b
Melittidae			
Melitta	butyrate esters (DG)	germicides?	Tengö & Bergström 1976, 1977; Francke et al. 1984
Oxaeidae			
Oxaea	citral, heptanones (MG)	male territory markers? defense?	Blum, Wheeler, & Kerr, quoted in Blum 1981
Protoxaea	lactones, aldehydes, esters, acids (DG)	germicides?	Cane 1983b
Halictidae			
Dufourea	citral (MG) ketones, esters, acids (DG)	unknown germicides?	Duffield et al. 1984 Wheeler et al. 1985
Lasioglossum	lactones (DG)	sex pheromones	Smith, Carlson, & Frazier 1985
Halictus	lactones, hydrocarbons (DG)	antibiotics?	Cane 1983b
Nomia	lactones, esters (DG)	pheromones?	Duffield et al. 1982
Evylaeus	lactones, esters, hydrocarbons (DG)	discriminators?	Hefetz et al. 1986

Table 2.4 (*cont.*)

Bee family and group[a]	Volatile glandular substances[b]	Function	References
Megachilidae			
Megachile	lactones, esters, hydrocarbons (DG)	nutrition, germicide?	Francke et al. 1984; Williams et al. 1986
Coelioxys	spiroacetals, alkanols, ketones, alcohols (CG)	host defense evasion?	Tengö et al. 1982
Andrenidae			
Andrena	>50 compounds – farnesol, geranial spiroacetals, monoterpenes (CG)	male territory markers	Bergström & Tengö 1982; Borg-Karlson & Tengö 1986 Tengö & Bergström 1975, 1978
Panurginus	citral, ketones, alcohols, acetates butanoates, spiroketals (MG)	male territory markers	Duffield et al. 1984; Francke et al. 1980 Wheeler et al. 1984
Calliopsis	geranial, neral (MG)	defense?	Hefetz et al. 1982
Anthophoridae			
Centris	nerol, neral, geraniol, ethyl laurate, geranyl acetate, ketones, ethyl myristate, 2-heptanone, hydrocarbons (MG, HLG)	male territory markers	Vinson et al. 1982; Coville et al. 1986
	lactones, hydrocarbons, acids (DG)	germicides?	Cane & Brooks 1983
Eucera	not yet characterized	unknown	Shimron et al. 1985
Melissodes	butanoates (MG or LG)	attractants	Batra & Hefetz 1979

Xylocopa	benzoic acid, vanillin (MG) acid lactone, benzaldehyde, *p*-cresol	attractant or marker	Wheeler et al. 1976; Velthuis & Gerling 1980 Frankie & Vinson 1977 (bioassay only); Hefetz 1983
	fatty acids, hydrocarbons (DG extract)		Vinson et al. 1978; Williams et al. 1983; Francke et al. 1984
	terpenes, geraniol, farnesal, fatty acids, farnesol (TG)		Williams, Vinson, & Frankie 1987; Andersen et al. 1988
Ceratina	terpenoid acetates, alkenes, farnesyl, geranyl & neryl acetates, nerolic & geranic acids, salicylaldehyde	defense?	Duffield et al. 1984; Wheeler et al. 1977
Pithitis	ethyl hexadecanoate, 1-hexadecyl acetate (MG)	ant repellents territory markers?	Hefetz et al. 1979b; Cane 1986
Holcopasites	geranyl acetone, acetones (MG)	defense?	Hefetz et al. 1982
Exoneura	ethyl dodecanoate, salicyl-aldehyde, benzoquinone, farnesyl hexanoate, geranyl octanoate (MG)	ant repellents	Cane & Michener 1983
Nomada	alkyl butanoates (MG)	attractants, defense?	Tengö & Bergström 1975, 1976, 1977; Duffield et al. 1984
Nomadopsis	neral, geranial (MG)	defense?	Hefetz et al. 1982
Epeolus	spiroacetals, alkanols, pyrazines (CG)	host defense evasion?	Tengö et al. 1982
Apidae			
Bombus	41 compounds, including farnesol, hydrocarbons, alcohols, aldehydes, acetates, ethyl esters, geraniol, citronellol, butanoic acid, geranyl geranyl acetate, ethyl dodecanoate (LG, MG)	male territory markers	summary by Duffield et al. 1984

Table 2.4 (cont.)

Bee family and group[a]	Volatile glandular substances[b]	Function	References
Psithyrus	citronellol, aldehydes, tetradecanol, farnesol, other alcohols, acids (LG)	territory marker, defense?	Kullenberg et al. 1970, quoted in Blum 1981; Cederburg et al. 1983
Euglossa	41 compounds: alkanes, alkenes, dienes (MG)	male territory markers?	Williams & Whitten 1983
Eulaema	acetates, diacetates, alcohols, and unidentified compounds (MG, CG)	attractants?	Williams & Whitten 1983
Melipona	skatole, 2-heptanol, undecane, nerol (MG)	alarm attractants, repellents?	Smith & Roubik 1983
Apis	2-heptanone (MG)	alarm attractants, repellents?	Morse et al. 1967; Gary 1974; Blum 1976, 1981
	fatty acids, hydroxybenzoate,	social regulators, discriminators? germicides, or unknown	Boch et al. 1979; Costa Leonardo 1980; Velthuis 1985
	9-ODA, 9HDA, 10-ODA, 10-HDA, other acids (MG) citral, geraniol, nerolic & geranic acids, nerol (MG,CG), isopentyl acetate, 2-heptanol, other acetates alcohols (SG)	attractants	Gary 1974; Pickett et al. 1980
		alarm and attractant	Collins & Blum 1983; Duffield et al. 1984
Oxytrigona	formic acid, hexedecanal, hexadecane, heptendione, nonendione, hydrocarbons, ketones, carboxylic esters, acids & acetates (MG)	alarm attractant, repellents, germicides?	Roubik, Smith, & Carlson 1987; Bian et al. 1985

Trigona	2-heptanol, 2-nonal, nerol (MG, CG)	alarm and attractants repellents, germicides?	Kerr, Blum, & Fales, quoted in Blum 1981; Johnson & Wiemer 1982; Johnson et al. 1985
Geotrigona	citral, neral, geranial (MG, CG)	alarm and attractants? repellents, germicides?	Blum et al. 1970
Hypotrigona	citral, neral, geranial (MG, CG)	alarm and attractants? repellents, germicides?	Keeping et al. 1982
Trigonisca	citral, neral, geranial (MG, CG)	alarm and attractants? repellents, germicides?	Roubik, Schmidt, & Buchmann unpub.
Lestrimelitta	citral, neral, geranial (MG, CG)	alarm and attractants, repellents, germicides?	Blum 1966; Blum et al. 1970
Meliplebeia	2-heptanone (MG, CG)	alarm and attractants? repellents, germicides?	Crewe & Fletcher 1976; Keeping et al. 1982
Scaptotrigona	undecane, tridecane, benzaldehyde 2-heptanol, other 2-alkanols, 2-heptanone, 2-alkanones, butanoates, tetradecenyl acetate, 9-hexadecenyl acetate, undecenone, pentadecane, dodecane, hexanol, hexyl-hexenoate (MG, CG)	alarm and attractants, repellents, germicides?	Francke et al. 1982; Luby et al. 1973 Blum 1981; Francke et al. 1983
Tetragona	deltalactones (MG, CG)	alarm and attractants, repellents, germicides?	Wheeler et al. 1975

[a] Either genus or subgenus.
[b] MG = mandibular glands; CG = cephalic glands (whole head extracts); LG = labial glands; DG = Dufour's gland; HLG = hindleg gland; SG = sting gland; TG = thoracic glands.

dersen et al. 1988; see also Table 2.4). Persistence of these odors in normal conditions is still unknown; some components are clearly volatile and short-lived, but others appear to last (Free 1987).

Opportunistic use of foraging trails of one species of meliponine bee by another closely related species has been reported for two South American stingless bees of the *Scaptotrigona* group (Kerr, Ferreira, and de Mattos 1963), but curiously, this behavior was not reported to be reciprocal. Similarly, I have observed males of several *Paratrigona* arrive at large male aggregations of a single species. Avoidance and defensive behavior in response to cephalic secretions of other species have been confirmed for a few stingless bees at artificial resources (Johnson 1980; Johnson and Wiemer 1982). Experimental bioassays at normal resources are lacking, but the present results indicate that foraging odors of aggressive or group-foraging species may be avoided by other foragers, hence averting direct contact, damage, and displacement.

2.3.3 Flight range

Bees are "central-place" foragers (Schoener 1979; Stephens and Krebs 1986), returning to a previously occupied site after foraging activity. Extensions of this theme exist, notably the hoarding behavior of many eusocial bee colonies and the emigration of colonies or individuals. The nesting female returns several times to the nest during a given day after foraging bouts. Male bees return either to mating territories, resting places (Fig. 2.21), or nests. Foraging habits of parasitic female bees are less known; but females likely display foraging behavior resembling that of males because each only gathers nectar from flowers and needs not have pollen sources. Females of parasitic species seem often to use the same nectar sources as their reproductive hosts and presumably remain close to their nesting areas. Bees of both sexes have been found to repeatedly cluster in a resting aggregation on a particular plant (Linsley 1962b). Dense resting aggregations of parasitic bees with their hosts have been recorded, such as *Anthophora* and *Melecta*, or *Colletes* and *Epeolus* (Evans and Linsley 1960). Both males and females also pass inactive periods within flowers, in holes in wood, or on plants at which they engage in mating or nectar feeding (Sakagami and Michener 1962; Eickwort and Ginsberg 1980). Social bees occasionally remain in the field during the night if prevented from reaching their colony (Michener 1974a; Morse 1982).

Bee flight ranges in tropical habitats have been studied in only a few species, and indirect methods such as release and recapture are usually employed. Several studies indicate that males and females of a given species do not have the same flight range and that there are straightforward morphological correlates attesting to this general pattern (Kapil and Dhaliwal 1969). Within a species, the larger sex

2. Foraging and pollination 83

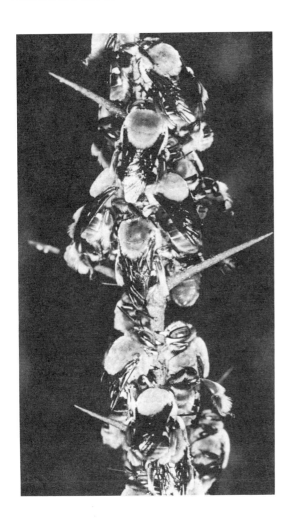

Figure 2.21. A resting or "sleeping" aggregation of male *Centris adani* (Anthophoridae) on a thorny branch in lowland Costa Rica. Original photograph by Stephen L. Buchmann.

might be expected to have the greater flight range, but this is not always the case. It is also sometimes expected that longer wing length and larger wing-area/body-weight ratios indicate greater flight range (Raw 1976; Casey, May, and Morgan 1985; Marden 1987). The wingbeat frequency is positively correlated with wing loading (g/cm^2 wing area), and heavier bees thus compensate for greater body weight by more rapid wing movement (Casey et al. 1985; Byrne, Buchmann, and Spangler 1988).

During flight, the fore- and hindwings are coupled by a series of hooks, the hamuli, on the anterior portion of the hindwing. Hamuli stabilize the wings and thus influence flight power and lift. The number of hooks on wings displays a broad positive correlation with bee size and flight range (Schwarz 1948). However, Lee (1974) shows that queens of *Apis mellifera* have fewer hamuli than workers and drones, and despite the larger size of drones and queens compared with workers, workers and drones have equal hamuli numbers. Drones of European *A. mellifera* ordinarily fly farther than virgin queens (Koeniger 1986). A different trend is evident in *Bombus* and *Psithyrus*. Reproductive females have more hamuli than either males or workers (Schwarz 1948). Among the many stingless bees examined by Schwarz (1948), the workers, males, and queens are similar, with some tendency of workers to possess more hamuli. In contrast to *Apis*, the newly emerged queens of *Melipona* are about the same size as workers and males although the queens of Trigonini tend to be larger than workers. No general hypotheses have been advanced to account for these contrasting trends.

One study by Kapil and Dhaliwal (1969) quantified both male and female flight ranges and hamuli number in two species of *Xylocopa* in India. Bees were released at six distances between 0.5 and 5 km from their nests. From 50% to 80% of the females returned to nests from the greatest distance, compared with only 25% of the males. Male hamuli number was less than that of females, although very slightly so, and the males were smaller and had shorter wings. Based on correlations between flight range and hamuli number and experiments with *Xylocopa* – for example, one in which 28% of *X. virginica* released 12 km from nests returned to them (Balduf 1962; see also Rau 1933) – Kapil and Dhaliwal suggest the Indian species might fly as far as 20 km. Hamuli number in all of the above studies was variable within species and sexes, ranging in number per wing from 3 in small Trigonini to 30 to nearly 40 in *Xylocopa* and Bombinae. It is relatively easy to encounter tropical bees of the last two groups flying across open ocean where the distance between the nearest island and the mainland is over 10 km, and their great flight ranges are thus suggested by the high number of hamuli and large body size.

The greatest flight range of a tropical bee was reported by Janzen (1971), who released 24 marked females of the large euglossine bee *Eufriesea surinamensis* at up to 23 km from their nests and later found females in the nests from which the released bees had been removed. Myers and Loveless (1976) studied bees in the same place and found that other females in the aggregation moved to unoccupied nests (usurping them) almost immediately, which led to the question of whether some of Janzen's bees were unable to return to their nest or were accidentally confused with other females. However, his study did not involve nest usurpation because the released females were individually marked. Therefore, some female bees plausibly had familiarity with a foraging area of over 1,770 km^2, and they certainly had the ability to navigate in such an area. This is the area within 23 km of a nest. It seems likely that the data represent actual flight capability; all 12 bees released at <6 km were recorded back at nests within nine hours, and 7 of 12 returned from

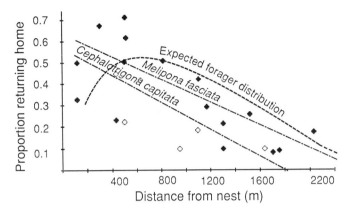

Figure 2.22. Expected foraging ranges resulting from a release–recapture study of stingless bees (Apidae: Meliponinae) in a tropical forest. The larger *Melipona* arrived at their nests from greater distances than the *Cephalotrigona*. Linear regression of the proportions of foragers from groups released at varying distances gave approximate maximum flight distances (after Roubik and Aluja 1983). However, studies using larger numbers of worker bees show that flight distances approximate a Poisson distribution (Figs. 2.24 and 2.25). Expected forager distributions must take into account the extended flight distances of some individuals (see text).

distances of 14–23 km during a full day. Decreasing numbers of foragers are likely to return to a nest from progressively greater distances (discussed later), as also indicated in the above experiment of Kapil and Dhaliwal. The specific cues used by these bees are unknown, but sun compass orientation and perhaps internal maps of the type previously described (Section 2.3.2), combined with landmark recognition, seem likely. Euglossine bees displayed both remarkable orientation ability and perception of learned orientation cues over an extremely large area.

The probability with which bees forage at a certain distance and whether they normally utilize their full flight range are subjects that invite further study. Little research has effectively dealt with these topics in forests, either in the tropics or the temperate zone. A mark-and-release study was performed with the relatively large stingless bees *Cephalotrigona capitata* and *Melipona fasciata* in the forest of Barro Colorado Island, Panama (Roubik and Aluja 1983). A magnetic recapture system (Gary 1971) was employed in this study; bees carrying a small, demagnetized ferrous tag glued to the thorax were released at a range of up to 3 km from their nests. Returning bees had their tags removed by magnets placed at nest entrances. The maximum return range was 1.5 km for the *Cephalotrigona* and 2.1 km for the *Melipona*. Linear regression analysis of the proportion of bees returning from various distances gave predicted probable maximum flight ranges of 1.7 and 2.4 km, respectively (Fig. 2.22). As will be clarified, these estimates are still conser-

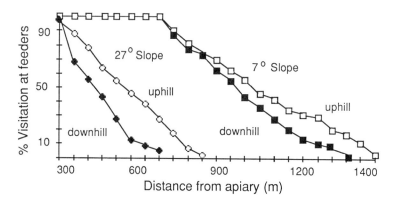

Figure 2.23. Foraging ranges and flight direction (uphill versus downhill) of *Apis cerana* (Apidae) in relatively flat and steeply inclined terrain (after Dhaliwahl and Sharma 1974). Bees trained to sugar solution feeders more readily traveled greater distances uphill than downhill when their return flight included considerable climbing.

vative because very few foragers of a colony travel more than a moderate fraction of their potential flight distance. Unpublished observations of *Melipona fasciata* foraging at flowers of *Hybanthus prunifolius* in the same area showed that *M. fasciata* tagged and then released at their nests flew up to 1.5 km to flowers (Roubik, Montalvo, and Ackerman).

The flight ranges in a variety of settings have been gauged for other stingless bees, honeybees, and halictids. *Scaptotrigona luteipennis* gathered pollen at a distance at least 1.3 km away from the nest in lowland forest in Panama, because this was the distance of the large trees it was visiting, *Cavanillesia platanifolia*, which I identified by pollen analysis. A similar technique revealed that a smaller Asian stingless bee foraged at 1.1 km from its nest, visiting a palm plantation outside the forest where it nested (Appanah 1982). Studies of the flight range of *Apis cerana indica* in India showed that this bee visited feeders at distances of up to 1.8 km (Dhaliwal and Sharma 1974; see also Fig. 2.23). Indirect study of meliponine flight ranges, incorporating small species (6 mm length), was performed with tagged foragers of observation colonies in western Sumatra (Inoue et al. 1985). Extrapolating from the known flight speed and collecting time at artificial resources, they calculated foraging ranges of 84–434 m for workers of *Tetragonula minangkabau*. In Costa Rica, Wille and Orozco (1970) and Wille (1976) reported that released, marked bees of the largest meliponine, *Melipona fuliginosa*, returned from 2 km to nests, and the small halictid *Lasioglossum umbripenne* returned from >100 m.

All of the above studies more or less confirm the flight range capability of the bees, but all are dependent on the bees' current foraging orientation, despite the possibility of navigating in varying patterns within known territory (Gould 1986). This type of information is inconclusive because if a very large number of individually marked bees were observed, it is likely that even greater flight ranges would be recorded. The studies of Dhaliwal and Sharma (1974) and Roubik and Aluja (1983) demonstrated that the chance of foraging at a given distance from the nest declined with increasing distance although rare foraging events at greater distances were probably not detected.

A more thorough approach to the question of flight and foraging range has been made with European *Apis mellifera carnica*. The flight range limit was determined by the co-workers of Frisch (1967), who positioned concentrated sugar solution feeders at increasing distances from honeybee hives. Their marked foragers from a known colony continued to visit feeders and then returned to the nest from 12 km. From this result, it appears that flight range and foraging range are equivalent, at least for bees gathering nectar. European honeybees gathering both nectar and pollen in arid, open areas in southwestern United States sustained their colonies adequately when forced to forage up to 8 km from the nest, and they continued to forage at a maximum distance of 13.4 km (Eckert 1933).

An ecological perspective on flight ranges in temperate forest was generated by Visscher and Seeley (1982) by using observation hives to record the communication behavior of returning foragers. After the essential step of calibrating the dance language, which varies among colonies in its tempo–distance correlation, foraging localities of returning foragers were mapped by simply reading their dances (Fig. 2.24). The resulting information showed that peak foraging range was 10 km on both nectar and pollen sources, but most foragers were working within 2 km of the nest, and the average forager flew 1.8 km to resources. Repetition of this technique with two colonies of Africanized honeybees in the lowland forest of Panama gave almost identical results (Vergara 1983; see also Fig. 2.24). The ecological range of foraging *Apis mellifera* may be similar in temperate and tropical forest, and it includes over 300 km^2 for any given colony, at least in flat areas. The amount of foraging tapers gradually away from the nest, tracing a negative exponential decline. These studies also show that honeybee colonies rapidly and at times radically shifted their foraging areas and ranges from one day to the next or over shorter time intervals (Visscher and Seeley 1982).

Such information allows construction of a useful theoretical model of foraging intensity, defined here as the proportion of total foraging trips at varying distances from the nest. If the maximum flight range of a bee is known, then the probability of foraging at a certain distance can be estimated by using the probability density function for a normal distribution. The only simplifying assumptions that need be made are (1) foraging occurs with equal probability in any direction from the nest

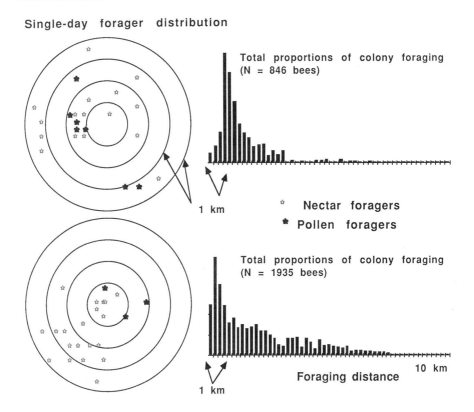

Figure 2.24. Foraging localities and flight distances for workers from colonies of *Apis mellifera* (after Visscher and Seeley 1982; Vergara 1983). Single-day forager distributions are shown in the circles on the left; larger stars represent several pollen foragers; smaller stars represent several nectar foragers. Histograms to the right show the proportions of colony foraging at a range of distances from the nest, determined from observation hive readings of the dance language of individual foragers. Both studies took place in a forest, those of Vergara in the lowland forest of Panama using observations colonies of Africanized honeybees. The study of Visscher and Seeley was performed with European honeybees in New York.

and (2) the *long-term* distribution of foragers approaches a normal distribution. If these assumptions seem reasonable, a statistical z-distribution table can be used to predict foraging intensity within a given distance band (Fig. 2.25). Visualized in three dimensions, the probability distribution of foragers around the nest assumes a bell-shaped pattern. The density of foragers or individual foraging trips is highest near the nest. Precise proportions of the area under the normal curve can be determined for differing distances from the origin (bee nest), expressed as standard deviations from the mean. According to statistical theory, 68% of a normally distributed population lies within one standard deviation, 95% within two, and 99%

2. Foraging and pollination 89

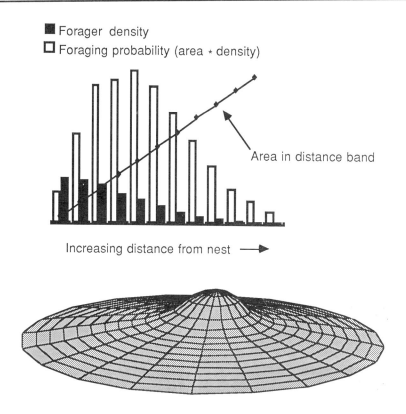

Figure 2.25. A theoretical distribution of foraging trips by workers of bee colonies or a nesting female bee. The bell-shaped figure represents foraging in three dimensions; its volume over a given area is equivalent to the probability of foraging there. However, when the direction from the nest is not considered, then the probability of foraging at a certain distance does not decline uniformly with distance. It is on the average most intense between one-third and one-half of maximum foraging range. The graph represents in two dimensions some concepts related to foraging distance, without regard to direction from the nest. The white bars show that the numbers of foraging trips increase up to the mean foraging distance and then decline. Despite the declining probability of foraging away from the nest, the area in a band of a given width increases with increasing distance from the nest. Foraging distance probabilities are the product of forager density and foraging area.

within three of the origin (a central place, or the bee nest in the present case). However, even though this accurately shows that the chance of foraging at a particular distance progressively declines away from the nest, it does not reveal how much foraging occurs at a particular distance from the nest irrespective of direction. Total foraging activity within a given section of foraging range is the product of both a probability density function and area. As a result, it is clear that most foraging does not occur close to the nest. In fact, most foraging takes place at a distance of one-third to one-half of the foraging range. The result of any study of flight range carried out for a sufficiently long time should reveal a peak of returnees coming from such an intermediate distance (Fig. 2.24)

As an example, consider the flight and foraging ranges of Africanized *Apis mellifera*. The hybrid African honeybees in Panama demonstrated mean flight range of about 1.7 km and ordinarily a range virtually limited to 7.5 km, with a few distances of 10 km (Vergara 1983; see also Visscher 1982, quoted in Seeley 1985, p. 92). Nonetheless, as shown by the white bars in the graph in Fig. 2.25, about as many honeybee foragers range 8–9 km from the nest as range less than 1 km from it. Although the number of foragers per unit area is indeed higher 1 km from the nest, the total area is far greater in a radius 1 km wide 8 km from the nest. Most foraging of at least these varieties of *A. mellifera*, over a season or relatively long time period, will occur between roughly 2 km and 5 km from the nest. The distribution of forage patches could, however, easily upset this theoretical prediction.

For large bees, such as some female *Xylocopa* or euglossines having a flight range of 20 km, most of the daily foraging will theoretically take place at between 6.7 km and 10 km from the nest. In addition, most competition between bees nesting close to each other will not occur in the immediate vicinity of the nests. If their nests are barely separated, it will occur in a circular band near the mode average flight distance, and if they are more distant, the probable amount of competition can be computed from intersecting volumes of the bell-shaped foraging distributions.

Habitat and resource topography influence bee foraging range and flight direction. Dhaliwal and Sharma (1974) studied three colonies of *Apis cerana* in which 3,000–5,000 foragers were marked and then recorded as they left their hives or arrived at feeding stations (Fig. 2.23). The plotted results show the approximately normal distribution of bee foraging distances, presumably determined by physical flight capability, energetics, and local resource depletion. The studies were performed with sucrose solution that was less concentrated than that which bees take from some flowers; thus the observed maximum ranges are probably conservative. However, a symmetrical distribution of foragers around the home base might only be expected in a relatively flat area. As shown in Fig. 2.23, foragers tend to fly shorter distances if the terrain is steep, and they fly the shortest distances to resources if they must carry them upward when returning to the nest. The authors claimed there were no statistical differences in these results, but I believe the data are stronger than they wished to suggest. Therefore, bees should prefer to forage uphill from the nest, and an estimation of foraging probability as a function of distance would have to take this factor into account.

2.3.4   Thermoregulation, metabolism, and response to weather

Meteorological factors influence bee flight activity and the cost of body temperature regulation. Explanations for flight behavior can include both the physiology of bees and that of plant nectar secretion, as well as the general quality of floral and nonfloral resources (Linsley 1978; see also Section 2.2). Besides determining the time when flight activity can be initiated or must be diminished, weather conditions affect bee foraging at flower patches. Bees sometimes move faster between inflorescences at higher temperature; and when metabolic heat production is expensive relative to caloric gains at flowers, large bees like *Bombus* may crawl rather than fly between them (Heinrich 1972). Once flight is initiated, about 80% of the energy metabolized by insect flight muscles is lost as heat (Kammer and Heinrich 1978). The activity patterns, timing, and initiation of flight activity of any species are likely to vary, and short-term characterizations are therefore tentative. Endogenous circadian rhythms exist for bees, and they can be further adjusted in colonial bees, possibly by contact pheromones (Southwick and Moritz 1987a).

Some of the basic physiological processes that come into play during foraging and general activity have been measured for a few bee species. The metabolic relationship between CO_2, O_2, and H_2O was found for flying male bumblebees to amount to the production of 24.5 mg CO_2 per hour, corresponding to respiratory oxygen consumption of 56.4 ml/g body weight per hour (Bertsch 1984). Bertsch shows that mass-specific oxygen consumption tends to be uniform in two of the genera of bees (*Bombus* and *Xylocopa*) studied thus far. Moreover, a male bumblebee that flies for four hours daily must void an amount of water equal to its total body-water content, about 60% of its weight, during 24 hours. This balanced water budget takes into account an observed evaporative water loss of 23% of the total metabolic water in addition to water ingested with nectar. The bumblebees in Bertsch's study, along with honeybees, euglossine bees, and some large stingless bees (Roubik pers. obs.), void water by "squirting." Water is jettisoned periodically while bees are in flight. Clearly, high metabolic water production due to extended flight would permit very limited feeding on dilute nectar without frequent elimination of excess water.

The anatomy of bees is modified to transfer heat to the flight muscles, retain it there, or carry heat away from the thorax and into the abdominal venter or the head, from which it is dissipated (Heinrich 1979a; May and Casey 1983; Casey et al. 1985; Heinrich and Buchmann 1986; see also Fig. 2.26). The path of the aorta is modified into a broadly arched loop in the thorax or a series of convolutions near the base of the thorax in some genera, such as *Bombus, Apis, Melipona, Euglossa,* and *Eulaema* (Wille 1958; May and Casey 1983). This structure, at least in

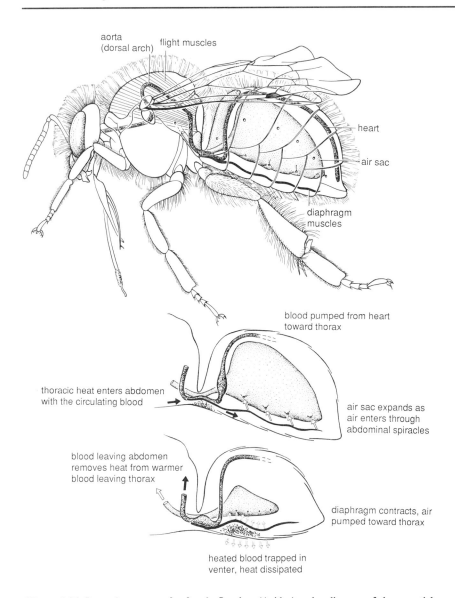

Figure 2.26. Internal anatomy of a female *Bombus* (Apidae) and a diagram of the essential mechanisms of losing excess heat by ventilation and allowing blood warmed by the heat generated from thoracic flight muscles to dissipate heat through the ventral abdomen. From Friedrich G. Barth (1985), *Insects and flowers: the biology of a partnership,* trans. M. A. Biederman-Thorson. Copyright © 1985 by Princeton Univ. Press (U.S.) and Allen & Unwin (U.K.). Reprinted by permission of the publishers.

some Bombinae and possibly in large meliponines with a similar thoracic aorta, allows them to shunt heated blood from the thorax into the abdomen, from which heat is lost as the cooler haemolymph of the abdomen is pumped into the thorax. During preflight warming of the muscles by flight muscle contraction, the pumping of the abdominal muscles and thus increased circulation from the heart in the dorsum of the abdomen, is halted. Thus no heat is transferred to the abdomen. The convoluted design of the honeybee dorsal vessel, however, prevents effective transfer of heat to the abdomen; instead, most of this heat is retained within the thorax, allowing incoming haemolymph from the abdomen to be warmed. Excess heat is shunted into the head of the honeybee, which is cooled by the regurgitation of a nectar droplet and evaporative cooling while in flight (Heinrich 1979b, 1980).

Cooper, Schaffer, and Buchmann (1985) found that honeybee foragers in the desert returned to the nest more often with a nectar or water droplet extruded on the proboscis at higher ambient temperature. In addition, thoracic temperatures of workers returning with liquid were at least 2° C cooler than those returning to the nest with only pollen. Both *Apis* and *Bombus* have a homeothermic response to cooling; they can raise body temperature without locomotor (wing) activity, possibly by contraction of the flight muscles but without the shivering behavior before flight seen in other insects (Morgan and Bartholomew 1982).

Little is known of physical mechanisms for temperature regulation in other bees, all of which may be susceptible to the lethal temperature range of about 45–50° C while in flight (Heinrich 1979a; Ono, Okada, and Sasaki 1987). Bee flight activity apparently requires a thoracic temperature of at least 25–30° C (Esch 1976; Heinrich 1981; May and Casey 1983). Hovering euglossine bees, similar in size to the honeybee, had average thoracic temperatures near 40° C during the lowland dry season in Costa Rica, where ambient temperatures are near 35° C (Inouye 1975). Their temperatures were lower in their normal environment within the forest (May and Casey 1983). *Xylocopa* fly with their thoracic temperatures reaching 46.5° C (Heinrich and Buchmann 1986). Honeybee thoracic temperatures are typically 10–15° C above ambient but are sometimes as much as 23° C above ambient temperature when returning from foraging in near-freezing temperature, and they reach 45° C in hot weather (Seeley 1985). Returning foragers of two *Melipona* species in the Amazon displayed ectothermy, having thoracic temperatures ranging from 32° to 38° C while ambient temperature was near 30° C (Roubik and Peralta 1983), as did large black colletid bees in the cloud forest of Panama when ambient temperature was 17° C (Roubik and Michener 1984). The abdomen may serve as a radiator for losing excess heat, and the insulating effects of abdominal air sacs (Heinrich 1979a) may also play some role in heat control. The wide and flat abdomen of *Xylocopa* may maximize exposure of its ventral surface to air current, thus enhancing heat dissipation. Recently, Heinrich and Buchmann (1986) confirmed that *Xylocopa* actively pumps heat generated from the flight muscles and then carried by haemolymph into its abdomen, from which the heat is lost while in flight.

Further conjecture might be made on the relative advantage enjoyed by social bee foragers with no need to protect reproductive organs or eggs from thermal shocks while foraging, or in the slight thermal inertia provided by a large nectar load. Unwin and Corbet (1984) suggest insects that frequently hover have wings with large lobes, which may increase air circulation and cause more convective heat loss. These authors offer a starting point from which to investigate other mechanisms for thermoregulation by foraging bees.

Comparative studies of heat production and regulation by foraging bees focus on variables such as mass-specific metabolic rates, flight speeds, and ratio of wing area to body mass (Dyer and Seeley 1987). Dyer and Seeley find some of these factors vary disproportionately among four species of the genus *Apis* and show that there are general implications for many aspects of behavior. Ruttner (1988, p. 71) mentions a correlate of *Allen's rule* in his finding that in *Apis* some body parts, such as the legs and proboscides, are shorter when heat dissipation is unfavorable; this occurs in cold climates. Study of bee activity during three flowering seasons led Boyle-Makowski and Philogène (1985) to conclude that *Apis mellifera*, andrenids, and halictids have differing responses to light intensity, humidity, and temperature. The solitary bees were active under a smaller range of conditions but apparently more active in the most adverse conditions, and the honeybees responded more to changes in humidity. Some examples from India show that light intensity was closely aligned with foraging activity of *Megachile* (Kapil and Jain 1980), but humidity and temperature were scarcely correlated with foraging either by this genus or by *Xylocopa* (Kapil and Dhaliwal 1969). Several euglossine bees were studied in Panama by May and Casey (1983). Like other bees, these were found to be strongly ectothermic (body temperature was higher than ambient). Heat loss was substantially retarded by pubescence on the thorax under conditions of high flight or wind speed, but hairy euglossine species lost heat from the thorax, to a large extent by pumping heated haemolymph into the head, substantially more so than the similar bumblebees of temperate areas. Further, at low wind speeds the loss of heat from the thorax was not greatly retarded by thoracic pubescence. Slow flight or calm weather, relatively poor insulation, or intermittent rests that allow loss of excess heat from the thorax may therefore be used to a greater degree by some tropical bees to avoid overheating. Relatively hairless bees, however, may increase their flight speed to enhance heat loss by forced convection, regardless of greater heat output by the flight muscles (Heinrich and Buchmann 1986). One means of increasing flight speed and avoiding increased heat production would be to maintain a downwind flight direction.

2. Foraging and pollination

Some of the highly eusocial apids adjust to changing temperatures in a manner that affects their current foraging behavior. *Apis,* the only bee genus known to regulate nest temperature by evaporative cooling from collected water (Section 3.1), may alter colony foraging activity in times of excessive colony heating (Lindauer 1967; Michener 1974a). Foraging bees are at such times directed by nest mates to harvest nectar that is more dilute or to switch to water collection (Lindauer 1954; Seeley 1986). Both male and female *Bombus* heat the brood cells within the nest (Cameron 1985) and thus may curtail foraging activity as they apply body heat to ensure brood survival rather than to generate body heat permitting food gathering outside the nest.

The degree of insolation at flowers influences foraging choice. Flowers in direct sunlight secrete nectar earlier than shaded flowers and at a higher rate (Percival 1965). Foragers experiencing cold temperatures may use flowers as a device for trapping solar heat, thus raising their body temperature (Kevan 1975), and foragers experiencing overheating may regurgitate nectar on part of the head, retreat to shaded areas, become inactive, or even preferentially visit shaded flowers (Heinrich 1979a, 1980; Willmer and Corbet 1981). Such behavioral mechanisms have been demonstrated in very few species and usually occur where there are environmental extremes. Their role in enforcing the partitioning of resources such as flower species is probably negligible, yet subtle differences in visitation within a flower patch might conceivably relate to thermal gradients (Willmer and Corbet 1981; see also Fig. 2.27). Although the work of these researchers was carried out for a very short period, it seems that the bees and hummingbirds gathering nectar from a single species concentrated most on flowers separated by differing thermal conditions.

Additional physical relationships between foragers and thermal environments provide evidence of the role that meteorological factors may have in shaping bee foraging activity. Body color is a primary feature for consideration although size and other characteristics are also correlated. The general hypothesis that dark color and large body size enhance heat retention and preflight warmup has been invoked repeatedly to explain bee behavior (May and Casey 1983; Plowright and Laverty 1984). It is questionable that color alone provides a good criterion. For instance, black bees are common foragers in widely differing habitats, such as the cool tropical cloud forest and hot, xeric lowlands (Cockerell 1926; Roubik and Michener 1984). The effect of dark body color is strongly dependent on the amount of hair on the bee's body, bee size, and the timing and location of foraging.

Heat may be rapidly lost through forced air convection or retained by hair on the body, which enhances warming (Heinrich 1979a; May and Casey 1983). A few tropical solitary bees (which lack nest temperature regulation) have both nocturnal and diurnal species (Linsley 1958; Janzen 1967b; Kerfoot 1967a,b; Wolda and

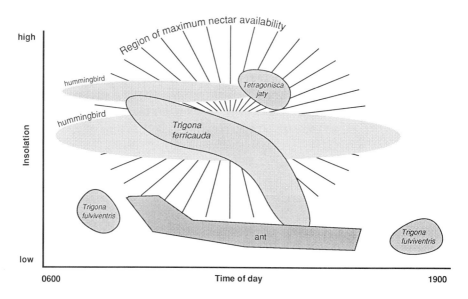

Figure 2.27. Spatiotemporal distribution of visitation by meliponine bees, birds, and ants to flowers of *Justicia* (Acanthaceae), plotted to show the possible relationship to sunlight and temperature (after Wilmer and Corbet 1981). The study was made during one day in lowland forest of Costa Rica.

Roubik 1986). Their body colors suggest that dark coloration is favored by diurnal activity. Some paleotropical *Xylocopa* (subgenus *Nyctomelitta*) and neotropical *Megalopta* are nocturnal and are light brown, and the former is quite hairy. Related crepuscular or diurnal species of each, in the same habitats, are often black (Roubik pers. obs.). Carpenter bees, *Xylocopa*, are mostly tropical and typically large, shiny black bees with sparse hair (Hurd 1978; see also Appendix Figs. B.1, B.5, and B.6). However, the differing flight behavior of the two sexes may explain why there is often a striking difference in coloration of male and female. Males are often light yellow or orange, contrasting with shiny black females. One study has compared their temperatures (Louw and Nicholson 1983). Foraging females of *X. capitata* in southern Africa had body temperatures near 39° while foraging on sunny days, and foraging males had body temperatures averaging 38° C; but territorial males had thoracic temperatures of 41° C. These bees foraged at ambient temperatures near 30° C but were inactive at temperatures lower than 23° C. The males of *Xylocopa* are somewhat smaller than the females, and some species spend hours in hovering flight at mating territories (generally in the shade). In contrast, males of neotropical *X. tabaniformis* hover in the late afternoon, and they are gray in color (Janzen 1964; see also Appendix Figs. B.1, B.5, and B.6).

In general, twilight-active or high-elevation bees tend to be dark-colored and have dense pile on the thorax whereas those of hotter areas tend to be relatively hairless. In the tropical highlands, many bees that forage through the day are black or dark gray; those that are active at twilight are lighter in color. Highland *Apis mellifera monticola* is darker and larger than closely related *Apis* in the central and south African lowlands (Southwick 1985a; Ruttner 1988). However, contrasting trends occur in other African *A. mellifera,* thus contradicting *Bergmann's rule* (Ruttner 1988, p. 191). The diurnal, large colletid *Crawfordapis* is black, and a bee nesting in the same area and of the same size, the matinal and crepuscular colletid *Ptiloglossa,* is lighter and has light abdominal stripes (Roubik and Michener 1984). The giant Himalayan honeybee *Apis laboriosa* is darker and much hairier than *A. dorsata*, which replaces it in adjacent lowlands (Sakagami et al. 1980; Roubik, Sakagami, and Kudo 1985).

Bees that are nocturnal, strictly crepuscular, or matinal (active at dawn) have enlarged ocelli whereas those that are at least somewhat diurnal have ocelli of an intermediate size compared with related diurnal species (Kerfoot 1967a,b). In Trinidad, Bennett (quoted in Rozen 1984b) found that the parasites of nocturnal bees also have enlarged ocelli; they were clearly necessary to locate the host nest and enter when the host forages and is away from the nest (Section 3.2.8). The function of the ocelli may differ for diurnal and nocturnal bees. Ocelli may be used primarily for detecting polarized light in diurnal species, and their enlarged size in nocturnal bees might suggest more importance as light receptors that help to detect the horizon, in turn permitting stable body position while in flight (Fent and Wehner 1985; Roessel and Wehner 1986). As noted by Dyer (1985), the mechanism by which a nocturnal *Apis* excludes the polarized light source of the moon, replacing it with remembered cues from the sun, is still a mystery. In *Apis mellifera* the passage of light through ocelli acts as a general stimulant to activity (Michener 1974a).

Mimetic complexes, automimicry, and sexual dimorphism complicate some general explanations of bee color, size, and pubescence based on temperature regulation alone and are discussed in Section 2.3.8. Furthermore, unambiguous correlations are unlikely to be found among all characteristics. For instance, Bergmann's rule has been applied to bees to explain why racial varieties of a given species are larger and darker at higher elevations or colder climates (Maa 1953; Sakagami et al. 1980). It has some 15 variations for bees (Pyke 1978b). Higher metabolic heat production, and presumably darker color, may be offset by smaller body size in bees (Heinrich 1979c) due to their more rapid convective heat loss. Because the ratio of surface area to volume is greater in smaller organisms, body temperature regulation at relatively low ambient temperature improves with a size increase. Bergman's rule was originally presented to explain why the largest individuals within a species's range occur in the coolest regions of its distribution. Maa (1953) gives the example of two tropical honeybees from southeast Asia (*Apis florea* and *Apis andreniformis*), stating that the darkest species is also the smallest, which he believed to inhabit the highlands. Wu and Kuang (1986), however, show that although these two species are generally found in the lowlands of southern China, the lighter colored *A. florea* has nearly twice the altitudinal range known for the smaller *A. andreniformis*. The same is true of *Apis cerana* in China; bees in southern China are smaller and darker than those in the north (Wu pers. commun.). A similar trend occurs in the stingless bee *Trigona fulviventris*. Its dark form, *guianae*, occurs in lowland Amazonia, and its lighter form is found at elevations up to 2,000 m. The lighter form is slightly larger; nonetheless, a hybrid colony that I observed in Muzo, Colombia, displayed later foraging activity by all black bees than by lighter, orange and black hybrids.

Seasonal differences in diurnal foraging activity may be indicated by bee color and pubescence. In the neotropics, centridine bees that forage during the hot, dry season are generally smaller, less pubescent, and lighter in color than those active in the wet season. This trend was evident in studies of several *Centris* and *Epicharis* in Central America, in which seasonal color dimorphism of *Centris inermis* is possibly related to flight of the lighter morph during the hot, dry season [Frankie et al. 1983; Snelling 1984; Wolda and Roubik 1986; see also Appendix B.1 (Nos. 13 and 14)]. Increasing melanism in polymorphic species of *Bombus* has been correlated with decreasing annual temperatures in their range of distribution (Pekkarinen 1979). However, exceptions are easily found in correlations among these features. The bumblebee that inhabits the southern tip of South America, *B. dahlbomi*, is light orange in color.

Adverse conditions for bee foraging in the lowland tropics are mist, high wind, heavy rain, and relatively low temperature. European honeybees are said not to fly when wind speeds exceed 24 km/hr. During gusts of wind, I have observed small stingless bees cling to inflorescences and commence foraging when the winds abate. Bumblebees that continue to forage in windy conditions move relatively slowly (Morse 1982). Flowers sheltered from the wind receive more visits than those in open, windswept areas in the tropical highlands, and I have noticed that larger bees foraging in the wind tend to keep close to the surface of the ground. The effect of wind is probably similar to the effect of rain (with the exception of a heavy tropical downpour) in causing bees to select resources closer to their home base. An approaching thunderstorm elicited cessation of foraging at feeders placed 6 km from honeybee hives, but bees continued to visit feeders 100 m distant until the rain began (Boch 1956). There is at least the possibility that detection of humidity and barometric pressure by bees allows them to sense the coming of wet weather and squalls some time before their arrival. Sudden darkening of the sky by heavy rain clouds appears to cause the rapid return of bees to their nests although foraging in light rain is observed in a number of bee genera (Linsley 1958; Morse 1982). It has been suggested that Africanized *Apis mellifera* continues foraging during light rain, whereas European honeybees seldom do so (Portugal-Araújo 1971). Further comparison of honeybee races from these two continents suggests that the African bees forage somewhat earlier in the morning due to generally higher metabolic activity, which permits an earlier warmup and exit from the nest (Heinrich 1979c). Preflight warmup has been shown in all bees tested thus far, including *Apis, Bombus,* and *Xylocopa* (e.g., Nicholson and Louw 1982; Heinrich and Buchmann 1986). The latter authors show that one species of *Xylocopa*, although hairless and poorly insulated, produces a preflight thoracic temperature of near 20° C when ambient temperature is only 12° C.

2.3.5 Social foraging

The *dance language* of European *Apis mellifera* has been described in detail by its discoverer, Frisch (1967), and certainly provides the best-known example of specialized adaptations in orientation and food harvest suited to a colonial species (Lindauer 1967; Seeley 1985; Wilson 1985b; Dyer 1987; Gould and Towne 1987). As a means of alerting nest mates and providing directional and distance information on the location of a forage source, honeybees perform a repeated series of movements on the comb surface. These are perceived by nearby bees and impart a symbolic representation of the direction a potential forager should take in relation to the position of the sun, and also the distance she should fly to locate the particular resource. The behavior that originally alerted nest mates to the physical location of resources, and led to evolution of stereotypic dance languages, may have its origin in the process of cooling off (Frisch, quoted by Gould and Towne 1987) by forced air convection. A forager arriving at the nest would take more time to lower its body temperature in this manner, the greater the distance it had just flown. Furthermore, returning foragers frequently shake their bodies dorsoventrally while grasping another potential forager, and this "vibration dance" might provide additional stimulus regulating colony activity (Schneider, Stamps, and Gary 1986).

The general ecological significance of honeybee communication behavior has been addressed, although many unexplained variations among the *Apis* species are known. These plausibly relate to the ecology of food gathering (Lindauer 1967; Michener 1974a; Nuñez 1979; Gould 1982; Dyer and Seeley 1987). For example, the minimum distance of forage from the nest that elicits the *waggle dance* of a returning honeybee forager differs among the geographic races of *Apis mellifera*. Distinctive behavior patterns during the waggle dance are enacted by arriving for-

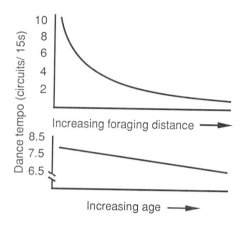

Figure 2.28. Variation in dance language tempos of worker *Apis mellifera carnica* according to distance traveled from the nest and the age of the worker performing the recruitment dance (Frisch 1967; Michener 1974a).

agers during the straight run on the comb surface (the portion of the dance that takes place between two loops that together form a figure eight). During the straight run, distance information is contained in the sum of the individual waggles (i.e., one waggle is equal to 10 m, etc.). Even more pronounced differences exist among four of the *Apis* species that have been studied (Gould and Towne 1987). Alterations in use of waggle dances are generally as follows: At short distances from the nest, a simple round dance is used instead of the figure eight and waggle dance. This alerts nest mates to the presence of a resource but gives no directional information. The waggle dance is performed when resources are at greater distances, from 2 m to about 100 m among the four common *Apis* (summary by Lindauer 1967). The smallest honeybee, *A. florea,* has the slowest dance. Because the communication of progressively greater distances requires increasing duration of the waggle dance (Fig. 2.28), an increase in the number of waggles by *A. florea* increases the distance covered by the waggle dance on the comb when it communicates forage distances greater than about 400 m (Seeley pers. commun.). Its waggle dances would otherwise require excessive time and possibly not even not fit within a dancing circuit of fixed size. Some variance in the tempo of movement on the comb surface, in relation to forage location, is evident even among bees of the same colony but of different ages (Frisch 1967; see also Fig. 2.28). The comparison of interspecies or racial variation in the dance tempo, as it relates to the distance of forage, thus corresponds to the age structure of a colony. It may be expected to change through the life of the colony, and it makes the accuracy of communication between foragers dependent on their ages.

Extrapolating the foraging range of *Apis* from the tempo of dances and the use of the round dance is complicated by the fact that one of the largest *Apis, A.*

dorsata, also performs waggle dances for resources relatively close to the nest, and at least one variety of the medium-size *A. cerana* communicates by using the waggle dance when resources are nearly as close to the nest as those that evoke this behavioral response in *A. florea* (Lindauer 1957, quoted in Frisch 1967; Michener 1974a). If resource patches normally harvested by these bees are small, or available for only a short time, then the greater accuracy of communication behavior would be advantageous. For some European *Apis mellifera,* the duration of the waggle dance increases about 75 msec for each 100 m in foraging distance from the nest. The less the frequency of waggle runs decreases with increased foraging distance, the greater the distances that foragers are able to communicate (Ruttner 1988). The giant honeybees (the *A. dorsata* group) must also increase the duration of the waggle dance to communicate greater distances, because they surely have a larger flight range than other *Apis.* Their nesting habit includes tall trees (see Fig. 4.14), while *A. florea* only nests near the ground (Seeley, Seeley, and Akratanakul 1982). This also suggests that perhaps another aspect of forager behavior (the range of heights at which foraging occurs?) differs among honeybee species or races, which might necessitate more precise communication of even nearby resources for *A. dorsata* or *A. cerana.* In European honeybees, Frisch (1967) found no indication of height communication regarding a forage source.

A positive correlation between the duration of the waggle dance and the duration of sounds produced by the flight muscles of a dancing forager, shows that two communication channels can be employed for distance information. The Johnston's organ in the base of the honeybee antenna detects the vibrations of a dancing worker and is maximally sensitive to their frequency, and the subgenual organs in the legs also detect vibrations (Michelsen, Kirchner, and Lindauer 1986). The view was held that possibly sounds were perceived through the air rather than exclusively through a solid substrate (Gould, Dyer, and Towne 1985). This was verified in detailed studies of *Apis mellifera* using a laser vibrometer (Michelsen et al. 1986). The measurements of Michelsen et al. demonstrated that messages relayed by a returning forager to attending bees are transmitted *exclusively* through the air. On the other hand, the comb itself is an extremely important medium for signals of both such attending bees and queens. The fundamental frequency of the begging signal given by attending bees to returning foragers is 320 Hz. Vibrations of this frequency borne on the comb cause the signal receivers momentarily to halt their movement. Their slight difference from the 280-Hz airborne 20-msec pulses emitted during the waggle dance of returning foragers evoke an entirely different behavior.

The sounds emitted during the waggle portion of the dance may have originated primarily to gain the audience of other bees rather than to provide distance information, and as already mentioned, the waggle run itself is possibly an extension of fanning of the wings during individual thermoregulatory behavior. Towne (1985) has found that *A. dorsata* and *A. florea,* species that do not nest in cavities but rather build a single hanging comb, appear to make no noise whatsoever while relaying forage information. Much of their behavior involved in drawing the attention of nest mates occurs in a visual mode. The striped abdomen is moved from side to side while raising it over the body, simultaneously jostling attending bees and performing a waggle dance in their midst. In addition, coincident sounds are emitted during the recruitment process only in species of *Apis* that nest in dark cavities, in which this acoustic information has probably replaced visual signals (Gould et al. 1985). The ways in which sound, touch, or vision are processed by the bees' nervous systems to yield the direction of forage from the nest are still unknown. The begging sounds of nest bees among *Apis* and stingless bees (Section 3.3.4) are presumably widespread, but only substrate-borne sounds are presently thought to function within nests of the latter.

Even a minimum of ecological interpretation suggests that the dance language variation seen among honeybee species and races need not primarily represent comparable means of communicating distance and direction. The tempo, correlation with distance, and duration of the waggle dance also reflect precision of communication concerning three dimensions, because the more accurate the dance, the greater the chances are for locating resources that may occur over a considerable height span. This aspect of forager behavior might be expected to hold greater significance where resources could be located anywhere over a vertical range of 60 m in tropical forests. Furthermore, the extent to which waggle dances or straight runs are used can relate to bee size and physiology, including variables such as flight range and olfactory sensitivity. The ecology of foraging, nest defense, and thermal regulation may have strongly influenced worker size in the genus *Apis* (Dyer and Seeley 1987). That selection pressures influencing size have the most pronounced effect in the more ecologically active caste of the honeybee is evident in the fivefold species size differences that occur among workers, compared with only twofold differences in queens and drones. Selection pressures for such size modification could secondarily produce variation in the dance language. The choice of nesting sites or other spatial and temporal resource characteristics might each contribute to the evolution of dance language behavior.

When a returning forager enters the nest of *Apis* or stingless bees, she not only imparts information that may describe her foraging location, she gives samples of the forage as well. The odor of food or nest construction material is dispersed among attending workers, and nectar samples are passed by *trophallaxis* (mouth-to-mouth food transfer) to nest mates (Frisch 1967). Pollen loads are deposited in storage areas in the nest, but nest mates are first allowed to taste, smell, and possibly measure the size of the pollen load with the antennae or mouthparts. In one species of the stingless bee *Melipona,* the returning forager will not dislodge her pollen load for storage in the nest before receiving nectar from a nest mate (Sommeijer et al. 1983). Information on the quality of the nectar source is immediately perceived by *Apis mellifera,* and undoubtedly other bees, detecting differences in sugar concentrations of less than 5% (Frisch 1967). Moreover, the availability of nectar at a resource, and probably pollen as well, is sometimes correlated with the amount gathered by the bee (Free 1956; Nuñez 1982). The best, most easily harvested nectar loads are larger, and returning foragers may have difficulty finding nest mates that will receive a small nectar load. Even if returning foragers reveal the size of their nectar load in an indirect way that is perceived by other bees, this has been shown not to affect the readiness with which it is taken by *receiver bees* within the nest (Seeley 1986). Nectar sweetness is apparently the most important proximate signal of nectar quality for the European honeybee. For a tropical honeybee, nectar availability could conceivably be more important (Section 4.3).

The size of a nectar load does not increase with increasing sugar concentration in European *Apis mellifera* (Wells and Giacchino 1968), and two of four *Melipona* species that have been studied collected similar amounts of nectar whether rich or poor in sugar (Roubik and Buchmann 1984). However, in two other species, the large *Melipona fuliginosa* collected larger loads of more dilute nectar, whereas *M. compressipes triplaridis* collected larger loads of more concentrated nectar. Some of the amino acids commonly found in floral nectars elevate the rate of solution imbibement by honeybees (Inouye and Waller 1982) and may thus alter the food transfer and recruitment behavior of foraging colonies. The same amino acids diminish colony feeding at sucrose solutions by *Melipona* yet have no effect in *Trigona, Tetragona,* and *Partamona* (Roubik and Aluja pers. obs.). Thus the details of resource communication behavior in the nest may include several as yet poorly understood variables, and these might vary considerably among bee or plant species. As discussed above, the distance of a resource from the nest may have particular significance in determining the extent to which it is harvested by a foraging bee.

Social regulation of worker recruitment to resources plays a key part in directing colony foraging effort, and less profitable foraging trips for individual bees may be a necessary outcome of following the information of nest mates to locate a particular resource (Seeley 1985; Seeley and Visscher 1988). A recruit of European *Apis mellifera* typically leaves the nest and returns a number of times with no foraging success – more often than do scout bees – until it locates the resource of exceptional quality to which it had been directed. When such behavior is summed over all foragers in the colony and over several days, however, the average return on foraging effort is considerably greater than if foragers worked independently in limited areas and with limited information on resource availability and quality.

Moreover, the yield per unit time is, overall, greater for recruits than for scouts. The reconnaissance activity of the entire colony is formidable and can in fact lead to discovery of most resources within the flight range (Seeley 1987).

A short comparative study of nectar foraging by colonies of European and Africanized honeybees in Venezuela showed that although both bee races gathered nectar of the same type and presumably from the same plant species, Africanized bees tended to have have high, low, or intermediate foraging success, and the European bees were either successful or unsuccessful (Rinderer et al. 1984). In light of a principle that appears valid – that less foraging success is more typical for a scout than for a successful recruit – it might be inferred that Africanized bees had a greater proportion of scouts in their colonies (Section 4.3). Scouts, responding primarily to floral cues, might at least have moderate success, but successful recruits and new recruits would display high or low success, depending on whether they had located a particular resource or were still in the process of searching for it. The distances from which the two races of bees in this study were collecting nectar, however, was unknown. This purely spatial aspect of colony foraging determines the ease with which a particular resource is located and thus must have an effect on the rate of successful foraging by recruits.

One species of stingless bee that has been carefully studied, *Trigona fulviventris,* normally forages solitarily at flowers and small, scattered resources; yet at large artificial feeders, or presumably the large inflorescences of tropical plants, bees are recruited rapidly and arrive by the hundreds (Johnson 1974; Hubbell and Johnson 1978; Roubik and Aluja pers. obs.). Among the stingless bees, this species has large colonies (>10,000 workers: Hubbell and Johnson 1977; Roubik 1983a). Colony size probably interacts with recruitment ability to determine when and if massive and directed recruitment is preferable to relatively independent foraging by workers (Johnson et al. 1987). Indeed, it is not realistic to expect the maintenance of very large colonies by solitarily foraging species or those foraging very near the nest (Gould 1982).

The components of social foraging behaviors, which are further influenced by recruitment ability of the particular species, can in theory be divided into the portion of foraging effort expended in location, harvest, and defense of a resource. Aggressive group foragers tend to locate new resource patches slowly but then focus a large colony effort on procuring the resource and defending it from other foragers (Hubbell and Johnson 1978; Roubik 1980a, 1981b; Appanah 1981). The degree of aggression expressed by foraging highly social bees can be discerned at three levels, all seemingly absent among other bees (Johnson and Hubbell 1974; Roubik 1980; Howard 1985). The lowest level involves threat, when a bee spreads its wings in a V position over the abdomen or turns to face a rival with open mandibles. The next level involves brief contact as a bee rapidly darts at a rival while in flight, possibly attempting to bite. At the highest level, there is sustained contact as the mandibles close on an opponent; bees have been recorded locked in combat, venter to venter, and dead on the ground near a rich resource. Level-2 and level-3 aggression in stingless bees are only displayed by medium-size or large trigonines that have strongly developed mandibular teeth (two or three more in the genus *Trigona* than in meliponines of comparable size). These species appear better equipped for combat. Such escalated aggression may occur at flowers but is more readily seen at artificial feeders or resin sources, and I have seen rival colonies of flesh-eating *Trigona* immediately engage in level-3 aggression at dead animals.

An additional unique behavior pattern of aggressive stingless bees is best regarded as a ritualized threat display because the end result is dominance of a resource by a forager or colony without escalated conflict. In this behavior pattern, rivals hover face to face and rise into the air repeatedly. Bees may lock their mandibles on a rival and fall to the ground, but both bees then fly again. For example, I have seen two competing colonies of *Trigona nigerrima* perform this ritual several times in Panama, on the leaves of trees offering no resources. The large workers of *Trigona silvestriana* either forage as marauding individuals or may initiate group foraging, hovering in a group of more than 100 bees and initiating threat displays that may result in massive attack and mortality among rival groups. The result is the eventual departure of one foraging group, either within minutes or days, although aggressively foraging *Tetragona* display the behavior when foraging singly (Johnson and Hubbell 1974; Roubik 1980a, 1982b). While no comparative studies have been performed on Old World meliponines, Appanah (1981) reports aggressive group foraging by *Lophotrigona canifrons* and aggressive pursuit of competitors apparently by *Apis dorsata;* I have observed aggressive foraging behavior at flowers and honey water by *Tetragonula drescheri,* directed both at conspecifics and at unaggressive *Heterotrigona itama,* in western Sumatra. This aggressive bee did not arrive at the resources in a group.

Unaggressive social foragers, which include *Apis, Melipona,* and most of the Trigonini, may recruit well or scarcely at all (Kerr et al. 1963; Lindauer 1967; Hubbell and Johnson 1978; Roubik 1980a). Further, even the geographic races of a species display large differences in recruitment. For instance, Kerr (pers. commun.) has noted that *Melipona compressipes fasciculata* in the eastern Amazon recruits rapidly to sugar sources, whereas *M. c. manaosensis* does not, nor does *M. c. triplaridis* in Panama (pers. obs.). Thus far, only a few *Melipona* appear able to recruit rapidly, and these are among species that maintain the largest colonies for their genus.

It was suggested by Esch (1967) that *Melipona seminigra merrillae* and *M. quadrifasciata* indicate both the distance and direction of resources to nest mates in a manner analogous to *Apis mellifera*. The distance of resources was negatively correlated with the frequency of sounds emitted by returning foragers in the nest, and direction was thought to be communicated by repeated flight (up to 30 guiding flights of up to 50 m) of a recruiting scout from the nest entrance toward a feeding station. Despite a complexity in the sounds and movements of returning *Melipona*, their behavior has alternatively been interpreted as a general alerting sequence rather than a detailed rendering of forage location equivalent to that given by *Apis* (Gould et al. 1985). Such precise communication behavior has not been confirmed at ordinary resources in any of the more than 40 *Melipona* species nor in the Trigonini. Sound communication within the nest is likely to be used in many species, in combination with receipt of forage, in eliciting foraging behavior (Esch 1967; Michelsen et al 1986). The majority of stingless bee species have apparently very effective recruitment abilities (Roubik et al. 1986), for which comparative studies are difficult. The uncontrolled variables of colony and forager size, pollen or nectar needs of the colony, and competing floral resources generally prohibit a meaningful comparison of colony recruitment by different highly eusocial bee species. Their comparison has been gauged by recording the number of new recruits arriving at a feeder after a single scout has visited it (Lindauer 1967). The simultaneous action of more than one scout on a larger resource therefore remains unassessed, which virtually excludes the type of resource upon which highly eusocial bees sometimes specialize (Chapter 4).

Occasional acts of aggression, in which a forager lands on another bee at a flower, are seen in both *Apis* and *Melipona* and probably many solitary and social bees (e.g., *Bombus* observed by Morse 1979), but these events are too rare to be considered normal foraging styles. Attack with the mandibles or sting does not occur in these instances. Intraspecific aggression among honeybees is often expressed by bees of different colonies; it is more readily seen at artificial feeders than at flowers, and it is far more common than is interspecific aggression (Roubik 1980a). Such behavior does include attempts to bite or even sting. At flowers, aggressive behavior that is an attempt to bite or chase away another forager may exist for the purpose of driving away parasitic flies that lay eggs on foraging bees (Section 3.2). Displaced-nest and food-defense behavior, or robbing behavior, are not normally expressed at small, scattered floral resources. The effects of occasional acts of interference, as they may relate to food competition and the physical size of normal resources, have not been studied. The most general correlate of aggressive foraging behavior that I have observed is that of extended stationary feeding or resource collection (Roubik 1980a; Howard 1985). Stationary feeding at sugary solutions frequently leads to fighting among foraging bee colonies, but the same species show no such behavior at flowers. Recruiting social bees that forage fruit, resin, sap, or carrion and thus spend several minutes collecting a foraging load, seem much more likely to interact aggressively with competitors than are other bees. In theory, expression of aggressive foraging behavior is likely in a highly eusocial bee when the species has limited foraging options or food reserves, relatively small foraging range, inflexible behavior geared toward massive recruitment, foragers that are of relatively large size or well equipped to fight with the mandibles (a minority of Trigonini), and little to lose by sacrificing bees in combat with other aggressive species. Precious foraging time or foragers are not squandered by solitary bees in fomenting such encounters, nor by colonies of small size or those having vastly superior scouting and recruitment systems such as *Apis mellifera*.

A further note can be added on the intriguing possibility of organized group foraging among large, solitary bees. These release mating odors near flowers (Table 2.4). Odors from netted bees are immediately discernible in some Colletidae and Anthophoridae. A number of bees, both male and female, fly toward the captured insect and even fly into the net. Frankie and Baker (1974) described a type of group foraging by *Centris aethyctera, C. adani* (see also Frankie et al. 1980), and *Gaesischia exul* in the crowns of flowering trees. Foraging groups of 10–300 foragers arrived in waves in a small area of the canopy ranging from roughly 0.5 m^2 to 12 m^2. These aggregations moved quickly and continuously between flower clusters; they were composed mostly of male bees in the *Centris,* but females predominated in the groups of *Gaesischia.* Arroyo (1981) reports the arrival of foraging solitary bee groups at trees, followed by apparent displacement of smaller foragers. Male *Centris* having foraging territories in trees or elsewhere individually pursue any object coming within their territory, as do many male bees. This aspect of mate selection and territorial competition will be reviewed in Section 3.3. Considering the apparent mobility of the solitary bee groups, their behavior is not comparable to the foraging style of aggressive groups of *Trigona,* which appear to remain in a small foraging territory bordered by the odors they place on plants (Roubik 1981b, 1982a; Johnson 1983a).

2.3.6 Foraging styles and syndromes

Considering the enormous range in physical parameters such as inflorescence size, flower size, and the distribution of rewards in flowers, there is ample opportunity for foraging specialization among bees. Consistent foraging styles seem to apply to some bee species that have received sufficient study. Small bees, for example, can crawl onto the anthers of flowers with long, exposed, widely separated filaments and extract pollen that a larger bee could not take without energetically costly hovering. In turn, large bees can force apart petals or depress a keel petal that exposes the anthers, an act impossible for a small bee to perform. Within

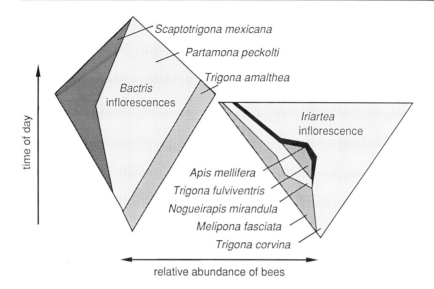

Figure 2.29. The relative abundances of bee species beginning at the time of dehiscence of two palms. The *Bactris* inflorescences were observed throughout the day (Johnson 1983a), and a single inflorescence of *Iriartea* was observed for 5 hours (pers. obs.). Studies were made in Costa Rica and Panama, respectively.

single species, the females and males sometimes forage nectar at widely different rates (Section 2.4.5). Moreover, in a study of two *Xylocopa* species visiting the cucurbit *Luffa*, females gathering both pollen and nectar still visited flowers at three to six times the rate observed in males (Kapil and Dhaliwal 1969).

The conformity between foraging style and resource characteristics is most evident in foraging by colonies of highly eusocial bees. These display a wide range of foraging behavior, colony and worker size, and worker morphology that can be related to measurable features of resources, such as their spatial arrangement. This is convenient for comparative studies and in marked contrast to most solitary bees, which rarely display overtly aggressive foraging behavior (Morse 1978; Eickwort and Ginsberg 1980) and are not readily studied in terms of group response at resources.

Hubbell and Johnson (1978), Roubik (1980a), and Johnson (1981) documented the succession or dominance of highly eusocial bee species at artificial feeders that presented rewards of differing quality, dispersion, or abundance. Hubbell and Johnson (1978) hypothesize that such bees, which visit many kinds of plant species, partition resources not so much according to taxon or flower size but by the spatial and temporal availability of a given flower species. Unaggressive foragers are displaced by large or aggressive species, and the smallest nonaggressive species may still attempt to harvest resources in the presence of aggressive bees. At flowers, these authors also showed that foragers of *Apis* and Trigonini responded differently to flower dispersion and foraged most effectively in distinct settings (Johnson and Hubbell 1975; Roubik 1981b; summary by Johnson 1983a). For the honeybee, detailed studies of pollen diet have shown that in tropical forest *Apis mellifera* specializes on large inflorescences or dense arrays of flowers (Roubik et al. 1984, 1986; see also Chapter 4). Figure 2.29 shows the succession of highly eusocial bees gathering pollen at large inflorescences of the palms *Bactris* and *Iriartea*. In one instance, the flowers were visited by at least three species throughout the day, but in the second study one species gradually dominated the inflorescence by aggressive group foraging.

The challenge and reward of detecting discrete foraging styles of highly eusocial bees is exemplified in the study of 14 sympatric Trigonini by Johnson (1983a), primarily at sucrose solution feeders but also at flowers. A total of eight foraging styles are given; more than one apply to some species. The principal foraging styles of these bees are either *solitary* or *group;* variations in these styles are indicated in the nomenclature applied by Johnson. Solitary foragers display, as a discernible foraging tactic during feeding, either *avoidance, displacement, gleaning,* or *insinuation*. The group foragers display *scramble, bustling, extirpation,* or *opportunism* in foraging style. Aggressive group foragers are only known among the *Trigona* that abruptly arrive at a resource in groups of 15–60 bees (Roubik 1982a; Johnson 1983a). These species have medium-size to large colonies, and a very few species have huge colonies (3,000–25,000–200,000). Despite justifiable skepticism regarding older literature reports on colonies of Trigonini exceeding a few tens of thousands of workers (Wille 1983a), Camargo (pers. commun.) has found normal forest colonies of *Trigona amazonensis* consisting of about 200,000 bees. An estimate of 150,000 workers for *Trigona spinipes* (in Michener 1974a) seems likely to be less accurate, because their nests are much smaller than those of *T. amazonensis* (see Fig. 3.1).

Worker *Scaptotrigona* and *Partamona* also appear to leave the nest in groups and feed in close proximity on flowers, but they do not forage aggressively. Group foraging is likely orchestrated by odor communication and scent marking of a resource (Johnson 1983a). Bees that recruit rapidly and in large numbers, such as *Apis* and some *Melipona,* attain a relatively high density on flowers, but no persistent role in the scent marking of flowers has been demonstrated. Solitary foraging has been defined as the independent foraging choice of a bee, irrespective of the decisions made by sister workers at the same flower species. This does not imply that group activity is unimportant in achieving recruitment to a resource. The odors of flowers or other resources are unavoidably imparted within the nest, and probably no bee occupying a nest with other bees is absolutely uninfluenced by the foraging experience of nest mates. Solitary and group foraging are generally identical among the scout bees or the first to discover a resource to which nest mates might then be recruited. The foraging patterns of highly eusocial bees may reveal still more variation in the hundreds of species that have not been studied.

In the mix of social and solitary species that occur on tropical flowers, several common patterns of flower exploitation can be distinguished. These can be predicted in a general way from resource characteristics. As noted above, the type and density of foragers will likely change over some period of time. The following foraging syndromes best apply to a relatively stable or equilibrium period in foraging. For all but extremely short pulses of resource availability, this period will be distinct from the initial presentation of resources and their final stages of depletion or senescence. I make a distinction between foraging syndromes, which apply to the behavior of bees at particular flower species, and the *floral syndromes* (Faegri and van der Pijl 1979; see also Section 3.4) that emphasize the principal pollinators and floral traits associated with them.

a. Generalized flower syndrome. At relatively small or open flowers, highly social species dominate the best resources by aggression or sheer numbers. Larger bees, and those displaying relative impunity to the bustling or aggressive pursuit of other foragers, move through patches in which social bees are dominant. Other small, "timid" or unaggressive bees forage on the scattered available resources unoccupied by dominant social bees and large, marauding foragers, either by insinuation or by removing what is left by other foragers. An example can be cited from foragers at large flowers of *Eugenia* in the Reserva Ducke near Manaus, Brazil. Here I saw individual trees dominated by aggressive *Trigona* of several species. While larger bees visited flowers of the various trees, the aggressive *Trigona* each remained on the trees they dominated and did not occur in mixed species groups. Small trigonines, however, were scattered among the several trees. Similar patterns in coexistence at resources have been recorded at artificial feeding stations (Hubbell and Johnson 1978). A variation of this syndrome is the arrival in early morning of foragers that remove pollen or nectar from nocturnally dehiscent flowers (Janzen 1968; Roubik 1988).

b. Specialized flower syndrome. Large bees or other animals predominate at large flowers to which they alone gain legitimate access, either by having proboscides or other resource-gathering appendages of the correct length or by having the power and physical bulk to force their way into relatively closed corollas (Frankie and Haber 1983). Smaller bees may glean residual pollen or nectar from the flowers and floral bracts (where pollen often falls after visits by buzz-collecting bees), gain access to floral rewards by perforating flowers, or perform as secondary robbers by using perforations made by larger foragers (Thorp and Estes 1975; Roubik et al. 1982).

Another type of specialized flower is only manipulable by a small bee that has exclusive use of very small flowers or those with a small space in which to enter the flower (Camargo, Gottsberger, and Gottsberger 1984; Mori and Pipoli 1984; Prance 1985; Gottsberger 1986). Restricted flowers of this type exclude visitors of certain sizes but do not exclude destructive flower visitors. In such apparently restricted flowers as the Lecythidaceae, which are normally pollinated either by tiny Trigonini that crawl down a long passageway to nectar or very long-tongued Euglossini, I have seen that many flowers are perforated by a few species of other Trigonini that remove floral resources without pollinating (Section 2.4.4).

c. Stratum fidelity syndrome. Bees forage at fallen flowers on the ground (stingless bees) and flowers at ground level (bees of all families), as well as at those in the uppermost reaches of the tallest trees. Stingless bees have been repeatedly noted for their wide stratum selection (e.g., Roubik et al. 1982; Wille 1983a), but solitary bees also nest in the ground or near ground level and may be active over the same height span as they travel to and from the showy floral displays of the forest canopy.

Stratum fidelity has been suggested for tropical forest bees (Frankie and Coville 1979; Perry and Starrett 1980) and for temperate bumblebees (Waddington 1979). As I interpret it, the basis for this behavior is the continuation of foraging within a profitable stratum. The same bees that appear to restrict their foraging to flowers in the canopy can probably be found visiting flowers at ground level at other times or places. The report of Perry and Starrett implied that the bees they saw in canopy flowers of the legume *Dipteryx panamensis* never visited flowers elsewhere in the forest and that they were exclusively canopy foragers. This claim could not possibly be verified without placing entire *Dipteryx* canopies at ground level and observing visitors to the flowers there. Furthermore, the implied dependency between the visitors of *Dipteryx* and these flowers is tenuous, because this tree does not flower every year where the study was made (Roubik pers. obs.). Bees will forage where they find rewarding resources. Their degree of specialization will usually depend on the relative abundance of different resource types both during their life span and over the course of many generations, if genetic variation at all affects such floral visitation patterns.

Foraging theory suggests that when a single type of several simultaneously available resources reaches a threshold abundance, foragers preferentially take that resource and ignore others (Pyke 1984). This is particularly applicable for flowers, because each species has a different shape and must be visited in a different way (Lewis 1986; Waser 1986). The notion that an individual bee exclusively directs its foraging activity toward a stratum, ignoring all other factors, may be true and is suggested by the results of Waddington's 1979 study. A learned preference for flowers in a certain area may occur, or during part of one foraging bout, just as within a more two-dimensional setting (Heinrich 1979a). In hilly terrain where the canopy is not of uniform height, the question of canopy foraging perhaps relates more to the openness of the foraging area and the ease with which it is surveyed than its height. Or, as discussed earlier, it may be less energetically demanding to forage in a horizontal plane or to carry a foraging load downward to the nest rather than upward. The experimental study of Frankie and Coville (1979) showed that centridine bees preferred flowers of a bush presented at a height of 4 m to one at a height of 1 m. A later study by Roubik et al. (1982) assessed visitation by some of the same bee genera and species to flowers in three patches of a flowering tree, revealing that bees showed no preference for flowers at a height of greater than 12 m to flowers under 7 m, and preferences for lower flowers also were documented. These behavior patterns had little relation to the number of flowers available in each of these strata and showed that these bees, in a population, could not be expected to forage at a particular height. In the three-dimensional foraging arena of the forest, height preferences probably arise when there is a vast difference in the availability of flowers at widely differing heights or where there is a mating advantage (Chapter 3) to patrolling and feeding at more open areas, such as the canopy. Both the studies of Frankie and Coville and of Roubik et al. concerned flowers that only offered pollen. This is a desirable characteristic for such work, because the sugar content in nectar may differ twofold on a single tree: Higher flowers often have more sugar (Crane, Walker, and Day 1984).

d. Trap-lining syndrome and the aggressive foraging syndrome. These two syndromes are mutually exclusive. For a large bee or a colony working collectively, many flowers are needed to obtain an adequate amount of nectar or pollen. If the amount of resource per flower patch or flower is small, then the foraging range of the bee will be larger. When the patch of many small flowers is worth defending, then it will be dominated by a stingless bee colony that possesses aggressive foraging behavior and at least some anatomical features appropriate for fighting (Sections 2.3.5, 2.4.3, and 2.4.4; see also Section 2.3.6a).

Some flowers or resource patches persist over several days or weeks, and others are far more ephemeral and may be available for less than a day. Botanists have described the endpoints in this broad spectrum of flowering phenology as "big-bang" and "steady-state" flowering (e.g., Gentry 1974a; Ackerman 1983a; Bawa 1983; Mori and Pipoly 1984). A trend that could well be generalized has been detected in a number of neotropical plants (Augspurger 1983a): Species whose individuals flower for a short period tend to have more synchronous flowering episodes. Janzen (1971) proposed that bees that visit plants that present few flowers but continue flowering for many days (steady state) should be visited by trap-lining bees. These move among many plants over a relatively large area. The trap-lining pattern has been demonstrated but is variable in a species (Heinrich 1976a; Ackerman et al. 1982). Inexperienced foragers are less likely to display the behavior, and even experienced foragers in habitats lacking suitable orientation cues are unlikely to trap-line in a predictable sequence. Big-bang flowering patterns can draw many territorial foragers, some of which may successfully defend resource patches. The aggressive stingless bee species tend to discover resource patches slowly, but after finding them to be worthwhile they will recruit massively to them (Section 2.3.5).

Trap-lining among flowers at different strata has not been demonstrated, but it may occur on the flowers of vines or lianas on forest edges, or among trees, treelets, or shrubs having a range of flowering heights or cauliflorous inflorescences (those that spread over the trunk and branches). Frankie et al. (1983) discuss a form of trap-lining on the flowers of a tree species with a vertically elongate crown; flowers at its base secrete nectar before those higher up. Foragers respond by gradually moving up the tree. Trap-lining appears profitable for a large bee visiting highly dispersed yet predictable resources, but there is no obvious reason why a small bee, either solitary or social, should shun limited local resources even though it does not use the trap-lining method. Larger resources, the tropical trees or large stands of a species, are potentially far more profitable. They are visited by a plethora of bee, wasp, fly, bird, butterfly, moth, and bat species that range greatly in size and foraging style (Section 4.1.1).

2.3.7 *Foraging patterns and optimality*

The changing daily profitability of a flower patch transforms it into several different resources; each one may transiently be exploited most effectively by a different bee species. The most energetically demanding foragers are often the first to arrive. Kerfoot (1967a), Schlising (1970), and Real (1981) observed the changing composition of bee species at flower patches during the day in Costa Rica, where Frankie, Opler, and Bawa (1976) also noted a succession of large to small bee species at a species of flowering tree, a pattern often found in temperate habitats (Heinrich 1976b; Linsley 1978). Linsley and Cazier (1963) and Schaffer et al. (1979), among others, provide detailed information on the turnover of bee species at flowers. In French Guiana, Mori and Pipoly (1984) and Roubik (unpublished data from *Mimosa pudica*) observed that after the peak foraging activity of bees at abundantly flowering resources, visits later in the day were primarily by small Trigonini. The causal explanation for change in the studies was incomplete, but foraging profitability clearly was perceived differently among species and could account for changes in their abundance. Changes in floral resource availability that corresponded to bee species turnover at flowers were documented by Murrell and Nash (1980). They carefully noted change in nectar volume per floret and the number of *Apis cerana* and smaller *A. florea* in a flower patch in Bangladesh. Both species foraged through the day. Although the larger bee peaked in abundance at 1000 hours, after this time the smaller bee increased abruptly in abundance, reaching a peak after 1200 hours. Nectar volume per flower steadily declined from 0.3 to 0.08 μl. Real (1981) followed the distinctive peak abundance patterns of 20 bee species visiting two species of *Ipomoea;* one flower was shallow and visited primarily by shorter-tongued bees, and one was deeper and visited primarily by bees having longer tongues. The largest bees (*Eulaema, Xylocopa, Bombus,* and *Thygater*) came first and the smaller bees later. The shallow flowers showed a considerably more staggered succession of the visitor species. Furthermore, their nectar sugar concentration rose from 20% to 40% sugar, but its quantity steadily decreased from the time visitors arrived until the flowers closed.

Species turnover at flower patches is driven primarily by resource depletion resulting from preemptive competition. It is rarely due to aggressive displacement. Very few bees attack other foragers, as is well known, although there can be many other mechanisms of displacement. In competitive situations, which are experienced to some degree by most foraging bees, the principle of an "ideal free distribution" (Fretwell and Lucas 1970) may often apply (Roubik et al. 1986). At an equilibrium, all competitors achieve equal gains by dispersing among resources, and good resource patches are harvested more intensively than poor ones. Thus a poor resource for large, energetically demanding or highly eusocial bees (Schaffer et al. 1979) may often be acceptable to smaller or solitary species, all other factors being equal. The process of reaching an equilibrium with resources and competitors at flower patches potentially orchestrates most of bee foraging dynamics.

The distribution of foraging bees from a colony and the patterns followed in flower visitation by bees in general continue to be difficult field problems. By using a novel yet extremely labor-intensive study method, Gary and co-workers (1977) demonstrated that honeybees foraging in homogeneous agricultural field settings were not particularly constant in foraging territory and ranged widely. These studies employed the magnetic recapture system in reverse. Bees were fitted at their hives with small ferrous tags, which were later retrieved with thousands of magnets placed over the umbels of onion flowers. Previous studies (review by Eickwort and Ginsberg 1980) had indicated repeated return by honeybees and bumblebees to small patches of flowers over several days. However, studies of Visscher and Seeley (1982) demonstrate that honeybees frequently shift their foraging sites. It is clear that until the foraging options of a bee are understood in relation to available flowers and other extrinsic features of the environment, as well as the intrinsic, stereotypic behavior of the species, foraging patterns are scarcely predictable (Waddington 1983). Indeed, without exact information on foraging conditions and the factors mentioned at the beginning of this section, there is perhaps no reason to expect most behavior of an opportunistic forager such as a bee, particularly learning behavior, to be predictable (Section 2.3.1).

Another direct method of determining foraging routes and day-to-day constancy of bees visiting particular resources is that of tagging bees at flowers and then monitoring the flowers to record returns. Ackerman et al. (1982) used this technique to study foraging euglossine bees in lowland forest. They found that males of *Exaerete frontalis* (a parasitic species) and *Euglossa imperialis* consistently foraged at the same patch or inflorescences of *Calathea latifolia* and ignored most of those nearby over a several-day period. Nonetheless, the same flowers were not always included in a foraging sequence, nor was the direction from which bees initiated and terminated their visits to 80 monitored inflorescences invariant. Convincing evidence of extremely accurate spatial orientation of returning foragers was shown by removing inflorescence stalks. Foragers arriving on the following day circled and searched in midair where there had previously been a flower, and this behavior was also noted by Janzen (1968) for the colletid bee *Ptiloglossa* and for *Euglossa* (Janzen 1971).

A further method to gauge foraging ranges of bees was employed by Frankie, Opler, and Bawa (1976). They placed foragers in containers with fluorescent powder of different colors (Frankie 1973); the powder cannot be groomed from between the coxae and some areas on the notum and head. Four groups of a flowering tree species were used to study bee movement between trees. A small minority of the 70 bee species and hundreds of marked individuals moved, over a several-day period, between trees separated by 0.4–1.9 km of open pasture. Whether such movement occurs in forest conditions is unclear. The bees found moving nearly 2 km were medium to large anthophorids (*Centris* and *Gaesischia*) and a megachilid *Chelostomoides*. Frankie (pers. commun.) has found marked male *Centris* moving 7 km between trees within a few days.

Once foraging at flowers, the movements of individual bees reflect the rules of cost–benefit assessment in a striking manner, and they also show that no behavioral trait is necessarily a stable one. Bees do not follow simple, self-defeating rules such as "go to the closest flower," and bumblebees gathering nectar demonstrate the flexibility of bee behavior (Hodges and Wolf 1981; Waddington 1981). The distance they move between flowers is inversely related to the volume of nectar removed during a flower visit (Fig. 2.30). In general, their movements while foraging are spatially restricted when foraging is profitable, but they fly increasing distances between flowers as foraging rewards decline. Furthermore, less of the

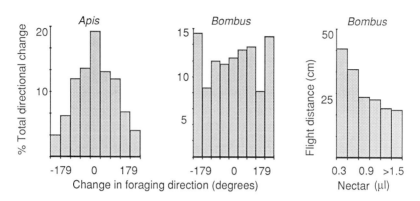

Figure 2.30. Diversity in flight distances and directions taken by foraging bees in resource patches. Variation in the direction of flight taken by foraging *Apis* and *Bombus* (Apidae), shown in the left and middle diagrams, reveals that foragers can tend to move in a straight line (*Apis* foraging nectar) or move almost randomly (*Bombus* foraging pollen). The *Bombus* shown at the right displayed increasing flight distances between successive visits to flowers in a patch when individual flowers contained relatively little nectar (Waddington 1981, 1983; Zimmerman 1979).

nectar in flowers is removed if bees have determined that nectar is locally plentiful. Change in direction in foraging movement, however, is random for bumblebees successfully removing nectar or pollen from flowers (Zimmerman 1979; Hodges and Miller 1981) but tends to occur more often when a profitable flower patch is entered (Pyke 1978a; Heinrich 1979a; see also Fig. 2.30). The reason is apparently uncomplicated: Bees in general forget from which direction they arrived after probing and shifting their position several times on a rewarding flower (Schmid-Hempel 1984). Foragers thus stay in profitable areas and increase the chance of finding better areas when foraging is less profitable. Honeybees also turn more frequently when nectar quality or quantity per flower is relatively high (Marden and Waddington 1981). Movement predominantly in a straight line has, however, been noted for other bees and is apparently more common (Waddington 1983). This may indicate that bees are attempting to avoid revisiting a flower that they have recently depleted, despite the variable rewards in flowers within a patch (Corbet et al. 1984; Marden 1984). It may also mean that bees have sometimes been observed in patches where foragers have depleted most of the food. The density of flowers and competitors can influence in other ways the choice a bee makes to continue visiting flowers near the one it occupies. Large bees such as *Xylocopa* and *Bombus* leave odor marks on flowers, and these may be used by other foragers to assess potential foraging profits without entering the flower (Pijl 1954; summary by Eickwort and Ginsberg 1980; Corbet et al. 1984). Honeybees may respond to odors left by other foragers at flowers (Free 1987). Still, no evi-

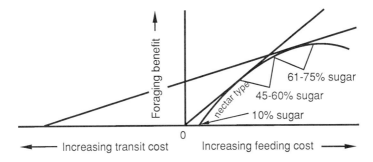

Figure 2.31. The concept of the marginal-value theorem applied to nectar feeding by bees (after Heyneman 1983; Roubik and Buchmann 1984; Stephens and Krebs 1986). Lines tangent to the forage profitability curve on the right show the predicted change in forage preference with increasing travel time to the resource patch. Feeding costs reflect the increasing difficulty of imbibing more viscous solutions (see Fig. 2.20). Explicit assumptions of the model are that bees maximize foraging rate (gain/time) and that the greater the cost of arriving at the resource patch, the more time is spent there and the more profitable it must be. The maximum rate of caloric gain for a number of long-tongued bee species occurs at nectar sugar concentrations of 45–60%. Thus when there are no transit costs to foraging, such bees prefer and benefit most from nectar of 45–60% sugar. The minimum nectar sugar concentration that is acceptable to bees is 10%, and the maximum is near 75%.

dence exists that this mode of communication determines foraging behavior in a significant way at natural floral resources.

If we propose that foraging bees try to maintain the highest rates of resource harvest, then we are applying the *marginal value* theorem. This caricature of bee behavior improves upon the rule "go to the closest flower," as mentioned. In brief, it poses the forager's problem as one of deciding whether to move on, based on its assessment of current and potential profits (Stephens and Krebs 1986, pp. 95, 103). A central element in the argument is that increased travel costs to arrive at a patch must correspond to prolonged foraging time and profit within it (Fig. 2.31). Whether a bee readily searches for new resource patches depends strongly not only upon the chances of finding them but also on the advantage of gaining information that will allow choosing among them, even if there is no serious depletion of the current foraging patch. The relative transit and feeding costs in foraging behavior have been evaluated for nectar-feeding animals by Heyneman (1983), who finds that as transit time increases between resources and the central base of the forager, the quality of the resource (in this case energetic value of nectar) should be greater (Fig. 2.31). This analysis included consideration of nectar viscosity as the feeding cost; viscosity and ease of imbibement are strongly dependent on temperature and sugar concentration (Section 2.2.7).

A relationship between nectar quality and foraging effort was indicated by studies of Roubik and Buchmann (1984) showing that 15 colonies of four species of

Melipona took higher-viscosity nectar as the day progressed. Because the nectar in flowers is gradually depleted by foragers and individual bees returned with similar amounts of nectar throughout the day, it was suggested that bees were flying further to resources and making more visits to flowers. Increasing sugar concentration in nectar should promote this behavior (Section 2.3.9). Each of these species as well as *Apis mellifera* showed peak rates of energy harvest at sucrose solutions between 45% and 60% sugar (Fig. 2.20). Yet distinctive preferences for nectars of differing sugar concentrations were found for some species. These could be predicted on the basis of net rate of energy intake at sugar solutions of a range of concentrations, taking into consideration the energy benefits of the time saved by gathering a load of lower-viscosity nectar. An exception to this pattern was found for the largest bee studied, a colony of which gathered large loads of dilute nectar (20% sugar) throughout the day. Heyneman's model suggests that bees should accept dilute nectar when it is plentiful and close to the nest, but location of the nectar source used by the bees in the Roubik and Buchmann study was unknown.

Within a given flower patch or proximate series of patches, it appears to be an accurate statement that bees work in small areas when foraging is profitable and then extend their foraging area, eventually abandoning the flowers altogether. Regarding foraging bumblebees, Heinrich (1976a) proposes that bees "major" on the most profitable flowers they encounter, while "minoring" on other less profitable flowers in the same patch or area. Maintaining one or several minors provides a means of assessing changing resource profitability and then, based on a reward scale of the major resource, switching to another resource as it becomes the more rewarding. There is little doubt that it is adaptive for a bee occasionally to sample flowers other than those at which it is specializing. This behavior might be accentuated in tropical habitats where hundreds of flower species can be present at one time (Chapter 4). For example, Raw (1976) documented an average of 4–5 and up to 10 pollen species contacted during a single foraging bout by two *Exomalopsis* species in Jamaica.

The weight of the foraging load, particularly nectar, has quite definite significance in determining foraging behavior, regardless of the distance of the resource from the home base. An absolute limit in the foraging load of bees, and apparently for all insects achieving free flight with the wing muscles alone, is that the weight of the bee's flight muscles cannot be less than 15–16% of its total weight (Marden 1987). Marden's rule indicates that Hymenoptera can carry a load equivalent to their body weight because flight muscle acounts for about 30% the weight of winged species. Applying Marden's rule to nectar foraging by bees implies that full nectar loads must be smaller, on average, when the nectar sugar is more concentrated. The nectar load of a bee also gathering pollen must also be smaller than that of another bee foraging only nectar. A further prediction is that the largest nectar loads will tend to be more dilute, for which there has been some supporting data in the large stingless bee *Melipona fuliginosa* (Roubik and Buchmann 1984).

There is significant variation among views of how individual bees and colonies assess foraging profitability. Differences seem to correspond well to the basic dichotomy described at the introduction, that of rate maximizers versus efficiency maximizers. On one hand, Heinrich (1979a) summarizes the commuting and foraging behavior of large and energetically demanding bumblebees: "The costs of carrying large loads and flying long distances are insignificant, in terms of energy expenditure, when compared with the cost of food energy not collected because of lost foraging time." Hence, even if greatly rewarding resources are present, less rewarding ones might be preferred if these are more available. On the other hand, Schmid-Hempel and co-workers (1985) found that honeybees often returned to the nest without a full nectar load of 50% sugar solution. Their interpretation was that achieving a high net energy intake rate, as well as a high rate of colony energy acquisition through recruitment of other foragers to the resource, were less important than the conservation of workers by diminished flight effort. The argument is basically that certain irreversable physiological costs of prolonged flights outweigh potential energy gains (Sohal 1986).

Individual foragers of European honeybees and presumably solitary bees seem to gather more food energy per foraging trip when nectar is readily accessible but flight expenditure is considerable. Nuñez (1982) found that *Apis mellifera* gathered a larger load of 50% sugar solution at 2 km from the nest than at shorter distances. He also found that bees imbibed less than their full capacity when the food source was close to the nest or when it was relatively more difficult to harvest. When nectar was easily harvested and close to the nest, bees also returned to the nest without a full load. Later work by Kacelnik, Houston, and Schmid-Hempel (1986) provided evidence that this type of foraging behavior is explained better as energetic efficiency than as rate maximization. The partial crop filling by honeybees had been interpreted to show that colonies maximized rate of resource harvest when foragers returned to the nest with less than a full load, thereby potentially enhancing colony recruitment. However, the load that a foraging bee carries in her crop may be less than that of bees imbibing sugary liquids while feeding stationarily. This was tentatively confirmed for both honeybees and stingless bees (Roubik and Buchmann 1984) and brings into question the definition of maximum crop capacity. Bees feeding without taking flight seem to imbibe a greater amount of sugary solutions than do normal foragers.

For insects in general, increased activity tends to result in a shorter lifespan (Sohal 1986). To survive through long inactive periods, colonial bees in the temperate zone adjust their behavior apparently to strongly favor efficiency over maximum intake rate. Recently, a foraging-efficiency hypothesis has received experimental confirmation for European *A. mellifera* (Kacelnik et al. 1986; Seeley 1986; Houston et al. 1988). The first authors show that individual European honeybees that forage at a range of distances from their nests maximize efficiency and not harvest rate. This supports the hypothesis that the individual worker bees are less likely to perform energetically demanding foraging activities. Extending the efficiency approach to the entire colony, Seeley (1986) concludes that the colony assesses resource patches in such a way that its forager population and nest storage areas in their combs (Rinderer 1982) are allocated by minimizing total cost to the forager rather than by maximizing the net rate of energy harvest of the colony. The contrasts with tropical *A. mellifera* that display colony behavior that may place emphasis on rapid discovery of resource patches and maximum rate of intake are discussed in Section 4.3.2.

The bumblebees do not recruit nest mates to food sources, and a contrast in their foraging behavior compared with highly eusocial apids reflects this in some startling ways. Wells and Wells (1983) set up experiments to test the hypothesis that individual European honeybees respond optimally by selecting the most rewarding or predictable resources, rather than simply remaining faithful to a particular color of flower. The bees were found to be very constant to previous resources, which resulted in individual behavior that was undeniably suboptimal. They constantly visited inferior flower types. Fortunately, such behavior is normally regulated and corrected by the colony for honeybees and the other permanently eusocial apids. As Seeley (1986) emphasizes, all resources carried by returning foragers are ranked by the response of the bees receiving it in the nest. First, a bee returning to a colony cannot unload the nectar she carries if it is substantially less rewarding than other nectar collected by colony members. In fact, if she does not transfer her nectar load to another bee within 100 sec, she does not even communicate the location of the resource (Seeley 1985, 1986; see also Section 2.3.5). Second, if the bee is collecting pollen, cells that have been prepared to receive it are necessary, and a similar lack of communication about the food source will occur if the returning forager has difficulty finding a place to store her pollen load (Free 1967).

Potential foraging profitability is not adequately assessed by measuring flower density alone, or even by measuring the distribution of the resources that flowers contain. Patch selection, which is perhaps the common denominator of a wide range of behavior (Wiens 1978), provides examples of varied cost–benefit assessment and the principle of ideal free distribution, or the dispersal from crowded or depleted resource patches. Often the same characteristics of a flower patch can be correlated with contrasting forager behavior. In studies of foraging *Melipona* at a lowland tropical shrub, both Augspurger (1980) and Roubik and Buchmann (1984) found that one common species most often visited the flowers of plants presenting many flowers. However, another common *Melipona* showed no foraging preference related to flower number on the plants. In a study of foraging solitary bees, Waddington (1976) observed positive correlations between the number of halictid foragers and the number of *Convolvulous* flowers in patches near the ground. In studies of trees, Stephenson (1979) also found that larger floral displays received more visits on a per-flower basis. In contrast, other studies of a tropical tree and shrub species showed visitation no more frequent to flowers in large displays than to flowers in smaller displays, and flowers in smaller displays even received more visits in some patches (Roubik 1982b; Roubik et al. 1982). Developing an empirical concept of what group of flowers, or even what size of an individual flower, is treated as a patch by a bee is an area that deserves more investigation. Furthermore, when the distribution and abundance of resources are changing constantly due to foraging competitors, it is likely that floral characteristics such as number, size, and arrangement of flowers, or cues indicating pollen and nectar availability, may be superseded in importance by chemical, visual, or other signals from the foragers themselves.

Many more studies of bee foraging patterns could be cited here, but it should be evident that forager behaviors are determined both by extrinsic and intrinsic variables that, although basically similar in all bees that have been studied, vary among species and particular ecological settings. Fortunately, it is likely that the studies and models of foraging choice and pattern in honeybees and bumblebees (Oster and Heinrich 1976; Heinrich 1979a; Waddington and Holden 1979) apply equally well to solitary and social species. Foragers of solitary bees as well as the scout bees of colonies work independently rather than react specifically to the needs and demands of a colony (Seeley 1985). From the studies reviewed here, it is at least probable that larger bees or, in the case of *Apis,* those with exceptional communication and recruitment systems travel a few to several kilometers to resources. Small bees, which could be designated to be those less than 10 mm in length (e.g., Frankie et al. 1983), will rarely travel more than a kilometer during foraging (Section 2.3.3). Male bees are thought to be less constant in central foraging area than females because they are not often associated with a nest site. Stable central-place foraging by some males is likely, and transient nesting habits are known among females such as *Megachile brevis* (Michener 1953). I have seen the males of *Centris analis* return at dusk, apparently throughout their lives, to a particular hole in the wood in which the females nest and possibly where the males emerged as adults, and males of *Eulaema nigrita* return to a leaf of a particular plant at which they pass inactive periods hanging from its underside, holding the central vein with the mandibles. The vagility of both males and females during their life spans are at present difficult to characterize for most bees.

Observations in the field still reveal seemingly anomalous behavior that might be due to the varied means bees have for measuring forage quality, their response to forager odors, or the way that a flower presents its nectar or pollen. Because of this, perhaps it is still too early to decide the extent to which bees exhibit optimality in their foraging choices – the predictions for one set of circumstances can easily be the opposite of the predictions for another. Hodges and Miller (1981) found that pollen-gathering bumblebees apparently did not detect the amount of pollen they collected from a single flower, because the time they spent on a flower that had just been visited was the same as that on an unvisited flower. Although bees tended to visit plants with many flowers, they only visited a small number of flowers on a plant. Foragers at buzz-pollinated flowers may spend more time on anthers that have previously been visited (Section 2.1), whereas *Melipona* visiting the ball-shaped flowers of *Mimosa pudica*, which they vibrate to gather pollen, spend more time on flowers that have received less visits (Roubik 1978). During nectar foraging, some bumblebees probe only a few of the several dozen florets of flowers before moving to another inflorescence (Free and Butler 1959; Hodges and Wolf 1981). This could mean that the flowers have been largely depleted or that the bees are minimizing their chances of overexploiting flowers to which they will shortly return, in the absence of any persistent competition. The bumblebees studied by Hodges and Wolf (1981) left nectar in flowers that had large standing crops and tended to deplete it from flowers that had lower rewards. Another possible explanation for very brief visits or seemingly very superficial sampling of a flower patch involves scent-marking of flowers by bees, at least by *Bombus* (Cameron 1981; Free and Williams 1983) and other bees such as *Xylocopa* (Frankie and Vinson 1977). Honeybees often mark artificial feeders or water sources with odors that are partly volatile and partly adherent to the substrate (Nuñez 1967; Free and Williams 1971; Ferguson and Free 1979). As Free (1987) comments, the chemical nature, glandular origin, and significance of odors involuntarily placed on substrates by honeybees are largely unknown. The possibility that sporadic release of volatile honeybee Nasanov gland odor attracts other bees to flowers is less likely, but conclusive studies have yet to be made. The use of forager odors is far clearer where truly aggressive foraging takes place by groups of stingless bees (Sections 2.3.5 and 2.3.6).

A characteristic of the few instances in which bee odors elicit demonstrable responses from foragers is that foraging bees are either repulsed or attracted to the flowers and odors upon them (Corbet et al. 1984; Free 1987). Corbet and her co-workers postulate that the same odor may initially have a deterrent effect, preventing revisitation of a flower just depleted. But as the odor spreads over the surface of the flower, its concentration may then correspond to a lower threshold of response for a foraging bee, one that indicates that the flower is a rewarding one given that nectar secretion continues. Such temporally reversing odor cues might also occur at pollen resources, where pollen dehiscence occurs gradually and resembles the secretion of nectar. One situation in which odor marking of flowers would be of advantage to female bees concerns buzz-pollinated flowers, for which there are no visible cues to pollen availability, and apparently (Buchmann 1983) gradual dehiscence. Short-lived deterrent odors that later become attractive should have little benefit for bees visiting flowers that have most of their potential standing crop removed in a single visit.

Social bees with effective communication systems can more readily abandon a resource and shift to a better one, and solitary bee individuals are less likely to abandon resources entirely. Aside from the awesome information-gathering ability of socially foraging colonies, this is suggested because the social species, with ample food stores in the nest, enjoy additional flexibility over the solitary bees. They alone endure periods of resource scarcity without obligately moving to a new site or lowering activity and metabolism. The migratory and cold-weather behavior of some honeybees is an exception (Section 4.3.5), but local migration of some larger bees having powerful flight might be caused by resource shortage, and this behavior may be a common phenomenon in the tropics (Janzen et al. 1982). These authors interpreted the variable number of euglossine bee species with tattered, worn wings and those with perfect wings at chemical baits as an indication that during certain times of the year the bees were dispersing over a distance of a few tens of kilometers. The habitat in which this study took place was tropical lowland deciduous forest, which experiences pronounced seasonal drought. Wetter mountainous areas were within 20 km of this forest and may have provided refugia or other resources for bees.

Different flexibility in foraging area is also possible in male and female bees of the same species. Dressler (1979) and Janzen (1981a) both suggest that males of the euglossine bees that have to collect fragrances in addition to nectar and that need to seek mating areas having pollen resources used by females may range over a wider area than the females. The studies of Ackerman et al. (1982) of bees in this group did not substantiate these differences in foraging behavior at flower patches; both sexes repeatedly arrived at the same resources. Undoubtedly, the simultaneous availability of all required resources in a small area is more likely in wet forest where the latter study was made than in the more xeric area studied by Janzen (1981a).

2.3.8 Predation and mimicry

At tropical flowers and other resources, bees are commonly captured by ants, reduviid, phymatid, and other bugs, dragonflies, jumping and crab spiders, web-spinning spiders, mantids, robber flies, vespid wasps, frogs, lizards, small snakes, toads, and birds. Many of these animals were listed among the natural enemies of bees at flowers in recent reviews by Knutson, DeJong, Morse, and Ambrose [in a volume edited by Morse (1978)], and reviews have also treated predation on foraging bumblebees (Morse 1982; Plowright and Laverty 1984). Because many observations focus upon introduced honeybees, some of the predators may be facultative, particularly near large aggregations of honeybee colonies. Available information does indicate that the number of taxa attacking bees in the tropics is undoubtedly large. Many such predators of honeybees at flowers take any insect prey within their grasp, but little is known regarding their relative preference for bees. Avian specialist predators of flying Hymenoptera occur both in the Old World tropics (the family Meropidae, or "bee eaters") and in the neotropics (Galbulidae, or jacamars, and Tyrannidae, or flycatchers). Predation at flowers and its consequences constitute a poorly developed area of bee ecology. I will focus on the predators that appear specialized on bees and discuss the mimicry sometimes involved in this predation and in predator deception. A discussion of predators attacking the nests of bees is given in Chapter 3.

Mimics of bees are composed of *aggressive* (i.e., predatory) mimics; *Müllerian* mimics that are distasteful to predators, as are bees they resemble, and palatable, *Batesian* mimics that gain protection by appearing where unpalatable bees have "educated" the local predators. In a given assemblage of species, there can be found organisms including all three types, frequently wasps and flies, and sometimes also Hemiptera or true bugs. When the mimetic relationship with bees is Batesian, the mimics are unlikely to be active as adults during the peak abundance period of the bee "models." Similarly, they are likely to be scarce when many potential predators are naive and are in the process of learning to avoid the models (Huheey 1984; Waldbauer and LaBerge 1985). Migrant birds in the tropics can plausibly be a reason for temporal separation of the appearances of model and mimic. Although no studies have shown that such a relationship exists with tropical bees, the population dynamics of fledgling birds clearly shows that mimics in the temperate zone have adjusted their phenology to avoid periods of abundant naive predators (Waldbauer and LaBerge 1985).

Reduviid bugs pierce foraging bees captured at nest entrances or at flowers (Fig. 2.32). Mimicry between the stingless bee *Trigona fulviventris* and the slightly larger reduviid *Notocyrtus vesiculosus* was noted by Jackson (1973), but despite its presence at the same flower species, the bug was not seen with this prey. Johnson (1983b) described the persistent but rather clumsy behavior of *Apiomerus pictipes* at the nest entrances of *Tetragona dorsalis* and *Trigona fulviventris* but suggested that substantial colony mortality could accrue. This bug has elongate forelegs that it covers with resin or other sticky substances, but its legs move too slowly to catch flies although it often ensnares small bees. Several other genera of reduviids, including both neotropical and paleotropical species, capture bees at re-

Figure 2.32. Hemipteran predators and mimics of bees. Top: a mimetic *Apiomerus* and the potential prey it resembles (after Jackson 1973), *Trigona fulviventris* (Apidae). Bottom: *Apiomerus* with prey *Augochlora* (Halictidae) captured at flowers; original photograph by Sean Morris, Oxford Scientific Films.

sources and attempt to capture honeybees and stingless bees at their nests. Miller (1956) and Usinger (1958) report that two Oriental genera capture resin-collecting Trigonini at trees, applying the same resin to their foretibia. Johnson (1983b) believes that the bugs perceive and are attracted to the mandibular gland odors of stingless bees. I have watched neotropical genera appear at resin sources in deep forest where stingless bees are foraging and capture both Trigonini and the larger *Melipona*. Neotropical *Apiomerus* arrives in early morning at flowers of *Mimosa*

Figure 2.33. Top: Asilid fly mimic *Hyperechia* that is also a predator of the model, Asian *Xylocopa*, and a parasite of its nests (Yoshikawa et al. 1969); original photograph provided by S. F. Sakagami. Bottom: Neotropical asilid fly mimic of the euglossine bee *Eulaema* with a honeybee prey; original photograph by James Marden.

pudica and *M. pigra*, where there is no nectar or strong floral odor but where bees intensively harvest the pollen. Because bugs do not feed on the flowers, it appears they actively locate resources at which bees are found.

Asilid flies are both bee predators and bee mimics in Asia, Africa, and tropical America, but are not restricted to the tropics (Fig. 2.33; Appendix Fig. B.22). There are few literature references to asilid predation on native tropical Hymenoptera including bees (Hurd 1978; Fisher 1983; Shelly 1984, 1985; Moure and Hurd 1987); observations deal largely with introduced *Apis mellifera*. Linsley (1960) and Powell and Stage (1962) showed that asilids feed primarily on other insects, and most of the taxa listed by Fisher (1983) were Hymenoptera but did not include bees. Neotropical asilids, when they mimic bees, are strikingly close mimics of the large, hairy euglossines *Eulaema* and *Eufriesea* (Dressler 1979). As shown in Figure 2.33 and Appendix Fig. B.22, some species are not large enough to subdue larger bees; but some are sufficiently large to capture them, as has been noted with *Eulaema*, a large euglossine bee (Shelly 1985; Janzen pers. comm.). I have seen an asilid (not an obvious bee mimic) capture and attempt to kill several *Euglossa,* but they were released by the flies. The asilid *Hyperechia* in Sumatra is a mimic of a large, black *Xylocopa* (Yoshikawa, Ohgushi, and Sakagami 1969; see also Fig. 2.33), which is probably one of its prey items. Poulton (1924) discusses a *Hyperechia* that in Africa is both a predator and mimic of adult *Xylocopa,* having larvae that are parasitoids of its immatures. The resemblance of some asilids to *Apis* or *Bombus* has been noted (Waldbauer, Sternburg, and Maier 1977), and smaller bumblebees may be taken by large, *Bombus*-like robber flies (Brower, Brower, and Westcott 1960). Linsley (1960) offers the view that a beelike appearance, if not entirely coincidental due to introduction of the honeybee, affords the fly protection from some of its natural enemies, a view shared by Dressler (1979). Unfortunately, natural enemies of the adult flies in the tropics are not well known, nor have they been tested for the response of naive foragers to bee models and fly mimics (Huheey 1984). In the temperate zone, several studies give credibility to the hypothesized protection of flies from two bird species and a toad, which were unable to distinguish flies from *Bombus* or which avoided flies after experiencing a bumblebee sting (Brower et al. 1960; Waldbauer and Sheldon 1971; Evans, quoted in Plowright and Laverty 1984).

The resemblance of flies and other insects to large stinging bees is probably a Batesian mimicry complex, as the large bees mimicked by them possess painful stings. Syrphid and bombyliid flies mimic both *Euglossa* and *Bombus* (pers. obs.). Female bees might also be considered Batesian models for male bees (Stiles 1979). A close resemblance of male to female bees thus potentially involves Batesian mimicry, also called automimicry within a species (Brower, Pough, and Meck 1970). Widely diverging appearances of males and females of the same species might indicate a general lack of predation on foragers.

At a given latitude and thermal regime, larger bumblebees show less sexual dimorphism than smaller species, and males of dimorphic species have longer but less dense pubescence and lighter body color (Stiles 1979). Sexual dimorphism also increases with increasing altitude (Plowright and Owen 1980). An explanation for the variable color patterns of the males is that those that hover during mating flights, thus generating more body heat than females, have shifted toward pubescence and color characteristics that radiate more light of the visible spectrum but nonetheless serve to maintain body temperature. It should be noted that this characteristic does not imply reflection of heat per se, because infrared radiation is not within the visible spectrum. Sakagami and Itô (1981), however, point out that in tropical bumblebees, the subgenus *Fervidobombus,* there are minor differences between the sexes, whereas in the similarly large temperate subgenus, *Pyrobombus,* there are pronounced sexual differences in color pattern and pubescence. They say that this bias within subgenera prohibits analyses that lump them together, which was done by Stiles to suggest that dimorphism is not favored in the tropics. Sakagami and Itô imply that there are fundamental phylogenetic constraints to expression of dimorphism and demonstrate that the degree of dimorphism in two large subgenera is not associated with the thermal environments of their species.

A noteworthy uniformity exists among the over 160 euglossine species (Kimsey 1987); males do not diverge markedly in appearance from the females, although the color patterns on the face do sometimes differ between the sexes. The males remain stationary for a brief time on tree trunks or branches at which they collect fragrances (Ackerman 1985) or in courtship displays (Kimsey 1980; Schemske and Lande 1984; see also Section 3.3.1), where they might easily be captured by arboreal lizards. I have seen one instance of such predation by a lizard that consumed a male *Euglossa imperalis* collecting a chemical fragrance. Other territorial male bees that perch while seeking mates rather than hover might also gain protection from predators if they closely resemble female bees. The large perching male *Xylocopa* and *Bombus volucelloides* that I have seen in the neotropics have the same coloration and general appearance as the females, whereas some hovering male *Xylocopa* are light orange or yellow, in contrast to the black females (Appendix Figs. B.1, B.5, and B.6). Stiles (1979) notes that females of the crepuscular bee *Ptiloglossa guinnae* are darker than males (Roberts 1971), yet it is not clear to what these differences are related. The ecology of automimicry, sexual dimorphism, predation pressure on foragers or reproductively active males, and thermoregulatory properties of bee color and structure is an intricate subject that needs further study.

Müllerian mimicry among sympatric bees is easily seen in the tropics; not surprisingly, the mimicry patterns often involve the abundant and ubiquitous social species. In the Old World tropics, mimics of *Apis* include melittids, megachilids, and halictids (Appendix Figs. B.8, B.11, B.16, and B.21). The New World tropics has many close mimetic complexes of two neotropical genera, *Paratetrapedia* and *Tetrapedia* (Anthophorinae, and Exomalopsini and Tetrapediini, respectively) and some wasps, along with many species of Trigonini (Appendix B).

Mimics of some *Melipona* occur in the Megachilidae and the *Euglossa decorata* group (Appendix Figs. B.9 and B.20). Mimicry involving stingless bees is intriguing because, aside from strongly distasteful and foul-smelling mandibular gland substances like skatole (Nogueira-Neto 1970b; Smith and Roubik 1983) and possibly resins carried on the legs, the stingless bees are not equipped to discourage predators. Johnson (1974) found several *Trigona* and *Scaptotrigona* to be unpalatable, presumably due to their pungent mandibular gland chemicals. Female bees that mimic meliponines have painful stings. Furthermore, both the stingless bees and their co-mimic *Paratetrapedia* forage together on the same flowers. In Panama, *P. calcarata* and *T. fulviventris* forage together on various species of

Psychotria (Rubiaceae). This *Trigona* normally forages individually and unaggressively at flowers (Johnson 1983a), and I have found it to be outnumbered by the *Paratetrapedia* at inflorescences of *Psychotria*. In contrast, some other mimetic solitary bees resemble large and aggressive *Trigona* that attack other foragers (Appendix Figs. B.2 and B.18). These include black bees, bees that have an orange-and-black-striped abdomen, and orange bees. Because neither black nor orange alone is among the classical warning color patterns of unpalatable insects (Rettenmeyer 1970; Huheey 1984), these mimicry complexes might have some significance in foraging. For example, competing foragers might not approach bees that appear to be aggressive *Trigona*. Other plausible interpretations await study. For the moment, the least complicated evaluation of this particularly widespread mimicry between stinging and nonstinging female bees suggests a Müllerian relationship between distasteful and stinging foragers. Polybiine wasps, cleptoparasitic bees of the genus *Osiris,* worker *Trigona,* and *Paratetrapedia* are involved in neotropical mimicry complexes (Shanks 1986). Other co-mimicry complexes of stinging bees, to name but a few, include *Bombus volucelloides* and a strikingly similar euglossine, *Eulaema leucopyga,* and also the Amazonian *Bombus transversalis* and both the males and females of *Ptilotopus americanus* (Centridini, Appendix Fig. B.1). Seabra, Campos, and Moure (1961) mentioned the close mimicry in coloration found between several Centridini of two differing taxonomic groups, depicted in Appendix Figure B.1.

Crab spiders, phymatid bugs, and mantids mimic flowers (Wickler 1968; Ospina 1969; Rettenmeyer 1970; Yong 1976), and each presumably can often capture small bees (Morse 1982, 1979b; Moure and Hurd 1987; see also Fig. 2.34). One quantitative study of bee predation by a crab spider in the temperate zone revealed that these predators probably captured 1–5% of the bumblebees entering a flower patch during the flowering season (Morse 1979). Web-spinning spiders of many kinds frequently build webs close to flowering plants and bee nests; their impact on foraging bees is probably large. The extent to which flower mimics are mobile in their choice of flowers or flower patches is unknown. It would not be surprising if their foraging and dispersal occurred among the flowers of the closed canopy of mature tropical forest rather than in the understory.

Large vespid wasps forage for bees at flowers and near bee nests in much of Asia as well as in Africa and some of the Mediterranean and Middle East (Matsuura and Sakagami 1973; Starr pers. commun.; review by DeJong 1978). A distribution wholly overlapping with that of *Apis* in subtropical, warm temperate, and tropical areas strongly suggests that widespread predation of at least *A. mellifera* and *A. cerana* foragers is perpetrated by these wasps. The species *Vespa tropica* is the largest and the most common of two southeast Asian species, however, and is also present in Melanesia, where there are no native honeybees. *Vespa tropica* and

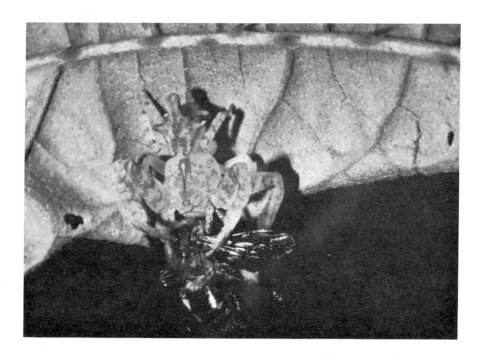

Figure 2.34. Spider predator of bees: a crab spider shown with a worker *Melipona fasciata* (Apidae) captured while foraging on *Solanum*. Original photograph by Sean Morris, Oxford Scientific Films.

related species live in large carton nests made from wood fibers that the wasps gnaw from plant stems, and a *Mimosa* shrub is a species used for this purpose by foraging groups of a few dozen wasps, at least in northern Thailand. Honeybees gathering pollen from this plant may occasionally be killed by the pulp-foraging wasps. I watched approximately 25 *Apis cerana* feeding at honey water baits in this area disperse immediately when a forager released the alarm pheromone after capture, a behavior that may be associated with wasp predation at the flowers.

Large neotropical ponerine ants feed at extrafloral nectaries and drive away, attack, or even occasionally kill approaching small bees. Their defense of plants from defoliators has led them to be dubbed "pugnacious bodyguards" by Bentley (1983). Whether bees have chemical or behavioral defenses that repel or thwart attack by Hymenoptera or other animals could provide ecologically relevant insight into their foraging. Some chemical components in alarm pheromones and odor marks used in foraging or mating also have a function in predator deterrence, as appears to be the case for mandibular gland odors of anthophorids such as *Exoneura* that repel ants (Table 2.4).

2.3.9 *Activity patterns of social and solitary bees*

A large literature exists on the activity of entire colonies of honeybees, bumblebees, and some stingless bees, or nesting aggregations of solitary species, but there are few data that include activity during an entire day. A problem with studies of colonial species is that, unless careful attention is paid to the presence of nectar or pollen loads, bees making afternoon orientation flights may be confused with foragers. This is a pattern that is quite consistent in honeybees (Frisch 1967). Factors that affect bee development and phenology are discussed in Section 3.1.2. Although the mechanisms producing such patterns are understood in considerable detail (Tauber, Tauber, and Masaki 1986; Danks 1987), there is relatively little information on the proximate factors that influence the activity of tropical adult bees. This section presents largely descriptive data, rather than analyses, concerning daily activity patterns that have been recorded for social and some solitary species.

All bees appear to forage in bouts that may last from a few minutes to a few hours, and pollen-collecting trips generally take less time than those for nectar. Further, the European honeybee has the ability to remember up to five separate foraging localities and the times of day they must be visited, but memory of such patterns fades after 24 hours (summaries by Michener 1974a; Barth 1985). The daily number of foraging bouts for a bee may vary from roughly 5 to 15, with a maximum of a few dozen trips observed for worker honeybees visiting flowers. So far as is known, there is a lack of time specificity in the foraging of individual bees of a colony (Roubik, Parra, and Holbrook pers. obs.). That is, workers that reach foraging age fly to and from the nest throughout the day. Peak foraging periods in social or aggregated species have been documented for a number of tropical bees in natural forest conditions and will be reviewed below.

Besides demonstrating graphically the flux of returning foragers, the colony activity depicted in Figures 2.35–2.39 shows the similarity in foraging patterns over a broad range of conditions. The influences of the weather appear in the performance of three Sumatran stingless bee species on one day during squalls compared to performance in more favorable weather a few days later (Inoue, Salmah, and Yusuf 1985) and in the cessation of foraging during rains or early in the morning during cool weather.

In most cases, the figures show the type of resources being gathered at different times of the day. Peak pollen collection occurs in the early morning, peak nectar collection follows later in the day, and a second, smaller peak in nectar and pollen collection may occur in the late afternoon. Gilbert (1973) noted a maximum in colony pollen collecting in the early morning in a short study of *Trigona fulviventris* during the dry season in Costa Rica; a second, smaller peak in flight activity occurred near sunset, but it is questionable whether this activity involved foraging.

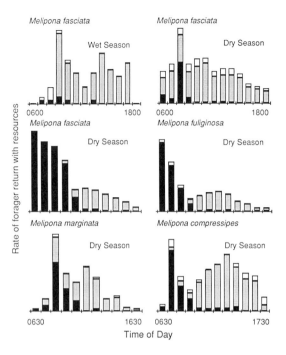

Figure 2.35. Colony activity patterns and relative harvesting intensity of pollen (black bars), nectar (gray bars), and resin (white bars) by *Melipona* species in lowland Panamanian forest (Roubik, Parra, and Holbrook, unpublished). The first two graphs of wet and dry season activity were taken for the same individual colony, the remaining data from an earlier study (Roubik and Buchmann 1984).

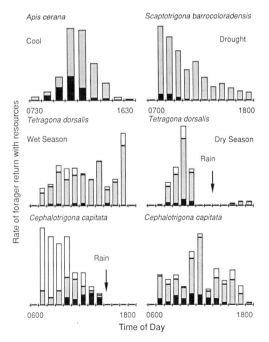

Figure 2.36. Colony activity patterns during local climatic extremes and relative harvesting intensity of pollen (black bars), nectar (gray bars), and resin (white bars) by *Apis cerana* in Bangladesh (Murrell and Nash 1981) and stingless bees (Meliponinae) in lowland Panamanian forest (Roubik, Parra, Holbrook, and Wittmann, unpublished). Comparative data of colony activity during wet and dry seasons were taken for the same individual colonies.

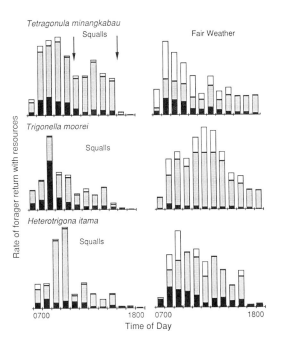

Figure 2.37. Colony activity patterns and relative harvesting intensity of pollen (black bars), nectar (gray bars), and resin (white bars) by stingless bees (Meliponinae) in agricultural areas and secondary vegetation in lowland Sumatra (Inoue et al. 1985). Comparative data of colony activity during fair weather and squalls were taken for the same individual colonies.

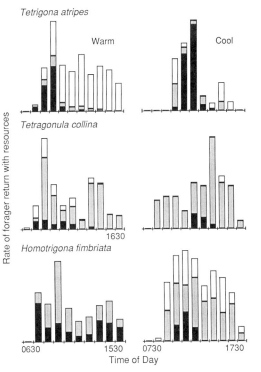

Figure 2.38. Colony activity patterns and relative harvesting intensity of pollen (black bars), nectar (gray bars), and resin (white bars) by stingless bees (Meliponinae) in secondary vegetation and forest of lowland northern Thailand (Roubik pers. obs.). Comparative data of colony activity were taken from the same individual colonies during the last day of the warm wet season and the first days of the following cool, dry season.

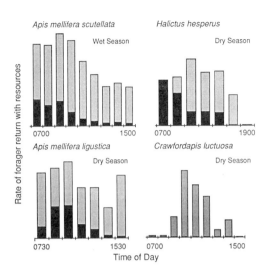

Figure 2.39. Colony activity patterns or general activity in nesting aggregations and relative harvesting intensity of pollen (black bars) and nectar (gray bars). Three colonies of descendants of African *Apis mellifera scutellata* were observed during 40 days in lowland forest of Panama (Vergara 1983), and four colonies of European *A. mellifera ligustica* were observed in secondary vegetation and forest at 900 m elevation in New Guinea (Roubik pers. obs.). Several colonies of *Halictus hesperus* (Halictidae) were observed in lowland forest of Panama (Brooks and Roubik 1983). General flight activity of *Crawfordapis luctuosa* was observed in Costa Rican cloud forest (Otis et al. 1982).

Gilbert quoted older literature by von Ihering, Peckholt, Schwarz, and Lutz, each of whom noted changes in the first flights of various tiny stingless bee species from their nests. Such species did not begin foraging until at least one hour after daylight. Initiation of flight was delayed in cloudy, rainy, and cool weather. Lutz (1931) suggested that such variation might be due to the interaction of light intensity and its perception by the bees with an internal clock that prompts bees to emerge from the nest. Species of meliponines that start to forage early in the morning assemble outside the nest entrance during darkness, and they fly as the sun reaches the horizon. As indicated in Section 2.3, flowers may open and begin to receive visits at any time, and meteorological conditions also affect nectar, pollen, fragrance, and resin production by plants. Seasonal variation is evident through the year, particularly in the amount of pollen gathered by bee colonies (Roubik et al. 1984; Appanah et al. 1986). It is an open question whether the preferred resources of particular bee species are closely tied to the time at which bees begin to forage or whether delays in initiating foraging are related to physiological constraints on the flight and orientation performance of the bees.

Among the species that collect pollen, nectar, and resin or mud, the nesting materials are often collected slightly less frequently than is pollen. Nectar is usually collected more intensively than any other resource, but a small portion of foragers return empty (Rinderer et al. 1984; Roubik and Buchmann 1984; Inoue et al. 1985; Seeley 1985).

Relatively cold weather retards the foraging activity of bees, even on clear days (Fig. 2.39). During subtropical winter prolonged cold weather apparently leads to

2. Foraging and pollination 135

cessation of brood rearing by stingless bees (Ihering 1903; Nogueira-Neto 1970a). At this time they forage little for pollen. The pattern is probably due not only to the problem of generating a body temperature high enough to permit flight but also to the sluggish nectar secretion of most flowers at low temperature. Three species of stingless bees studied in northern Thailand (21°N latitude) were active only after mid-morning during relatively cold winter weather in late November, when ambient temperature at dawn was near 10° C and maximum temperature at the nest entrance reached 22° C (Fig. 2.38). At a similar latitude, five colonies of *Apis cerana* foraging primarily on one species of flower were observed through the day in Dacca, Bangladesh, during the cool season (Murrell and Nash 1980). These colonies also delayed until mid-morning to reach maximum activity. The activity of the large solitary bee *Crawfordapis* is given for the middle of the dry season in the cloud forest of Costa Rica, where temperatures range between 10° C and 27° C and winds are stronger than in many tropical habitats, generally 15–20 km/hr (Otis et al. 1982). These bees were most active between 1000 and 1400 hours.

Meteorological factors have an obvious relation to foraging activity. Again, these factors are scarcely discernible in all but environmental extremes, and many of the same factors affecting nectar secretion or other resources can also be correlated with bee activity. Thus Fowler (1979) was able to detect positive correlations between ambient humidity and temperature during seven months of the year and the foraging activity of two colonies of *Tetragonisca angustula* in semideciduous secondary forest of Paraguay (25°S latitude). Foraging intensity did not change over the months of the study, nor did the timing of peak foraging activity, which was in the morning. The cold-weather performance of the Thailand stingless bees at 21°N latitude is interesting and is seen clearly in the activity of *Homotrigona fimbriata,* the largest Asian meliponine (Fig. 2.38). After the morning ambient temperature dropped 5° C between November 20 and November 25, colony foraging commenced a full two hours later. Resin was a major resource collected by all three species during the cold weather although they still collected some pollen.

Early morning foraging peaks are typical of bees in the tropical lowlands. Activity of the eusocial *Halictus hesperus* is given for the late dry season in forest of lowland Panama, when ambient temperatures reach 37° C and ground temperature reaches 52° C (Brooks and Roubik 1983). Among these few hundred colonies, peak pollen collection occurred at 0700 hours, followed by peak nectar harvest at 1100 hours. In the same habitat, three colonies of Africanized *Apis mellifera* were censused through the day on 42 days during the rainy season, between June and October. The temperature remained between 25° C and 33° C, yet the timing of pollen and nectar harvest peaks was much like that of the halictid bees. Four colonies of European *A. mellifera* were studied for one day during the wet season in second-growth middle-elevation forest near Wau, Papua New Guinea, in ambient temperatures of 21°–27° C. Like the honeybees in Panama, despite introduction into an area outside their natural range, the colonies displayed peak foraging in midmorning, and almost all pollen was gathered before noon.

Stingless bee colonies in mature forest on Barro Colorado Island, Panama, are compared in different years and seasons in Figures 2.35 and 2.36. Their pollen-gathering peaks are generally followed by peak nectar harvest, with the latter occurring at few hours later. The patterns of dry-season (January–April) and wet-season colony activity appear similar between years, but the amount of activity varies considerably. Prolonged drought appeared to affect the colony of *Scaptotrigona* (and other colonies not shown here) not only by shifting all foraging activity toward the morning but also by eliminating a midmorning peak in nectar gathering.

What shapes the diurnal trends of bee foraging activity? Regardless of the particular plant taxon or bee species involved, pollen appears to be removed first; the nectar is depleted and foraged more intensively later in the day. The general foraging activity of tropical bees indicates peak pollen and nectar gathering in the morning, a pattern that is frequently learned by bee collectors studying the tropical fauna. Considering that sugar concentration in nectar tends to increase through the day in loads carried by tropical bees (Roubik and Buchmann 1984; Inoue et al. 1985), it is likely that increasing caloric return per unit foraging time, combined with the preferences of bees for nectar having 60% sugar, entices bees to fly further searching for nectar sources as the day progresses. No such change in pollen quality occurs. Once pollen is depleted, there may be little reason for bees to fly further afield seeking new resources. Therefore, pollen is depleted while nectar quality increases, and at some point individual bees or colony foraging effort switches from pollen gathering to nectar.

In a habitat relatively depauperate of foraging bees, other behaviors are likely. In this situation, bees probably forage nectar and pollen over the same area, and they may concentrate on nectar when it is energetically the most profitable and abundant, which should occur within a few hours of anthesis. Individual foragers would thus be expected to shift from pollen or nectar-and-pollen collecting to more specialized nectar collection, and then back to pollen collection after the nectar standing crop is depleted. This behavior pattern would explain minor afternoon foraging peaks of highly social bees, at least when floral resources do not first become available in the afternoon (Sommeijer et al. 1983; see also Section 2.4).

Linsley (1978) discerns trends of larger solitary bees foraging before smaller species in arid habitats of the southwestern United States, where relatively large crepuscular bees appear at many flower species. The generality of this phenomenon, its applicability to groups of bees that are not closely related, and its ecological consequences should help to clarify the foraging ecology of bees in tropical habitats, where similar trends may exist. The succession of large to smaller bees at flowers, noted by numerous authors, was mentioned in Section 2.3.7. Exceptions are known among the euglossine bees, in which a relatively high energy metabolism (per unit body mass) is due to the elevated wing-beat frequency while in flight (Casey et al. 1985). All euglossine bees apparently display a very high power output in each wing stroke, but differences in body heating among different species may not be predictable on the basis of their relative sizes (Pyke 1978b; Dyer and Seeley 1987).

Closely related species also show that ecological factors, as well as the physical parameters of body size or coloration, influence foraging activity. The above *Melipona* studied by Roubik and Buchmann give one example of apparently rigid differences among related species. Two dark species, *M. fuliginosa* and *M. fasciata*, began foraging at dawn, whereas a species intermediate in size between these two bees but light-colored, *M. compressipes*, repeatedly showed no foraging activity until about an hour later. These patterns were observed during the dry and wet seasons (Roubik et al. 1986). The three trigonine species studied by Inoue, Salmah, and Yusuf showed no predictable foraging patterns based on color. But the largest species, *Heterotrigona itama,* was more active later in the day than the smallest, *Tetragonula minangkabau;* both bees are black and almost hairless. Behavior and physiology, rather than external morphological features, help to explain differences detected by Heinrich (1979c) in early-morning foraging by two races of *Apis mellifera.* African *A. mellifera,* although smaller, initiated foraging decidedly earlier than European bees, and it is only slightly darker in color than the lightest races of European bees. Pronounced early-morning foraging activity by tropical highly eusocial bees, particularly at pollen sources, has been noted repeatedly (Roubik and Buchmann 1984; Roubik et al. 1986; Danka et al. 1987; Roubik 1988). Asian *A. dorsata* forages to some extent at night (Dyer 1985); and although it is relatively large, its body coloration does not differ markedly from other *Apis*. Within these principal groups of highly eusocial bees – *Apis, Melipona*, and the Trigonini – a prediction might be that the larger bees will be likely to forage earlier or later in the day or to display some nocturnal flight. Such trends are easier to detect in the highly eusocial bees because statistics on their flight activity are readily obtained at colonial nests. The vast majority of bees, the solitary species, likely display some diurnal staggering in their foraging activity that is related to size within their taxonomic group.

Notwithstanding some apparently general trends, single colonies vary from day to day in both the timing and the extent of resource harvest. The most striking feature of colony activity patterns of 12 stingless bee species studied in lowland forest in Panama was a rare burst of foraging, seen on average during 4% of hourly observations made during 13 consecutive mornings (Roubik et al. 1986). These short bursts of incoming resource lasted for about an hour. Nevertheless, despite their short duration and sporadic occurrence, they accounted for an average of over one-third the total resources harvested by colonies during nearly two weeks. For species capable of recruiting, the occasional discovery of an extremely rich resource produces completely unpredictable surges in foraging activity. This pattern of sporadic resource discovery and harvest may be one factor permitting the coexistence of many highly social bee species in tropical forest, despite broad similarity in the resource taxa (Section 2.4.4 and Chapter 4).

2.4 Pollination ecology

Traits associated with floral resources and feeding by bees have been introduced in the preceding sections. Pollination ecology (or anthecology) attempts to bring such insights to bear upon the relationships between bees and flowers, in both the short and long term. Tropical anthecology provides much impetus for the study of plant reproductive biology through the diversity of interactions (and sometimes the concomitant lack of information) it reveals (Baker 1983; Janzen 1983; Meeuse and Morris 1984).

2.4.1 General considerations

Plant reproductive fitness is influenced by bees that are
1. *pollen dispersers* and *donors,*
2. *pollen consumers,*
3. *destructive visitors,* and
4. *nectarivores.*

These tendencies are not always independent. For example, a bee might alight on the stamens and stigma of a flower pollinated by a vertebrate, walk to the base of the outside of the corolla, make a hole in it, remove nectar, and then fly to another flower of the same species. It may thus fulfill three of four roles. A bee might arrive at a nectarless, buzz-pollinated flower, vibrate the anthers and thereby collect pollen, and then fly to another flower of the same species. If it is a small bee visiting a large flower and does not contact the stigma, then it is nothing more than a pollen consumer. Similarly, a bee that contacts stigmas while imbibing nectar of pistillate flowers (the female flowers of dioecious species) can only be called a nectarivore unless it also carries the pollen of staminate flowers and if that pollen is genetically compatible with the maternal (gamete). These are common tropical flower-visitation patterns for bees. They show that careful observation and experimentation are necessary to establish the effect of visitors on flowering plants and also on other visitors of the same plant.

Distinguishing among the alternative relationships that bees have with their host plants requires the analysis of populations. When a bee's visit to a flower does not result in pollination, even though the bee uses no novel or destructive foraging techniques to gather the floral resource, it is sometimes given the label *thief* (Inouye 1983). The connotations of this term are misleading because many if not most bee visits to flowers do not result in pollination. A clear example is provided by flowers that are obligately pollinated by bees. Among neotropical orchids that depend solely on euglossine bees for pollination (Ackerman 1983b), an individual bee that carries no pollinia and visits flowers that have already dispersed their pollinia, is usually a commensal. However, if its presence at the flower lowers the probability that another pollinator will deposit pollinia, the bee is in fact a parasite. Nonetheless, the bee's *population* is the only means by which the plant reproduces; thus the bee *species*, at least locally, is clearly a mutualist of the plant.

Bees are often commensals of flowering plants, a trend that seems more widespread in rich tropical floras that are pollinated through diverse mechanisms and by a wide variety of animals. The concept may seem novel at first encounter, but bee visits to certain flowers often make absolutely no difference to plant fitness (Janzen 1977). This is true when a bee makes its visit only when the stigma is no longer receptive or after most of the pollen has been removed from the flower. Bees that are truly scavengers visit flowers that have already been available to pollinators and have either dispersed their pollen, received pollen from other flowers, or have fertilized themselves. It seems likely that most flower-visiting bees are commensals at one time or another. However, the relationships of some bees to particular types of flowering plants are likely to be entirely commensal. Bees that visit nocturnally dehiscent flowers in the early morning on the following day most likely fit into this category. The bats, sphingid moths, vertebrates, or beetles that provide the pollinating visits have left a food source that is later opportunistically taken by bees (Lobreau-Callen, Darchen, and Thomas 1986; Roubik et al. 1986; Eguiarte, del Rio, and Arita 1987). Eguiarte and co-workers pointed out that continuation of nectar production during the day implies further visits and pollination by diurnal visitors, provided that stigmas continue to be receptive at this time.

Legitimate visitation of a flower by bees has no exact significance for fertilization of its ovules, transfer of pollen, or successful maturation and growth of seeds. How important is bee visitation of a flower that is autogamous (self-pollinating) or automictic (able to produce seeds and generate new genotypic combinations with no contribution from pollen), and what is its importance to a plant that reproduces both vegetatively and sexually? Limited observation certainly suggests whether a bee is capable of pollinating a flower, but the ways in which the bee influences reproductive success and evolution of the plant are difficult to deduce. Even less obvious are the ways that bee populations influence flower populations, because both vary constantly due to several principal factors. Weather and natural enemies affect bee abundance and activity (Sections 2.3 and 3.2), as does the availability of flowers (Bawa 1983; Reich and Borchert 1984). Bee–flower interactions also vary due to competition with other species (Eickwort and Ginsberg 1980; Bawa 1983; Bullock, Beach, and Bawa 1983; Rathcke 1983; Waser 1983; Roubik 1988). Pollination biology texts have addressed such themes with increasing alacrity (Proctor and Yeo 1973; Faegri and van der Pijl 1979; Feinsinger 1983; Jones and Little 1983; Kevan and Baker 1983; Real 1983; Meeuse and Morris 1984). Their generality as applied to tropical flora depends largely on data yet to be presented, making any summary unstable. Indeed, even bees that are unequivocal pollinators can differ substantially in the amount of pollen they remove relative to the amount they deposit on stigmas (Snow and Roubik 1987). Finally, and perhaps most significantly, the ways in which bees influence male fitness (successful dispersal of pollen to stigmas) and female fitness (successful reception of pollen and maturation of seeds from ovules) are complex and varied. Thus the number of visits a hermaphrodite flower (one possessing both anther and stigma) receives can have differing consequences for its male and female functions. Much of the following discussion will be devoted to elucidating basic patterns in this rich interplay, emphasizing tropical studies.

A new generation of tropical pollination studies takes on the questions of ultimate consequences from visitation by bees. Horvitz and Schemske (1988) hand-pollinated the tropical herb *Calathea* (Marantaceae) in experiments designed to test the limiting factors of fruit initiation and ultimately of seed maturation. Their hypotheses, extracted from various recent studies, led them to assess

1. fruit-to-flower percentages;
2. pollen quantity limitation to the initiation of fruit and production of seeds;
3. selective plant abortion of fruits;
4. bet-hedging (adjustments to the temporal abundance and activities of various animals interacting with the plant, including ants that guarded the fruit);
5. pollinator attraction (nonlinear increase of visitation to individual flowers as a function of total flower number in the floral display); and
6. male function (whether the flowers not producing fruit were useful primarily in donating pollen).

The authors found that fruit initiation was limited by the amount of pollen received by stigmas but that mature fruit and seed production were limited by plant resources. These variables were also affected by nonpollinating animals at the plant during different phases of seed maturation. In addition, the low percentages of flowers producing fruit may have enhanced reproductive success through male function.

Pollinators might be viewed as primarily pollen donors or pollen receivers, but a large part of their effectiveness depends ultimately on the number of visits they make and *how, where,* and *when* they distribute pollen, both as individuals and as populations. To offer one example, suppose a bee delivers a large pollen load to a flower's stigma during a single visit, completly covering the stigmatic surface. If the plant discriminates even only slightly among pollen parents, then this "efficient" pollinator may have canceled the female reproductive success of that flower. One pollen donor type might visit many flowers and deposit little pollen on each; another might visit fewer flowers before grooming or returning to its nest, but it might deposit a large amount of pollen at each flower. There are certainly advantages and disadvantages, to the plant, corresponding to each type of behavior. If a small amount of pollen is distributed to many stigmas, the fitness of the pollen parent might be increased if there are few or no compatibility barriers and little resource limitation of seed maturation; but in other circumstances, deposition of less than a relatively large amount of pollen may not result in seed set (Snow 1982; McDade and Davidar 1984; Schemske and Horvitz 1984; Bertin 1985; Sutherland 1986).

Plants can often be characterized as primarily self-pollinating or primarily outcrossing, relatively few species having intermediate breeding systems (Schemske and Lande 1985). The value of outbreeding or inbreeding and the value of long- or short-distance dispersal of pollen effected by a given bee vary widely among plant species (Harper 1977; Levin 1979; Handel 1983; Waser and Price 1983). Populations of tropical trees and shrubs seem to maintain high genotypic diversity, at least in preliminary electrophoretic studies carried out in one area (Hamrick and Loveless 1986). For natural plant populations and local groups within them, adaptation to soil, light, or moisture conditions, pathogens affecting seedlings (Augspurger 1983b), and many other traits can depend strongly on genotype. Therefore, pollen coming from too far away can result in *outbreeding depression*. A few experimental studies have shown that pollen transported from an intermediate distance provides greater pollination and seedling success than pollen originating either very close to the parent plant or relatively distant (Waser and Price 1983; Schemske and Pautler 1984). Similar results might be expected for tropical forest trees when genetic variability is slight or particular genotypes are closely associated with discrete habitat patches within a forest. Moreover, pollen viability can decline precipitously in a few hours or less, and it can decline more rapidly if exposed to direct sunlight and ultraviolet radiation (Stanley and Linskens 1974; Mulcahy, Mulcahy, and Ottaviano 1986). This could greatly diminish the effective pollen flow between individual plants, making pollinator movement patterns even more important. Such features may directly influence plant evolution and lead to traits that favor visitation by certain types of pollinators. This underscores desirability of viewing plant reproduction with a conceptual flexibility that has proven useful in studies of animals (Janzen 1977, 1983; Waser 1983; Willson and Burley 1983; Wyatt 1983). Merely by encompassing so many thousands of sexually reproducing plant species, tropical anthecology is likely to be one of the largest biological frontiers.

Dependence of a bee population on a particular plant has scarcely been established anywhere, and the significance to bees of a flower species they visit is probably just as variable as are the roles of bees to the plants they visit. For instance, during one foraging bout, a bee can visit a staminate flower of a dioecious species, then visit the pistillate flower and find no pollen, visit another flower that mimics the first species, visit another flower species for nectar, and then return to its nest. What would happen to the bee population if one type of plant were absent? Does the bee visit each flower type throughout that flower's range? Questions like these may appear too difficult to be resolved, but they are often neglected altogether. Both are particularly relevant in the tropics. Regarding the first question, flowering periodicity in tropical forest seems a priori to prohibit extreme dependence of a bee on a flower species. Many forest trees, shrubs, and lianas do not flower each year, and this is true for populations as well as for individuals (Gentry 1974a,b; Appanah 1982; Baker et al. 1983; Bawa 1983; Whitmore 1984; Whitten et al. 1987). Entire Southeast Asian dipterocarp forests experience widespread flowering episodes or "mast" years at unpredictable intervals of 2–21 years, and such flowering vegetation is visited heavily by bees (Ashton 1969; Medway 1972; Appanah 1985). Thus the choice of a particular flower by a bee should in some areas be strongly influenced by community-wide flowering patterns and yearly variation in flowering. Flower visitation by bees should almost invariably reflect evolved preferences tempered by the availability of flower species at a given time, even though a few solitary bees reportedly delay their reproduction until the flowering of a preferred host (Danks 1987, p. 184; Cane and Eickwort in press).

One type of information often gathered on flower visitation by bees is straightforward – the species they use in a given area and time (Section 4.1). A drawback of such studies is their limited ecological scope and usefulness. Short-term studies or inventories of bees at flowers may explain little of the efficacy of pollination by different bees. The relative importance of a plant to bee fitness remains unknown, as does the contribution of the plant to adult bee metabolism, nest building, or reproduction. However, given the potential variability in available flower species between years, extensive or comparative studies of this kind potentially generate more accurate pictures of bee ecology than short-term detailed descriptions of species' pollen and nectar harvest. Ideally, both extensive comparative studies and long-term attention to the diet (e.g., proportion of pollen consumed of different taxa) of bees could provide desired ecological data. Further, data tend to have more to say about evolution and ecology if they come from habitats that are not artificially disturbed.

Disturbances immediately create conditions favorable for invading species (see, e.g., Fox and Fox 1986), which in pollination ecology are analogous to opportunistic flower visitors that appear when the normal pollinators are absent (Roubik et al. 1985). The added difficulty of studying natural bee–plant interactions should be compensated by its capacity to clarify the difference between opportunism and more persistent relationships, which may lead to coevolution or close adjustments between bees and plants (Section 2.4.3).

2.4.2 What makes a pollinator effective?

Flowers manipulate bees in a variety of ways, perhaps most of which depend upon stable apoid behavioral and physiological traits, parallel to those of physiology and foraging energetics reviewed in Sections 2.2 and 2.3. Bees are attracted to flowers by perceived color, size, smell, shape, and texture (Baker and Hurd 1968; Kevan 1978; Kevan and Baker 1983; Dafni 1984). The odors are learned perhaps more rapidly than the other features (Barth 1985), and close-range cues also involve the reflection of sunlight upon or even the fluorescence of nectar.

The color vision of bees is trichromatic, as it is for humans, but the three primary colors are shifted toward shorter light wavelengths (Daumer 1958; Frisch 1967; Silberglied 1979; Kevan 1983). Thus bees perceive ultraviolet as a distinct color, but the longest wavelengths, in the deep orange-red band, are not perceived. The European honeybee, which has been studied in depth, detects a maximum contrast between ultraviolet and green. Thus flowers with ultraviolet nectar guides are maximally attractive against a background of green foliage. Nectar guides are highly ultraviolet-reflectant areas on floral corollas that generally attract bees and direct them to nectar (Jones and Buchmann 1974; Faegri and van der Pijl 1979). Other attractive floral colors for bees are those that humans perceive as white, yellow, and blue-violet or purple.

It has recently been suggested, with experimental support, that increased attractiveness of flowers serves as a means to enhance pollen dispersal and male fitness. And in some cases it also favors increased female reproductive success, even when maternal resources place a limit on the number of fruit and seeds produced (Stanton, Snow, and Handel 1986). These workers show that the greater preference of bees and other insects for yellow flowers over white in a population of the self-incompatible wild radish *Raphanus raphanistrum* has two contrasting results. First, maternal success (seeds produced per ovule) does not differ for the two color forms, despite substantially greater visitation to yellow flowers. Second, paternal success (determined by the phenotype of resulting seedlings) is 50% higher for yellow flowers, in agreement with the greater number of visits they received. Therefore, female reproductive output reaches its maximum regardless of flower type, but a greater number of visits to yellow flowers augments dispersal and fitness of their pollen. A possible subtle advantage to female reproductive success could accrue from the same greater visitation – more competition among pollen tubes could occur (Mulcahy 1979) – but experimental confirmation is meager. Even though not evident by resulting seed number, seed quality could be improved via the influence of superior male (pollen) parents. The degree of such pollen tube competition should be apparent in an enhanced rate of fruit and seed growth and later in germination success. It is thought to be a direct consequence of superior genotypes produced by the fastest-growing pollen tubes (summaries by Stephenson, Winsor, and Davis 1986; Mulcahy and Mulcahy 1987).

Waser (1983) conducted experiments with *Bombus* and *Diadasia,* finding that differences in color alone caused foragers to discriminate among flowers of similar morphology (and presumably smell). Both types of foragers most often visited the type of flower from which they had just departed, but they also regularly visited flowers of the other type. If Stanton et al. (1986) are correct, then these examples show that both learned and inherited color preferences of bees have influenced the evolution of hermaphrodite flowers via their differential success as pollen-donating organs.

Repeated visits by an individual bee to flowers of a given color, size, and shape on plants of a given stature, termed *floral constancy*, is thought to increase directly in proportion to effective pollination (Heinrich 1976a; Faegri and van der Pijl 1979; Laverty 1980; Waddington 1983). The degree to which such behavior is beneficial to the forager or plant has been discussed by Waser (1983, 1986), but the general concept is vague. For instance, it is uncertain whether interspecific pollen transfer diminishes plant fitness, relative to the pollination that would be achieved if the visitor could visit no other flower species. Floral constancy can be a means of economizing flight and foraging time by searching for flowers of the type that a bee knows are rewarding and has learned how to manipulate (Darwin 1876). This implies that bees are unable to learn and to *remember* the correct way to manipulate many different types of flowers at the same time (Lewis 1986; Waser 1986). Floral constancy makes sense in light of foraging theory only to the extent that flowers are sufficiently rewarding (Heinrich 1979a; see also Section 2.3). However, several types of pollen are often carried by bees at a given time (Raw 1976; Absy and Kerr 1977; Absy et al. 1980; Appanah 1982). Whether this poses a disadvantage to any of the plant species is an unresolved question. Two studies of neotropical *Cochlospermum* demonstrated that although the large solitary bees pollinating this nectarless flower foraged elsewhere for nectar, no pollen of the nectar species was deposited on stigmas of *Cochlospermum* (Pereira and Gottsberger 1980; Snow and Roubik 1987). The first authors showed that pollen transfer in the opposite direction (from *Cochlospermum* to the stigmas of the nectar flower) did occur. From these examples, and considering the very large proportion of nectarless tropical flora (Section 2.2), it seems that strict constancy to flowers of only one species might be rare among tropical bees. Presumably, less-constant behavior would be likely where many flower species are present and floral morphology is not complex.

Lack of floral constancy need not diminish pollination potential. Large pollinators such as hummingbirds sometimes carry pollen from more than one type of flower on different parts of the body and thus are *partitioned* by flowers visited simultaneously (Feinsinger 1983). Similar differences in sites of pollinaria placement are easy to observe for orchids; euglossine bees visit up to several orchid species on a given day (Dressler 1981, 1982; Ackerman 1983b; Roubik and Ackerman 1987). Bodily partitioning of pollen from different plant species is known for temperate-zone bumblebees (Heinrich 1979c); it is likely true for tropical bees, such as large anthophorids and euglossines that I have watched visit papilionaceous legumes. Some flowers deposit pollen strictly on a bee's notum or sternum by "plunger" dehiscence after the forager depresses the keel petals to reach nectar. Bees with larger bodies may thus potentially be better pollinators than smaller bees. If both regularly visit several plant species on a foraging bout, then the larger insect provides greater chances for spatial segregation of pollen from different plants. Schemske (1981) found that *Euglossa imperialis* visiting two species of *Costus* (Zingiberaceae) carried pollen of each in the same position, but the plants have strong hybridization barriers and might benefit in some way by jointly supporting a larger pollinator population than could either species individually. Armbruster (1986) saw that associated species of *Dalechampia* (Euphorbiaceae) shared pollinators but at the same time avoided negative interspecific effects, possibly through evolution of incompatibility between related plant species.

Deception of naive bees and mimicry between co-occuring flowers may have tremendous influence on pollination (Baker, Cruden and Baker 1971; Baker 1976; Wiens 1978; Bawa 1983; Bullock et al. 1983; Little 1983; Dafni 1984; Ackerman 1986). A good pollinator is often one that makes mistakes. Such mistakes are sometimes costly, when food deception mimicries are concerned, usually the case in orchids (Ackerman 1986), or they merely result in visitation of flowers that are rewarding and similar to another species in appearance, smell, or texture. Apparently, based upon a single study, bees that are diet specialists are less likely to make mistakes than are generalists: A study of two sympatric flowers of the same genus showed that individual *Centris* mistook one species for the other at half the rate of *Apis mellifera* (Jones 1978). However, since most of the foraging by European *A. mellifera* results from recruits directed to high-quality resources (Seeley

1985), even slight differences in reward between flowers similar in appearance would likely result in large differences in total visitation. Deception and mimicry systems can be cataloged as follows:

1. *Automimetic systems.* These apply to plants with dioecious or monoecious/ dioecious breeding systems. The female flowers have no pollen and may also be nectarless; they mimic male flowers with which bees have prior experience. The male flowers are often more abundant than female flowers and last longer (Bawa and Opler 1975; Baker 1976; Absy and Kerr 1977; Bullock and Bawa 1981; Appanah 1982; Little 1983). Tropical examples include Arecaceae (palms), Euphorbiaceae, Sapindaceae, Cucurbitaceae, and Annonaceae.

2. *Phenological tracking.* Episodic flowering (two to five pulses of flowering within a year) may be timed to occur after rewarding flower species; it may also be related to unpredictable weather or pollinator activity (Gentry 1974b; Schemske et al. 1978; Frankie et al. 1983; reviews by Bawa 1983 and Little 1983). Bawa (1983) notes that episodic flowering in Costa Rican forests was more common among shrubs and treelets than among trees, as shown in studies of Frankie et al. (1974) and Opler et al. (1980a,b). This sequential mimicry is common among some large and brilliantly colored flowers of the forest canopy (Frankie et al. 1983).

3. *Shared pollinator mimicry.* This category includes several finer distinctions (Jones 1983); it can be deceit mimicry, where one of the co-mimics does not offer a resource to pollinators, as well as mutualistic mimicries, where resources are present in all flower types. Both of these mimicry systems are found among, for example, orchids and Malpighiaceae. Malpighiaceae display floral conservatism, and most are yellow and the same shape and size (Anderson 1979); no nectar is provided, only oil. *Oncidium* and other orchids that produce oil look like the malpighs (Buchmann and Buchmann 1981; Buchmann 1987). Several neotropical orchids are food- or fragrance-deception mimics of orchids or other flowers (Boyden 1980; Ackerman and Chase 1986).

4. *Strong attraction.* Strong attraction to color or other floral features may be sufficient to allow deception of a naive visitor, leading to brief but pollinating contact between the visitor and the flower (Dafni 1984). The bee eventually learns to avoid unrewarding individuals. In this case, mimicry is sometimes invoked but there is apparently no model (Jones 1983). Concealed floral rewards, which include both nectar and poricidally dehiscent pollen, seem likely to function in this way.

Variable nectar availability and quality, sometimes combined with automimicry, are primary mechanisms that may encourage bee interplant movement and outcrossing in the tropics (Appanah 1982; Frankie and Haber 1983; Roubik and Buchmann 1984). For the mechanism to function, bees must respond to changes in resources by shifting their foraging area or learn and follow sequential resource production peaks at different individuals, but not abandon the plant species. Clearly, floral constancy is a requirement, but it is strongly reinforced by the individual plants. Although the trap-lining behavior described in Section 2.3.6 shares much in common with resource tracking, there is a difference in the size of the resource. Trap-lining involves small resources that are relatively long-lived, and the tracking of variable nectar availability applies to the many flowers of larger woody plants, which may not flower for as long. Frankie and Haber (1983), as well as Bawa and Opler (1975), discuss the manner in which mass-flowering tropical trees were found either to display asynchrony in timing and peak of nectar production among individuals in a local population or to produce nectar at a variable rate, such that two peak flow periods occur, one in midmorning and one in the afternoon. Either pattern could be tracked by bees, and it would be profitable for them to move between trees if they were capable of finding the rewarding flowers. Movement between treelets or shrubs could also be favored, as appears to be the case for the shrub *Hybanthus prunifolius* visited by a species of *Melipona* (Augspurger 1980). These bees did not visit all of the flowers on one individual shrub before moving to another, but appeared to forage profitably. One tree with decidedly bimodal nectar production, *Andira inermis* (Papilionoideae), seemed to receive more visits during nectar production peaks (Frankie and Haber 1983) although the nectar standing crop may not have changed substantially. Memory and past rewarding experience of the foragers seem necessary to encourage such interplant movement. Large bees conceivably monitor several trees in this way, and small highly eusocial bees might receive information from nest mates allowing them to switch among plants in a large area; however, bees of small size but lacking a social communication system would be unlikely to move often between trees.

Changes in nectar quality could very plausibly promote interplant movement and continued visitation of flowers, despite uncertain nectar reward per flower. Appanah (1982) observed several species of small Trigonini, numerically the major visitors, concentrate their foraging first on male individuals of *Xerospermum* (Sapindaceae) and then on trees having pistillate flowers. Pollen was carried by bees arriving at the pistillate flowers. Roubik and Buchmann (1984) and Inoue et al. (1985) each found a gradual daily increase of about 100% in the sugar content of nectar harvested by *Melipona* and Trigonini in Asia and Central America, a pattern seen also by Real (1981) in *Ipomoea* visited by other types of tropical bees in Costa Rica. Roubik and Buchmann documented this change in *Hybanthus prunifolius,* a shrub pollinated primarily by *Melipona,* and they also demonstrated that the most concentrated nectar harvested (about 60% sugar) was that from which the bees derived the most caloric reward per unit time (Section 2.2). Appanah (1982) found that pistillate flowers of *Xerospermum* had much higher amino acid concentration in their nectar than did staminate flowers and contained predominately sucrose (instead of hexose) sugars. The nectar to be found at a new plant, after abandoning a depleted foraging area, must sometimes be unavailable due to competitors already feeding there. It is probable that bees inclined to move between plants require qualitative as well as potential quantitative gain in order to readily travel to another foraging area. On the other hand, the few exploratory visits to a new foraging area before learning that it, too, is unprofitable may provide sufficient outcrossing to massively flowering trees (Frankie et al. 1974).

Whether or not a certain pattern of bee visitation and movement between plants results in few or many seeds per plant ovariole is perhaps irrelevant in qualifying the success of a particular pollination system. In fact, the reproductive physiology, breeding system, and flowering habit of plants seem to adjust readily to the probability of pollinator visits (Sutherland 1986). Low fruit production in one year may be offset by accumulated resources and subsequently greater fecundity in the next flowering season.

Synchronous flowering of many individuals in a population is particularly suited to attracting pollinators that do not emerge at the time a host plant flowers. A Panamanian shrub receiving most visits from two species of *Melipona* attracted far fewer pollinators and produced less seed when experimentally induced to flower out of synchrony with the main population (Augspurger 1980, 1983a). Pollinators of this plant are active throughout the year (Roubik and Buchmann 1984), but for reasons that have not been discovered, they heavily use *Hybanthus* when it flowers, as do a number of less abundant anthophorid and halictid bees.

Frankie et al. (1983) describe a large contrast between the breeding systems of flowers pollinated by small bees and those pollinated by large tropical bees. Many of the small, white-green, or yellow (inconspicuous) flowers attractive to small bees were dioecious. That is, in order to realize male or female reproductive success, pollen *must* travel between plants. Why should small bees, having small flight ranges and foraging requirements that can surely be filled in the flowering crown of a single tree, be depended upon for outcrossing? Bawa (1980) clarifies the value of dioecious breeding systems by comparing them with self-incompatible hermaphrodites:

> For hermaphroditic plants, restricted foraging should increase self-pollination in self-incompatible species and decrease outcrossing in self-compatible species. The evolution of dioecy in such cases could increase the amount of pollen flow from one plant to another. The male plant could disperse more pollen than the hermaphrodite because its pollen would not be trapped by its own stigmas. The female plant in turn could be more efficiently pollinated than the hermaphrodite because its stigmas would not receive its own pollen and the pollen brought by the flower visitors would not be displaced or contaminated.

Arroyo (1976) makes similar remarks on the evolution of self-incompatibility systems in the angiosperms, which may encompass about 70–76% of flowering tropical tree species (Sobrevila and Arroyo 1982; Bullock 1985). Isolated individual trees would benefit, in particular, by limiting self-pollination. However, tree distribution in the neotropics appears often to be aggregated (Hubbell 1979), such that flowering individuals would rarely be isolated from conspecifics. Dioecy and the consequent obligate outcrossing is perhaps the most available means of avoiding excessive self-pollination. When outcrossing predominates, intrapopulation genetic variability should exceed interpopulation variability, with the result of disproportionately high genetic variability among tropical plants (Hamrick and Loveless 1986; Hubbell and Foster 1986).

The preceding arguments suggest that a need to escape a potentially high rate of inbreeding is accomplished by evolution of dioecy. Geitonogamy (transfer of pollen from anthers to stigmas on the same plant) may be frequent even for large bees visiting a small flowering tree crown, and outcrossing is sometimes obligatory in such species as well (Roubik et al. 1982). However, from studies in several tropical forests, it seems that the visitors of obligately outcrossing species that also are dioecious are usually stingless bees (Meliponinae) and honeybees (Apinae) (Bawa and Opler 1975; Appanah 1982; Roubik et al. 1984, 1986). Other social bees, such as halictids and exomalopsines, also arrive in large numbers at dioecious species. It is therefore possible that pollen transfer through a series of flower visitors, rather than just one, might be mediated through such social bees. Studies with European honeybees visiting obligately outcrossing varieties of apples show that pollen from different plants is acquired by bees within the hive (Degrandi-Hoffman, Hoopingarner, and Klomparens 1986; Dicklow et al. 1986). Even male honeybees, which never visit flowers, acquire pollen from foragers that pass them in the nest. Intranidal pollen transfer is especially likely in the natural nests of stingless bees and the burrows of parasocial species because there is direct contact between returning and exiting foragers in the nest entrance area. If this type of pollen transfer occurs, interplant movement is not the only means by which outcrossing can be promoted by small, tropical bees.

2.4.3 Specialization and coevolution

Interpreting the history of plant–pollinator evolution suffers from lack of fossil evidence, but the data point to increasing specialization in flowers and bees beginning in the late Cretaceous period (Crepet and Taylor 1985; Taylor and Crepet 1987; Cane and Eickwort in press; see also Section 4.1). Coevolution between flowering plants and bees undoubtedly occurs, but specifying which species have evolved in this way is extremely difficult. Widespread associations between genera of bees and flowering plant genera have been documented. Linsley, MacSwain, and Raven (1963) provide examples of pollen specialists (*Andrena, Anthophora, Sphecodogastra, Megachile,* and *Tetralonia*) on *Oenothera* (Onagraceae); Soderstrom and Calderón (1971) show repeated associations of meliponine bees with rain forest grasses that depend on animal pollinators (*Pariana* and *Olyra*); Ackerman (1983b) describes the neotropical distributions of orchids possessing solely euglossine bee pollinators; Armbruster and Mziray (1987) document the relation of Trigonini and two tribes of the Megachilidae (Megachilini and Anthidiini) with resin producing plants of the genus *Dalechampia* (Euphorbiaceae); Buchmann (1987) summarizes the close distributional and pollinating relationship of oilflowers of the genus *Lysimachia* (Primulaceae) with *Macropis* (Melittidae); Neff and Simpson (1981; see also Section 2.1.3) provide detailed information on the associations of various anthophorid bees and many neotropical Malpighiaceae; and Steiner (quoted in Buchmann 1987) is documenting the association between African oil-gathering bees of the genus *Rediviva* (Melittidae) and a few dozen species of *Diascia* (Scrophulariaceae) and orchids.

The broad associations noted between plant families such as cactus, palms, cucurbits, composites, and mallows and certain bees (Simpson and Neff 1985; Henderson 1986) indicate that some type of coevolution has been likely, perhaps only awaiting more complete documentation to emerge on a basis comparable to that of the examples just cited. Exact congruence in the distribution of such plant–pollinator pairs in pristine habitat would be strong evidence for coevolution, especially when the pair in question are species rather than a larger taxonomic unit. However, even studies presenting such data are unlikely to be complete, and there are always doubts where species' absolute distributions are under consideration. Even though the postulated strict coevolution should not obliterate variation of pollination relationships within populations, near distributional congruence may be acceptable as a demonstration of coevolution (Simpson and Neff 1985; Buchmann 1987).

Flowers may be as likely to evolve traits that protect floral resources (Section 2.4.4) as to encourage pollinator specialization. But how do bees specialize? The general matches between bee and flower size, dimensions of the floral corolla and a bee's resource-gathering appendages, pollen size and structure or floral lipid sources and bee hairs that collect the pollen or oils, and nectar characteristics and bee imbibing traits have been considered in Section 2.2. However, none of these matching traits proves that any type of coevolution has taken place (Janzen 1980, 1985; Schemske 1983). They are, however, expected correlates of specialization, as is a seasonal abundance pattern coinciding with that of preferred flowers (Chapter 4). Persistence of these associations can lead to coevolution. Given the unlikely existence of 1:1 coevolution (i.e., exclusively influencing gene frequencies of two interacting populations), bee and flower coevolution is most likely to be *diffuse* (Janzen 1980; Schemske 1983; Howe 1984; Manning and Brothers 1986; Buchmann 1987; Roubik and Ackerman 1987). Diffuse coevolution occurs when several species interact, and the outcome is usually unbalanced. That is, some of the species seem scarcely, if at all, dependent on their counterparts (Ackerman 1983b).

The reliance of a plant population on a particular type of pollinator or a single bee species is often seen as conducive to plant speciation, but the result for bees seems to consist mainly of behavioral and physiological specializations rather than speciation events following those of floral hosts (Feinsinger 1983; Montalvo and Ackerman 1986; Cane and Eickwort in press). For instance, the specialist bee will harvest pollen and nectar faster than other bee species, nest close to the host plant, and respond, by adult emergence, to possibly some of the same environmental cues that trigger flowering. Floral fragrances may also have differential attractiveness to different pollinators; further changes in the fragrances that narrow the attractiveness of the flower might culminate in reproductive isolation of plant subpopulations (Williams and Dressler 1976). Although these traits in extant populations suggest *preadaptation* as much as coevolution, additional physiological traits provide nearly unequivocal evidence of bee evolution in response to particular flowers.

Not only do bees that restrict their pollen feeding collect this food more rapidly than some less specialized bees, but their larvae sometimes digest and assimilate it more efficiently (Wightman and Rogers 1978; Strickler 1979; Louw and Nicholson 1983; Schmidt and Buchmann 1985). Considering the toxic substances found in both pollen and nectar (Section 2.1), larval assimilation efficiency can depend upon detoxification if hindrances to feeding or digestibility are present. One species of *Andrena* is known to specialize on pollen toxic to other bees (Tepedino 1981), and various types of pollen are toxic to bees (Schmidt, Buchmann, and Glaiim in press; see also Section 2.1.6). Furthermore, honeybees prefer a mix of high-quality pollen to a single species of comparable nutritional value (Schmidt 1984), suggesting the possibility that generalization may help to dilute the specific toxins of flower species. Comparing the highly generalized honeybee to solitary bees with presumably more specialized diets, it can assimilate 83% of consumed nitrogen, whereas nearly 90% assimilation was found for *Xylocopa* in South Africa and *Megachile* in New Zealand (Wightman and Rogers 1978; Louw and Nicholson 1983; Schmidt and Buchmann 1985). The *Megachile* and *Xylocopa* also incorporated 54–58% of all energy ingested during their development, far more than that observed in other phytophagous insects (Waldbauer 1968). Thus dietetic specialization seems to promote high reproductive efficiency in bees. Evidence favoring adaptive evolution was provided by controlled experiments with different pollen diets. These verified that larval bees, including the species of *Megachile* studied by Wightman and Rogers, grew more slowly and attained smaller maximum weight when fed pollen differing from their normal diet, or they did not feed at all (Guirguis and Brindley 1974; Bohart and Youssef 1976; Tasei and Masure 1978). Although these data are the first of their kind, they suggest the possibility of close dietetic evolution of bees in response to their host flowers. The abundance and stability of bee populations is another feature linked to the potential for coevolution (Chapter 4).

Few pollen determinations have been made for the nest provisions of solitary tropical bees; thus statements about their degree of specialization are premature. Pollen from perhaps several dozen buzz-pollinated taxa (Melastomataceae and Solanaceae) predominated in cells of large highland *Ptiloglossa* and *Crawfordapis* (Colletidae) studied in Costa Rica and Panama (Roberts 1971; Roubik and Michener 1984), and the pollen of another buzz-pollinated plant constituted all of the stored pollen in cells of *Centris heithausi* in Costa Rica (Coville et al. 1986).

Are the highly eusocial, perennial bee species that abound in tropical forests strictly generalists, or do they, too, display specialization and the potential for evolution with certain plant species? More evidence exists for their persistence as plant enemies than as potential specialist pollinators (Section 2.4.4), but preliminary data from pollen studies hint at specializations. Absy et al. (1984) collected some of the nest pollen from 37 nests of 24 meliponine species during two months in the central Amazon. Of 122 pollen species that were identified, fully 75% were used by three or less bee species. Another preliminary study in a Panamanian forest showed that Africanized *Apis mellifera* and 10 stingless bee species, including the largest and smallest meliponines, were very similar in the pollen they gathered (Roubik et al. 1986). They shared most of the 48 species harvested during two weeks, heavily utilizing six of them, all of which were functionally dioecious. Neither of these studies assessed the importance of a pollen type in the colony diet or its role in colony nutrition. This is indicated as the next step for such work. New information might reveal that a few of the shared species are of paramount importance to all of the bees.

In the tropics, colony diets have been studied quantitatively thus far only with introduced tropical *A. mellifera* in Panama (Roubik et al. 1984). This study was made by maintaining pollen traps on hives, which continuously removed a portion of the incoming pollen loads of foragers. Combined with similar but less quantitative studies, also utilizing pollen traps (Villanueva 1984; Roubik et al. 1986), the pollen diet of this species in lowland to upland forests in Panama and southern Mexico was found to include 135 to 240 taxa (Roubik 1988). Of these, one or two tree species might provide over 10% of the total *volume* of pollen harvested by colonies during an entire year.

Some meliponines having large colonies may be just as generalized (or specialized) as honeybees in the tropics, although no complete data exist for colonies in natural habitat (Cortopassi-Laurino 1982). Some broad patterns are seen in scattered studies of varying duration and sampling intensity. Lists of species visited by meliponines have been provided by study of pollen in some relatively disturbed neotropical habitats (Absy and Kerr 1977; Iwama and Melhem 1979; Absy, Bezerra, and Kerr 1980; Engel and Dingemans-Bakels 1980; Moreno and Devia

1982; Sommeijer et al. 1983; Ramalho et al. 1985). The last authors took monthly samples from nests of the small *Plebeia remota* and found that only 7 of 64 pollen species recorded during an entire year were abundant in the colony stores (over 10% of total pollen grains counted). This type of study did not assess pollen volume as did the above study of *Apis,* due to differences among pollen grain sizes. Similar species lists of plants visited for nectar and pollen have been compiled from bee collections made in secondary and primary forest, as well as semiagricultural areas [Heithaus 1979a,b; Roubik 1979a and studies by Ducke (1901, 1902, 1906, 1925) reviewed therein; Cortopassi-Laurino 1982; Camargo and Muzucato 1986]. The largest numbers of pollen and nectar sources associated with single species were obtained from a year's field data on floral visitors and thus are surely incomplete. Nevertheless, *Tetragona clavipes* visited at least 61 species in lowland French Guiana (Roubik pers. obs.), *Trigona fulviventris* visited 95 in lowland Costa Rica, and *T. spinipes* visited 104 in a botanical garden in southern Brazil. Africanized *Apis mellifera* at the last site visited 105 plant species. Other, stingless bees, including three species of *Melipona* observed for one year in French Guiana and Brazil, were found to visit about 50 species (Absy and Kerr 1977; Roubik 1979a; Absy et al. 1980). This number is as high as was recorded for most stingless bees in the literature cited above, and it is similar to that of a eusocial halictid bee, *Halictus hesperus,* studied in Panama (Brooks and Roubik 1983). These numbers represent close to 15–20% of the local angiosperm flora.

Two additional hypotheses regarding specialization by bees deserve special mention, and neither has been tested for an assemblage of tropical bees. The first hypothesis concerns proboscidal dimensions as they relate to corolla length. In experiments with bumblebees, both Inouye (1980) and Pyke (1982), among others, demonstrate convincingly that bumblebees tend to specialize and forage more profitably upon flowers that match their functional tongue length. Similar data failed to yield statistically convincing results in another bee–flower assemblage (Ranta 1982). This approach should be promising where many flowers have long, fused corollas, such as those used by euglossine bees (Kimsey 1982a; Ackerman 1985) and bumblebees. The second notion involves potentially both nectar and pollen preference and has been termed *Hopkins host-selection principle* (Feinsinger 1983; Cane and Eickwort in press). This interesting but unproven idea suggests that chemical stimuli associated with larval diet or food odors in the nest guide host-seeking behavior of the adult. Dobson (1984) has shown that lipids from the surface of pollen can be used by bees to discriminate flower species. These and other odors in the nest might facilitate learning by foragers, much in the way that it might for the highly eusocial bees that are demonstrably poor recruiters to sugar baits (Lindauer 1967; see also Section 2.2.5).

2.4.4 Bees as persistent plant enemies

Some bees regularly perforate flowers to obtain nectar or pollen, giving them an apparently unique role in pollination ecology. Little destructive activity by bees has been documented, but it may not be fortuitous that the same very limited number of genera, and even species among the meliponines, cause more than one type of damage. The megachilid bees that make nests from leaf sections may defoliate cultivated fields of plants (Batra 1984), and xylocopines that nest in living plant stems could also cause some damage (Chapter 3). A few neotropical *Trigona*, such as *T. spinipes* and *T. silvestriana*, have long been noted as orchard pests due to their resin- and sap-mining activities at branches, buds, and flowers (Ihering 1903; Wille 1969; Freire and Gara 1970). But the vast majority of bees that damage plants are foraging females with modified mouthparts that facilitate robbing from flowers. In these cases, maxillae or galeae are fused or pointed, and apical mandibular teeth are present (reviews by Schremmer 1972; Barrows 1980; Roubik 1982b, in press b; Inouye 1983). So far as is known, robbing bees are opportunists that forage at some flowers without robbing nectar and pollen. That is, they are not obligate floral parasites but rather legitimately visit flower species that they do not rob. However, they can consistently, year after year, be the primary or sole robber of a given plant species (Roubik, Holbrook, and Parra 1985; Renner in press).

Tropical bees known to damage flowers in order to collect pollen are exclusively meliponines, which usually perforate anthers of nectarless buzz-pollinated flowers visited only by bees (Renner 1983; Roubik in press b; see also Fig. 2.40). Nectar is often removed from flowers pollinated by hummingbirds (Table 2.5). As indicated in the table, it is very likely that a bee that robs nectar also robs pollen; thus the robbing stingless bees frequently compete with both bees and hummingbirds (see following). The small subset of stingless bees that rob nectar make large holes in flowers, which are then used by other bees such as *Ceratina, Paratetrapedia, Apis, Trigona,* and Halictidae as well as hummingbirds (Barrows 1980; Inouye 1983; Roubik, Holbrook, and Parra 1985; Roubik in press b). The phenomenon is not uniquely tropical by any means; it has been noted in temperate areas (Proctor and Yeo 1973; Faegri and van der Pijl 1979). Such foragers are called *secondary robbers* even though they are not destructive. Known nonmeliponine nectar robbers include *Megachile* (Schremmer 1941), *Pseudaugochloropsis* (Sakagami and Moure 1967), *Xylocopa* (Schremmer 1972; Barrows 1980), and *Oxaea* (Camargo et al. 1984). Thus very few bees besides highly eusocial species are destructive flower visitors. A *Perdita* (Andrenidae) robs pollen by entering an unopened flower through a hole cut with the mandibles (Hurd and Linsley 1963), but no similar behavior has been recorded for other small, solitary bees, temperate or tropical. Nec-

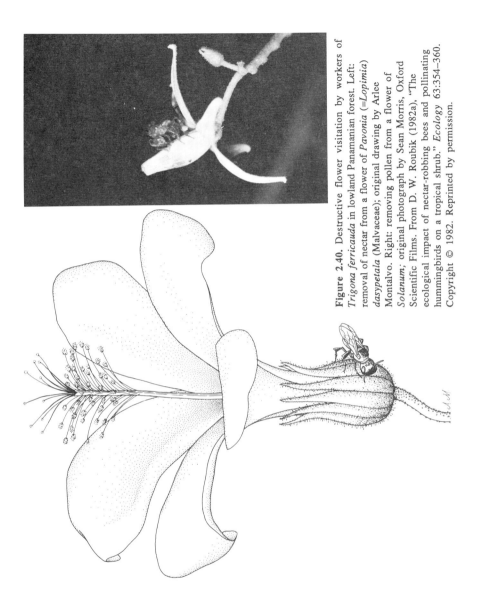

Figure 2.40. Destructive flower visitation by workers of *Trigona ferricauda* in lowland Panamanian forest. Left: removal of nectar from a flower of *Pavonia* (=*Lopimia*) *dasypetala* (Malvaceae); original drawing by Arlee Montalvo. Right: removing pollen from a flower of *Solanum*; original photograph by Sean Morris, Oxford Scientific Films. From D. W. Roubik (1982a), "The ecological impact of nectar-robbing bees and pollinating hummingbirds on a tropical shrub." *Ecology* 63:354–360. Copyright © 1982. Reprinted by permission.

Table 2.5. *Known destructive apid flower visitors in the tropics and the resources illegitimately removed from flowers of plant genera*

Bee	Resource removed[a]
Bombus transversalis	*Isertia* P_9
Trigona fuscipennis[b]	*Hamelia* P_{14}, *Cochlospermum* P_{11}, *Cassia*[c] P_1, *Tabebuia* NP_{13}
Trigona corvina	*Tabebuia* NP_{13}, *Erythrina* N_{13}, *Pavonia*[d] N_{10}, *Jacaranda* N_{10}
Trigona pallens[b]	*Asystasia* N_9, *Hamelia* NP_{13}, *Faramea* N_9, *Dioclea* N_9, *Cassia* P_6 *Vigna* N_9, *Cochlospermum* P_{11}, *Stromanthe* N_{13}, *Isertia* N_{13}
Trigona ferricauda	*Pavonia* N_{10}, *Heliconia* NP_{143}, *Solanum* P_{13}, *Aphelandra* N_{13}, *Stromanthe* N_{13}
Trigona fulviventris[b]	*Passiflora* N_{14}, *Quassia* N_{12}, *Cassia* P_1, *Dioclea* N_{13}, *Lantana* N_4, *Tabebuia* NP_{13}, *Asystasia* N_9, *Aciotis* P_{15}, *Clidemia* P_{15}, *Comolia* P_{15}, *Miconia* P_{15}, *Isertia* N_{13}, *Myriaspora* P_{15}, *Nepsera* P_{15}, *Rhyncanthera* P_{15}, *Sandemania* P_{15}, *Psychotria* N_{13}, *Tibouchina* $P_{2,15}$, *Tococa* P_{15}, *Bellucia* P_{15}, *Stromanthe* N_{13}, *Justicia* N_8
Trigona silvestriana	*Cassia* P_1, *Aphelandra* NP_7, *Pavonia* N_{10}, *Musa* N_{13}, *Passiflora* N_{14}, *Ochroma* P_{13}
Trigona dallatorreana	*Bellucia* P_{15}, *Desmoscelis* P_{15}
Trigona amazonensis	*Bellucia* P_{15}, *Tococa* P_{15}
Trigona hyalinata	*Bellucia* P_{15}
Trigona sesquipedalis	*Rhyncanthera* P_{13}
Trigona spinipes	*Tibouchina* P_2, *Citrus* NP_3
Trigona nigerrima	*Hedychium* P_6
Trigona williana	*Passiflora* N_9, *Mouriri* P_{15}, *Rhyncanthera* P_{15}, *Tococa* P_{15}
Parapartamona zonata	*Fuchsia* N_{13}
Paratrigona impunctata	*Cassia* P_6
Paratrigona subnuda	*Tibouchina* P_2
Heterotrigona itama	*Melastoma* P_{13}

[a] N = nectar, P = pollen.
[b] More than one geographic race or apparent sibling species are included (taxonomic names have been updated from the original source when necessary).
[c] *Cassia* = *Senna*.
[d] *Pavonia* = *Lopimia*.
References: [1] Wille 1963, [2] Laroca 1970, [3] Freire & Gara 1970, [4] Giorgini & Gusman 1972, [5] Barrows 1976, [6] Michener et al. 1978, [7] McDade & Kinsman 1980, [8] Willmer & Corbet 1981, [9] Roubik 1979a, [10] Roubik 1982b, [11] Roubik et al. 1982, [12] Roubik et al. 1985, [13] Roubik unpublished, [14] Gill et al. 1982, [15] Renner 1983.

tar robbing, but not pollen robbing, is common for some temperate species of *Bombus* and *Alpigenobombus*, which perforate flowers by executing a short, powerful bite or by piercing with the galeae (Alford 1975; Sakagami 1976; Morse 1982; Inouye 1983). Although only female bees generally are robbers, Camargo et

2. Foraging and pollination 157

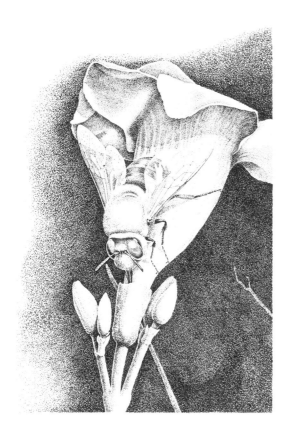

Figure 2.41. Destructive flower visitation by the male of a solitary bee, *Oxaea flavescens* (Oxaeidae), removing nectar from a flower of *Arrabidaea* (Bignoniaceae). Original drawing provided by J. M. F. Camargo (Camargo et al. 1984).

al. (1984) give a detailed account of one nectar-robbing tropical bee (*Oxaea flavescens*) in which both males and females perforate flowers. They have modified galeae that are pointed, not fused like the maxillae of nectar-robbing *Xylocopa* (Schremmer 1972). Furthermore, the male showed unusual reduction in pollen-filtering structures in the alimentary tract (at the proventriculus), which suggests that it has become dependent on nectars that are rich in amino acids and contain little pollen. These authors showed that a tropical bignoniaceous flower produced more amino acids in its nectar shortly after being pierced by this robber (Fig. 2.41).

The general behavior of tropical destructive bees varies markedly. Large bees usually make slits; those made by *Xylocopa* are up to 12 mm in length and are executed rapidly simply by inserting the proboscis through the corolla (Barrows 1980). A few destructive stingless bees dissect the flower and slowly remove stamens and pistils (McDade and Kinsman 1980). I find these to be the species that typically damage plants for sap and resin. Pollen robbers gradually dissect anthers by clipping off sections of the anthers and then inserting the proboscis to remove pollen (Renner 1983; see also Section 2.2). The nectar-robbing stingless bees also perform their work gradually and were seen taking 10–30 min to complete a perforation before harvesting nectar (Roubik 1982b; Roubik, Holbrook, and Parra 1985). Holes left in flowers by these robbers are round and a few millimeters in diameter. Large nectar robbers do not defend flowers from other visitors; their visit to a flower is too short to make this behavior useful. In contrast, some of the stingless bees defend the flower vigorously from all other visitors, excepting those originating from the same colony. This is often part of their normal foraging behavior, even when not robbing. They spend a long time making a flower perforation, and they can potentially harvest several loads of nectar from a flower producing the copious amount associated with visitation by hummingbirds; thus aggressive behavior is advantageous to some (Roubik in press b). The markedly aggressive robbers are, with a few known exceptions, those that arrive at resources in a small group (Section 2.3.6).

Although data are far from complete, strong phylogenetic constraints appear in destructive flower-visiting behavior among bees. Thirteen of the 16 neotropical stingless bees in Table 2.5 are of the subgenus *Trigona,* a group of about 30 species. Compared with other Trigonini, workers of these bees have two or three additional apical mandibular teeth, used also in contesting aggressively with other foragers, mining resin from woody plants, harvesting animal flesh, and deterring colony predators at the exposed nests built by some species (Roubik 1982a,b, 1983a; Johnson 1983a). The remaining three neotropical species also have modified mandibles: The small *Paratrigona* and *Aparatrigona* have four mandibular teeth, and *T. (Parapartamona) zonata* has two enlarged basal mandibular teeth. The Sumatran *T. itama* has two slightly enlarged basal mandibular teeth, as do many neotropical and paleotropical Trigonini that are not known to rob flowers (Sakagami 1978; Wille 1979b). A few stingless bees of other groups have enlarged basal mandibular teeth (e.g., *Ptilotrigona, Melipona, Meliponula,* and *Meliplebeia),* but none is known to use them for robbing. Comparative data are incomplete because no observations have been made in the Old World tropics except in Malesia, where there are more local stingless bee species than are found in much of the neotropics. Neotropical data come from Nicaragua, Costa Rica, Panama, Colombia, French Guiana, and Brazil.

Why are flowers perforated by small Trigonini and not the many solitary small bee species that have strong mandibles and are used to burrowing in soil or wood? Bolten and Feinsinger (1978) suggested that hummingbird-pollinated flowers produce dilute nectar so that bees do not take it, but this does not help explain why some bees are secondary nectar robbers of these flowers. Furthermore, some stingless bees gather nectar having 9–65% sugar, which completely encompasses the nectar sugar concentration of bird-pollinated flowers, 10–35% (Roubik and Buchmann 1984; Roubik et al. 1985; Roubik in press b). However, a closer look at the actual sugar concentrations of hummingbird- and bee-pollinated flowers in the same tropical habitat shows that the "bird flowers" often have nectar of 11–17% sugar and that most bees studied, except for the destructive *Trigona*, rarely take nectar below 19% sugar (Section 2.1). If amino acids become more concentrated in the nectar of robbed flowers (Camargo et al. 1984), then this may provide an additional attractant. This has yet to be tested for complex amino acid mixtures, but extensive experiments with four common robbing species of *Trigona* showed no preference for sugar plus single, common floral amino acids over a sucrose solution standard (Roubik, Aluja, and Inouye pers. obs.). Different ratios of glucose, fructose, and sucrose in floral nectar have been noted (Baker and Baker 1983b; see also Section 2.2.1), but it is not clear whether such combinations matter to bees of sizes similar to *Trigona*. The primary attractant for the highly eusocial bees is probably the large amount of carbohydrate that can be harvested from a single robbed flower by making several trips to the nest, which is an observed pattern for robbing *Trigona* (Roubik 1982b). Because most small solitary bees cannot use nectar at a comparable rate, they should not pursue such a laborious foraging method. Neither should *Apis* or other highly eusocial bees when other carbohydrate sources are available. It might be added that none of the above observations preclude coevolution between nectar robbers and chemical deterrents in nectar, such that deterrents discourage nectarivory by all but the most persistent nectar robbers.

In contrast to most temperate flowers that receive destructive visits (Barrows 1980; Inouye 1983), many of the flowers affected by robbing *Trigona* are large, simple (not composite) flowers, often presented individually on terminal inflorescences. Compact multiple-flower inflorescences of some flowers may prevent nectar robbing in some of the central flowers, and this is commonly found in tropical Bignoniaceae, which are heavily robbed by bees and birds (Barrows 1976; Gentry 1978; Roubik in press b). It seems clear from a recent analysis by Plowright and Laverty (1984) that this type of floral arrangement combined with nectar robbing may force pollinators to visit more flowers rather than merely abandon them and thereby enhance outcrossing (as postulated earlier by Heinrich and Raven 1972). However, no proof has been given that such forced outcrossing occurs or is beneficial for tropical species.

The negative effect of nectar robbing, aside from occasional aggressive behavior on the part of the destructive visitor toward pollinators, is the result of diminished nectar availability. Two studies, in which hummingbirds were the sole pollinators, showed considerable reduction in visitation and seed set due to robbing by *Trigona*

(Roubik 1982b; Roubik, Holbrook, and Parra 1985). Another study with a flower-destroying *Trigona* also showed its negative effect on plant reproduction (McDade and Kinsman 1980), and a study with *Xylocopa* implicated diminished plant reproductive success due to nectar robbing (Dorr and Martin, quoted in Hurd 1978). There was either less visitation or a postulated shorter visitation time to flowers containing little nectar. In general, foragers transfer less pollen at flowers containing relatively small amounts of a resource that is typically removed in a single visit by the pollinator (Thomson 1986). The destruction of anthers, like destruction of entire flowers, should have no subtle outcrossing benefits to flowering plants.

From 64% to 100% of all flowers were robbed on at least some individuals of five tropical forest tree and shrub species studied in Panama (Roubik in press b). The result of nectar removal by robbers and the secondary robbers they support can be seen not only as an added nectar production cost for the plant. Its end result appears to be a diminished amount and quality of visits received and seeds produced. This is particularly evident when normally outcrossing plants evolve closed breeding systems. For instance, temperate-zone flowers pollinated by hummingbirds are thought to be largely self-incompatible or to require outcrossing (Kodric-Brown et al. 1984). In contrast, two hummingbird-pollinated plants studied in lowland Panama were self-compatible, and one was also autogamous. The pistils of the latter gradually curve downward until anthers contact the stigmas (Roubik 1982b; see also McDade and Davidar 1984). Tropical hummingbird-pollinated flowers in the highlands, where there are fewer stingless bees, appear to be frequently self-compatible (Feinsinger et al. 1986), although there is no study suggesting the degree that destructive flower vistation might influence this trait.

The ecology of nectar and pollen robbing is a strong selective force in many tropical communities. Any organism can visit a flower as a thief, taking nectar or pollen without damaging the flower but providing negligible pollination. Moreover, a destructive hummingbird or bee visitor sometimes also visits a flower as a legitimate pollinator during the same foraging bout (Waser 1979; Roubik, Holbrook, and Parra 1985). The general distinction between a thief, a robber, and a pollinator is therefore blurred. However, the ecology of destructive flower visitation is distinct because

1. it involves unusual structural adaptations among bees and persistent rather than occasional interactions with plants;
2. it can alter pollinator behavior by uncoupling pollen availability in anthers from nectar availability in flowers (pollinators deplete *both* simultaneously); and
3. it supports secondary robbers, which then obtain otherwise unavailable resources.

3 Nesting and reproductive biology

3.1 Bee nests

Female bees construct nests to receive food that will be transformed into larvae, pupae, and then adult bees. To allow this process, the basic components of most bee nests are similar. They have an *entrance,* a *main burrow* or nest cavity, and *cells* or a *brood chamber* in which immature bees are reared in closed or protected compartments. At one extreme, the small, simple nests of solitary bees are shallow excavations. An unornamented hole terminates in a single cell having an inner surface barely different from the substrate. The first bees probably burrowed in the soil and nested in this way. Such nests, each composed of a single cell at the burrow's end, are now found among some primitive wasps. Through their elaboration, the nest architecture widespread among the sphecoid wasps and presumably the ancestors of bees has developed (Malyshev 1935; Michener 1964b; Brothers 1975; Bohart and Menke 1976; Iwata 1976). Evidently, intense pressure from natural enemies can force tropical bees to construct nests that consist of only one brood cell (Vinson, Frankie, and Coville 1987). However, solitary female bees usually build and provision at least five or six cells in each nest and then may go on to construct additional nests (Malyshev 1935).

Different sections of the burrow serve different purposes. The main burrow penetrates the substrate, separates the larvae from potential enemies, allows bees to develop in suitable conditions, and provides an exit when they emerge as adults. The *laterals* of a nest are formed by short branches of the main burrow that connect brood areas to it, usually separating brood cells from each other and allowing emerging bees to leave the nest without damaging cells closer to the entrance. Bee nest entrances are frequently narrowed but immediately within the nest they may widen to form a *vestibule;* both features can be useful in nest defense (Section 3.2). The entrance, which in some species projects as a tube, also serves for nest recognition and for orientation by adults. These attributes usually apply to both solitary and social bees, encompassing the nesting biology of *digger bees* that excavate nests in the ground or another firm substrate, *lodger bees* that use and modify preformed cavities, and *mason bees* that construct nests that are not enclosed in a nesting substrate and consist mainly of bee cells. The mason bee has no nest burrow, and its nest entrance is simply the cap of the cell, closed upon completion by the nesting female.

3.1.1 General considerations

The nesting habit of solitary species is extended by bees that live in colonies. A subtropical exomalopsine bee, *Exomalopsis aureopilosa,* forms extremely large communal nests. Zucchi (1973) found a nest that reached 5.3 m below the soil surface, occupied by over 900 adult bees. Soil-nesting *Halictus lutescens,* a primitively eusocial bee species in Guatemala, had a similarly large nest, inhabitated by almost 600 adult females (Sakagami and Okazawa 1985). Even these very populous nests contrast greatly with those of contemporary highly eusocial bee colonies. The largest of all bee nests is that of a South American meliponine, *Trigona amazonensis.* Its colony makes an elaborate resin nest up to 6 m long (Camargo, pers. commun.) and over 1 m across, weighing a few hundred kilograms and cemented to a tree trunk (Fig. 3.1).

The origin and structure of brood cells can be particularly complex (Figs. 3.2 and 3.3). Some cell walls are made completely from leaf material or plant lipids or resins, and from materials as diverse as pebbles and plant hairs, the frass from beetle burrows, spider webs, or flower petals; others are of glandular derivation or a mixture of bee and plant secretions (Malyshev 1935; Eickwort et al. 1981; Cane 1983a). Part of the cell lining secreted by the Dufour's gland of some anthophorids, halictids, and possibly other bee families even provides food for the larva (Norden et al. 1980; Cane 1983b; Duffield et al. 1984; Williams et al. 1986; Hefetz 1987).

In contrast with the digger bees, the lodger and mason bees utilize less building material that is of glandular origin (Malyshev 1935). This author proposes that the lodger bees display a much wider range of nesting sites than do other bees. They should require more building material to seal off part of the nesting cavity or to otherwise modify it, and the needed quantity probably could not be produced by the glands of an individual bee. Digger bees may have the advantage of using a nesting substrate that can be modified to fit the exact specifications of the bee. In contrast, for the lodger bee the high risk from predators or competitors at potential nest sites is probably offset by the diminished cost of building a nest. One of the consequences of investing relatively little in establishing a nest might be that protective glandular chemicals are unnecessary. Although there is not sufficient information to show whether the generalization holds for the tropics, Malyshev's theory is that more refined chemical adaptations exist in bees that must build their entire nests, and such bees are often highly specialized in their nesting preferences.

Nesting substrates, nest architecture, and to some extent placement of the brood cells differ greatly among species, and some variability within species is not un-

Figure 3.1. The exposed nest of *Trigona amazonenesis* (Apidae) attached to the trunk of *Ceiba pentandra* (Bombacaceae) in the Colombian Amazon. Original drawing by Sally Bensusen.

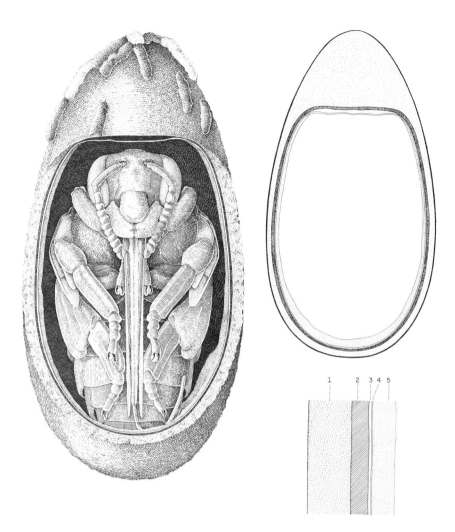

Figure 3.2. The brood cell and pupa of *Eulaema nigrita* (Apidae), showing layers of the cell and deposits on its surfaces. Resin and collected animal feces are deposited on the top of the cell. Original drawing by J. M. F. Camargo (Zucchi, Sakagami, and Camargo 1969).

common (Figs. 3.4–3.6). For instance, nest entrance turrets (Fig. 3.7) lined with secretions from the Dufour's gland were found in nests of *Halictus hesperus* in Panama and Costa Rica but were lacking for this species in southern Mexico (Wille and Michener 1971; Brooks and Cane 1984; Packer 1985). The obvious correlate of this difference was a vertical nesting substrate in Mexico and a horizontal sub-

3. Nesting and reproductive biology   165

Figure 3.3. Cell partitions constructed by *Xylocopa* (Anthophoridae) from the wood pulp chewed from walls of the burrow (Anzenberger 1977). Original photograph provided by Gustl Anzenberger.

strate at the other sites. Turrets were not sufficiently rigid to project at right angles from vertical surfaces.

Nest characteristics unique to particular stingless bee species often include the type of internal nest entrances they make. However, the stingless bee *Frieseomelitta savannensis* in French Guiana built the three types of nest entrances shown in Figure 3.8, and two strikingly different entrances were built by the same colony in different nesting cavities. The food-storage containers made by stingless bees are of a certain size and shape: spherical, egg-shaped, or cone-shaped. The size but not the general shape varies with the nest cavity.

The size, thickness, and shape of brood cells made by bees also show traits unique to each species. Other features of a bee's nest are more variable and include the length and number of tunnels and other chambers that connect the entrance to immatures and stored food. The nest entrance heights and the directions they face are variable for bees nesting in woody plants. In addition, soil-nesting bees often produce a mound of earth around the burrow entrance, called a *tumulus,* which has variable shape and dimensions. It is sometimes used to fill portions of the nest, particularly by bees nesting in loose sand or light soil (Malyshev 1935; Buchmann and Jones 1980; Roubik and Michener 1980; Vinson, Frankie, and Coville 1987).

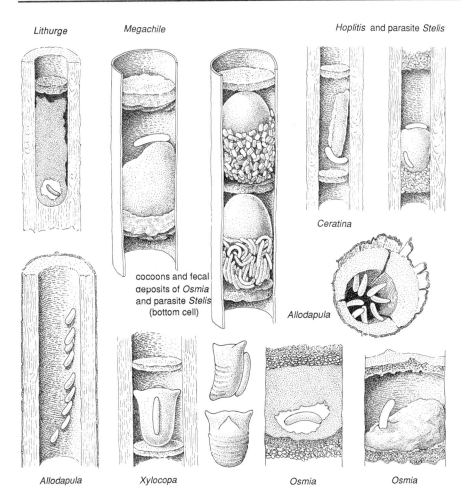

Figure 3.4. Nests and cells of xylophilous (twig- and stem-nesting) bees. So-called composite cells of *Allodapula* are hollows where eggs are laid. Eggs and/or pollen provision masses are shown for each type of bee nest, except the middle drawing showing cocoons surrounded by feces of *Osmia* and its parasite *Stelis*. The shaped pollen mass peculiar to a few species of *Xylocopa* is shown to the right of the nest, with and without the large egg (adapted from Stephen et al. 1969; Michener 1974a). Families of the taxa are Megachilidae (*Lithurge, Megachile, Osmia, Stelis, Hoplitis*) and Anthophoridae (*Ceratina, Xylocopa, Allodapula*). Original drawing by J. M. F. Camargo.

Construction patterns of bee nests are summarized by Malyshev (1935) and are

1. *progressive* if the main burrow is lengthened as laterals and cells are added;
2. *regressive* if cells and laterals are built gradually approaching the entrance from the burrow's deepest point;
3. *stationary* if laterals extend at a depth even with that of the burrow's end; and
4. *dispersive* if cells are scattered along the main burrow in no particular sequence.

3. Nesting and reproductive biology 167

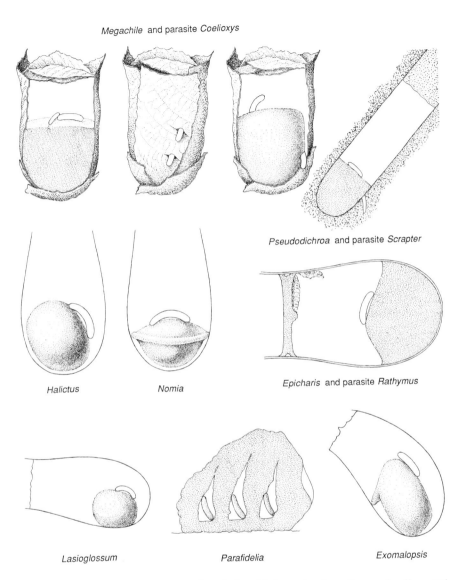

Figure 3.5. Cells of hypogeous (ground-nesting) and hole-nesting bees. Eggs and pollen provision masses are shown for each, except the middle right drawing showing the larva of *Rathymus* near the cell cap of *Epicharis*. The hole-nesting bees are represented by the leaf wrapped cells of *Megachile*, shown in the top row with the eggs of various parasites inserted through the cell wall or lying on top of the egg. The egg of the parasite *Scrapter* is shown inserted in the outer surface of its host cell (adapted from Rozen and Michener 1968; Stephen et al. 1969; Michener 1974a; Camargo et al. 1975; Rozen 1977b, 1984a). Families of the taxa are Megachilidae (*Coelioxys, Megachile*), Anthophoridae (*Pseudodichroa, Scrapter, Epicharis, Rathymus, Exomalopsis*), Halictidae (*Nomia, Lasioglossum*), and Fideliidae (*Parafidelia*). Original drawing by J. M. F. Camargo.

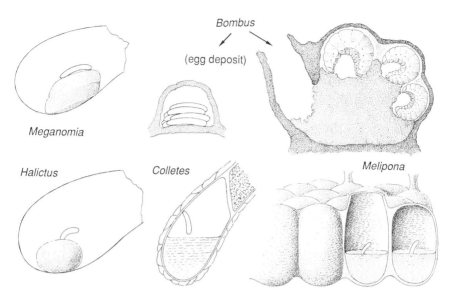

Figure 3.6. Cells of hypogeous or wax-making bees. Eggs or developing larvae are shown for each type of bee. A cluster of eggs placed in a cell to which no provisions have been introduced is shown for *Bombus*, in addition to a pollen pocket used to provision several larvae simultaneously by some bees of this genus (adapted from Michener 1974a; Rozen 1977b; Batra 1984). Families of the taxa are Melittidae (*Meganomia*), Halictidae (*Halictus*), Colletidae (*Colletes*), and Apidae (*Bombus, Melipona*). Original drawing by J. M. F. Camargo.

Figure 3.7. A forager leaving the hypogeous nest and a guarding female at the entrance of a colony of *Agapostemon* (Halictidae). Original photographs provided by G. C. Eickwort (Eickwort 1981).

3. Nesting and reproductive biology 169

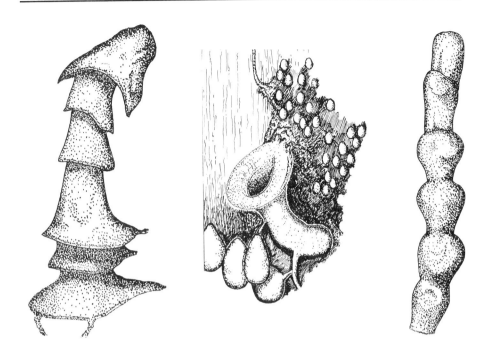

Figure 3.8. Internal nest entrances of cerumen built by colonies of the stingless bee (Meliponinae) *Frieseomelitta* (after Roubik 1979b). Storage pots for honey and pollen, along with the small brood cells, are shown in the middle drawing. The two nest entrances shown at the left and middle were constructed by the same colony in two different nesting cavities.

The burrows of digger bees are ramified as cells are placed either in separate, coalescent, or linear arrangement, and some halictid bees (*Halictus, Augochlora*) construct brood chambers that are filled with earthen cells (Figs. 3.3–3.11). As bee nests are enlarged from single cells to their completed size, the arrangement of cells can become planiform, nearly symmetric, or even spiral. The stingless bees, in contrast with all other bees, complete the entire nest and seldom change its different areas (Section 3.3.4) before broods are produced.

What excavation procedure is followed by the ground-nesting species that use one burrow for more than a generation or through the emergence of more than one brood? For *Halictus* and *Exomalopsis*, excavation of the main burrow is not continuous (Michener and Bennett 1977; Raw 1984). The general pattern is regressive. The main burrow is extended until the stratum of soil appropriate for cell construction is penetrated. After this, cells are made from the bottom of the burrow toward the entrance. If broods continue to emerge, however, or if the nest is used

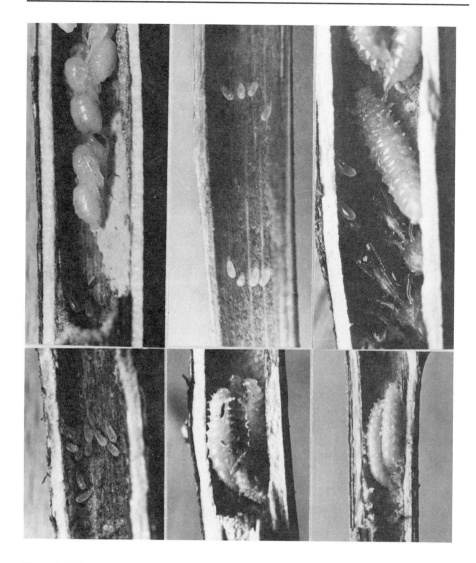

Figure 3.9. Nests and small colonies of allodapine bees, *Allodapula melanopus* (Anthophoridae), showing eggs, larvae, and pupae (Michener 1971). Original photograph provided by C. D. Michener, courtesy of the American Museum of Natural History.

for more than one generation, then work on the main burrow is resumed. Cells are built, once again, beginning at the lower extremity of the burrow. The nest then extends further downward, and used cells are found near the nest entrance, or the youngest immature bees are found at both extremes of the nest with older immature bees between them.

3. Nesting and reproductive biology 171

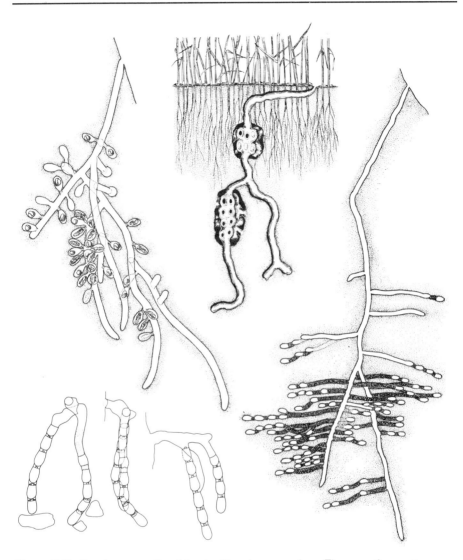

Figure 3.10. Complete nests of social and solitary hypogeous bees. The upper three nests are made by the various Halictidae and are the work of several females (Sakagami and Michener 1962). The nests shown on the lower left consist of two or three burrows made by individual *Centris* (Coville et al. 1983). Original drawings provided by S. F. Sakagami and R. E. Coville.

If the immediate environment can be modified to accommodate the needs of adult and immature bees, then nesting is successful. If it cannot, then the nest site will be abandoned; the nesting bees or their offspring will die. Mating precedes nest construction by most solitary bees; but for stingless bees, if nest construction

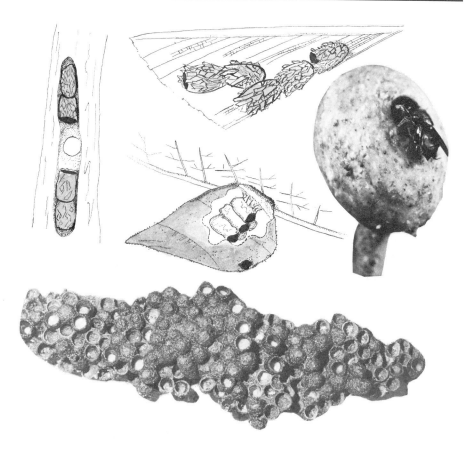

Figure 3.11. Diversity of nests and cells constructed by euglossine bees (Apidae). The large aggregation of cells at the bottom is a complete nest of *Euglossa imperialis* in Panama, found by H. Harlan, taken from a small cavity in limestone (photo by author). The photograph on the upper right is the nut-shaped hard resin nest of an individual *Euglossa dodsoni*, shown entering the nest (photo by author). Above left are resin and wood chip cells of *Eufriesea* placed in the abandoned nesting tunnel of *Xylocopa* (after Kimsey 1982a); above center are cells of *Eufriesea* attached to the underside of a palm frond (after Sakagami and Strumm 1965); and at center is the brittle resin nest of *Euglossa turbinifex*, attached under a leaf, showing open and completed cells within the hollow interior (after Young 1985).

is hampered, the queen does not even mate (Section 3.3). Unmated female Halictini and probably most bees (Michener, pers. commun.) will occasionally build nests and produce only male bees if mating is unsuccessful (Section 3.5). Selection of a nest site is determined by availability of resources and protection from the elements and natural enemies. The extent to which nesting bees can perceive and measure these and other pertinent factors will enhance their reproductive fitness. Such features affect the physical conditions experienced by immature bees, which in turn influence their rate of development and the timing of adult emergence.

An intriguing illustration of bee nesting behavior is found in a study of *Xylocopa* nesting in equatorial Africa (Anzenberger 1977). Burrows and cells are excavated in hard wood, where Anzenberger found several trial bores made to only a few millimeters' depth. At an observation nest, a newly emerged female began to construct a burrow in a timber crowded with the burrows of five other females. The small new burrow was abandoned for nearby unoccupied substrate. The bee was evidently able to judge that there was insufficient space without breaking into the burrows of its neighbors. This author also found that bees never stayed away from the nests for more than 25 min, possibly to prevent raids by doryline ants. This phenomenon has a strong connection to resource availability within flight range. Anzenberger surmises that tropical *Xylocopa* that impregnate their burrow entrances with potentially repellent secretions (Skaife 1952; Kapil and Dhaliwal 1968a) probably make longer foraging trips. Other studies of this conspicuous bee genus have shown that female bees can interact aggressively at crowded nest sites (Janzen 1966) and that conspecific pillaging of provisions or usurpation of nest burrows is common when the burrow is left unguarded (Bonelli 1976; Ben Mordecai et al. 1978). These are additional factors that could regulate nest site selection by females.

Most female bees lay eggs in fairly well-separated individual cells or in cells grouped into clusters within the nest (Figs. 3.4–3.6). An egg may be buried in a chamber in the provisions, it may rest on one or both tips or lie flat on liquid provisions, or it may be cemented to the upper portion of the cell. After the egg is laid, the cell is sealed, usually terminating further contact between the nesting female and developing larva. However, notable exceptions do exist. Some female halictids and xylocopines open cells while their larvae feed, to remove their feces; and allodapines also remove feces from their nests, in which the main burrows contain all of the immature bees (Michener 1974a; Raw 1984). An unusual subtropical species of the widespread genus *Lithurge* (Megachilidae) often lays two eggs within a single pollen mass, with no cell partitions (Garófalo et al. 1981). Deposition of more than one egg in a cell has also been found in African fideliids (Rozen 1977a; also see Fig. 3.5) although it may occasionally occur in the communal nests of other bee families that are used by several reproductive females (Raw 1984). Such bees, as well as *Heriades* and *Sayapis* (Megachilidae) make no partitions between cells. The allodapines make no cells but excavate nests (composite cells) in the soft pith of stems and herbaceous plants, where the offspring are then cared for within the tubular cavity (e.g., Michener 1974a; Maeta et al. 1985; see also Fig. 3.9). In some bee families, cells contaminated with fungi that kill larvae and may spread to other cells are reopened by females and filled with earth, along with laterals leading to them (Batra 1984).

Some social bees progressively feed their larvae. These include bumblebees, probably all of which progressively feed the larvae of queens, directly feed regurgitated nectar and pollen to developing larvae, and feed honey directly through perforated or permanently open orifices in cells (Sakagami 1976). Other *Bombus,* although apparently not the lowland tropical species (Sakagami, Akahira, and Zucchi 1967) deposit pollen at the bases of a larval cell, lay a batch of eggs on it, then occasionally feed larvae directly with regurgitated nectar and pollen (Fig. 3.6). Female allodapine bees (Anthophoridae) of the Old World tropics regurgitate nectar and pollen to their broods, and honeybee workers feed larvae in the same manner (Michener 1974a). Michener also points out that the allodapines and most *Bombus* deposit eggs in batches within a single chamber (Figs. 3.4, 3.6, and 3.9).

Nest location or the building material used may strongly influence the potential for social development (Section 4.3). There are no visible differences between the nests of closely related solitary and social species in the Halictidae (Sakagami and Michener 1962; see also Fig 3.10), a fact that leads these authors to suggest that nest structure had no obvious correlation with the development of bee sociality in this particular group of bees (Michener 1964b, 1974a). Recent analysis of two bee groups displaying broad similarity in size variation within species and within nests (*Ceratina* and Halictini) showed that, although larger bees make effective nest guards and are beneficial to both groups, only the Halictini have evolved fixed size differences (Michener 1985). These bees nest in the ground and are more exposed to nest predators than are the twig- and stem-nesting *Ceratina*. Thus predation pressure, as a result of nest substrate selection, may help promote evolution of specialized *castes,* some of which serve primarily as guards or foragers rather than as reproductives. Other authors have suggested that the evolution of wax production and of nest structures containing wax has had considerable importance in allowing continual colonial living (Duffield et al. 1984). Free-standing cells and storage containers for food may be constructed in tight aggregations, providing some of the conditions for social evolution and sustained colonial living. This subject will be taken up again in discussion of social evolution in tropical bees (Chapter 4).

All bees, except some of the *Apis* and mason bees, carry out much of their egglaying or social functions in darkness. In the darkened nest, materials are manipulated by bees using their legs, mouthparts, and metasoma; diverse chemical and tactile signals are perceived with the antennae. Among social species, the substrate may be a medium through which the chemical cues are transmitted that lead to many behavior patterns (Free 1987). For example, the chemicals bees encounter in certain parts of their nests may cause them to forage, clean the nest, sting intruders, feed a nest mate, or prepare to abandon a nest site, or they may be prompted to release pheromones. Fresh feces of queen *Apis mellifera* can act as repellents and evoke self-grooming behavior of worker bees in the nest (Post, Page, and Erickson 1987). As Free (1987) recognizes for the honeybees, there are likely to be numerous undiscovered bee pheromone systems. In the nesting cavity, chemical, acoustic, and tactile stimuli regulate bee behavior, and the context for all interactions is an artificial construct made by the bees. In turn, bees carefully regulate their nest environment.

The behaviors involved in nest construction differ slightly among the solitary bees, as they construct and line cells for provisioning. Megachilids differ from the halictids, melittids, andrenids, and colletids by using the forelegs and mandibles to remove soil from the burrow (Eickwort et al. 1981). Adult bees possess one or two spurs on the apex of the middle or hind tibiae, projecting at right angles to the body. These provide purchase against the burrow walls, as likely do the one or two spines on apices of the tibiae of many bees. Bees pivot on these spurs, somersaulting to change the direction of their head-first movement in their tunnels (Anzenberger 1977; Batra 1984). Because the spurs also are present in males, it is doubtful they are important in nesting success, although the emergence of male bees from burrows may be aided by the spurs. Soil-dwelling bees, excluding some Colletidae (*Colletes*), possess a pygidial plate at the tip of the abdomen that is used to compact the soil of burrows and cells (Batra 1980a). Digger bees of the Anthophoridae and Halictidae that work the soil often possess pygidial and basitibial plates specialized for this purpose, also used to apply Dufour's gland secretions to the cell walls. Basitibial plates help brace the legs of most soil-dwelling bees that move, with legs bent, within their burrows (Batra 1964).

Bee mandibles are often modified to deal with a certain type of nesting substrate. Large or sharp series of mandibular teeth serve to perforate or cut either hard or soft materials (Fig. 2.12) or to excavate in hard substrate. Apids that extensively use wax have smooth, rounded mandibles that are used to shape waxen structures (Fig. 2.12, *Bombus, Apis, Melipona*). A radical alteration of mandibular type is apparent in the *Xylocopa* subgenus *Gnathoxylocopa;* females have broad, spoon-shaped mandibles suited to excavate the soft, moist stems of the African *Aloe* plant in which they construct their burrows (Hurd and Moure 1963). Behavioral and chemical traits of bees nesting communally also appear to have undergone modifications, perhaps allowing the burrowing activity of one female to proceed without damaging established cells or burrows. The cell and burrow linings of some ground-nesting species are thought to cause a nesting conspecific to avoid them (Duffield et al. 1984). Buzzing and stridulatory sounds made by adults may have the same function (Batra 1980a; Rozen and Jacobson 1980).

3.1.2 Physical conditions, bee cells, and life cycles

Atmospheric temperature, gas, food, photoperiod, natural enemies, population density, and moisture affect the activity of adults and the survival and development of immature insects (Tauber, Tauber, and Masaki 1986; Danks 1987). A fascinating body of information exists on ecology of the insect microhabitat (Gates 1980; Unwin 1980), which has yet to be applied to full advantage in studies of nesting bees. Optimal temperature conditions clearly exist for pollinators such as *Megachile rotundata* (Tepedino and Parker 1985). Developmental times of bees are influenced by temperature, and smaller individuals tend to be produced at higher temperatures within a single nest (Jay 1959; Stephen et al. 1969; Kamm 1974; Undurraga and Stephen 1980; Garófalo 1985; Tepedino and Parker 1985). Undurraga and Stephen (1980) discuss "developmental threshold," which they show to be 19° C for *Megachile rotundata*. At slightly lower temperatures, its development from late pupa to adult and also adult emergence are delayed. Diapause of mature larvae (prepupae) and of pupae appears to be widespread in bees as one means of surviving through unfavorable environmental periods (Danks 1987). Tauber et al. (1986) outline the range of adaptive responses involved in insect dormancy and their chief mechanisms:

1. *Quiescence* is an adaptive response to unpredictable, aseasonal changes and is not mediated by neuroendocrine changes.
2. *Nondiapause dormancy* involves a more complex set of physiological adaptations that evolve due to seasonally recurrent environmental conditions; metabolism is suppressed, yet there remains a high degree of responsiveness to immediate environmental conditions.
3. *Diapause dormancy* is the most common form of dormancy that evolves in response to persistent environmental changes; it relys upon the operation of a biological clock and neurohormonal processes under the control of token stimuli that signal the approach of unfavorable conditions.

Diapause dormancy does not end until certain physiological processes are completed. Denlinger (1986) reviews the subject of diapause in tropical insects and finds a general trend that tropical insects and related temperate-zone species diapause at the same stage of their life cycle. In his survey, 7% of the insect species diapaused as eggs, 32% as larvae, 30% as pupae, and 31% as adults.

The nest-cell humidity of bees is usually near 100% and must be kept from diminishing to assure larval survival, yet innundation and mold growth also must be avoided (Malyshev 1935; Michener 1974a). To this end, largely liquid feces may be completely partitioned from the pupa by a cocoon, and the lipid cell linings made by most ground-nesting bees are hydrophobic (Shinn 1967; May 1972; Cane 1983b; Batra 1984; Roubik and Michener 1984; Rozen 1984a,b; Hefetz 1987). Hydrocarbons produced by the glands of ground-nesting bees must have properties similar to wax. Dufour's gland secretions of some bees, among them certain *Calliopsis* and *Xylocopa*, are made up solely of hydrocarbons (Kronenberg and Hefetz 1984); so their primary function seems to be that of manufacturing waterproof cell linings. Solitary bees have no further control over cell conditions after the cell is sealed, and thus nest temperature and humidity are determined by components of the cells and its contents.

Little is known of the degree to which bees control atmospheric gases such as CO_2 or O_2 in their nests (Rozen 1984b) or the effect that these may have on immature bee development. Experimental study of *Apis mellifera* showed that workers responded by fanning their wings in the brood area when normal levels of CO_2 were elevated, irrespective of O_2 concentration (Seeley 1974). This bee can detect differences in CO_2 on the order of 2% (Lacher 1967). Nest CO_2 concentrations were kept between 0.44% and 0.78%, or at about 10–20 times normal atmospheric concentration (Seeley 1974). At sea level there is approximately 20% combined CO_2 and O_2 in the atmosphere. The changes in O_2 and CO_2 concentration in the nests of temperate-zone honeybees were found to be tidal, rising and falling a few times each minute due to the coordinated fanning behavior of worker bees (Southwick and Moritz 1987a). Removal of CO_2 from the nest environment likely prevents narcosis of adults and immatures.

One recording of atmospheric CO_2 in the lowland forest of central Panama was 0.034%, whereas the concentration in arboreal nests of six stingless bees there was 0.44–0.90%, reaching 0.93–1.38% when these nests were artificially closed for 24 hours (Roubik and Kursar, pers. obs.). The watertight cells containing pupae of a large anthophorid bee *Epicharis rustica* had 1.32% CO_2 in the same environment. These fragmentary data indicate that low levels of CO_2 narcosis are probably avoided by the fanning behavior of highly social bees and resisted by solitary bee larvae. The cells containing larvae of the largest meliponine bee, *Melipona fuliginosa,* have a small hole in the middle (personal observation), which may exist to allow escape of metabolic CO_2. The level of O_2 is perhaps unlikely to have much effect on immature bee metabolism, because it varies less than several percent.

Roberts (1971) and Coville, Frankie, and Vinson (1983) discuss cell modifications of tropical ground-nesting bees, such as an open cell closure or one with a small tube leading from it (Fig. 3.10), as modifications made to allow gas exchange. On the other hand, Rozen (quoted in Michener 1978) suggests that excessive liquefaction of larval provisions may occur in open cells due to the hydroscopic properties of the pollen provision mass. Both factors could come into play in regulating the timing of larval development and adult reproduction necessary to ensure larval or pupal freedom from excessive CO_2 and humidity. Certain tropical and temperate andrenid bees may further reduce liquefaction of provisions by coating them with a secreted water-resistant lipid layer, similar to that of cell linings (Rozen 1967; Cane 1983a).

Details of larval food composition are given in Section 3.3, and the consistency of provisions suggests pronounced differences in larval environments of bees. The mass of provisions can basically be described as relatively dry and firm, pasty and smooth, or liquid. The doughlike provisions of some digger and carpenter bees (bees that nest in excavated galleries in wood) are molded in various shapes that minimize their contact with the cell walls; if the provisions are in permeable cells, they absorb water and expand, due to their relatively large exposed surface area (Malyshev 1935). Provisions in the cells of stingless bees are roughly one-half to two-thirds water (Roubik and Buchmann, pers. obs.), but the egg stands upright on this soupy mixture, apparently due to surface tension and a reticulate egg surface that prevents the egg from sliding (Sommeijer, Zeijl, and Dohmen 1984). Egg adhesive is apparent on freshly deposited honeybee eggs, on the eggs of *Colletes,* and on eggs of some apoid cleptoparasites, none of which place eggs on top of food provisions (DuPraw 1967; Batra 1984; Torchio and Trostle 1986; see also Fig. 3.6). Larval provisions of many colletids are liquid. In fact, the euryglossine and hylaeine colletids carry all collected floral resources in their crop (see Table 2.2); thus their provisions must be liquid when regurgitated into cells. The egg floats on the surface or is attached to the upper portion of the cell (Roberts 1971; Iwata 1976; Batra 1984; Roubik and Michener 1984; Rozen 1984b). Various South American, African, and Australian anthophorids, including some *Thygater, Melissoptila,* and *Anthophora,* have similar but less liquid provisions (Iwata 1976). Most bees make a provision mass that is smoothed by the female and somewhat sticky, but as shown in Figures 3.4–3.6, its form varies much among taxa.

Bees protect their nests in various ways from water that can enter the burrows and damage cells, provisions, and immatures. The laterals from the central burrow of a ground-nesting bee are often arched slightly upward before meeting the top of a cell, which would prevent rainwater from entering it. Laterals are filled by soil from the main burrow, also protecting the cell. In addition, ground-nesting bees of 13 groups of Meliponinae (Portugal-Araújo 1963; Wille and Michener 1973; Camargo 1974) may construct a basal drainage tube so that the nest cavity does not collect water (Wille 1966). The subterranean nests are built in well-aerated soil (Nogueira-Neto 1948). Further traits of exclusively ground-nesting groups of meliponines such as *Schwarziana* and *Geotrigona,* include an entrance tube that diverts surface water around the nest as well as furrows in the involucrum, which prevents water from entering the brood area (Camargo 1974). Workers of *Melipona fasciata* frequently eject a brown liquid from the nest by regurgitation outside the nest entrance (pers. obs.). This probably comes from both condensation of metabolic water within the nest and liquid excretions of adult bees. Many tropical ground-nesting bees select nesting sites that are sheltered from the rain (Rozen 1984a,b, pers. commun.). Innundation is avoided by bumblebees that coat the ceiling of a nest with their wax (Michener 1974a). The waterproofing afforded by resin and cerumen nest linings, cells, or nest envelopes is undoubtedly important in allowing meliponines and some megachilids and euglossines to nest in wet environments. Leaf and leaf-pulp cells of megachilid bees (Fig. 3.5) presumably have similar properties (Williams et al. 1986). Iwata (1976) points out that the ground-nesting habit of xylocopine and megachilid bees has largely been lost; the bees now utilize preexisting cavities and wood or pith. Their existence in the wet tropics was aided by this shift in the nesting site.

Differential permeability of cells to atmospheric moisture, respiratory gases, and water have been ascribed either to properties of a cell lining or to the cocoon (Malyshev 1935; Michener 1964b; Houston and Thorp 1984; Rozen 1984a,b). Specialized cell linings are made by many of the older bee groups, such as colletids, stenotritids, oxaeids, andrenids, and melittids, but also by more derived forms, such as anthophorines. Some groups utilize both plant lipids and Dufour's gland secretions in cell construction (Section 2.2.3). Larval fecal material may also be pressed over the entire cell wall, forming a separate layer upon the cocoon (Rozen and Jacobson 1980; see also Fig. 3.2). Alternatively, feces remain within the cocoon or are mixed with cocoon silk, usually pushed to one end of the cell, or are removed as the mature larva defecates (Rozen 1984a; Maeta et al. 1985). A complete fecal cell lining or one on the inside of the cell closure is unusual. Whether such variations might aid regulation of the bee's physical environment, besides separating it from contaminants, is unknown. Cocoons are generally not made by many bees that construct waxy cell linings, and Michener (1964b) suggested that either structure may exclude water and yet allow humid air to flow into a cell and respiratory gases to escape. The combination of a cocoon and lipid cell lining is advantageous in wet nesting environments. Only one melittid larva is known to make a cocoon, and the adult female also lines the cell with a waterproofing lipid gathered from plants (Rozen and Jacobson 1980; Cane et al. 1983). The diphaglossine colletids, some of which nest in the wet ground in tropical cloud forest, line the cell with a glandular secretion, and the larva also makes a cocoon (Rozen 1984b). Furthermore, the extremely liquid defecation of the mature larva is sometimes allowed to drain from the base of the cell. To accomplish this, the larva bites a hole through the cell lining before defecating (Rozen 1984b).

Cocoons are not made by *Xylocopa* and other xylocopines (Fig. 3.9) that nest in wood, nor do they generally coat the cell walls with a water-resistant lining. This may restrict nesting to drier substrates (Hurd and Moure 1963; Anzenberger 1977). Kronenberg and Hefetz (1984) found that ground-nesting *Proxylocopa* line their nest cells with secretions from the Dufour's gland, but *Xylocopa sulcatipes,* which nests in wood, does not. Many ground-nesting bees pass several months, or even years, in the prepupal stage, and their cocoon or cell lining probably helps to avoid desiccation or drowning during this period (Michener 1974a; Roubik and Michener 1980; Raw 1984). In the extremely dry or shallow nest habitats of some andrenids and stenotritids, the mature larva secretes a protective coating that covers its integument and promotes water retention (Rozen 1967; Houston and Thorp 1984). It makes no cocoon. In contrast, wood- or pith-nesting anthophorids, even in relatively cool subtropical and temperate areas, survive prolonged periods as immatures without the protection of a cell lining or cocoon (Maeta et al. 1985).

Cocoons provide little insulation, although they are often impermeable (Sakagami, Tanno, Tsutsui, and Honma 1985). Bee species that complete two reproductive cycles per year in the temperate latitudes have been found to construct cocoons only prior to the coolest season (Rozen 1984a; Torchio and Trostle 1986). A similar trend was noted in the Brazilian megachilid *Lithurge* (Garófalo et al. 1981), which builds a thicker cocoon in which prepupae of the second yearly generation diapause. However, this occurs during the warmest and wettest time of the year. Increased protection from water (or enemies?) apparently favors cocoon building by such tropical bees.

Bees that nest communally in a single cavity but have no organized social regulation of nest temperature are likely at the mercy of environmental conditions. I can provide one observation of mortality due to high ambient temperature in an observation colony of *Euglossa imperialis*. After six months of activity during the wet season and early dry season, when humidity and temperature averaged near 80% and 28° C, respectively, the entire immature population (roughly 20 cells provisioned by 6–12 females) may have died during the middle to late dry season when temperatures exceeded 34° C and humidity declined to 55%. Female bees abandoned the nest (Fig. 3.11), which was in an open, elongate box. The nest had originally been found in a small, wet limestone cave.

Highly eusocial species, with the exception of *Apis* (Southwick 1987) and possibly some of the stingless bees (Roubik pers. obs.), do not closely regulate their nest temperatures. They are therefore extremely dependent on local climate and suitable nest sites within cavities. Nogueira-Neto (1948) studied the air movement produced by fanning workers in nests of 10 stingless bee species and concluded that they were able to draw warm air into the nest as well as expel moist air probably created by evaporation of water from nectar. The result of these rudimentary controls over the nest environment is protection of the immature and newly emerged bees from environmental shocks. When subjected to artifical cooling of the nest, *Melipona* elevates metabolic rate to reduce temperature variation in the brood area (Roubik pers. obs.).

3. Nesting and reproductive biology   181

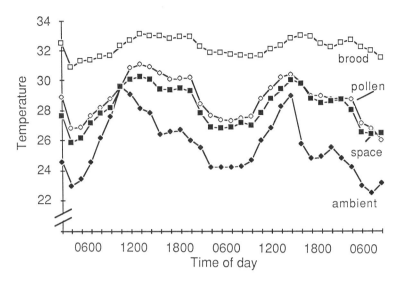

Figure 3.12. Nest temperatures during two days in three distinct regions of the nest of a colony of *Melipona seminigra* in the central Amazon (after Roubik and Peralta 1983).

Adult behavior and nest architecture are closely related to the production of a relatively stable nest environment. For instance, colonies of two species of Amazonian *Melipona* displayed relative temperature stability and higher than ambient temperature in the brood area because of an envelope that is built to surround the brood (an *involucrum;* see Fig. 3.19), although temperatures there and elsewhere in the nest followed ambient temperature change (Roubik and Peralta 1983; also see Fig. 3.12). The brood area was the warmest part of the nest and linked directly to the external environment by an internal nest entrance tube. A thermal gradient was maintained by the large brood area, such that temperature diminished in nests of the two *Melipona* by roughly 1° C for each 6 cm above it. Sakagami and coworkers (1983a) recorded the temperature changes within bamboo occupied by a colony of *Tetragonula laeviceps* in Sumatra. The nest and an empty bamboo internode were quite similar, remaining 2–3° C above ambient temperature. Only one stingless bee from Asia, *Lepidotrigona terminata,* nests exclusively where the climate is cool (Sakagami and Yamane 1987; Salmah, Sakagami, and Inoue pers. commun.).

Preliminary study of a southern Brazilian species, *M. quadrifasciata,* showed that brood temperature varied between 26° C and 35° C in ambient temperatures of 20–37° C (Kerr and Laidlaw 1956). Wille (1976) presents information on the relatively stable brood nest temperatures of *M. fuliginosa* (25.5–29.5° C) at ambient temperatures of 14–20° C, and Wille and Orozco (1975) recorded brood nest temperatures of 23–29° C in shaded ambient temperature conditions of 15–34° C. Such species probably could not survive in colder areas, although it is not clear to what extent their climatic limitation arises from difficulty in utilizing sufficient honey to permit the metabolic activity that ensures adequate brood nest temperatures (Section 4.3.5). Kerr (1972) has shown that polyploid spermatids were produced by males of *M. marginata* that were exposed as pupae to temperatures of 18–21° C. Terada, Garófalo, and Sakagami (1975) believed that larval mortality in the nests of *Plebeia droryana,* a small neotropical stingless bee with small colonies, was caused by extreme temperature fluctuations in the environment. Nonetheless, colonies that have minimal control over nest conditions survive temperatures as low as 1° C and as high as 37° C (Kerr 1972). The temperature control of termite colonies (Wilson 1971) is shared by the stingless bees *Scaura, Partamona, Trigona, Ptilotrigona, Aparatrigona, Apotrigona, Axestotrigona;* see Darchen 1972a; Camargo 1980; Roubik 1983a), euglossines (*Eufriesea;* see Kimsey 1982a), anthophorids (*Centris, Ptilotopus;* see Coville et al. 1983), and megachilids (*Chalicodoma;* see Messer 1984) that nest there.

Nests of stingless bees have been found at elevations of 1,500 m in Taiwan and at up to 3,400 m in Colombia (Nates pers. commun.), where average temperatures during the coolest part of the year are 9° C or lower (Sakagami and Yamane 1984). Stingless bees in Sumatra, Borneo, and New Guinea also inhabit regions above 2,500 m elevation (Schwarz 1939; Salmah pers. commun.). Possible homeothermy of the bees, which would regulate brood nest temperature in a stricter manner resembling that of *Apis,* has not been studied. Another alternative, induced torpor of both adults and immatures, also seems plausible. The initiation of diapause or quiescence in such species has not been studied.

The ability to regulate nest temperature by a combination of behavioral and architectural mechanisms is most evident in meliponine bees that build exposed or "aerial" nests, often exposed to full sunlight and protected by hard containers of resin, mud, vegetable fibers, and feces (Nogueira-Neto 1962; Wille 1983). Study of the Brazilian stingless bee *Trigona spinipes* in a single nest showed that this colony kept brood temperatures at 34–36° C in ambient temperatures of 16–28° C (Zucchi and Sakagami 1972; Sakagami 1982). Darchen (1966) measured temperature near a nest of African *Dactylurina staudingerii* and found that as ambient temperature rose from 23° to 28° C, the bees increased ventilation of the nest by augmenting the number of 1-mm-wide openings in its surface. They apparently closed them again with resin at night. Nests of both species and the other dozen or so meliponines that build large exposed nests have populous colonies that maintain several layers of envelopes around the core brood nest, with space for bees between them. The outer layer is often thin or soft, and holes can readily be opened or closed to regulate the flow of air. Within the nest layers, worker movement and activity can regulate temperature with precision. Other stingless bees that build small exposed nests, especially *Paratrigona* (Wille and Michener 1973; Roubik 1983b), have three or four involucral layers surrounding the brood. The closed brood area presumably maintains temperatures higher than ambient, as it does for *Melipona* that build involucral sheaths around the brood (Roubik and Peralta 1983). Possible homeothermy in these bees was shown by the elevated thoracic temperatures of workers resting between combs. Elevated body temperature and the insulation of involucra combined to produce a relatively stable brood nest temperature.

Apis dorsata in its exposed nests maintained a brood temperature of 30–34° C when ambient was near 24° C (Viswanathan 1950) but was unable to maintain activity or, presumably, viable brood at much lower temperatures. Over 80% of the colony's adults may remain on the nest, which allows the colony to maintain sufficient brood temperature (Morse and Laigo 1969). Dyer and Seeley (1987) found that this bee cannot maintain body temperature sufficient for flight at ambient temperatures below 17° C. In contrast, the related *Apis laboriosa* nests at elevations of over 3,000 m in Nepal and Tibet on exposed cliff sides and displays a good deal of cold tolerance in addition to superior insulation derived from its substantially hairier body (Appendix B). Its colonies are at an elevation where temperatures often drop below freezing (Roubik, Sakagami, and Kudo 1985). During the coldest part of the year, however, entire colonies leave the high Himalayas to overwinter as a resting cluster of bees, without wax comb, near the subtropical lowlands (Grimaldi and Underwood 1986; Underwood, pers. commun.).

Bumblebees are among the most cold-resistant bee species in tropical, temperate, and arctic areas. Nesting is observed at up to 4,300 m in the Andes and at 5,500 m in the Himalayas (Liévano and Ospina 1984; Williams 1985). Individual bees in the nest climb upon a brood cell and press closely to it, transferring body heat that is generated by thoracic flight muscles (Heinrich 1979a; Cameron 1985). The lethal temperature for the immatures is not known, nor have there been studies of temperature change within the nest cavity. It is likely that the nest sites are carefully selected and deep below ground (Itô et al. 1984).

Movement into protected nest cavities may have occurred as an evolutionary response to cold climates (or predators inhabiting cold regions) in both honeybees and stingless bees. Whereas the smallest honeybees build exposed nests and are completely restricted to tropical climates, honeybees of intermediate size, *A. cerana* and *A. mellifera,* inhabit both the tropics and temperate zones as far north as northernmost Japan (excluding Hokkaido) and Norway (Ruttner 1988). The Saban honeybee, *Apis vechti,* is the third species of cavity-nesting *Apis* (Tingek et al., 1988). These species may nest exclusively in protected cavities, although *A. mellifera* also builds exposed nests in warm climates. Such bees have the largest range of temperature tolerance of any bee species. Southwick (1985b, 1987) provides a full discussion of cold resistance by honeybee colonies, showing that they cooperatively maintain temperature differentials of up to 114° C between the core nest and the external environment; this performance is comparable to the most cold-resistant warm-blooded animals. The trigonine bee group *Partamona* is neotropical and contains approximately 40 species (Camargo 1970, 1980). Many build nests that are partly exposed, inserted under the rootlet masses of epiphytic bromeliads or constructed within termite nests; only one related group of them is known to nest in tree cavities. This is *Parapartamona,* which in Colombia and Ecuador nests between 2,000 and 3,400 m (Roubik pers. obs.; Camargo pers. commun.). Perhaps not coincidentally, this area also has native bears that should prey upon social bee colonies, and bears also occupy the natural ranges of *Apis mellifera, A. vechti,* and *A. cerana* (Section 3.2.3).

Cold tolerance by *A. mellifera* and probably *A. cerana* is accomplished by architecture and behavior that maintain the brood nest temperature near 34° C. Immature bees are deformed if their temperatures fall below 31° C or rise above 37° C, although adults endure temperatures as low as 6° C (Seeley and Heinrich 1981; Southwick 1985b). In general, the cavity-nesting honeybees have higher metabolisms and shorter lifespans and generate more heat during flight than any of the *Apis* that build exposed nests (Dyer and Seeley 1987). The most important differences in the cold-weather behavior of temperate honeybees compared with tropical races seem to involve heating efficiency, since both tropical and temperate *Apis mellifera* cluster tightly within their nest and increase metabolism to offset cooling in the environment (Southwick and Roubik pers. obs.). Stored honey is utilized to maintain worker or immature-bee temperatures within the cluster (Seeley 1985). The cold-weather clustering behavior of honeybees compresses their workers in a space only 12% of the volume that they otherwise would occupy, and the optimum number of bees in the cluster is approximately 16,000 workers for temperate-zone *A. mellifera* (Southwick 1987). Temperate honeybees maximize exposure of the nest entrance to the sun (entrances face south in northern latitudes) and restrict the size of the entrance to less than 60 cm^2, also preferring an entrance near the base of the nest and a total nest volume close to that filled by the combs (Seeley and Morse 1978; Seeley 1985). All such traits are useful in heat economization during cold weather. The opposite of these traits might be expected for tropical *A. mellifera* if loss of excess heat is promoted by nesting site. There is some indication that Africanized *A. mellifera* prefer larger cavities for nesting than do temperate honeybees (Rinderer et al. 1981). These experimental findings are tentative, because swarms were sometimes given choices of cavities previously used by other colonies that may have been chemically attractive.

3. Nesting and reproductive biology

Water is apparently used only by *Apis* to bring about cooling in the brood area. No stingless bee or bumblebee is known to lower nest temperature by evaporative cooling of the nest from water or nectar droplets placed within it, although both groups fan the wings while on brood to lower brood nest temperature and may also open holes in the roof of the nest (Nogueira-Neto 1948; Hasselrot 1960; see also the summary by Michener 1974a). One species of *Scaptotrigona* does gather water more frequently when its nest is heated (Roubik, Farji, and Sierra pers. obs.) but their use of the water within the nest has not been studied. The honeybees fan their wings over water droplets placed on the edges of cells, causing a draft that disperses the water-cooled air (Lindauer 1961; Seeley and Heinrich 1981). These authors, along with Michener (1974a) and Seeley (1985), provide many details of this unique social behavior. Its origin can perhaps be traced to relatively constant collection of water by a small number of bees. Water is generally used to dilute honey that is too viscous to be imbibed by bees – for example, by the small workers of *Apis florea* (Akratanakul pers. commun.). Water is often used to dilute the provisions given to larvae (Section 3.3.2). However, when the nest temperature increases, adult bees in the nest eagerly receive water and less readily take other types of forage. Foragers are thus forced to specialize more upon water collection (Section 2.3.5).

3.1.3 Nest sites

Natural nests of tropical bees are often cryptic and distributed in a way that is only slightly understood. Most bees are known only as adults from collections made on flowers. Nesting biology, however, seems to follow patterns indicated broadly by the higher taxonomic groups. General nesting characteristics of bee families are given in Table 3.1, which shows that some families are far more variable than others. For example, it would be difficult to describe a typical nesting habit of an anthophorid bee, for within the family, nesting occurs in hard substrates such as tree branches or open ground, in the soft pith of herbaceous stems, and in preexisting cavities such as the nests of other bees, beetle burrows left in tree trunks after adult emergence, and within bamboo internodes. Apids and megachilids primarily nest in preexisting natural cavities in wood and living trees and thus are lodger bees, but some also build exposed nests or cells and thus can be termed mason bees, or dig holes in termite and ant nests as digger bees; most other bee families comprise digger bees that excavate nests primarily in the ground.

For the biologist, success in finding the nest of a particular bee will often depend more on luck than experience, even if the type of nesting substrate is known beforehand. If the most suitable nesting sites for a bee species are usually occu-

Table 3.1. Nesting habits of bees

Bee family	Solid substrates						Preexisting cavities			
	Open ground	Termitaria	Ant nests	Exposed	Twigs	Foliage[a]	Tree hollows	Used nests[b]	Beetle burrows	Caves, sub-terran. holes
Stenotritidae	~exclus.									
Colletidae	freq.							rarely		
Oxaeidae & Andrenidae	~exclus.									
Halictidae	freq.				rarely			rarely	rarely	
Melittidae & Ctenoplectridae	~exclus.									
Fideliidae	~exclus.									
Megachilidae	occas.	rarely		rarely	occas.	occas.	rarely	occas.	freq.	rarely
Anthophoridae	freq.	rarely	rarely		occas.	rarely		rarely	occas.	rarely
Apidae		rarely	rarely	rarely		rarely	freq.	rarely	rarely	occas.

[a] Within appressed leaves, epiphytic rootlets, or plant galls.
[b] The nest of the parent or the nest of another hymenopteran.
Abbreviations: occas., occasionally; freq., frequently; ~exclus., almost exclusively.

pied, then bees possess remarkable specializations in their nesting requirements and the ability to locate potential sites and to discriminate among them. Nests in felled trees, in brush that has been burned or that has been trampled or cut by man or other animals, in the earthen banks of road cuts, railways, paths, athletic fields, and in banks made by rushing water have frequently provided the only opportunities to study bee nesting biology (e.g., Michener and Lange 1958a,b; Sakagami and Michener 1962; Maeta et al. 1984). This may bear some resemblance to the manner that hypogeous (ground-nesting) bees find suitable nest locations. Good clues to site preference for xylophilous (wood-nesting) and lodger bees are provided by surveying dead trees, branches, and cavities in trees and under surface roots. Both the bees that nest in wood and their parasites can be seen flying close to these surfaces as though searching for small holes, and they briefly fly into larger cavities. On the other hand, study of the highly eusocial bees in intact tropical forests is usually undertaken successfully only with the help of local guides and aboriginal peoples, who to some extent depend upon the honey, brood, or resin contained in these nests for their own livelihood (Section 4.3.4).

a. Nesting microhabitat. Bees build nests at a wide range of heights, nearly everywhere but in small dry branches at the very tops of trees. As shown in Table 3.1, many bees nest in exposed, firm ground. Most bee families nest only in this microhabitat, which often has little vegetation, whether temperate or tropical. These areas only exist stably in arid or semiarid conditions. Suitable natural nest sites are therefore likely to be scattered and transient in places such as treefalls or glades with sparse vegetation. In the mature forest, bee nests are made in flat surfaces on the ground or in inclined or vertical banks, and they are also pressed against tree roots, under tree bark, or in crevices between tree buttresses. Even for a given species, nests are found from a few centimeters above ground to some 20 or 30 m high in living trees. Tropical lowland bumblebees, for example, nest near the surface of the ground under vegetation or roots, whereas those in temperate and tropical highland areas frequently nest in ground cavities (Dias 1958; Sakagami and Zucchi 1965). Highly eusocial species rarely make their nest entrances level with the ground or in the tallest branches of forest trees. A height of 30 m seems to be their normal limit (Roubik 1979a, 1983b; Inoue pers. commun.). Some euglossine bees build a nest within bromeliads, under leaves or in the woody exocarp of fruits (Dodson 1966; Zucchi, Sakagami, and Camargo 1969; Young 1985), which could well be high in the canopy. The unpublished work of Zucchi, Sakagami, Camargo, and Dias in Belém, Brazil, which entailed examining some 300 nests of 20 euglossine species, shows that nest sites are consistent for each species, corresponding to several strata from below ground to relatively high in the canopy (Camargo pers. commun.). The intense desiccation and insolation to which the forest canopy is subjected, however, might prohibit the construction of nests in small, dead branches or other relatively exposed substrates high above the ground. Allodapine bees in the Old World tropics tolerate exposure and highly variable nesting temperatures (Michener 1971), as do *Ceratina*, small megachilids like *Heriades*, and the small colletid *Hylaeus* (Michener pers. commun.). Exposed twig and stem nest sites are highly ephemeral. Dead, dry tree branches are broken by winds, and the upper branches of large trees are perhaps unreliable as nest sites for other reasons; the upper trunk or branches of tropical trees frequently break (Foster 1980).

Foraging and nesting sites may to some extent be related. In a study of tropical bees arriving at light traps, Wolda and Roubik (1986) found that a consistent majority of ground-nesting species usually came to a light operated at 3 m above the ground rather than to another light 27 m higher. On the other hand, wood-nesting bees arrived most often in the upper trap. Two of these, purely nocturnal species of *Megalopta* (Halictidae), arrived in stable but different proportion at this trap, about 60% and 80% of all catches, in different years. Because nocturnal bees are active for up to several hours during periods of moonlight (Kerfoot 1967b), it is unlikely that the bees arriving at light traps were mostly those nesting nearby. This raises the possibility that different species are most often active at different strata above the ground, but it is not clear from this information whether the reason is more closely related to food or to nest sites.

Selection of nearly vertical earth banks for nesting is common among diverse ground-nesting bees (Michener and Lange 1958a,b; Iwata 1976), but there are no experimental and little comparative data on whether such sites are preferred by these bees or on what the consequences of nesting in other conditions might be. Vertical slopes might, for example, provide better protection from predators or rain. Bees nesting in a vertical surface often do not extend their nesting area to include intergrading horizontal substrates that are apparently of the same soil type. But the exclusive use of horizontal substrates is also notable for some bees.

The slope of nest burrows initiated in a vertical surface is almost always downward or horizontal rather than upward and may simply indicate positive geotaxis. Earth pushed out of the main burrow is more easily moved downward, and therefore the added energy expenditure needed to push debris up and out of a burrow appears to be of relatively little importance. A female bee pushes excavated nest material toward the entrance usually by moving forward, pushing debris with the head, and cradling it next to the venter with the legs. Near the burrow entrance, the bee may somersault, rotate 180° on its longitudinal axis, and then push debris out with the tip of the abdomen and the legs. Once the main burrow is made in soil, excavated laterals are made, and often are later filled. The downward slope of the nest, which would prevent the loss of substrate material needed to fill these spaces, is one explanation for the tendency to make burrows that are horizontal or slope downward. Nests excavated in wood are often slanted upward, like those of *Xylocopa* both in hardwoods and bamboo, whereas those in pithy stemmed plants can extend downward or upward toward the entrance (Sakagami and Laroca 1971; Iwata 1976; Anzenberger 1977). The upward slant of a bee burrow could be related to protection of the nest from rainwater or direct sunlight. In species that open several interconnected burrows and have more than one provisioning adult in them, both above and below the nest entrance, finishing construction first in the upper portion of the nest is advantageous, for then excavated materials do not fall into brood cells in lower chambers. Malyshev (1935) observed that the degree to which cells of a particular bee were inclined had no correlation with the consistency of its provisions. Even bees that fill their cells with liquid food masses can build cells that slant strongly upward toward their bases.

Marked preferences that exist for sand or clay soils seem related to the depth at which cells are built by nesting bees. Tropical bees, like those in temperate climates, excavate burrows from a few centimeters to over 3 m in depth (Camargo 1974; Michener 1974a; Iwata 1976; Rozen 1984a,b). A given species digs deeper in loose soil than in hard ground (Malyshev 1935). The anthophorid *Melitoma* is an example of a solitary tropical bee limited in distribution by its preferred hard nest substrate, and its nests reach only 5 cm depth (Linsley et al. 1980). In French Guiana, small patches of clay substrate such as adobe walls, introduced by man near large expanses of sandy soil, were utilized intensively and over many generations. *Melitoma* is rare in the Amazon valley, where most soil is sandy, although one flowering malvaceous plant that it visits for food is abundant in this and other regions of sandy soil (Linsley et al. 1980). Another case of soil stratum limitation in a ground-nesting species was described by Roubik and Michener (1980). They found that cells of Amazonian *Epicharis zonata* were made in sandy soil at depths of up to 52 cm, at a time when the water table was below 56 cm. Shortly after the dry season and the period of adult bee activity, the cells became submerged for up to nine months, during which time prepupae remained in watertight cells. Rather than the usual single cell closure, *E. zonata* sealed its cells with two. Camargo (pers. commun.) has found that other centridines having nests submerged under water for several months also make a double cell closure. Such species and perhaps most bees that nest in the ground fill in the laterals, and sometimes the central burrow, as cells are completed. The tumulus soil surrounding the nest entrance is crumbled and then pushed into the main burrow. This behavior would reduce predation and parasitism of the cells (Vinson et al. 1987).

b. Nest dispersion (solitary bees). Solitary and facultatively social bees nest in aggregations or in scattered, predictable localities as well as those that

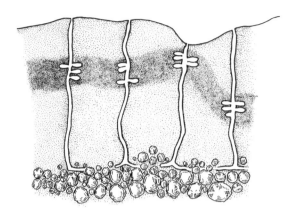

Figure 3.13. Extension of halictid nest burrows through soil, showing construction of cells in a particular layer of soil type and moisture and termination of burrows in coarse gravel (Sakagami and Michener 1962). Original drawing provided by S. F. Sakagami.

are seemingly unusual and unpredictable. Iwata (1976) mentions plant galls and snail shells as preexisting sites used occasionally for nesting by lodger bees. Michener (1970b), in a survey of the twig- and stem-nesting bees of Africa, also noted nesting in galls made by wasps and in the thorns of *Acacia*. He found many more nests of *Heriades, Ceratina,* and allodapine bees in the introduced plant species of gardens than in normal vegetation. The enormous aggregations of *Centris* and *Epicharis* along rivers in the Amazon basin have long been noted (Camargo pers. commun.). Nesting areas in the vertical riverbanks extend several kilometers, encompassing tens of thousands of active female bees. The same sites are submerged during part of the year. Solitary bees may also be wholly dependent on disturbed sites or the nests constructed by other species (which frequently include wasps, ants, termites, or birds), even the burrows made by dung-provisioning scarab beetles (Raw 1984, 1985; Rozen 1984a) The giant Moluccan megachilid *Chalicodoma pluto* nests within arboreal termite nests, as do some centridine anthophorids (Coville et al. 1983; Messer 1984; Snelling 1984; see also Section 3.1.2). Several megachilid, euglossine, and anthophorid genera include many lodger bees that utilize the abandoned nests of other bees or even wasps, which are then filled with nesting material and cells (Vesey-Fitzgerald 1939; Hurd and Moure 1963; Sakagami and Laroca 1971; Michener 1974b; Kimsey 1982b; Garófalo 1985). Many tropical bees are dependent on the burrows excavated by wood-boring beetles (see Crowson 1981), as are certain stingless bees (Schwarz 1948; Kerr et al. 1967; Michener 1974a; Roubik 1979a).

The criteria used by solitary female bees to find ideal nesting areas have yet to be elucidated. Ground-nesting solitary bee species use open ground with few or no plant roots (Sakagami and Michener 1962; see also Fig. 3.13), but this may be an artifact considering the ease of finding the nests and making studies in such an area. The root mat in a tropical forest tends to be extensive and dense, but it is

3. Nesting and reproductive biology

usually very thin since soil nutrients are on or near the surface of the ground, and tree roots or rootlets thus have large horizontal spread but slight vertical extension (Whitmore 1984). A firm or hard surface in combination with softer soil in deeper strata with moisture content sufficient to prevent collapse of excavated burrows is the preferred substrate of ground-nesting bees. One study of such bees in Brazil showed that clay and decomposed gneiss were used for cell and gallery substrate, irrespective of depth below the surface within range of a few meters. Aside from the apparent preference for decomposed rock, minute details of soil chemistry or pedology had little importance (Michener et al. 1958). Similar nesting patterns have been found in halictid bees (Fig. 3.13) and are probably widespread. A frequent requirement of ground-nesting bees of diverse families is that there be soil available for the filling of laterals. A separate chamber near the nest entrance is sometimes excavated in the process of obtaining this material, provided it cannot be obtained from other laterals excavated in the same nest (Malyshev 1935; Sakagami and Michener 1962; Michener 1974a). However, hard substrates of lateritic, sunbaked soils are nonetheless used for nesting. Bees that nest in hard earth may soften the substrate with water they carry to their nesting site, often requiring many trips to a water source in order to excavate the main tunnel (Malyshev 1935). Without readily available water, or perhaps nectar, some ground-nesting species would have little success. Rozen (1977b) reports several incomplete nests of the fideliid *Parafidelia pallidula* in southern Africa. Very compacted or hard clay soil encountered in the process of burrowing was evident and had caused the female to shift her burrow site.

There is an indication that the distribution of solitary bee nests is often locally aggregated or clustered; perhaps it is determined by the location of existing successful nests, often the parental nest, or by chemical attractants, discussed in the next paragraph. The central burrow of a nest made in a hard substrate is likely to be reused by adult females emerging from the nest, which then construct new laterals and cells. The nest may be extended downward, ramifying from the central burrow, until the soil layer suitable for nesting is exhausted.The few detailed studies that have been made of soil chemistry and nesting responses of bees suggest that soil texture and compaction are important. Hard clays are usually unsuitable, although they can be worked by some species, and soil moisture may determine the depth at which new cells are built (Sakagami and Michener 1962; Stephen et al. 1969; Packer 1985; Williams et al. 1986). Michener (1974a) mentions that some tropical halictids nest at the same level below the ground throughout the year, but other bees make their nests deeper during periodic hot and dry seasons.

Nest aggregation pheromones originating in the Dufour's gland have been postulated for solitary bees such as ground-nesting colletids, andrenids, halictids, and anthophorids (Bohart, quoted in Michener 1974a, p. 49; Cane 1983a; Duffield et al. 1984; Hefetz 1987; see also Table 2.4). The work of several authors suggests diverse biological functions of these secretions in helping females recognize their nests and cells currently being provisioned, attracting male bees to nest sites where they may mate with females, and even preventing female bees tunneling in the ground from breaking into completed cells. Forthcoming experimental studies of these proposed functions may provide a key to nest site selection. In addition, Cane (1983b) has shown that female anthophorid bee parasites of *Andrena* are strongly attracted to the Dufour's gland secretions of its host.

If a species reuses part of a parental nest, this is likely to be due to very limited nest site availability. In addition, dependence on a hole in the ground left by a defunct ground-nesting ant colony, the existence of hard or relatively long-lasting nest substrate like the solid wood or the hard substrate of an abandoned termitarium, and continuation of some freedom from depradations of predators or nest parasites could promote continuous nest occupation. Beneficial symbionts are also more available at an active nest site (Section 3.2.4); they conceivably are passed by the fecal meconium frequently ejected by a callow bee as it emerges from its nest (Malyshev 1935). Reuse of the parental nest, except by eusocial colonies, seems relatively rare in tropical bees, but data are too fragmentary (Hurd and Moure 1963; Stephen et al. 1969; Zucchi et al. 1969; Michener 1971, 1985; Gerling et al. 1981). *Xylocopa* nesting in thick-walled bamboo such as Asian *Bambusa stenotachya* reuse the nest cavity for four to five generations, which can be as long as five years (Iwata 1976). Occasional reuse of nests, cells, or burrow entrances has been seen in some solitary, gregarious, and cooperatively nesting Anthophoridae (Rozen and MacNeill 1957; Daly, Stage, and Brown 1967; Sakagami and Laroca 1971; Raw 1977; Rozen 1984a), Halictidae (Sakagami and Michener 1962; Michener 1974a), Colletidae and Megachilidae (Stephen et al. 1969), and Euglossini (Zucchi et al. 1969; see also Section 4.3). Cell shape and the diameter of burrows made in wood can reveal this phenomenon. Barrel-shaped cells, rather than cells with nearly parallel sides, can indicate reuse of cells excavated in wood, but they also exist in nests of soil-nesting species without having the same implication for these bees.

c. Nest dispersion (eusocial colonies). In a study of stingless bee nest distribution within a neotropical dry forest, Hubbell and Johnson (1977) located 26 trees with nests within 13 hectares of a forest; the presence or absence of nests was scored for plots 10 m across. Their analysis of 67 nests in this and adjacent areas revealed that four of nine resident species had relatively uniform (overdispersed) distributions, four were slightly aggregated or random, and the remaining one was probably randomly scattered, being rare among the surveyed trees. Furthermore, it is interesting that 16 stingless bee species occupy this forest; thus many nests are rare or cryptic (Janzen and Roubik pers. obs.). A similar study by Fowler (1979) of *Tetragonisca angustula fiebrigi* in Paraguay showed random nest dispersion for 149 nests, 83% of which were made in abandoned nest sites of leafcutter ants, *Acromyrmex*. Neither study revealed a preference of a bee for a particular tree species. Perhaps not surprisingly, the species that nested at even intervals in the first study were not restricted to nest locations that are obviously very limited. One builds exposed nests on branches or tree trunks, one nests in the nests of arboreal termites, one lives in cavities anywhere between 3 and 25 cm in diameter, and the fourth nests among the root systems at the bases of large trees (Wille and Michener 1973; Roubik 1983a).

Each of the three species of *Trigona* included in the study of Hubbell and Johnson (1977) was found unlikely to be close to another nest of this subgenus, which may indicate that these closely related bees respond to each others' pheromones and space their nests accordingly (see Table 2.4). The four bee species displaying intraspecifically uniform nest dispersion had among the largest colonies of the local species. The workers of one species, normally a completely unaggressive forager, are intensely aggressive toward bees from rival colonies at potential nest sites being investigated for future nest establishment (Hubbell and Johnson 1977). The corresponding hypothesis was that foraging odors used by some species are similar to those used to mark potential nest sites (Johnson 1980; Johnson and Wiemer 1982). This seems not, however, to readily account for the dissimilar behavior displayed in the two different contexts. A further adaptive reason for dispersion of nests may be to avoid confusion between conspecific communication trails, whether or not related to food availability.

The work of Camargo in the Amazon basin has revealed that suitable trees may be used by up to 25 nesting colonies of both trigonines and *Melipona,* and one tree infested with termites had 37 stingless bee nests built within the termite galleries (Camargo pers. commun.). Darchen (1972b) also found up to 22 nests of African stingless bees of two genera in a single tree. Diverse stingless bee species, with the exception of the largest colonies comprising tens of thousands of workers, such as *Lophotrigona canifrons* in Malesia (Inoue pers. commun.) and many neotropical Trigonini, occasionally nest in the tree occupied by another colony (Hubbell and Johnson 1977; Roubik 1979b). Large trees with many cavities of suitable diameter are often occupied by several colonies of species of *Scaptotrigona, Nannotrigona, Frieseomelitta, Tetragonisca,* and *Melipona,* for example, in *Enterolobium cyclocarpum* in Panama. In Colombia, Salt (1929) found the nests of *Trigona, Cephalotrigona,* and *Melipona* in a single tree branch. The requirements for nest defense, the indirect benefits of nesting near aggressive colonies, and the utilization of a suitable nest substrate or system of cavities, which from available data may be the deciding factor in nest location, combine to produce these aggregating tendencies. Data from natural forest habitats are few, but the Amazon region typically has widely varied bee nest density and very patchy dispersion even of species that do not aggregate nests, with sometimes dozens of colonies in a hectare (Camargo pers. commun.). If careful studies are made on a large geographic scale, aggregation could prove to be common.

At present, there are no data to confirm that different colonies of a given species detect differences in their respective trail pheromones. Even species not known to use odor trails in foraging may employ odors to lead nest mates to a new nest site. Posey and Camargo (1985) state that the Kayapó indians of the Amazon forest recognize odors used by stingless bees, including *Melipona,* during the founding of new nests. Foraging by *Melipona* apparently does not involve odors (Michener 1974a), and trails to new nest sites have not been investigated. As suggested by Duffield et al. (1984), some biologically active substances of the stingless bees might be produced by the Dufour's gland in the abdomen. Establishment of a nest by *Trigona fulviventris* is mediated by workers that use one or more glandular odors to mark potential nest sites. This is followed by threat displays and fighting until one colony wins the competition or discourages establishment of a nest (Hubbell and Johnson 1977). Nest initiation will therefore occur when not too close to a healthy established colony. More than one colony may persist in the same tree, but I have seen this only rarely.

Analogous nesting behavior may occur in the honeybees. Rinderer et al. (1981) reported fighting between Africanized honeybee workers that were investigating potential nest cavities. Honeybees also use pheromones to mark these sites and attract nest mates (Gary 1974; Michener 1974a; Free et al. 1984). Moreover, when given a choice, swarms of *Apis mellifera* are likely to select a nest site among the most distant of those encountered by scouts, and apparently stable racial variations in such nesting behavior exist (Lindauer 1967; Gould 1982).

The degree of spatial separation of honeybee nests has been analyzed by Gould (1982). He compared *A. m. carnica* to *A. m. ligustica* and found that the first race, from a colder climate, preferred to nest over four times as far from the parent nest as the Italian honeybee, *ligustica*. Gould also compared these two races to the Egyptian honeybee studied by Boch (1957) and showed that variation in dancing tempo among these three races reveals a trend. Bees from progressively warmer regions forage closer to the nest. Slower dance tempo for a given distance was interpreted to mean less likely communication of distant resources. An alternative interpretation is that a slower dance tempo provides more precise communication of foraging localities relatively close to the nest. It was shown in Chapter 2 that Africanized honeybees foraging in tropical forest fly as far from their nest as do foraging European honeybees in temperate forest. Gould (1982) suggested that preferred dispersal distance and the fixed behavioral differences or dialects of the dance language are evolutionary responses to food acquisition. Bees that need to store more honey in the nest, in order to survive the winter, need to forage over greater distances to survive. They may also require larger colonies to procure sufficient food. Whether this proposed explanation has some connection with competition with other colonies for resources, or with reducing competition with the mother colony, is an open question (Section 2.3.3).

The preferred dispersal distance of a honeybee swarm might also be related to breeding system. Specifically, the amount of inbreeding that can be tolerated by a species might influence the dispersal of daughter colonies. As Kerr (1969) has made clear, a haplodiploid population automatically regulates the deleterious recessive genes present in its populations, because the haploid males expressing these genes are eliminated (Section 3.3.5). Although there are few data that help give perspective, the distance between mother and daughter nests relates directly to the probability of inbreeding. The population consequences of frequent confamilial matings are known to be significant in *Apis,* where as many as one-half the offspring produced by queens mated with sons are infertile diploid males that are usually eaten while in the immature stage (Woyke 1963, 1986; Kerr 1972). Such diploid males also are produced following mother–son matings of neotropical *Bombus atratus* (Garófalo 1973; Plowright and Pallett 1979). Multiple matings may be favored in the honeybee so that occasional inbreeding does not greatly reduce colony fitness (Seeley 1985). Rejection of closely related females by males occurs in a eusocial halictid *Lasioglossum* (Smith et al. 1985). Despite various mechanisms that diminish the negative effect of inbreeding, separation in space greatly reduces the risk of mating between close relatives. Nonetheless, foundation of new nests by honeybees, as well as by stingless bees, often occurs within a few hundred meters of a parent nest (Lindauer 1955; Moure, Nogueira-Neto, and Kerr 1958; Wille and Orozco 1975; Seeley and Morse 1976; Inoue, Sakagami, Salmah, and Yamane 1984; Starr and Sakagami 1987).

A nest of permanently social bees near an established conspecific nest may benefit from increased protection, or it may have better access to resources. Only a few species of highly social bees regularly nest in aggregations, in which a large number of colonies are separated by only a few meters. Large and populous exposed nests of giant honeybees are usually built high off the ground and sometimes in densities of several nests and occasionally several dozen per tree (Singh 1962; Morse and Laigo 1969; Seeley et al. 1982; Roubik et al. 1985b). Some stingless bees of the *Partamona* group make exposed nests in aggregations of similar numbers on cliff sides, large tree trunks covered by the nest of termites, and on other hard substrates (Roubik 1983b; Camargo pers. commun.). Like the large species of *Apis* that naturally aggregate nests (Seeley et al. 1982), colonies of *Partamona* do not respond defensively to the destruction of nearby colonies. Molestation of the nest does lead to an impressive biting attack on the intruder (Roubik 1983b). Similarly, several nests of *Partamona* and *Scaptotrigona* are often built in large terrestrial termite mounds or, for the latter, in tree cavities in the neotropics (Camargo 1970, 1980; Roubik 1983b). Hundreds of workers issue from the nest to bite intruding animals. A small number of stingless bee species nest exclusively in the nests of arboreal and terrestrial termites or in abandoned terrestrial termitaria in the African tropics (Darchen 1972b). Some also build their nests in the nests of biting and stinging ants (Kerr et al. 1967; also see Section 3.1.4). Many of these colonies, however, are made singly and do not lead to aggregations.

Availability of a suitable nest substrate probably accounts for clumping of the southeast Asian *Tetragonula collina*, which nests in aggregations of from a few to a dozen colonies below the ground, among the dense roots of *Ficus* (Sakagami and Yoshikawa 1961). I have seen 16 nests of these and two other trigonine groups built in the trunk and among the roots of a single *Ficus* tree in Thailand. The bees are unaggressive when their colonies are disturbed. Aggregations of stingless bee nests in bamboo houses, under roofs, or in stone and adobe walls are known in many tropical regions (Michener 1946, 1974a; Schwarz 1948; Moure et al. 1958; Darchen 1972; Wille and Michener 1973; Sakagami et al. 1983b; Sakagami, Yamane, and Hambali 1983; Starr and Sakagami 1987). These species are best regarded as "weedy" and rapidly proliferate in the disturbed areas where nest sites are available and are not too distant from some forest and natural colonies. Loose colony aggregations of this type, involving species that do not display aggressive attack of intruders, suggests their nesting habit has nothing to do with defense.

Single nests of some highly eusocial bees are built in abandoned nests of birds or subterranean attine ants (Nogueira-Neto and Sakagami 1966; Imperatriz-Fonseca et al. 1972; Wille and Michener 1973; Michener 1974a; Fowler 1979). Africanized honeybees in the neotropics frequently nest in the excavations made in termite nests by nesting birds, such as parrots and trogons (Michener 1975; Roubik 1979a, 1983b). In equatorial Africa, nests of *Axestotrigona oyani* are made in cavities excavated by pangolins in nests of *Crematogaster* ants (Darchen 1972a), as are the nests of *Trigona cilipes* in the neotropics (Michener 1974a). Nests of some *Paratrigona* are built in the leaf nests of *Camponotus* (Wille and Michener 1973). In Malesia, the nests of *Trigonella moorei* are always built in arboreal nests of the highly aggressive ant genus *Crematogaster* (Sakagami, Inoue, Yamane, and Salmah 1983a,b, and in press; see also Fig. 3.14). These authors made the first observation of how such nests are initiated with resin.

The long-term distribution of social bee nests should be influenced by the robbing tendencies of bees such as *Apis* that frequently raid the nests of weaker congeneric colonies (Sakagami 1959; Michener 1975). *Lestrimelitta*, an obligate cleptobiotic genus of stingless bee distributed throughout the neotropics, may regulate

3. Nesting and reproductive biology   197

Figure 3.14. Nest entrance and host ant nest of the Southeast Asian stingless bee *Trigonella moorei* (see Sakagami, Yamane, and Hambali 1983). The right side shows the emergence of stinging *Crematogaster* from the ant nest some 20 sec after a slight disturbance. The entrance to the bee nest is the small hole near the middle of the ant nest. Photo by author.

its own numbers effectively by intraspecific robbing (Sakagami, Roubik, and Zucchi in press). In experimental settings I have seen this lead rapidly to the death of an introduced colony. The degree to which other highly eusocial bees practice intraspecific robbing is unknown but appears slight (Nogueira-Neto 1970a). The Kayapó Indians of the southern Amazon basin report that *Scaptotrigona* may raid other colonies for food (Posey and Camargo 1985). However, these include semi-domesticated nests that are opened and then repatched in their original tree cavity. This might produce the conditions for an opportunistic foraging raid. If extensive intraspecific robbing in natural circumstances is discovered for a particular species, this behavior would certainly produce additional explanations for nest distribution.

d. Nest site preferences of stingless bees. Colony selection of a suitable cavity for nest construction in trees seems very dependent on the local nest predators. However, bees of long-lived colonies nesting in tree hollows may not be able to differentiate between hardwood nest sites in relatively long-lived trees and those in more ephemeral secondary vegetation. Murillo (1984) noted that most nests of stingless bees in a relatively forested area in southern Mexico were found in second-growth and introduced tree species rather than within large native forest trees. The implication is that the bees are opportunistic in selecting a nest site and use whatever tree species presents a cavity of the correct dimensions (see also Hubbell and Johnson 1977; Fowler 1979). Other data, however incomplete, indicate some degree of nesting site selectivity. Kerr (1984b) states that only 12 of the over 200 local tree species at one site in Mato Grosso, Brazil, were used for nesting by stingless bee colonies. Hubbell and Johnson (1977) and Johnson and Hubbell (1984) show that the stingless bee colonies in Costa Rican dry forest were found in 20–35% of the local large tree species, and in Panama, 15 species of trees contained 22 stingless bee nests, also primarily in larger trees. All three studies were made in areas having 16–50 local stingless bee species, and none succeeded in locating more than several of them. This fact may mean that many nests were missed or that the nest dispersion patterns and nest sites of many species are aggregated or random.

The age of bee colonies found in trees is difficult to determine and is seldom documented, but Murillo (1984) found that among 35 semidomesticated colonies of *Melipona beecheii* in southern Mexico, the oldest colony was 61 years old, and the average colony age was 10–19 years. All of these nests had been harvested by local residents and then kept in protected areas. Thus colonies were abnormal in at least two ways. First, the nesting cavity was no longer part of a standing tree, and second, attack by predators and parasites was prevented.

With very few exceptions, cavity-nesting stingless bees only build nests in the cavities left after the core of living tree has rotted from rainwater that enters through the scar left by a branch (secondary branches fall from a tree during growth). The inside of tree cavities is often wet; water sometimes gushes from them when cut to reach a bee colony. The bees do not enlarge the cavities or make the hole necessary to enter them; they are entirely dependent on their natural occurrence. Why only living trees? One reason is that colonies of stingless bees rarely find and establish new nesting sites, and they have a complicated reproductive behavior (Section 3.1.4). A dead tree in tropical forest will be devoured by termites and decompose within a few years (Lieberman et al. 1985). Moreover, as termite colonies begin to consume the tree, they enter and kill the smaller colonies of eusocial bees. Dead trees are nonetheless used frequently by some neotropical meliponines with relatively small colonies, *Frieseomelitta*, *Duckeola*, and *Trichotrigona* (Camargo and Moure 1983). Such colonies are very abundant in a periodically flooded region of the Amazon, and the trees they nest in, such as *Dipteryx*, *Tabebuia*, and *Coumarouna*, are of hard woods that last for many years even when dead. Here many of the standing trees are dead, and termite pressure is presumably lessened. In other habitats, dead trees offer little resistance to the nest predators of stingless bees – predators that can claw or gnaw through several centimeters of wood to reach and destroy the nest. I have seen this happen to dozens of nesting stingless bee colonies kept in the forest, placed there in sections of tree trunks they originally occupied. With a breach in the tree wall and disturbance of the nest, small parasites such as flies rapidly invade. Their larvae completely consume the brood and stored food of neotropical stingless bees in a few days (Section 3.2.6).

Stingless bees of greatly differing worker and colony size build nests in tree cavities of roughly the same dimensions. Cavity-nesting meliponines can significantly reduce the size of a tree cavity by blocking part of it. Cavities tend to be cylindrical in natural tree hollows, having an inside diameter of 3–45 cm. Cavities average about 9 cm in diameter and 8 l in volume for tree-nesting stingless bees (Roubik 1979a, 1983b) and have a larger diameter, with a volume of 50 l, for temperate *A. mellifera* (Seeley and Morse 1976; Seeley 1985).

One factor apparently leads to discrimination against cavities by many meliponine species, and it is related to potential exposure to colony predators. Although the internal dimension of the cavity overlaps among species, the size of the natural opening leading to it varies markedly. A cavity connected to a large hole leading to the external environment is shunned by all but the species able to make thick, hard walls (Roubik 1983b). These *batumen* or wall-like plates (Nogueira-Neto 1970b) are built by *Melipona* and neotropical *Cephalotrigona* to plug all but the single, small hole leading to the nest cavity. The patched fissure is large enough to be entered by vertebrate colony predators or their muzzle and claws; it is frequently 10 cm across. These bees exhibit the remarkable behavior of working outside the nest to patch openings leading to the tree cavity. Bees carrying resin arrive outside the nest and then deposit their resin loads on irregularities on the surface of the nest substrate, effectively patching or augmenting an existing partition made from within the nest. The result of this behavior is easily appreciated on wooden hives made for these bees. Thick ridges of resin are deposited on cracks and the edges of the nest boxes, as well as between sections of the hive, along with small pebbles and mud. Regurgitated substances or mandibular gland secretions are added (Roubik pers. obs. of *Melipona compressipes*). Such behavior makes cavities unused by other stingless bees suitable to these neotropical meliponines.

3.1.4 Nest construction

The richness of bee nest architecture and construction far exceeds those prominent traits introduced in the first section of this chapter. A few additional examples, largely concerning tropical species, are given here. Highly complicated wax nest structures are made by the apine bees that build double-sided combs of hexagonal cells (Darchen 1962; Michener 1974a; Seeley 1985; Hepburn 1986). Their vertical combs have cells that slant slightly upward, and each cell is placed with its middle directly opposite the intersection of three cell corners, providing maximum comb rigidity. All other apids make round cells, spherical in some of the Trigonini, and elongate to egg-shaped in the remaining meliponines, bombines, and euglossines. Only one wax gland exists in the female Euglossini, placed at the dorsal extremity of the abdomen; dorsal and ventral wax glands are found in Bombini. Wax glands are all dorsal in the worker Meliponinae and all ventral in the Apinae (Cruz-Landim 1963; Michener 1974a; see also Fig. 3.15). Thus euglossines and to a lesser extent the bombines construct nests with little wax. Comb-building meliponines initiate a comb with a single cell built atop the middle of an existing horizontal comb and then build around it. During this process, the initially round cells are compressed at the sides, resulting in hexagonal cells. Combs of some meliponines are also made vertically (African *Dactylurina*), or the cells are scattered in small groups or are spaced regularly but not contiguously in clusters. Their combs or cells are supported and connected by pillars of cerumen (summary by Michener 1974a; Sakagami 1982; Wille 1983a). The brood area of meliponines usually does not change. It is delimited by a surrounding layer of involucrum, as well as storage pots. The bees continuously add new cells in the space left by the emergence of adults. Thus new cells are constructed essentially in waves passing through the brood area.

In addition to the obvious limitation imposed by availability of nest sites, the materials used to build a bee's nest often come from the external environment and subsequently determine reproductive success. If a specific nesting material is unavailable, most bees cannot reproduce. Exceptions occur among bees that excavate burrows into pith, wood, and the culms of large grasses like bamboo. In such instances, the finishing materials for constructing the entrance or closing a cell come from the nest substrate (Fig. 3.3) or are produced by the bees. A small amount of resin is gathered by some honeybee species, but it may not be an essential architectural component of the nest. Their nests are made solely of wax that the bees produce from honey and pollen. Secreted cell linings of bees also require pollen as a starting material, to be metabolized into the glandular product applied to cell walls (Batra 1980a; McGinley 1980; Cane 1983b; Torchio 1984). As outlined by Batra, the materials are transferred to the mouthparts from Dufour's glands located in the abdomen (Fig. 3.15) and are later applied to cell and burrow walls. The Dufour's gland size and shape vary greatly among bee taxa. Its secretion has several possible functions and produces a waxy material whose main purpose is moisture balance (Cane 1981; Hefetz 1987). Witnessing only the application of the material with the mouthparts had misled some authors, who suggested it came from

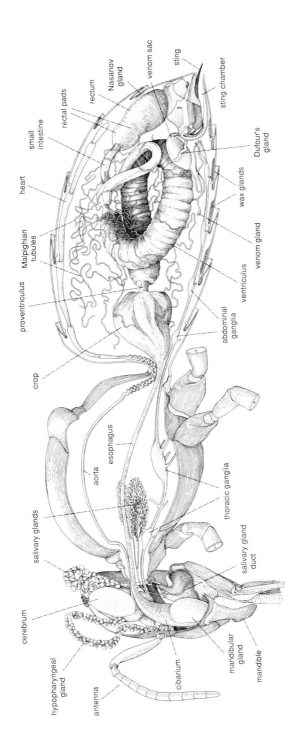

Figure 3.15. Internal anatomy of a bee showing the locations of exocrine gland systems. A worker *Apis mellifera* is depicted (Camargo 1972). Original drawing provided by J. M. F. Camargo.

glands in the head or thorax. Most other nonparasitic bees must have access to a raw material that is incorporated directly into the nest.

Resins are gathered extensively by certain megachilid, euglossine, and meliponine bees. Although also applied in the nest to a lesser degree by *Apis*, resin is used in all major nest structures by meliponines, euglossines and some megachilids (Chapter 2). Various megachilid bees are leaf cutters and cut at least a few leaf sections from living leaves to make part of each cell (Iwata 1976, p. 391; Batra 1984; see also Fig. 3.5). Leaf sections of a uniform ovoid or round shape and size are attached with an adhesive made by the bees; round leaf sections form cell partitions or the closure of a completed cell. Either leaf material chewed to a pulp or mud constitutes the primary nesting material for other megachilid bees. A few meliponines like *Trigona nigerrima* and *Trigona dallatorreana* collect rotten wood to build a "paper" or vegetable fiber and resin nest, and others like *Melipona fuliginosa* make their nest entrance exclusively from small seeds and the resin produced by the fruit of *Vismia* (Clusiaceae) (Roubik 1983b). Fecal material, loose dirt, and small pebbles are collected by some anthophorids, megachilids, euglossines, and meliponines and serve to fill spaces or to make nest or cell walls. The anthidiine Megachilidae use plant hairs and pappi in nests, and even soft animal hairs (Michener 1968a; see also Fig. 3.16). Finally, some anthophorids and melittids use the secreted oils of flowers, both for making part of their cells and for sealing the nest when a series of cells is completed (Cane et al. 1983; Roubik and Buchmann pers. obs.).

Bees measure potential nesting cavities as well as the size of cells. A worker *Apis mellifera,* assessing a cavity in which its colony may nest, can discriminate cavity volumes differing by 15 liters by integrating the information on the distances and directions in which it moves while walking throughout the cavity (Seeley 1977). Queens of *Apis* (excluding the *A. dorsata* group) determine whether to oviposit a fertilized egg or a drone egg by directly measuring cell size and detecting the larger opening of drone cells (Koeniger 1970). Similar abilities are known in solitary bees (Raw and O'Toole 1979). Frohlich (1983) described the processes by which a female solitary bee *Osmia* (Megachilidae) measured the size of the cell being constructed and the placement of partitions between linear series of cells. The bee used its mandibles to deposit a small clump of leaf pulp where the cell partition would be built after cell provisioning. This was done by touching the base of the cell with the head and then curling so that the head met the tip of the abdomen maintained in the position to which it extended when the cell base was contacted with the head. Reference points can thus be made during the process of cell construction by solitary bees to ensure proper cell dimensions. In this manner, the bee builds part of the cell base and cell cap before depositing the food and egg (Fabre, quoted in Malyshev 1935).

3. Nesting and reproductive biology   203

Figure 3.16. The nest of African *Serapista denticulata* (Megachilidae) from Cameroon, constructed with plant hairs or trichomes (Michener 1968). Mature larvae are shown in the cutaway photograph at the right. Original photographs provided by C. D. Michener.

How much time does a bee invest in nest building, cell construction, and provisioning? Much of the information on temperate and some tropical species indicates that one or a very few cells are completed by a solitary female bee in a single day, weather and floral resources permitting (e.g., Kapil and Jain 1980; Cane 1983b). Availability of flowers and nest material influence this reproductive tempo. Bees that dig extensive burrows in soil or wood, as well as those requiring large amounts of resin to complete the nest, complete cells slowly. About as much time is invested in preparing the nest site as gathering food. For example, by working most of the day and night, a nesting *Xylocopa* extends a burrow in hard wood by about 4 cm each 24 hours (Anzenberger 1977). The burrow must be completed before cells are initiated, because they are made from the end of the burrow toward the entrance. Average burrow length among four species was about 7 cm; thus nearly two days of continuous work was required to make the central burrow. Construction and provisioning of a cell required one to two days during peak flowering, and most burrows led to two completed cells. Thus about five days were required to produce two offspring. A similar observation of two to four days to complete cell provisioning during 20 trips was reported for *Xylocopa* in India (Kapil and Dhaliwal 1969). My observation of small *Centris analis* in Panama indicates they require from one to two days to complete and provision a cell in a preexisting cavity. They gather dirt, pollen, nectar, and plant oil. Roughly one-half of the time they spend in this activity involves working with the nest material.

Extensive use of resin in cell construction slows the production of offspring. Dodson (1966) established that the relatively large *Euglossa ignita* in the lowland forest of Iquitos, Peru, completed an average of about 12 cells in 45 days, within an established nesting cavity already containing some of the resin needed for cell construction. Here three or four days were usually required to produce one offspring. Study of a nest built by a single female of *Euglossa turbinifex* in Costa Rica suggested that five days were required to construct and provision a cell (Young 1985). Solitary *Euglossa townsendii* nesting in my observation hives in Panama require two to four days to complete a brood cell. The extensive study of Garófalo (1985) with observation hives of *Euglossa cordata* on a college campus in Brazil revealed production of a batch of 6–10 progeny by a female, followed by a pause until the adults emerged between 46 and 83 days later. A dominant female in one of these eusocial or semisocial colonies participated in up to two reactivations with the emergence of new adults, and lived from 39 to 191 days, while her total reproductive output was 2–50 bees. Small euglossine colonies, or those of one active female, thus may display a reproductive tempo of construction and provisioning, followed by a relatively long inactive period.

By comparison, larger semisocial euglossine colonies such as those of *Euglossa imperialis* sometimes complete a cell each day (Roberts and Dodson 1967). Highly eusocial species (*Melipona* and Trigonini), which also construct a new cell for each egg, produce from a few dozen to several hundred offspring in a day (Roubik 1979a, 1982a; Sakagami 1982; Sakagami, Inoue, Yamane, and Salmah 1983b; Sakagami, Yamane, and Inoue 1983; Sommeijer, Houtekamer, and Bos 1984; Sommeijer, van Zeijl, and Dohmen 1984). Their work involves much recycling of cerumen within the nest, with some additional foraging required for resin as well as for food. Honeybee colonies, with their permanent wax brood cells, can produce over a thousand eggs per day (Seeley 1985). Of course, these cells are not provisioned prior to egg laying. Honeybee net reproductive efficiency probably resembles that of stingless bee colonies, but compilation of total time and energy budgets to compare such bees will be a significant research accomplishment.

Meliponine colony reproduction and nest building apparently require from a few days to over three years for completion (Nogueira-Neto 1954, 1970b; Wille and Orozco 1975; Darchen 1977a; Inoue et al. 1984; Camargo pers. commun.). Colony reproduction by *Apis dorsata* may also be performed gradually (Ruttner 1988, p. 117). As indicated, stingless bees are unable to reproduce before they find a new nest site. The process is initiated by a group of workers that searches for a site, and about one-third of the older workers of the mother colony eventually switch to the daughter nest (Inoue et al. 1984). The changes in colony behavior preceding this event have not been analyzed. Once the nest site is chosen, workers begin to make a nest entrance, but even prior to its completion the inside of the nest is prepared. Debris is cleared, or soil and termite or ant galleries are removed. Some of the ground-nesting species make tunnels in excess of a meter to the nesting cavity, which can require many months. Tree-nesting species may block the nest cavity at the extremities that will encompass the new nest, also a time-consuming process. A resin deposit is always made near the entrance, and a few species line the entire cavity with resin.

Next, honey is brought from the mother meliponine nest and stored in containers made from cerumen also from the parent colony (Nogueira-Neto 1954). Several honey storage pots and a few pollen pots are provisioned. Pollen is usually freshly collected. The ability to pack pollen in the corbiculae that has previously been stored in a nest is an extremely rare trait in meliponines, apparently confined to bees such as *Partamona* that have a large corbicula (Kerr 1959; Roubik pers. obs.). After the food provisions are in place, a small group of workers departs from the mother colony with a virgin queen. She then mates and loses her flight capability, and about half the accompanying workers remain at the new nest site. The method by which a new nest site is communicated to nest mates that then become the workers of the new colony is probably chemical. Thus many of the species appearing to lack chemical foraging trails may at least use their characteristic mandibular gland pheromones (Table 2.4) or possibly other glands to guide nest mates to a new site.

When Wille and Orozco (1975) studied colony establishment by *Partamona* in an urban setting in Costa Rica, workers continued to transport food and materials between the mother and daughter nest for six months. By far the longest continuing preparation of an incipient bee nest was noted by Camargo (pers. commun.), who witnessed nest preparation by an Amazonian species of *Ptilotrigona* in an extremely hard sand substrate at the bank of a forest river. The nest was in construction for over three years. In contrast, the widespread neotropical *Tetragonisca angustula* requires only two weeks from nest site selection by scouts to the first oviposition of the new queen (Roubik pers. obs. in second-growth forest in Panama and Colombia). This is certainly not a fixed trait, because Nogueira-Neto (1954) recorded an interval of a half-year for this species to complete swarming in a disturbed habitat in southern Brazil. Rapid nest preparation also occurs in *Frieseomelitta* in southern Brazil, which completed a nest in the remarkably short period of three days in one instance (Terada 1972). The Southeast Asia species *Tetragonula laeviceps* and its races complete the swarming process in two to three weeks (Inoue, Sakagami, Salmah, and Yamane 1984; Starr and Sakagami 1987).

3.2 Natural enemies, associates, and defense

Bees ranging from soil-dwelling solitary colletids to permanently eusocial apids close their nest entrances at night to prevent entry by natural enemies. Even following a short disturbance, certain meliponine bees such as *Scaura* pull together the sides of their soft nest entrance tube, blocking the direct entry of an enemy and apparently relying further on chemical repellence. However, many bees never completely close their nests, and some partly close the nest even when leaving for a short time (Malyshev 1935). Chemical and behavioral defenses show much diversity but are not readily correlated to physical protection of the nest, or the lack of it. Trends in nest defense behavior have been the subject of few comparative studies, and nearly all (except Cane 1986) have considered only eusocial bee colonies (Kerr and Lello 1962; Kerr et al. 1967; Nogueira-Neto 1970b; Michener 1974a; Collins et al. 1982; Seeley et al. 1982; Nates and Cepeda 1983; Roubik 1983a; Wu and Kuang 1986; Ruttner 1988). Solitary and eusocial bees are broadly similar in their individual defensive equipment. Yet its employment and organization are markedly developed in species that maintain long-lived or relatively large colonies.

3.2.1 *General considerations*

Bees defend their nests with the mandibles or the sting, at times pursuing an intruder outside of the nest and grappling on the ground. Loud buzzing accompanies defensive attack, within burrows and outside the nest. As could be expected from the lack of a sting and weak mandibles, males do not often participate in this activity. Some male xylocopines apparently block the nest and do respond by biting if potentially harmful ants attempt to enter (Linsley 1966; Camargo and Velthuis 1979; Scholz and Wittmann in press). Among female bees, the persistence of a threat causes them to block the nest entrance completely, either with the head or the abdomen, or with hard or sticky resin (Michener 1974a; Eickwort 1975). In relatively inactive periods, females remain near the entrance, often with their abdomen and sting facing outward. Defensive secretions used by some bees are produced within mandibular or abdominal glands. Liquid feces also are ejected with force to defend the nest of certain bees (Janzen 1966).

Apid bees display varied additional nest defense mechanisms, which include covering an intruder with honey (*Bombus* and some Trigonini) or resin (trigonines). A list of 17 stingless bee nest defense tactics presented by Kerr and Lello

(1962), Michener (1974a), and Nates and Cepeda (1983) includes brief descriptions of such behaviors. These defensive acts serve to immobilize or deter both large and small predators and some parasites. Honeybees have at least one dissimilar defensive behavior, that of harassing or flying rapidly at an intruder without biting or stinging (Free 1961; Collins 1987). After disturbance of the nest, workers of *Apis cerana* and *A. florea* simultaneously produce a snakelike hissing sound, by shivering the wings, that evokes rapid escape behavior in a marauding bear or smaller predators (Fuchs and Koeninger 1974; Ruttner 1988). Snakes are considered to be the most formidable natural enemies of bears in tropical and subtropical areas. Bees guarding nests of *Apis cerana* and *A. dorsata* have a distinctive shimmering behavior when approached by large vespid wasp predators. While facing a hovering wasp, guarding bees rapidly jerk the raised wings several times, doing so in staggered succession, which produces a wavelike effect. For completely unknown reasons, these movements discourage the wasp from entering a honeybee nest (Seeley et al. 1982). However, Ono et al. (1987) have documented the effective thermal defense applied by *Apis cerana* against *Vespa;* the bees literally cook the wasp by covering it with their bodies and raising its temperature to a lethal level. Fanning of the wings is a behavior that could indicate elevation of body temperature in preparation for this defensive tactic. This behavior is absent in *Apis mellifera* which has no evolutionary history with large Asian *Vespa*. A defensive behavior has been seen in colonies of *A. mellifera syriaca* that regularly experience the attack of Middle Eastern *Vespa* (Ishay et al. 1967; Butler 1974, in DeJong 1978). The workers quickly cease foraging activity if wasps attack nest mates near the nest (Ruttner 1988, p. 188). Scouts of *Vespa* colonies that arrive to assess potential for a colony raid are attacked en masse by *A. cerana* if they approach the nest too closely. Introduced European *A. mellifera* has no such behavior, and its colonies are readily killed by *Vespa* (Matsuura and Sakagami 1973). This is one reason that *Apis mellifera* does not maintain feral populations in most of Asia.

The chemical arsenal of the apids is similarly developed beyond that known in other bees, and this implies that important selective pressures have been posed by larger animals. Schmidt, Blum, and Overal (1980) compare toxicity of venoms taken from several aculeate Hymenoptera. Of the three bee species they survey, venom of *Apis mellifera* was twice as toxic as that of *Bombus,* which in turn was three times more toxic than venom of *Xylocopa*. Schmidt (1982) discusses the uniqueness of peptides in the venom of *Apis*. Melettin is a hemolytic peptide that constitutes approximately half of the venom; apamin is considerably more lethal and is the smallest known neurotoxic peptide molecule.

A very unusual defense employed by the several neotropical species of *Oxytrigona* is the production of caustic mandibular secretions containing formic acid

(Bian et al. 1984; Roubik, Smith, and Carlson 1987; Table 2.4). The secretion, which has an unpleasant smell, induces a burning sensation when placed on human skin – but only if it is placed on an exposed membranous area, in a wound, or on the skin when pores are opened during sweating (Nogueira-Neto 1970b). Blisters caused by the secretion do not appear until a day after it has been deposited (these bees do not bite with sufficient force to cause open wounds in humans). However, the biting behavior of defending bees and unique mandibular gland components of *Oxytrigona,* including novel diketones (Bian et al. 1984), may enhance penetration of the acid solution (Roubik et al. 1987).

Individual colonies of apids show variation in their defensive behavior that can obscure the potential aggressive behavior of a species. Thus the very small *Plebeia jatiformis* was not found to display aggressive nest defensive behavior by Roubik (1983a), but some of the colonies used in this study later revealed intense biting and defensive recruitment after a disturbance. Similarly, Nates and Cepeda (1983) describe *Melipona fuliginosa* as a "very gentle bee," although this species normally attacks any large object moving by the nest and is very persistent in harassing a potential natural enemy (Roubik 1981a, 1983a).

Notwithstanding the aggressive techniques used to repel predators, the most common protection for nesting bees is perhaps concealment of the nest and partial blockage or restriction of the nest entrance (Fig. 3.17). One survey of stingless bees showed that half of the 40 species studied in natural forest habitat displayed absolutely no aggressive nest defense behavior when the nest was molested (Roubik 1983a). Although crypsis prevents detection by large predators, and coordinated attacks may drive them away when a nest is molested, smaller parasitic and predatory invertebrates regularly locate and enter the nests of bees or arrive there by attaching to bees at foraging sites or outside the nest (Section 3.2.6).

3.2.2   Guard bees and kin recognition

Effective nest guarding and social behavior share a common mechanism: the perception of genetically controlled odor traits, called discriminators (Hölldobler and Michener 1980; reviews in Fletcher and Michener 1987), as well as that of smells from the environment of colonies or individuals. Pheromones perhaps similar to those allowing individual recognition have significance in mate selection, discussed in Section 3.3. While at a nest entrance, bees assess the odors of other bees by slight touching with the antennae or near-approach with the antennae extended. A guard bee at the nest entrance will not let another individual pass if it lacks the correct smell. This subtle cue in individual recognition is of paramount importance in preventing nest usurpation by bees of the same species. The bees of small colonies can even remember the odors of individual nest mates. Within large

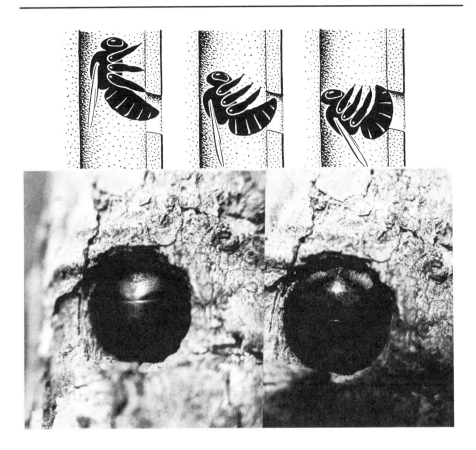

Figure 3.17. Defense of the nest by a female bee, *Xylocopa* (Anthophoridae), showing the stages of entrance blocking by the sting and dorsum of the abdomen (partly from Anzenberger 1977). Original drawings and photographs provided by Gustl Anzenberger.

colonies of *Apis mellifera* there are discriminating odors of differing genetic origin among the workers (Michener 1982b; Breed 1983; summary by Gadagkar 1985).

Knowledge of the role of odors in guard behavior has been provided in detail through studies with a primitively social halictid bee, *Lasioglossum zephyrum*, in laboratory colonies (Barrows 1975; Kukuk et al. 1977; Greenberg 1979; Buckle and Greenberg 1981; Michener 1982; Smith 1983; summary in Fletcher and Michener 1987). This species lives in groups of morphologically identical workers and queens, numbering from 2 to 24, in an association of a mother and daughters that lasts for a single breeding season. The bee odors are genetically controlled and recognizable by the bees at the level of the individual. Responses to such odors are learned from the bee's nest mates. Thus artificial colonies composed of five sisters from an inbred line and one unrelated nest mate showed that when the unrelated bee guarded the nest, it would generally not allow entrance of its own sisters from another inbred line. When one of the five sisters guarded, it allowed the unrelated nest mate to pass as well as her sisters (Greenberg 1979; Buckle and Greenberg 1981). In large colonies of *Apis, Bombus,* and meliponines, it has long been thought that the mixing of individual odors, in addition to the nest environment, produce in the guarding bees a general template of nest mate odor (Free 1987), which provides the basis for discrimination.

Responses of guard bees to unrelated individuals involve more than merely preventing entrance with the head or body; attack of an intruder can be elicited by additional physical, behavioral, and chemical features of another species. A full defensive reaction of a guarding bee at the nest entrance is sometimes the simultaneous opening of the mandibles and exposure of the sting as the bee curls its body in a C position, resting on the top of the abdomen. Sakagami (1976) summarizes the work of earlier authors and reports that female *Bombus*, when repeatedly harassed in the nest, assumes a defensive posture by flipping upon its dorsum and exposing the sting while extending the legs.

Relatively small colony predators such as *Azteca* ants and phorid flies evoke an immediate biting response and exit from the nest by stingless bees (and some halictids, Wcislo pers. commun.); some preliminary evidence suggests that this is in part an instinctive aggressive response toward small insects. The guards of *Melipona fasciata* respond by biting if approached by a single ant or phorid fly. Although this bee displays no aggressive foraging behavior at flowers, while at artificial feeders or gathering resin on tree trunks (Howard 1985) its reponse to very small stingless bees is immediate biting. Repeated attack by ants or bees at a nest evokes complete blockage of the nest by this species. This and other *Melipona* maintain several hard resin balls on the floor of the nest, which are somehow carried up to the inside of the nest entrance hole and fixed in place (Roubik 1983a; Portugal-Araújo, quoted in Kerr 1984b). Similarly, defending females of primitively eusocial and solitary Halictidae, Megachilidae, and Allodapini bite and sting. Repeated attack leads to blockage of the nest with the upper surface of the abdomen (Sakagami and Laroca 1971; Michener 1974a; Eickwort 1975; Iwata 1976; Anzenbeger 1977; Sakagami and Maeta 1984; see also Fig. 3.17).

Guard bees in small colonies tend to be the larger colony members and sometimes the reproductively active females. However, Houston (1970) reported extreme cephalization in some males of an Australian halictid bee; mate competition is intense because unlike related halictids, the bees mate within the nest (Kukuk and Schwarz 1987), which appears to favor large males (Section 3.3). The large males do not necessarily function as guards from natural enemies (Knerer and Schwarz 1976). Graded cephalization and associated size gradation among females in a single nest has been noted in several tropical halictid bees (Sakagami and Moure 1965). The largest individuals of eusocial halictids are the queens, and in a few species such colonies contain two morphs – queen and worker – with no intermediates (Sakagami and Fukushima 1961; Brooks and Roubik 1983; Packer 1985). Presumably, extreme cephalization with the absence of intermediate sizes is rare because most bees constrict their nest entrance to a size that can be filled by the head of a single, ordinary worker; guards having larger heads would be unnecessary (Michener 1974a; Maeta et al. 1985).

The small colonies occasionally formed by adult *Ceratina* or made by forcing individuals to live together in artificial nests, studied by Sakagami and Maeta (1984), display a division of labor. The larger female is the reproductive individual, guards the nest entrance, is less likely to forage, and may solicit regurgitated nectar from smaller returning foragers before allowing them to enter the nest. However, smaller individuals can guard in this genus and solicit food from the larger bees (Michener 1985). Single females or pairs of *Ceratina laeta* were observed nesting in French Guiana (Michener 1985); in one colony the smaller individual showed tattered wings and other signs of being a forager, although her role was not clearly established. Presumed food transfer by trophallaxis between allodapine bees of the same size has been reported in nests of Australian *Exoneura* and African *Braunsapis* (Michener 1972b). Because food transfer is lacking in eusocial halictid bees but likely occurs in a communal halictid *Chilalictus erythrurum* (Kukuk and Schwarz 1987), it may have no special importance in the evolution of interaction between guards and bees entering nests in that bee family. Females of *Chilalictus* tolerate non–nest mates, but macrocephalic males might guard the nests. In addition, some *Xylocopa* seem to organize nest defense by having the youngest bee remain in the nest while the reproductively active females forage (Barrows 1983; Gerling et al. 1983).

Feeding of guard bees by returning foragers probably played a significant part in formation of a natural mixed colony formed facultatively by two *Melipona* (Roubik 1981b). The larger species guarded the nest entrance and was the bee that built the nest; returning foragers of the smaller coexisting species were approached by the guards and fed nectar to them before passing. Within the nest, queens of *M. fuliginosa* and *M. fasciata* had oviposited in brood cells made by their respective workers, separated by only a few centimeters. Artificial mixed colonies are readily made by introducing brood from other species to an established nest (Nogueira-Neto 1950; Michener 1974a). Moreover, occasional drifting of returning foragers to nearby nests of other species does happen, as I have observed a few large foragers of *M. compressipes* become incorporated in the colony of smaller *M. favosa*. Mixed colonies of *Bombus* are common, often involving species of the same subgenera (Sakagami 1976). However, eventual nest usurpation is the result of these associations. They may arise through the feeding of resident bees by another bee species, although this is by no means certain. Although feeding between adult *Bombus* does not take place, returning foragers will regurgitate their nectar loads if harassed by bees guarding the nest entrance (Brian 1952; Alford 1975).

The age of a bee influences its ability to effectively guard the nest. In honeybees, as well as in stingless bees, the age of a worker partly determines the type of tasks – for example, nest building, brood provisioning, and foraging – that it is likely to perform (summaries by Michener 1974a; Sakagami 1982; Seeley 1982, 1985). Studies of Bego and of Simões in Brazil showed that stingless bee workers guard the nest toward the end of their life, which occurs before they reach the median age of foragers (Sakagami 1982). It was noted that the relative number of guards increased in very small colonies, which was probably due to lessened foraging activity; Sakagami suggests this may also represent an attempt to augment the defensive capacity of a weak colony. Workers of *Melipona favosa* guard the nest and begin to forage at approximately the same time, which corresponds to an abrupt decline in survival rate (Roubik 1982d; Sommeijer 1984). In colonies of *Apis mellifera,* which have been scrutinized in depth, three periods of behavioral revision occur in the lives of workers (Seeley 1985). Roles are clearly related to the portion of the nest in which bees spend their time and, statistically, to their age. Guard bees are those associated with foraging and the handling of incoming nectar and pollen; their work sites are on the nest periphery and entrance, in which they spend the last portion of their life. Sekiguchi and Sakagami (1966) reported that guard bees of *A. cerana* and *A. mellifera* are normally a very small portion of the colony, only guard for a few days, and occasionally forage and guard on the same day. Breed and Moore (1988) found that individuals of *A. mellifera* guard for up to seven days in a row, and the time interval spent guarding during a day increases slightly with experience. Furthermore, guarding bees occasionally take patrolling flights, possibly in order to detect enemies, *outside* the nest (Breed pers. commun.). When the colony is disturbed by natural enemies, the number of guards increases dramatically. Additional guarding bees are probably recruited from bees near foraging age rather than from the general colony. In agreement with the temporal behavioral analysis of *Apis mellifera* by Seeley, guard bees and foragers of other *Apis* and stingless bees have separate behavioral roles. All indications are that guards are younger than most foragers.

The adaptive significance of guarding behavior that occurs at an advanced age in highly eusocial bees has at least two components. First, since guarding and foraging are the most dangerous activities of a worker, colony survival might be enhanced if the bee has already contributed substantially to the colony before becoming exposed to extensive risks. Second, glandular maturation may be required before guards can produce sufficient repellent secretions against nest predators or liberate the quantity of alarm pheromone necessary for colony defensive recruitment. In addition, as shown below, the foraging pheromones used by some species are not present in young workers. Any bee that has a glandular defensive secretion must be constrained in defensive capacity by the amount available. Glandular development by females probably depends on consumption of pollen – for example in *Xylocopa* (Gerling and Hermann 1978), in which "yellow glands" then develop on the ventrolateral region of abdominal segments; the authors cited mention similar glands in female *Anthophora*.

Glandular secretions repellent to ants have been noted in a variety of solitary and social bees, but their chemical nature has rarely been identified (Cane and Michener 1983; Cane 1986; see also Table 2.4). Dufour's gland secretions including citral and linalool have long been thought to function as repellents of at least small invertebrate predators of bees (summary by Batra 1980a) but are not proven repellents of these or larger nest predators. Allodapine bees produce a brownish secretion, likely of mandibular gland origin and resembling the repellent discharges of tenebrionid and carabid beetles, which might protect the larvae of their subsocial colonies from ants (Michener 1965a). The mandibular glands of *Scaptotrigona postica*, which produce a strong-smelling secretion (Table 2.4), do not develop until the bee reaches foraging age (Cruz-Landim and Ferreira 1968). This group uses alarm pheromone and foraging recruitment trails, neither of which are produced by young bees. Workers of all species in the genus *Apis* appear to possess a similar primary alarm substance, isopentyl acetate, which is absent in young adult bees (Boch and Shearer 1966; Free 1987).

In summary, bees that guard using chemicals seemingly need to forage pollen, or be fed within the nest, in order to ensure an adequate supply of the glandular substances required for effective guarding. An important role of pollen as a nutrient needed for glandular development seems at least likely. Guard bees that repeatedly use glandular secretions may be forced to forage, unless there is considerable social development and feeding. Nonetheless, the social halictids rely on nonchemical nest defense from larger predators, and highly eusocial apids work as guards for apparently only a few days. Delayed production of foraging or defensive chemicals in highly eusocial bees may encourage their permanence within the nest during much of their life. West-Eberhard (1981) makes a case for individual selection as a determinant of temporal caste behavior patterns in eusocial insects. Young workers stay near the brood area for the purpose of contributing directly to colony offspring while potential fertility is high. Later in their lives, with the decline in fertility, they work to defend the colony and brothers or sisters, having lost the potential to contribute personal offspring. In the stingless bee *Melipona favosa*, workers that lay male eggs are younger bees (Sommeijer and Velthuis 1977), thus supporting West-Eberhard's suggestion.

3.2.3 Colony defense, pheromones, and predation

Pheromones are the chemical signals exchanged between members of a given species; these signals in turn release behavior patterns of a predetermined type. In addition, chemicals triggering the prey's defensive behavior – *kairomones* – may be released by predators, in which case the odor is useful to the defending species that perceives it (Wilson 1975; Bell and Cardé 1984). *Allomones* are a third class of odors that elicit predictable responses in other species. The organism producing the odor benefits from its effect, as in the case of repellents or "propaganda substances" (Wilson 1971). In this section the role of odors in the defense of eusocial bee colonies will be discussed, along with the attributes of colonies that increase their chances of surviving predator attack.

Volatile odors that regulate bee defense behavior have different functions on the outside and on the inside of a nest. One instance of a chemical alarm kairomone has been recorded for a small stingless bee, *Tetragonisca angustula,* and another stingless bee, *Lestrimelitta limao,* an obligate robber of stingless bee colonies. *T. angustula* is unaggressive but will, under attack by *L. limao,* release an alarm pheromone (Francke, Wittmann, and Smith pers. commun.). However, *Tetragonisca* is extremely sensitive to the blend of citral and its isomers produced by raiding *Lestrimelitta. T. angustula* is one of few species that always has a group of workers hovering near the nest entrance (Fig. 3.18). Studies in Panama demonstrated that these workers are guards (Wittmann 1985). The hovering bees maintain a fairly rigid alignment on either side of a corridor leading up to the nest; when a live *Lestrimelitta* passes within the corridor, it is attacked from behind on both sides. If the robber bee releases its odor near the nest entrance tube, several dozen additional workers issue from the nest and join the hovering defenders. Dead *Lestrimelitta* and live or dead stingless bees and other insects produce no such reaction. Mandibular gland volatiles of *Lestrimelitta* and possibly the similar African *Cleptotrigona* are lemon odors and serve to coordinate raiding parties by forming odor droplet trails from their nest to that of another stingless bee (Kerr 1951; Portugal-Araújo 1958; Sakagami and Laroca 1963; Blum et al. 1970; Michener 1974a). Many neotropical meliponines have a strong reaction to presentation of live *Lestrimelitta* outside the nest; they rapidly bite the bee and draw it into the nest, or fly from the nest while holding it in the mandibles. A few species such as *Melipona fasciata* (Fig. 3.19) hover in large numbers outside the nest following release of *Lestrimelitta* odor, but these bees also produce an alarm pheromone (Nogueira-Neto 1970a; Smith and Roubik 1983). The odor of *Lestrimelitta* is a kairomone outside the nest and becomes an allomone, debilitating the defense system of another bee colony, after release inside the nest. At termination of a raid, the lemon odor of *L. limao* triggers the mass exit of the robbers and massive return of host

3. Nesting and reproductive biology 215

Figure 3.18. Nest defense by a colony of *Tetragonisca angustula* (= *jaty*). Defending workers are hovering while facing the entrance to their nest (Wittmann 1985). Original drawing provided by D. Wittmann.

workers that remain outside the nest during during the raid (Sakagami, Roubik, and Zucchi in press).

Alarm pheromones have been characterized in the three major groups of highly eusocial bees, *Apis, Melipona,* and Trigonini (Table 2.4). Their synthesis is likely confined to exocrine glands in the head or terminal abdominal regions (Fig. 3.15).

Figure 3.19. Defending workers of *Melipona fasciata* (Apidae) hovering while facing the nest entrance (in a wooden hive) after a disturbance by the robbing stingless bee *Lestrimelitta limao* in lowland Panama. Photo by author.

Such odors can be called alarm attractants because they mark the object being attacked and subsequently guide defending nest mates to it. Production of an alarm attractant by all the hundreds of taxa in these groups is improbable; many stingless bees appear to have no alarm pheromone whatsoever. However, the specific context of pheromone release, or even production by certain workers but not others, may make potential use of alarm pheromone difficult to detect.

Worker stingless bees that produce odors attach them by biting, maintaining the mandibles closed on the target even as the head is torn from the body. In honeybees the odor is kept on the target by anchoring the barbed sting and its attached poison and pheromone sacs. The stinging bee dies in most honeybees, although workers of *Apis cerana* sometimes work the sting free of skin without causing evisceration that would in a short time kill the bee (Sakagami and Akahira 1960). In contrast, stingless bees die only if they are physically damaged by the animal they attack. Defending stingless bees display a notable lack of escape behavior after closing the mandibles on a rival stingless bee. As Schwarz (1948) has written, pairs of bees lie on the ground and are carried away by ants while still alive.

The honeybee *Apis mellifera* bites as it stings; alarm recruitment odors issue from the mandibular glands, but they are only 5% as effective in attracting stinging nest mates as is the acrid sting chamber pheromone, isopentyl acetate (Boch, Shearer, and Petrasovits 1970). The amounts of isopentyl acetate in the four common *Apis* species range from 0.2 μg per worker in *A. florea* to 58 μg in *A. dorsata,* decidedly associated with their respective defensive behavior patterns (Morse et al. 1967). *Apis cerana* defends its colonies primarily by biting (Ruttner 1988). Other honeybee alarm pheromones have been identified as 2-heptanone (produced by the mandibular glands), isobutyl acetate, and 1-pentanol (Collins 1981). *Apis florea* and *A. dorsata* each possess 2-decen-1-yl-acetate in their sting glands, which is absent in other *Apis* (Ruttner 1988). Collins and Blum (1983) found that 15 of 20 known compounds associated with the honeybee sting have demonstrable influence on the alarm response of worker bees. A total of 25 alarm pheromone components, one from the mandibular glands and the remainder from the sting chamber, have been assayed for *Apis mellifera* (Free 1987, p. 138). These compounds, tested individually, can strongly affect worker behavior patterns ranging from stinging to attraction or repellence.

The alarm pheromones of honeybees, like those of stingless bees (see Table 2.4), are a blend of molecules differing in molecular weight and volatility. Consequently, colony defensive reaction is influenced by ambient temperature and humidity affecting the rate at which these substances are dispersed. Within the nest, honeybees exposed to their alarm pheromones respond by fanning their wings; some also raise the abdomen and expose the Nasanov gland (Figs. 3.15 and 3.20). Nasanov glands produce an "alert and assembly" pheromone containing citral, geraniol, nerolic, and geranic acids (Collins 1981; Collins and Blum 1983; Free et al. 1984), and other, as yet unidentified chemicals (Free 1987). An attractant pheromone associated with neither the sting nor mandibles may thus enforce a defensive reaction. The tests of Collins and co-workers showed that the probability, speed, and intensity of worker response to controlled presentation of an alarm pheromone were augmented by increased temperature. Intensity of bee response was also more pronounced at higher humidity. After the genetic variation in colonies' defensive characteristics was eliminated statistically, 92% of the variation in defensive responses of European honeybees was shown to be predictable as a function of temperature, wind speed, solar radiation, relative humidity and barometric pressure (Moritz, Southwick, and Harbo 1987; Southwick and Moritz 1987b). It is reasonable to expect that the continued flight of workers around an intruder, after loss of the sting, further influences stinging behavior of nest mates.

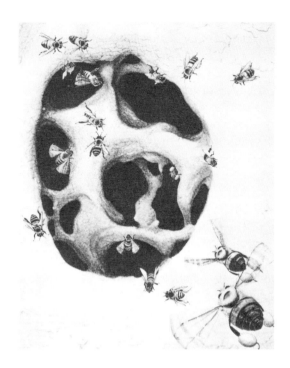

Figure 3.20. Foragers returning to the nest and guarding or fanning workers of a natural Africanized honeybee colony in lowland French Guiana. The pillars shown inside the nest entrance are constructed entirely by the bees from resin and wax. Original drawing by Sally Bensusen (from photograph).

This behavior is highly developed in an African race of the honeybee, now well established in the Americas (Chapter 4). Its descendants are known for continuing pursuit of a sting victim up to a kilometer from the nest.

The stingless bees possibly make use of mandibular gland products as allomones, to allow them to construct nests within active termite or ant nests and then benefit from the physical and behavioral defenses of these species (Kerr et al. 1967; Wille and Michener 1973; Roubik 1979b, 1983a; Nates and Cepeda 1983; Sakagami, Yamane, and Inoue 1983; Camargo 1984b). Most of these stingless bee species are unaggressive toward intruders and contribute nothing to the defense of compound nests, making them unequivocal usurpers of other social insect nests and defenses. However, some neotropical species of *Trigona, Partamona,* and *Ptilotrigona* are extremely aggressive and only reside in termitaria (Fig. 3.21). They can be regarded as mutualists with the termites. In Malaysia, *Trigonella moorei*

Figure 3.21. Nest and the details of nest structure of *Partamona vicina* (Apidae), an Amazonian bee that nests solely in active nests of nasute termites (see Camargo 1980). Original drawing provided by J. M. F. Camargo.

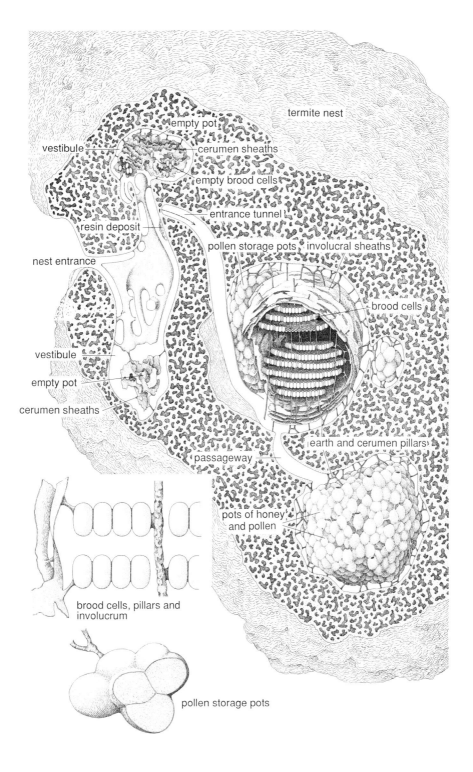

only nests within nests of aggressive ants, *Crematogaster*, whereas in the neotropics the nests of unaggressive *Trigona cilipes* and aggressive *Trigona pallens* are sometimes built in nests of stinging ants such as *Azteca* and in the nests of aggressive nasute termites (Kerr et al. 1967; Roubik 1979b, 1983a; Nates and Cepeda 1983; Sakagami et al. 1983a,b; Posey and Camargo 1985). This relationship is likely facultative in *T. cilipes* and *T. pallens* because they nest in unoccupied tree cavities in some regions. It is plausible that their ability to establish nests associated with ants or termites depends on the various chemicals produced by the host species and the bees in different geographic populations. The other neotropical bees that nest in occupied termite nests are *Aparatrigona isopterophila* and three bees of the subgenus *Scaura*. A close correspondence between the odor of an ant and the bee nest usurper is likely for neotropical *Nannotrigona mellaria,* which nests within the nest of myrmecine ants. In the nests I have opened it seems impossible for the bees to avoid direct contact with the ants.

Establishment of protected nests by usurping species has been observed in detail by Camargo (1984b) and Sakagami, Inoue, Yamane, and Salmah (in press). Camargo found an incipient nest of *Scaura latitarsis* in the Amazon forest. The bee always nests within active termite nests, where it had been seen in the early stages of colony founding but not observed over a period of time (Wille and Michener 1973; Roubik 1979b). This species initiates the nest by depositing a resin "beach head" on the surface of the termitarium. Bees then bring cerumen from the mother nest to construct an entrance tube, after which small groups arrive carrying more resin, gradually excavating and extending a series of small pockets within the galleries of the termite nest (Fig 3.22). The nest is lined with a continuous shell of resin. There is little contact between the termites and bees, except when the bees' nesting cavity is expanded. In the event of attack, bees working within the incipient nest seal themselves within and die there (Camargo pers. commun.). In the process of nest cavity expansion, the area of contact between the termite and bees is subsequently sealed with resin. Chemical deterrents are probably employed by the bees when penetrating further into the host nest, to prevent attack by termites.

The odor of army ants (*Eciton*), predators of social insect colonies throughout the neotropical forests (Franks and Bossert 1983), is also produced by bees. Skatole is an acrid and foul-smelling odor that is released by army ant raiding parties and is also used as a nest disinfectant (Brown, Watkins, and Eldredge 1979). The odor likely delimits the areas being raided and may deter other such raiding columns. Skatole permeates the nests of the neotropical *Eulaema* and those of at least one *Melipona* (Zucchi et al. 1969; Camargo pers. commun.). Females of the bees collect and store animal feces on cells of immatures (Fig. 3.2) or in containers, permanently establishing the odor of an area freshly raided by army ants. *Melipona compressipes triplaridis* in Panama envelopes its nest entrance with the smell when

3. Nesting and reproductive biology   221

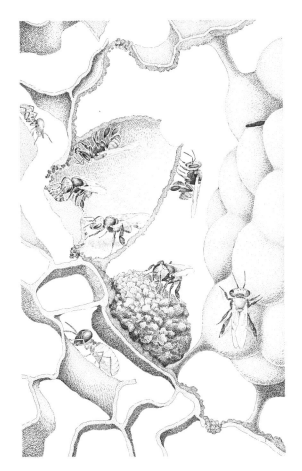

Figure 3.22. Nest initiation by workers of *Scaura latitarsis* that nest obligately in active termite nests (Camargo 1984a). The gradual process of nesting begins with the construction of an entrance tube from resin. Bees are shown working with the deposit of resin; storage pots filled with honey (right) have also been added to the excavated cavity. Original drawing provided by J. M. F. Camargo.

skatole mandibular gland contents are released by a biting guard bee (Smith and Roubik 1983).

Bees that make empty cells or chambers near the nest entrance may deceive nest predators or parasitoids (Malyshev 1935). However, the structures have additional defensive functions in the eusocial bees. Stingless bees of the subgenera *Partamona* and *Ptilotrigona* (Camargo 1970, pers. commun.) and of the subgenus *Trigona* (Wille and Michener 1973; Michener 1974a) build a series of pillars and connectives immediately inside the entrance. The former groups do this in an antechamber or vestibulum, separated by nearly a meter from the brood and food-storage area of species nesting in the ground or termite nests. A complete false nest is made by *Partamona pseudomusarum* and *P. vicina* that is as large as the real nest and complete with empty brood cells, empty storage pots, and an involucral sheath around the brood area (Fig. 3.21). This separate nestlike cavity is normally filled with foraging and guard bees, so that these aggressive bees have many potential defenders near the entrance. A relatively naïve predator attacked with sufficient force might depart upon finding little in what appears to be the nest. In addition, Kerr (1969) suggests that *Tamandua* and other edentate predators may be deceived after probing the vestibule with the tongue and finding no liquid food. Cavity-nesting *Eulaema* make several brood cells with nothing in them or make tunnels to empty chambers immediately inside the nest entrance (Zucchi et al. 1969; Ackerman and Roubik pers. obs.). Although behavioral studies are lacking, these consistent and unusual aspects of nest architecture might prevent some parasitoids from reaching the open brood cells. Random placement of empty cells in nesting areas seems to discourage parasitoids of wasps (Tepedino, McDonald, and Rothwell 1979).

A study of pronounced differences in the nesting biology of three sympatric Asian species of *Apis* by Seeley et al. (1982) leads readily to an appreciation of how colony and bee size, nest material, associations with conspecifics, and long-term association with predator species might interact to shape the evolution of defense behavior and nesting habit. In the study area, the forest of northeastern Thailand, there are many natural enemies of honeybees. Animals attacking nests include honey buzzards *Pernis apivoris,* Malay honey bears *Helarctos malayanus,* giant wasps *Vespa tropica,* tree shrews *Tupaia glis,* macaque monkeys *Macaca mulatta,* and lizards. Ant predators include *Oecophylla, Diacamma, Polyrachis, Odontoponera, Pheidologeton, Monomorium,* and *Dolichoderus.* The honey guide *Indicator archipelagicus* also exists in the forests of southeastern Asia, as do civets, *Arctictis* and *Arctogalidia,* that may prey upon nests of honeybees (Wong 1984).

One dwarf honeybee in Thailand, *Apis florea,* has small colonies and ineffectual stings. Its single-comb nests are concealed in foliage in the forest understory but are protected from ants with a sticky resin barrier on either side of the branch supporting the nest (see Fig. 4.14). The occasional chance encounter between a large colony predator and the nest has but one result: The bees rapidly emigrate. Later, after the nest has been plundered, some workers return to collect wax, food, and other salvageable materials and then build a new nest in another concealed site. The giant honeybee, *Apis dorsata,* builds large, single-comb nests that are conspicuous to any visually hunting predator (see Fig. 4.15). Colonies are as large as those of any other *Apis,* composed of more than 40,000 workers. Nests are often clustered, often on the underside of branches, rock ledges, or leaning tree trunks. If a predator begins to scale the tree, a powerful onslaught of stinging bees is immediately directed at the animal with a rapidity that astounds observers familiar only with temperate-zone *Apis mellifera* (Starr et al. 1987). The species of intermediate size, *A. cerana,* builds protected multiple-comb nests in tree cavities and small caves (see Fig. 4.13), has small colonies, and readily stings intruders. The last two species make limited use of resin barriers but appear to receive relatively few attacks from ants. All three species abscond if predator attack is successful.

The differing defense behaviors of the three *Apis* surveyed in Thailand appear to be effective means of applying the gross features of the bees and their nests to solve some problems of survival. Further comparisons of these species show strong physiological similarities between small, unaggressive *A. florea* and large, rapidly attacking *A. dorsata* (Dyer and Seeley 1987). These authors provide some evidence that workers of these two exposed-nest honeybees live longer than do cavity-nesting *A. mellifera* and *A. cerana*, probably to offset the cost of producing extra bees to protect the nest, which is largely protected by a tree cavity in the last two species. Thus nesting habit can direct the evolution of defense behavior and correlate with physiological traits. In addition, comparison of the two dwarf honeybee species found in southern China, *A. florea* and *A. andreniformis*, reveals contrasts in defensive behavior that are correlated with nest exposure (Wu and Kuang 1986). The more exposed bees, *A. andreniformis*, are more defensive. More subtle differences may exist between the open- and cavity-nesting species. Seeley (1985) and his co-workers suggest that a high worker/brood ratio is necessary for *A. florea* and *A. dorsata* to maintain the curtain of bees three to six layers deep over the nest for both defense and thermal control. A means of providing the larger number of workers for guarding seems to involve extended adult longevity rather than a shift in the schedule of age-related guarding behavior (Dyer and Seeley 1987). In contrast, the relatively low survival rate of European honeybees introduced into nests of Africanized honeybees (Winston and Katz 1982) suggests that the defensive capacity in colonies of the latter are augmented by earlier participation in foraging and nest guarding. We do not know if this change is due to pheromonal interactions with other workers or brood within the nest. Danka et al. (1987) switched brood between honeybee colonies of African and European ecotypes and found that the bees introduced into another colony foraged in the manner of the host bees. Foraging behavior was in some manner regulated by bees or the nest environment. Thus in the honeybees there appear to be both shifts in worker population size as well as changes in the statistical age–behavior schedule that can produce differences in colony defensive ability and other behavior.

African *Apis mellifera,* of which there are several behaviorally distinctive races (Ruttner 1988), has a suite of formidable predators (Walker et al. 1975; Fletcher 1978). Those known to interact strongly with honeybees in the lowland areas permit an attempt to deduce an evolutionary explanation for development of highly defensive behavior present in the some forms of African honeybees. The honey badger *Mellivora capensis* is an omnivorous predator that can climb trees, excavate subterranean nests, and rip apart the nests of African *Apis* to devour their immatures with relative impunity. The avian honey guide *Indicator indicator* has a mutualistic relationship with this mammal. It can digest wax, and after it locates a honeybee nest, it vocalizes loudly, calling to it any honey badger that might be in range. Aboriginal peoples and other primates have long profited by this interaction, also locating bee nests by following the calls of *Indicator* (Crane 1975). Four species of this bird are found, two in Africa, one in the Himalayas, and one in Southeast Asia (Friedmann 1974; Wong 1984).

The Africanized honeybee of the American tropics stings very readily following a disturbance to the nest. Its response to volatile honeybee alarm pheromone is attained at lower pheromone titers than temperate *A. mellifera,* and it recruits an overwhelmingly superior number of stinging bees to defend the nest in a matter of minutes (Collins et al. 1982). The colonies and queens are mobile and evasive, rapidly emigrating after a predator breaches colony defenses (Chapter 4).

Mammals are the significant large colony predators of contemporary social bees. None of them digest wax, and resin mixed with wax could certainly hinder their digestive process. Nonetheless, large animals throughout the tropics break into bee nests when they are not protected by thick tree walls. Humans pursue honey-making bees throughout the world (Section 4.3.4). There are entire tribes of aboriginal peoples that specialize in robbing and destroying colonies, but other groups have mastered the techniques of maintaining bees in living trees (instead of removing the entire colonies and placing them in hives, they open and then reseal the natural nest cavity) and providing protection from other predators (Schwarz 1948; Nightingale 1983; Posey and Camargo 1985). Since the evolutionary history of *Homo* occupies a small fraction of the time that meliponines and apines have existed (Chapter 1), it is doubtful that human molestation (or protection) of bee nests has much influenced their evolution. Relatively strong selection pressures from human populations might coincide with recent explosive human population growth and the advent of selective breeding techniques for honeybees (Seeley 1985). Both processes are unlikely to have influenced more than a small fraction of honeybee populations prior to the last century.

The largest contemporary bee nest predators in the tropics are the bears *Helarctos, Selenarctos*, and *Melursus* in Asia (review by Caron 1978b), related Eurasian bears, and *Tremarctos* in South America. The sloth bear *Melursus* has developed a buccal structure that is highly modified for scooping up small termites, and it attacks the nests of bees (Laurie and Sidensticker 1979). Although the South American bears arrived across the isthmus of Panama about three million years ago, there are large nest predators with a longer history in South and Central America (Simpson 1980). A member of the mustelid group, the tayra *Eira* (=*Tayra*) *barbara,* is locally called the honey robber in Panama and South America. This 6-kg animal energetically excavates nests of neotropical stingless bees in the ground and in tree cavities (Posey and Camargo 1985; Roubik pers. obs.). Other mustelids often raid honeybee hives (Caron 1978a) or nest aggregations of solitary bees (Stephen et al. 1969). The originally South American xenarthrans, particularly the giant armadillo *Priodontes,* apparently excavate terrestrial nests of stingless bees such as *Trigona fulviventris* (Kerr 1969; Roubik pers. obs.). Other edentates – *Dasypus, Myrmecophaga, Cyclopes*, and *Tamandua* (armadillos, anteaters, and tamanduas) – regularly attack honeybee hives, and *Tamandua* is also locally referred to as a honey robber (and therefore a predator of stingless bees) in South America (Röhl 1942; Caron 1978a). These are normally predators of termite colonies, and *Cyclopes* is arboreal (Simpson 1980). The claws of edentates are powerful and may be used to mash the honey, brood, and liquid provisions of stingless bee colonies, which can then be drawn into the mouth with the tongue. Stingless bee colonies in termite nests should be readily consumed by anteaters. The pangolins (*Manis*) of the Old World tropics probably display such behavior. Chimpanzees and baboons regularly attack the natural nests of *Apis,* as do gorillas and macaques (*Langur*) (Botha 1970; Caron 1978). Crane (1975) writes that baboons follow the guiding calls of *Indicator.*

Numerous vertebrates that dig in the ground, such as common skunks, armadillos, and rats, consume nests of bees. Opossums (*Didelphis*) are omnivores that occasionally consume kilograms of honeybees at the hive entrance (Stonehouse and Gilmore 1977; Caron 1978a). Considering the wide tropical distribution of marsupials and the nectar-feeding habit of Australian *Tarsipes,* it is possible that other occasional bee nest predators have existed among the marsupials on that continent and elsewhere. An amphibian predator of stingless bees and honeybees worth mentioning is the toad genus *Bufo,* which consumes hundreds of bees when it can reach the nest entrance (e.g., Malagodi, Kerr, and Soares 1986). Birds such as woodpeckers that feed on xylophilous insect larvae occasionally eliminate small social bee colonies (Gerling et al. 1983).

3.2.4 Parasitism, mutualists, and commensals

Following the definition given by Price (1980), a *parasite* is "an organism living in or on another living organism, obtaining from it part or all of its organic nutriment, commonly exhibiting some degree of adaptive structural modification, and causing some degree of real damage to its host." Another definition needs preliminary clarification: A *parasitoid* is a type of parasite that inexorably causes the death of its host (Iwata 1976). The larvae of some parasites of bees often kill their host larva and will be called parasitoids here, yet the adult parasite, even if it kills the host larva, is not a parasitoid. The biology of most bee natural enemies is inadequately known but at least can be described in a general sense, based mainly upon studies of pests of honeybees or a few solitary bees employed as pollinators in the temperate zone. An unavoidable inadequacy is created by emphasizing these particular relationships, because many natural enemies of solitary tropical bees may be overlooked. Even for *Apis* and the solitary bees managed for pollination, the details of natural history of associates often come from research directed at their effect once in the bee nest rather than prior to arrival. Thus the process by which the host is located or selected is often obscure.

Roughly 60% of all insects are parasites of plants or animals, and large numbers of microorganisms, nematodes, and mites are also specialized in this way of life (Price 1980). Considering the food stores rich in protein, lipid, and carbohydrate concentrated in the nests of bees, it would be remarkable if there were not several natural enemies potentially affecting each species. I have not attempted an exhaustive list of the organisms that cause the death of bee adults and the immatures in their nests nor mention of every protozoan or mite that may be a specialized parasite of a particular tropical bee. Advances in the knowledge of the often subtle interactions of bee parasites and their hosts are developed rather slowly, with many possibilities for misunderstanding along the way. Bailey (1981) discusses the way that a common mite, the trachael mite *Acarapis woodi,* was said to cause the "Isle of Wight" disease simply because the infestation of this acarine was first studied there. It was far more likely that the mite was widespread; and the mite was not demonstrated to produce the symptoms attributed to this honeybee malady. Other trachael mites are known – for example, *Bombacarus,* which infests *Bombus* and *Psithyrus* (Alford 1975). Like *Acarapis,* they seldom cause the host extensive damage. The large mite *Varroa* is a serious pest of honeybees (Section 3.2.7) introduced to South America from Japan and to Europe from Sri Lanka and China, tranferring from its original host *A. cerana* (Needham et al. 1988).

Various generalizations regarding parasitoids seem to apply well to parasitic bees and also to the interactions of other bees with their particular parasites. Further, parasites need not represent the termination of a lineage.Their populations can speciate concomitantly with the speciation of their host species (Section 3.2.6). No compelling evidence exists that extinction more often terminates the lineage of a highly specialized parasite than it does that of a generalist predator, or more often that of a specialist compared with a generalist (Simpson 1953; Flessa, Powers, and Cisne 1975; Price 1980). However, fossil evidence from ants preserved in amber shows that even though extinction rates were no more accelerated in highly specialized or parasitic groups among continental ant fauna, extinction in these groups was proportionately greater on the island of Hispañola, Greater Antilles (Wilson 1985a). This suggests that local extinction is indeed more likely, but it could also be offset by subsequent immigration and colonization from surrounding populations.

Relationships that are parasitic continue to be modified by natural selection, such that some degree of mutualism can result. If coevolution occurs, rather than merely a match between the biology of independent host and parasite organisms (Janzen 1980, 1985), the host and its parasite can then become the unit of selection. A variety of definitions can be used for coevolution, but it is generally thought to produce extraordinary specializations. The biology of parasitoid and host both evolve to control growth, reproduction, or lethal influence of the corresponding host or parasite (Price 1980; Jones 1985). Both mites and microbes have repeatedly evolved a commensal or mutualistic relationship with bees (Batra et al. 1973; Eickwort 1979; Gilliam et al. 1985), which may be the result of coevolution following a parasitic relationship.

Microbes and other nest symbionts discussed in this chapter are perhaps essential to some bees. However, there is little indication whether the bees actively seek such organisms. One visible example of a highly developed symbiosis is the specialized "acarinarium" of bees that transport mite hypopae or adults. In halictids, anthophorids, and stenotritids, this is formed by smooth dorsal cuticular areas fringed with hairs or with chitinous invaginations on the dorsal surface of the metasoma (Cross and Bohart 1969; Eickwort 1979; Houston 1984, 1987; McGinley 1987).

Mutualism may well be illustrated by some spore-forming bacteria that live in stored pollen, but not necessarily in bee cell provisions. Machado (1971) demonstrated that elimination of bacteria from pollen stores of *Melipona quadrifasciata,* using an antibiotic, led to colony death. Furthermore, at least one species of *Bacillus* appeared in the larval food of 13 stingless bee species. Gilliam (1979) and Gilliam et al. (1984, 1985) studied the provisions of *Apis, Anthophora, Centris,* and an obligate necrophage of the genus *Trigona*. All harbored one to five *Bacillus*. The bacteria produce enzymes that metabolize proteins, lipids, starches, sugars, and esters. More importantly, their metabolic activities include the production of antibiotics. Microbes of the same type were found in the glandular cell provisions made by the *Trigona* in lowland tropical wet forest and in the pollen of European honeybees living in the North American desert. The bacteria grew in an acid pH and tolerated high osmotic pressure, allowing them to survive in the physical environment of bee provisions. Beside the metabolic conversion of bee food, which may aid digestion by the bees, these bacteria could compete successfully with spoilage or toxic bacteria that would otherwise proliferate in bee nests. It should be noted that one of the most virulent diseases affecting honeybees is *Bacillus larvae* (Bailey 1981; Seeley 1985). Considering these divergent characteristics of microbes belonging to the same taxonomic group, an interpretation of their association might be that the ancestors of the present-day potential mutualists had a very negative effect on bee fitness (Brooks 1985). Strong selection pressures then modified the effect of bacteria on bee populations. However, the benign or beneficial bacteria are associated with many bee species, and this indicates that coevolution with each of them was unlikely (see Janzen 1985). The harmful *B. larvae* is apparently restricted to *Apis,* with which it has possibly coevolved. However, the case for microbes found in diverse bee nests might be one of functional convergence rather than phylogenetic similarity, because microbe taxonomy has been based upon the metabolic activity of the organism rather than solely upon morphological traits.

No adequate study has thus far been made of the mode in which microbes may enhance digestibility of pollen. The exine wall of pollen is not broken during the bee's digestive process, but it is somewhat reduced, allowing the contents of pollen to be extracted through the specific areas from which pollen tubes grow (Klungness and Peng 1984). Gilliam and Prest (1987) identified a number of molds, bacteria, and yeasts from feces of larval honeybees but could not identify a consistent symbiotic mycoflora in all the larvae they examined. Bacteria associated with bees certainly cost the host some of its stored resources, but at least for some species, the presence of microbes is beneficial or may even be essential. Jones (1985) assesses the role of microbes as mediators of plant resource exploitation; from his summary, it seems that the roles of microbes could well include the detoxification of nectar and pollen used by bees (Chapter 2).

Commensal organisms in bee nests are for the most part beetles and mites that do no apparent harm and, as preliminary studies have suggested, could conceivably be mutualists (Flechtmann and Camargo 1974). These workers showed that fungi did not harm hived colonies of stingless bees inhabited by the mite *Neotydeolus therapeutikos,* because most of the fungus was eaten by the mites. Arthropod associates of stingless bees include unspecialized commensals such as Collembola, Coleoptera (Staphylinidae), and Dermaptera (Salt 1929; Nogueira-Neto 1970b; Roubik pers. obs.). In contrast, permanent residents of social insect colonies have undergone striking structural modification (Wilson 1971; Kistner 1982; Roubik and Wheeler 1982). They have flattened appendages and wing covers that fit into grooves or can be closely appressed to the body. Phoretic species have other unique traits, such as deep notches in the mandibles that grasp the hairs of bees (Roubik and Wheeler 1982). On the other hand, apparent phoretic commensals or even mutualists can be vectors of pathogens or create niches exploited by pathogens or parasitoids (Sections 3.2.6 and 3.2.7).

Cryptophagid and leiodid beetles are found in neotropical bumblebee and stingless bee nests (Crowson 1981; Kistner 1982; Roubik and Wheeler 1982); both groups have flightless species that attach to adult bees at foraging sites. Adult *Antherophagus* (Crypotophagidae) in Central and South America apparently transfer between foraging *Bombus* at flowers, clinging to a leg or hairs with their relatively long, curved mandibles. The leiodid genera *Scotocryptus, Scotocryptodes, Parabystus,* and *Synaristus* all may transfer between foraging *Melipona* and mud-collecting Trigonini at mud collection sites (Roubik and Wheeler 1982). Their presence in the nests of Trigonini and *Melipona* in the neotropics had long been noted, and their location of hosts is presumably due to refined chemical senses, because these beetles are not only flightless but blind. Nogueira-Neto (1949) reports transfer of beetles on *Melipona* engaged in intercolony food transfer (in this case, robbing). Within the bee nest, larvae and adults feed on pollen and feces accumulated by the bees, as well as on slime molds and fungi that grow on them (Roubik and Wheeler 1982). Therefore, their role is mutualistic as they reduce waste material and fungal invasion. A highly modified coccinellid beetle, *Cleidostethus,* inhabits African stingless bee nests, but its biology needs further clarification (Salt 1929).

Some ricinuleids, pseudoscorpions, and mites appear to feed on mites or microbes in the nests of bees. Large and active mites are predaceous on smaller mite species in the nests of stingless bees and honeybees (Section 3.2.6). I have seen them most often in the outer areas of the nest, often within narrow crevices in hardened resin, where bees are unable to enter. The pseudoscorpions inhabit nests of meliponines, *Apis,* and *Bombus* (Weygoldt 1969). They spin small, disklike silken nests, which probably afford protection from bees. Their adults are phoretic and have many more bristles than related nonphoretic taxa (Mahnert pers. commun.); the large chelae are used to grasp the host to disperse. The genera *Chelifer, Ellingsenius, Dasychernes,* and two other neotropical genera (Mahnert pers. commun.) are hosted most often by stingless bees. These animals are generalized predators and take any small arthropod in their habitat, often commensals such as Collembola and Thysanura that are found in stingless bee nests (Weygoldt 1969; Nogueira-Neto 1970b). I found an unusal ricinuleid, *Cryptocellus gamboa,* in the nest of *Trigona "hypogea"* in Panama. Adults are slow-moving and may live for several years; they also have grasping chelae, but their mode of dispersal and habit in bee nests are unknown (Platnick and Levings pers. commun.).

3.2.5 Pathogens and related symbionts

Bees have undoubtedly coevolved with some fungi and yeasts, found in the gut of adult bees, nests, and in larval provisions (Wheeler and Blackwell 1984). Such microbes are frequently not pathogenic and even provide nutrients to bees (Vecchi 1959; Batra et al. 1973; Gilliam 1978; Skou and Holm 1980; Bailey 1981). For example, a large portion of larval provision of neotropical *Ptiloglossa guinnae* is composed of the yeasts *Saccaromyces* and *Candida,* which were thought to provide more of the amino acids, vitamins, and sterols for developing larvae than the pollen grains (Roberts 1971; Batra et al. 1973). On the other hand, as the latter authors show, the action of the yeasts makes provisions susceptible to colonization by fungi. In one survey concentrated in a few temperate localities, Batra et al. (1973) provide records published on 9 genera and 124 species of fungi associated with bees. These authors also mention a tendency of gregarious, soil-nesting bees to abandon an established nesting site after fungal populations increase far beyond normal levels. In one detailed study of solitary bee associates, Batra (1980a) discovered provision-destroying fungi of the genera *Rhizophus, Fusarium, Cephalosporium,* and *Cladosporium* in nests of three North American *Colletes.* She also noted that *Cephalosporium* damages adult bumblebees.

Worker stingless bees inoculate stored food with the bacilli from their mandibular glands (Machado 1971), and newly emerged bees acquire bacilli from stored pollen or the pollen on flowers or by receiving contaminated food from nest mates (Gilliam 1979; Gilliam et al. 1984; Gilliam and Prest 1987). A large proportion of microbial and other honeybee diseases is harbored in the hypopharyngeal and mandibular glands (Bailey 1981). But as Bailey points out, pathogens are often less likely to be transmitted by debilitated bees of foraging age. They cease to gather pollen, which they would normally wet with infected secretions and then store, unmixed and relatively unmodified, in the nest (Section 2.2.2). As described in Chapter 2, all nonparasitic apid bees regurgitate nectar to secure the pollen they collect. Thus a bacterial strain may be passed from generation to generation within a nest, especially when it does not kill or seriously debilitate adults. Similar gut-to-provision transfer of microbes must also occur in solitary bees as they moisten provisions with nectar or as they defecate within the nest.

Fungi such as *Aspergillis, Plistophora,* and *Beauveria* grow profusely on the cells, provisions, and dead immatures of neotropical *Melitoma* in Mexico (Linsley et al. 1980). Specificity of fungal parasite and bee host is indicated by distinct *Ascosphaera* ("chalkbrood") species associated with some 40 native North American bee species, including *Anthophora, Nomia,* and *Megachile* (Gilliam 1978; Torchio pers. commun.). At least one fungus has an apparently mutualistic relationship with a neotropical stingless bee inhabiting wet lowland forests, because it may desiccate and retard spoilage of stored pollen in the nests of *Ptilotrigona* (Camargo pers. commun.).

Many microorganisms that adversely affect bees attack and destroy the larvae or pupae, sometimes differentially affecting the sexes, or queens and nonreproductives. Regarding honeybees, Bailey (1981) discusses the pathology of 15 viruses; several bacteria including *Streptococcus, Pseudomonas, Spiroplasma,* a *Corynebacterium* type, and *Bacillus;* three fungi including *Ascophaera* and *Aspergillis* (*Paecilomyces* also kills honeybee queens; Skou and Holm 1980); the yeast *Torulopsis;* and three protozoans, *Nosema, Crithidia* (apparently not debilitating to

Apis), and *Melpighamoeba*. Many other microbes are known to exist in bee nests, such as *Cladosporium* in the comb of honeybees (Batra et al. 1973), but their pathogenicity remains unestablished in natural host bees. At least one virus affecting *Apis mellifera* has been recorded in Australia, where there are no native *Apis* but potential meliponine hosts (Bailey 1981).

Disease organisms are transmitted through bee feces; through the cast exuviae (and gut linings) or cocoons of larvae or pupae; directly on the oocytes by deposition on eggs; by food exchange among colony members; or by ingesting spore-infested nectar, honey, pollen, honeydew, or water, via ectoparasites or by contact with diseased immatures. A fungal disease was shown to be transmitted from a male to a queen *Apis mellifera* during mating (Skou and Holm 1980). Secondary bacteria or fungi infect immature bees or their remains, following infection or death caused by a primary pathogen; these organisms are sometimes mistaken for primary pathogens (Batra et al. 1973; Bailey 1981). Disease organisms not transmitted to the egg during oviposition are acquired by ingestion or, for viruses, through pores left in the cuticle where bristles have been broken. It is probably significant that among bees, only apid immatures defecate after the cocoon is formed, thus potentially isolating spores or disease organisms from the colony. However, larvae infected by the bacterium *Streptococcus pluton,* European foulbrood, have poorly developed silk glands and produce flimsy cocoons through which bacteria can pass (Bailey 1981).

A primary line of defense against the spread of pathogenic microbes within a bee nest is the filtering device of the proventriculus (Section 2.2), which passes pollen into the digestive chamber of the midgut. Particles of a size greater than 5 µm are passed through the proventriculus, sequestering them from the crop and food that may be subsequently regurgitated into the brood cells and to nest mates. Spores, as well as protozoans such as *Nosema,* for example, are passed into the midgut and then ejected with feces outside the nest of apine bees (Seeley 1985). *Apis mellifera* also has the advantage of antibiotic acids produced in worker mandibular glands, which further reduce the chances of passing disease microbes to nest mates during trophallaxis.

Mold is one of the most significant natural enemies of nesting solitary bees. Plant resins, the chemicals of nest materials, and the products of Dufour's gland and cephalic glands of bee adults and even immatures are biocidal agents that are thought generally to be used to thwart development of such pathogens (Batra et al. 1973; Cane and Tengö 1981; Cane et al. 1983; see also Chapter 2). Plant resins are rich in triterpenes and provide antimicrobial isoprenoids. Fungi and bacteria certainly exist in the nests of bees and soil (Batra et al. 1973), readily growing on larvae and cell provisions if the cell is broken or inadequately protected. An estimated 20% of larvae are normally diseased in 12 temperate-zone species of halictids, andrenids, and anthophorids analyzed by Cane et al. (1983). The rate of such mortality among tropical bees seems very similar, but the data are meager. Ground-nesting tropical *Exomalopsis* that reuse the parental nest (Raw 1977) experience about 50% mortality as immatures and 40% in the egg and larval stages. The fact that mold grew in cells where immatures had died may indicate biocidal properties of eggs or living larvae. Vinson et al. (1987) found profuse growth of mold in cells of *Mesoplia* bees that had parasitized the nests of *Centris*. Total immature mortality in temperate and tropical bees may be comparable, but the available widely ranging figures are not readily compared. Hypogeous centridine bees in Costa Rica sustained 59% parasitism in a sample of 22 cells (Vinson et al. 1987), and in the same area Parker, Batra, and Tepedino (1987) found 50% parasitism among 16 xylophilous megachilid nests. The 1,900 separate cells of hypogeous *Melitoma* studied by Linsley et al. (1980) at five sites in central and southern Mexico revealed average mortality of approximately 40%; 20% of the cells examined were infested by mold. Considering only stem-nesting or xylophilous Hymenoptera, including one megachilid bee, Danks (1971) found 50–60% mortality in England. This author points out the considerably higher incidence of mortality in insects with less specific nesting requirements; and although over 40% of mortality for the bee species was due to simple mortality of egg or larva, whether due to microbes or genetics (Section 3.3.5), even this amount of success in survival is conceivably reduced by the difficulty of finding "safe" nest sites. Greater larval mortality of lodger bees, due to microbes or other factors, is a corollary to the theory of Malyshev (Section 3.1.1) on the general trade-off between opportunistic nesting or high selectivity.

The potential for microbial infection is likely to be relatively great for tropical bees nesting in moist, warm habitats. If these resemble the temperate species thus far assessed, volatile acyclic terpenoids and fatty-acid derivatives may be released from the mandibular glands of female bees during nest construction, initially disinfecting the cell and nest environment before the cell is sealed (Cane et al. 1983). Microbial growth is hampered by secretions of the Dufour's gland (Hefetz 1987), which are spread with the mouthparts on the inside walls of brood cells by some anthophorids and colletids. Microbes are also said to be inhibited by secretions of prepupae of the halictid bee *Nomia* (Bienvenu, Atchison, and Cross 1968). Fungicides and bactericides are present in mandibular gland secretions of all adult bees analyzed. Besides honeybees, they include *Protoxaea, Oxaea* (Oxaeidae), *Andrena* (Andrenidae), *Colletes, Hylaeus, Caupolicana* (Colletidae), *Ceratina*, and *Pithitis* (Anthophoridae). The oxidation product of a Dufour's gland secretion, farensal, is a patented germicide (Cane et al. 1983). Disinfectant chemicals produced by bees have been tabulated by Cane and co-workers. These include linalool, citral, geraniol, citronellol, octanones, and the acetates of geraniol and nerol. Regarding colony processes that affect the nests of highly eusocial species, Seeley (1985) reviews evidence that the elevated nest temperatures near 34° C maintained by *Apis* are sufficient to prevent development of fungi (Maurizio 1934) and viruses (Watanabe and Tanada 1972). Sustained temperatures higher than 35° C also inhibit multiplication of the protozoan *Nosema apis* in adult bees (Bailey 1981). Inhibition at low temperatures, or through variable nest temperatures experienced by most solitary bees and many meliponines, conceivably affect the development of pathogens (Section 3.1.2).

3.2.6   Insect natural enemies of bees

Knowledge regarding the apoid host–parasite relations among tropical species is extremely limited, but a large proportion of tropical and other bees are obligate nest parasites and use the food provisions of other bees (Section 3.2.8). Several insect orders, primarily Diptera, Coleoptera, and Hymenoptera, are widespread natural enemies of nesting adult bees and their progeny. This section is limited to nonapoid parasites and predators. References to temperate-zone relationships are included and may be sufficient to indicate the types of insects likely to be natural enemies of bees in the tropics.

a. Hymenoptera (wasps and ants). Among nonapoid Hymenoptera, the female mutillid wasps, ants, ichneumonid, and chalcidoid wasps consume bee eggs, larvae, prepupae, and pupae (Table 3.2). The wasps are parasitoids and the ants are predators. Adult *Philanthus, Palarus,* and *Trachypus* wasps capture at least foraging halictids, anthophorids, and apids (Evans 1955; Alford 1975; Packer 1985; Moure and Hurd 1987; Ruttner 1988). Foraging *Vespa* of several species capture *Apis cerana* and introduced *A. mellifera* at nest entrances (Matsuura and Sakagami 1973; Ono et al. 1987). Ant genera that invade bee nests and occasionally kill even highy eusocial colonies are *Camponotus, Azteca, Eciton, Iridomyrmex, Monomorium, Pheidole, Crematogaster,* various formicines like *Solenopsis* and *Dorylus,* and potentially the Southeast Asian species previously noted in the study of Seeley and his collaborators (Schwarz 1948; DeJong 1978). I have seen the first three genera invade and kill colonies of Trigonini and *Melipona* in natural forest settings in the neotropics. Gerling et al. (1983) show the result of an attack by ants of a small colony of *Xylocopa:* The raiding ants entered the tunnels, removed the cell partitions, and killed 9 of the 16 immature bees in the nest within two days. Linsley (1966) reports that the minute workers of *Pheidole* often kill nesting *Xylocopa* in the Hawaiian islands.

Although the ants are generalized predators, other Hymenoptera appear more specialized though they are frequently natural enemies of both bees and wasps

Table 3.2. *Wasp parasitoids of bees*

Wasp parasitoid	Host	References
Bethylidae, *Scleroderma*	Apoidea	Iwata 1976, p. 63
Braconidae, *Syntretus*	*Bombus, Psithyrus* (AP)	Iwata 1976
Chalcididae, *Tetrastichus*	*Megachile*	Krombein 1967
Chrysididae, *Chrysis*	*Anthopora, Pithitis* (ANTH); *Megachile*	Krombein 1967, Kapil & Jain 1980
	Anthidium, Dianthidium (MEG); *Hylaeus* (COL)	Hurd 1979
Chrysididae, *Chrysura*	*Osmia* (MEG); *Anthophora* (ANTH)	Iwata 1976, Hurd 1979
Encyrtidae, *Coelopencyrtus*	*Hylaeus* (COL); *Ceratina* (ANTH)	Krombein 1967, Daly & Coville 1982
Encyrtidae, *Encyrtus*	*Hylaeus* (COL)	Hurd 1979
Eulophidae, *Aprostocetus*	*Ceratina* (ANTH)	Daly et al. 1967
Eulophidae, *Melittobia*	*Ceratina, Paratetrapedia* (ANTH); *Bombus* (AP)	Krombein 1967, Daly et al. 1967, Roubik unpub.
Eupelmidae, *Calosta*	*Megachile, Osmia, Heriades* (MEG)	Jayasingh & Freeman 1980
Eurytomidae	*Ctenoceratina* (ANTH)	Daly 1988
Eurytomidae, *Axima*	*Megachile* (MEG)	Kapil & Jain 1980
Eurytomidae, *Eurytoma*	*Ceratina* (ANTH)	Daly et al. 1967
Gasteruptiidae, *Crassifoenus*	*Ceratina, Ctenoceratina* (ANTH); *Hylaeus* (COL)	Daly et al. 1967, Parker & Bohart 1966
Gasteruptiidae, *Gasteruption*	*Ctenocolletes* (STEN)	Houston 1984
	Ceratina, Ctenoceratina (ANTH); *Hylaeus* (COL); *Dianthidium*	Daly et al. 1967, Skaife 1954, Hurd 1979
Gasteruptiidae, *Hyptiogaster*	*Ctenocolletes* (STEN)	Houston 1984
Ichneumonidae, *Aritranis*	*Ceratina* (ANTH)	Daly 1983, Daly et al. 1967
Ichneumonidae, *Grotea*	*Ceratina* (ANTH)	Daly et al. 1967
Ichneumonidae, *Hoplocryptus*	*Ceratina* (ANTH)	Daly et al. 1967
Ichneumonidae, *Kaltenbachia*	*Ceratina* (ANTH)	Daly et al. 1967
Leucospididae, *Leucospis*	*Chalicodoma, Hoplitis, Osmia*	Parker & Bohart 1968, Iwata 1976
	Megachile, Parevaspis, Anthidium, Paranthidiellum (MEG)	Parker & Bohart 1966, Torchio pers. commun.
	Centris (ANTH)	Roubik unpub.

Host	Associates	References
Leucospidae, *Micrapion*	*Ctenoceratina* (ANTH)	Daly 1988
Mutillidae	Melittidae	Iwata 1976, p. 75
Mutillidae, *Dasymutilla*	*Dianthidium, Paraanthidium* (MEG); *Diadasia, Anthophora, Melitoma, Ptilothrix* (ANTH); *Halictus, Nomia* (HAL).	Brothers 1972, DeJong 1978, Mickel 1928, Hurd 1979; Linsley, MacSwain, & Michener 1980, Plateaux-Quénu 1972
Mutillidae, *Ephutomorpha*	*Stenotritus* (STEN)	Houston & Thorp 1984
Mutillidae, *Hoplognathoca*	*Thygater* (ANTH)	Roubik unpub.
Mutillidae, *Hoplomutilla*	*Eufriesea, Eulaema* (AP)	Lenko 1964, Roubik unpub.
Mutillidae, *Mutilla*	*Bombus, Apis* (AP); *Ptilothrix, Ceratina, Peponapis* (ANTH); *Osmia* (MEG); *Lasioglossum, Augochlorella* (HAL).	Iwata 1976, Eickwort & Eickwort 1973; Brothers 1972, DeJong 1978
Mutillidae, *Myrmilla*	*Lasioglossum, Halictus* (HAL)	Mickel 1928, Krombein 1967
Mutillidae, *Myrmilloides*	*Augochlorella* (HAL)	Hurd 1979
Mutillidae, *Neomutilla*	*Corynura* (HAL)	Brothers 1972
Mutillidae, *Pappognatha*	*Euglossa* (AP)	Yanega pers. commun.
Mutillidae, *Photopsis*	*Chalicodoma, Megachile, Hoplitis* (MEG)	Mickel 1928, Parker & Bohart 1968, Torchio pers. commun.
Mutillidae, *Polistomorpha*	*Diadasia, Anthophora, Melissodes* (ANTH)	Mickel 1928, Torchio pers. commun.
Mutillidae, *Pseudomethoca*	*Euglossa* (AP)	Dodson 1966, Roberts & Dodson 1971
	Lasioglossum, Augochlorella, Nomia, Halictus (HAL)	Batra 1965, Brothers 1972, Mickel 1928
	Pseudagapostemon (HAL); *Xenoglossa, Svastra* (ANTH)	Hurd 1979, Brooks & Roubik 1983
Mutillidae, *Sphaeropthalma*	*Dianthidium, Anthidium, Callanthidium* (MEG)	Krombein 1967
	Diadasia, Anthophora (ANTH)	Hurd 1979
Mutillidae, *Stenomutilla*	*Osmia* (MEG)	Krombein 1967
Mutillidae, *Timulla*	*Svastra* (ANTH)	Hurd 1979
Perilampidae, *Echthrodape*	*Braunsapis* (ANTH)	Michener 1969
Pteromalidae, *Dibrachys*	*Osmia, Stelis* (MEG)	Torchio pers. commun.
Pteromalidae, *Habitrys*	*Ceratina* (ANTH)	Daly et al. 1967
Pteromalidae, *Merisus*	*Ceratina* (ANTH)	Daly et al. 1967
Sapygidae, *Eusapyga*	*Dianthidium* (MEG)	Pate 1947

Table 3.2 (cont.)

Wasp parasitoid	Host	References
Sapygidae, *Huarpea*	*Xylocopa* (ANTH); *Megachile* (MEG)	Pate 1947
Sapygidae, *Polochrum*	*Xylocopa* (ANTH)	Hurd & Moure 1961
Sapygidae, *Sapyga*	*Osmia, Heriades, Chalicodoma, Dianthidium* (MEG)	Iwata 1976, Krombein 1967, Hurd 1979, Torchio pers. commun.
	Megachile (MEG); *Anthophora* (ANTH)	
Tiphiidae, *Myrmosula*	*Lasioglossum, Nomadopsis* (HAL)	Iwata 1976, Hurd 1979
Torymidae, *Diomorus*	*Ceratina* (ANTH)	Daly et al. 1967
	Dianthidium (MEG)	Krombein et al. 1979
	Osmia, Megachile, Anthidium, Dianthidium (MEG)	Krombein 1967, Zucchi et al. 1969
	Euglossa (AP)	Kapil & Jain 1980, Hurd 1979
Torymidae, *Microdontomerus*	*Ceratina, Anthophora, Melitoma* (ANTH)	Daly et al. 1967, Linsley et al. 1980

Abbreviations: ANTH, Anthophoridae; AP, Apidae; COL, Colletidae; HAL, Halictidae; MEG, Megachilidae; STEN, Stenotritidae.

(Table 3.2). Iwata (1976) lists 33 wasp families that are parasitic; 22 of these have known hymenopteran hosts, including bees that are cleptoparasites of other bees. Their depredations include consumption of some larval provisions but are almost exclusively directed at the immature host. If a parasitic female also takes some of the bee provisions, then there is the possibility for chemical protection of the nest by inclusion of toxic and repellent substances in bee food, but this possibility has not been investigated.

Hymenopteran natural enemies of bees arrive within a nest by stealth as well as by force, although at least some mutillids lay eggs at the host nest entrance. Mutillid wasp females have disproportionately large heads and mandibles, a large sting, and a thick exoskeleton, and their larvae have large mandibular teeth (Brothers 1972; Alford 1975). The wingless female (Appendix Fig. B.22) locates an active nest and moves directly to the nest entrance and within, unless repelled by a guard bee (Batra 1965). Once inside the burrow, mutillids are only weakly resisted by adult bees in the nest; thus the parasitism within nests can reach 90% of all cells while adjacent nests are unharmed (Brothers 1972). Both Brothers and Batra note that *superparasitism* by mutillids is common – multiple parasite larvae are found in one bee cell. The adult parasite does not kill the host and attacks only pupal cells; and like some parasitic bees (Section 3.2.8) the parasite larva emerges with mandibular structure and behavior that are useful in combat. However, because its host is always in the pupal stage and therefore defenseless, the aggressive traits may play a more important role in competition with rival parasitoids. Rau (1926) found that *hyperparasitism* is permitted in anthophorid nests invaded by mutillids. When the mutillid wasp tunnels beside bee cells, it excavates a pit in each to deposit eggs; the pit provides an entrance for torymid wasps (*Monodontomerus*) that can produce up to a few dozen wasps in each host cell. Two-thirds of the cells of containing mutillids were also hyperparasitized by torymids.

Many parasites specialize in sneak attack. Some wait until a nest is completed by a nesting solitary bee, then laboriously chew through the cell cap and move within the nest. One such parasite in Panama (*Leucospis*) penetrates a sticky, resinous barrier placed on the last cell in a series completed by *Centris analis*. The notion that sneaking is the best tactic for these parasites may not be accurate. Their very slow movements may in fact be highly adaptive if they are to conserve energy while searching for hosts that have been selected for deceptive ability, such as making cell caps at entrances to cavities where no eggs or provisions have been placed (Tepedino et al. 1979).

The extremely small size of chalcidoid wasps and some ants probably allows them to slip by guards of many bees and into the nests. Over 48% of the natural nests of *Megachile nana,* and up to 100% in some places, were parasitized by the chalcidoid *Monodontomerus obscurus* in northern India (Kapil and Jain 1980). Michener (1969) actually found social *Braunsapis* caring for larvae of parasitic chalcidoids that had penetrated their nest. However, it appears that hymenopteran parasites are restricted for the most part to solitary bee species (Table 3.2). Entry into the nest of a solitary bee is always possible while it is on a foraging flight. Furthermore, it would be difficult for a returning forager to detect the presence of a small parasite egg within the nest. Observation of a megachilid bee, *Chalicodoma*, suggests that females can detect cells parasitized by another megachilid species and remove their contents (Parker et al. 1987). Similar deductions were made by Rozen (1977a) after observing soil-filled cells in active nests of the andrenid *Psaenythia* in southern Brazil. Often, however, the egg of the parasite is embedded in the host provision or cell wall, effectively eliminating possibility of discovery (Iwata 1976; Tepedino and Parker 1984; see also Fig. 3.4).

b. Coleoptera (beetles). A few beetles have attained remarkable larval specializaton that allows them to attach to an adult bee, either as it emerges from its natal nest or while foraging. The meloid, rhipiphorid, and stylopid larvae of the first developmental stage are triungulins, highly active and mobile larvae equipped with grasping claws and relatively well developed senses (Crowson 1981). The adult beetles are usually short-lived and often do not feed; many are flightless, blind, or weak fliers. Their host specificity seems to be low although particular beetle species parasitize single bee genera (Linsley et al. 1980). Female stylopids are flightless as adults; they develop and also mate while within the bee's body, protruding slightly between abdominal segments. However, the beetles do not seem to be host-specific. Jones, WIlliams, and Jones (1980) saw one species parasitize two species of *Andrena*. *Stylops* infests honey bees in India (Adlahka and Sharma 1976), after which they can neither fly nor sting, and halictid hosts also exist (Caron 1978b). Adults of the rhipiphorid genus *Rhipiphorus* oviposit on *Sida* (Malvaceae), which is a pollen source for the anthophorid bee *Diadasia;* apparent restriction of the bees to this pollen source in areas studied in California led to a 15–30% parasitism of bee cells by the beetle (Linsley 1958). Raw (1977) found two species of the tropical anthophorid *Exomalopsis* parasitized by rhipiphorids of the genus *Macrosiagon*.

Meloids are perhaps the most widespread and diverse coleopteran bee parasites, especially the Nemognathinae and genera *Hornia, Cissites, Synhoria, Tricrania, Zonitis, Tetraonyx, Nemognatha, Pyrota*, and *Meloe*. All known triungulin larvae develop in the nests of bees, although there are probable exceptions in the Old World (Selander 1987). Adults often feed on leaves and flowers; the larvae feed on nectar and pollen, and most do so in the nests of bees. A toxic substance, cantharadin, is present in the adults, eggs, and larvae (Blum 1981). Such a chemical might deter a host bee from attacking this parasite. A list of meloid plant and bee hosts, mostly restricted to the nearctic region, includes 109 species of host bees: 5 Andrenidae, 41 Anthophoridae, 52 Megachilidae, 2 Melittidae, 3 Halictidae, and 6 Colletidae (Erickson, Enns, and Werner 1976). Records of tropical bee genera that are meloid hosts include anthophorids (*Epicharis, Centris, Xylocopa, Melitoma, Melissodes, Diadasia, Ptilothrix, Anthophora, Amegilla*), bombines (*Eufriesea, Eulaema, Bombus*), megachilids (*Megachile, Chalicodoma, Heriades, Hoplitis, Anthidium*), halictids (*Nomia, Lasioglossum*), andrenids (*Calliopsis*), stenotritids (*Stenotritus* and *Ctenocolletes*), and colletids (*Colletes, Crawfordapis*) from varied nest-site substrates (Vesey-Fitzgerald 1939; Linsley 1958; Linsley and MacSwain 1958; Erickson et al. 1976; Linsley et al. 1980; Selander 1983, 1985; Houston 1984; Houston and Thorp 1984; Roubik and Michener 1984). A tiny larva, or group of larvae, attaches to a foraging bee on the hairs, legs, or antennae. Adult *Hornia* are wingless, and the female mates and lays eggs within the bee cell where she develops; larvae of *Hornia* disperse with newly emerging bees leaving the nest (Linsley et al. 1980). Alternatively, meloids locate the open nest of a bee and then deposit eggs near the burrow (Erickson et al. 1976; Crowson 1981; Kistner 1982). Within the bee nest, the beetle larva moults through a series of successively more grublike and sedentary larvae, as it feeds on pollen stores. The resemblance between meloid larvae and those of bees helps to explain the fact that they are even fed by honeybee workers while in the nest (Bailey 1981). The blind and flightless meloid *Meleotyphlus* parasitizes euglossine bees (Selander 1965; Dodson 1966), and the adult appears to spend its life in or near the nest. Related *Tetraonyx* are winged and sighted parasites of ground-nesting colletids and anthophorids (*Melitoma, Centris, Epicharis*), and the flightless but sighted *Opiomeloe* is likely a parasite of *Bombus* (Linsley et al. 1980; Roubik and Michener 1984; Selander 1985).

The large meloid *Cissites maculata* is a parasite of one of the largest bees, *Xylocopa frontalis* (Hurd 1958, 1978; see also Fig. 3.23). It apparently enters bee nest galleries to search for cells, where I have seen it attacked and damaged by female *X. frontalis*. The separate tunnel in which it pupates may be made by the larva (Sakagami and Laroca 1971), but normally only adults have strong mandibles (Selander 1985). Adult beetles remain outside the xylocopine nests and are slow-moving; there the female mates and oviposits close to the nest entrance, leaving the triungulin larva to locate its host. Such meloids are likely to be relatively host-specific (Linsley 1958), but larvae may nonetheless disperse to flowers to locate a potential host. Evidently, all members of the pantropical tribe Horniini parasitize *Xylocopa* (Selander pers. commun.).

Burrowing beetle larvae consume multiple bee cells and their provisions. The clerid genus *Cymatodera* was reported to consume eight bee larvae slightly before

Figure 3.23. Host bee and parasite beetle. *Cissites maculatus* (Meloidae) shown with its host in lowland Panama, the large *Xylocopa frontalis*. Original photograph by Sean Morris, Oxford Scientific Films.

it pupated, indicating that it normally attacks many more (Linsley et al. 1980). Furthermore, the larvae bored through hard adobe walls in Mexico and were able to find the cells of *Melitoma,* evidently the result of a specialized searching behavior for these bee cells. The clerid *Trichodes* parasitizes both solitary and social bees. Very generalized predaceous beetles, cicindelid larvae and adults, prey upon bee adults and larvae at nesting sites (Linsley 1958). These make burrows in the soil in the same places as nesting bees, sharing parasitic flies of the genus *Anthrax* (Palmer 1983). Dermestid beetle larvae such as *Trogoderma* attack cells of megachilids and anthophorids (Linsley et al. 1980; Raw 1984).

c. **Diptera (flies).** Flies are among the natural enemies of foraging bees (Chapter 2) and significant parasites and colony predators. Larval parasitoids such as *Anthrax* (Bombyliidae) consume the larvae of *Megachile, Chalicodoma, Centris, Ceratina, Xylocopa,* and *Melitoma* (Daly et al. 1967; Hurd 1978; Linsley et al. 1980; Coville et al. 1983; Raw 1984). The bombyliid parasites *Anthrax, Villa,* and *Heterostylum* usually oviposit while in flight; a projecting bee nest entrance tube can prevent the airborne eggs from landing within the nests (Linsley et al. 1980). *Bombylius* is also a parasite of *Colletes* (review by Batra 1980a).

The bombyliid *Villa* was found to be the most harmful parasite of the gregarious, ground-nesting anthophorid *Diadasia,* parasitizing 27% of its larvae (Linsley 1958). Another bombyliid, *Sparnopolus,* also parasitizes this bee genus (Eickwort, Eickwort, and Linsley 1977). They note that the larvae burrow through the soil and into bee cells. In northern India, Kapil and Jain (1980) report that *Argyramoeba distigma* parasitizes *Megachile.*

One genus of palaeotropical Asilidae, *Hyperechia,* shows extraordinary diversity as a natural enemy of bees. Its larvae are obligate parasites of larval *Xylocopa,* and the adults prey upon the adult bees in Africa and Asia (Poulton 1924; Yoshikawa et al. 1969). The host bee and adult fly are nearly the same size, and the fly also mimics the bee (see Fig. 2.33). Anthomyiid flies, *Hylemya* and *Leucophora,* are attracted to nests of ground-nesting genera such as *Colletes,* where they oviposit within the nest; their larvae consume provisions and immature bees (Batra 1980a). Calliphorid flies, *Caiusa indica* and *C. testacea,* parasitized 1–8% of larval *Megachile nana* throughout the active period of this bee in the northwestern Indian plains (Kapil and Jain 1980).

The Conopidae, Tachinidae, and Sarcophagidae include species that are bee endoparasites and parasitize bees in flight (Smith 1966; Stephen et al. 1969; Hüttinger 1974; Knutson 1978; Batra 1980a). Conopid parasites, including *Physocephala* and *Zodion,* parasitize *Apis* in Africa, and drones are not affected, indicating that eggs are attached to foraging workers. Hurd (1978) records parasitism of adult *Xylocopa macrops* by *Physocephala testacea* in South America. Other bee genera parasitized by conopids include *Megachile, Anthidium, Andrena, Anthophora, Eucera, Panurginus, Nomia, Halictus, Hylaeus, Apis,* and *Bombus* (Linsley 1958); Roubik pers. obs. for the last two groups). *Rondaniooestrus apivoris* is a tachinid parasite of *Apis mellifera* restricted to southern Africa. It is reported to hover near the nest entrance, where it deposits eggs on returning foragers (review by Knutson 1978). Other genera of this family that are believed normally to parasitize lepidopterans, *Crytophleba* and *Ictericophyto,* also arrive at terrestrial bee nests (Batra 1980a). The sarcophagid *Senotainia* attacks bees as they enter or leave the nest and immediately deposits tiny larvae between the head and prothorax. Larvae consume the flight muscles, haemolymph, and abdominal organs. The sarcophagid *Mitogrammidium* parasitizes megachilids in northern Africa (Krombein 1969). *Metopia* is a sarcophagid that is even attracted to nest tumuli removed from an active nest site, and its larvae parasitize colletids (Batra 1980a), the halictid genus *Lasioglossum* and ground-nesting wasps (Wcislo 1986 and pers. commun.). Stratiomyidae quickly locate damaged nests of neotropical meliponines, and their larvae destroy pollen stores and brood (Roubik pers. obs.).

Several species of *Braula* are commensals or ectoparasites of *Apis* in Africa and Asia, described chiefly as pollen feeders in the larval stage, although wax is also eaten (Grimaldi and Underwood 1986). The same authors also describe a new taxon, *Megabraula*, associated with the Himalayan honeybee *Apis laboriosa*. Because *Braula* also exists in Australia and South America (Örösi-Pal 1966), it is likely an introduced pest of honeybees. As an adult, it moves to the head of a honeybee, where it can intercept food transferred between bees at the nest (Bailey 1981). It is a large and mobile wingless fly; as it moves between hosts, microorganisms could well be transmitted, but its only known effect is to interfere with the activity of the queen, upon which over a hundred flies occasionally collect; normally there may be only one or two braulids on a bee. Braulid adults are 3–8% the size of their specific apid hosts.

Phorid flies of the neotropics, particularly *Pseudohypocera*, produce larvae that kill bees and entire colonies (Nogueira-Neto 1970b). This is especially evident when nests having strong odors, like those of the stingless bees, have been disturbed. The neotropical genus *Melaloncha* deposits its eggs on flying bees, after which the larva enters the thorax and kills the bee. Its known natural hosts include *Melissodes floris* (Anthophoridae) in Guatemala (Batra and Schuster 1977), *Scaptotrigona postica* in Brazil (Simões et al. 1980), and *Bombus mexicanus* in Costa Rica (Ramírez 1982). Following a raid of *Lestrimelitta* or another natural enemy of a stingless bee colony, and following breakage of the nest wall, I have seen phorids arrive by the hundreds in less than an hour. They apparently attack any stingless bee colony and are strongly attracted to the odor of bee provisions in general. A mating pheromone is probably released, and flies copulate while flying outside the nest shortly after they arrive. Their larvae consume provisions, pollen and honey stores, and immatures in all stages. Eggs are laid in batches on the surface of cells and between brood combs, in pollen storage pots, or on the involucrum, often in crevices and narrow spaces. Several authors have given partial accounts of phorid behavior at hives of stingless bees and honeybees (Pickel 1928a; Borgmeier 1971; Portugal-Araújo 1977; Reyes 1983).

d. Other insects. Termites are scarcely known as natural enemies of bees but may significantly affect nests of some species. At least some mortality due to termites is common for bees that nest in termitaria. Rozen (1968) reported that *Amitermes hastatus* penetrated the cell wall and killed larvae of *Anthophora* that nested in its termitaria, and Callan (1977) made a similar observation for *Centris*. I have seen termites invade the nests of small stingless bee colonies, kill them, and then forage within the tree cavity. Four colonies of *Trigonisca atomaria* were killed in this manner in Panama, in addition to two weak colonies of *Melipona fasciata*, each in natural tree cavities. On the Indian subcontinent, Batra (1980b) briefly describes the attack on ground-nesting *Anthophora* larvae and consumption of their provisions by *Microcerotermes*. Gerling et al. (1983) found larval bees killed by termites and report that nesting *Xylocopa sulcatipes* avoid woody substrates occupied by them.

Hemipteran predators of foragers have already been mentioned in Chapter 2; these insects also capture returning foragers at the nest entrance of stingless bee colonies. Several of the bugs can cause so many forager deaths that, over time, the colony could be exterminated (Johnson 1983b). Mimicry by these reduviid bugs is exceptional in the Asian species of *Ectinoderus* (Appendix Fig. B.22), and three of its *Trigona* prey species (Roubik pers. obs.). It lives and reproduces within the fallen nest entrance tubes of *Tetragonula collina*, which normally lay intact outside the nest. A reduviid has also been observed capturing *Ceratina* near the nest (Daly, Stage, and Brown 1967).

Recognition of mantispids (mantislike neuropterans) as bee predators followed from observing *Plega* enter open nests of *Melitoma* in Mexico and placing eggs in cells (Linsley et al. 1980). The young larva then may parasitize a prepupa, finally spinning its own cocoon within that of the host. I have found mantispids where *Tetrapedia*, *Megachile*, and *Centris* nest in trap nests in Panama, and Linsley (1958) noted that mantispids such as *Plega* lay eggs near bee nest aggregations in Ethiopia and in temperate areas.

A rather specialized type of parasite of apid colonies is the wax moth, a term applied to at least six genera of lepidopteran larvae that consume the wax or stored provisions and occasionally the immatures of *Apis* (Okumura 1966; Williams 1978). The genera *Galleria*, *Achroia*, *Vitula*, *Aphomia*, *Plodia*, and *Anagasta* are virtually cosmopolitan, some having been spread with the exportation of honeybees from the Old World. Principal food sources of these moth larvae often are bumblebee cells, immatures, and stored pollen and include nest contents of anthophorids and megachilids as well. Not all of them consume bee wax, but the two major honeybee pests, *Galleria* and *Achroia* (Pyralidae), consume large quantities of wax and are apparently native to the neotropics as well as the Old World, where they might readily display a preference for the comb wax of honeybees. *Achroia grisella* reportedly consumes the nesting material of two trigonine species in Australia (Hockings 1884). The mating system of this pyralid genus involves production of ultrasonic pulses by males that attract females, and sounds emitted by bees while ventilating their nests might also facilitate host location (Spangler 1984 and pers. commun.). Alternative hosts to *Apis* are necessary in the neotropics. I have observed larvae to eat the pollen stores of neotropical stingless bee colonies when artificially introduced into their nests but to do nothing to the cerumen building material. Nogueira-Neto (1970b), however, notes that their burrows are sometimes found within the cerumen of hived stingless bee nests.

3.2.7 Mites and nematodes

There is perhaps no nest of a tropical bee that is totally free of mites. The mites associated with bees are phoretic upon the adult bees (Cross and Bohart 1969; O'Connor 1982, 1988). The Mesostigmata mites consist of parasitic groups such as varroids and laelapids; Astigmata and Prostigmata consist mainly of scavengers and herbivores. Mites that are truly abundant in bee nests are usually facultative, scavenger mites. Many mites that are obligate bee associates do not readily transfer between solitary bee nests and those of eusocial species such as *Apis* (Eickwort 1988b). However, bee parasites such as mutillid wasps and cleptoparasitic bees may transfer mites between hosts (O'Connor 1988). Mites are ubiquitous in wet and dry habitats and benefit from the controlled nest conditions of many social insects. For example, they inhabit colonies of tropical ants and comprise 95% of all neotropical army ant ectoparasites (Rettenmeyer quoted in Kistner 1982). Mites inhabit bee nests and brood cells where they feed on surface exudations of adults, larvae, and pupae and probably some of the nesting materials or provisions; sometimes they consume haemolymph and may debilitate but seldom kill the bee (Krombein 1967; Cross and Bohart 1969; Eickwort 1979, 1988a; Bailey 1981; DeJong, Morse, and Eickwort 1982; Kistner 1982; Roubik 1987a). Bee eggs are eaten by some mites (Krombein 1967), but often there is no trace of bee eggs in infested cells, which implies either that mites kill young larvae or infest cells where no eggs are present.

Few mites are known to kill immature bees, and they are commonly beneficial to both solitary and social species. Many feed on fungi in nests or cells (Flechtmann and Camargo 1974; Alford 1975; Eickwort 1979, 1988b). The first authors found that 10 artificial nests of the social bee *Scaptotrigona* experienced an average 50% brood mortality before the mite *Neotydeolus therapeutikos* was introduced, after which brood mortality dropped to 1–6%. The level of mite infestation of cells in the bee nests was remarkable; dozens of mites were found in individual cells. Furthermore, the genital chamber of male *Scaptotrigona* was inhabited by the mites. Transmission of this beneficial symbiont plausibly takes place during mating. Mites on the venter of male halictid, xylocopine, and other bees are easily noted among pinned museum specimens (Appendix Fig. B.2, No. 27). They suggest possible phoretic transmission and sustained bee–mite symbioses.

Because fungal growth is extremely rapid after the death of an immature bee (Batra et al. 1973), it seems that its death would have a beneficial effect for fungus-eating mites. The mites, however, should often require a live adult bee to escape from the bee cell and to disperse to new nests. Their roles within bee nests should be to control but not to eliminate fungi. In response to immature bee mortality, possible reactions of the nesting female or workers include isolation of the cell or removal of its contents, thus potentially halting perpetuation of the mites or their spread to nearby cells within the nest. Beginning with a chance association of mites and fungi within bee cells, evolution of mites that regulate fungal proliferation might be expected.

African carpenter bees (*Mesotrichia, Koptortosoma, Platynopoda,* and *Cyaneoderes*) carry adult mites in pouches on the abdominal venter (review by Linsley 1958). This laelapid mite *Dinogamasus* lays its eggs on bee pupae, where they feed on exudations on the pupal skin and prevent fungal growth. A structure for carrying mites exists in both tropical and temperate halictid bees, as well as in the endemic Australian relatives of Colletidae, the Stenotritidae (Eickwort 1979; Houston 1984; McGinley 1986; see also Section 2.2). Houston (1987) has reared *Ctenocolletacarus* in the laboratory, finding close correspondence with host bee life cycles; the immature mites consume some of the host bee larva's feces as well as pollen.

At the risk of making a premature assessment of tropical mites associated with bees, most appear to feed on fungi, pollen, spores, and debris in the nest and are neither parasitoids nor significant parasites. Among soil-nesting bees, Norden (pers. commun.) has noted that mite-associates rapidly consume soil nematodes that enter the cells of bees, and this might be an additional advantage of bee symbioses with mites. Again, if the nematodes or other parasitoids killed the bees, the mites would have little chance of progeny dispersal to other bee nests.

With the possible exception of some Scutacaridae, Pygmephoridae, and other such Acari that feed on fungi growing on dead bee eggs or larvae, the Scutacaridae, Pygmephoridae, Anoetidae (= Histiostomatidae), Laelapidae, and Acaridae cannot be called natural enemies of halictid bees (Eickwort 1979). Eickwort found 15 genera in six families associated with the temperate halictids alone, which appeared similar to the types of Acari associated with the soil-nesting andrenids, colletids, and anthophorids. Cavity-nesting bees have associates of other Acari groups – for example, Saproglyphidae, Chaetodactylidae, and many Mesostigmata. These are found with bees like megachilids and xylocopines that are primarily xylophilous rather than hypogeous (Section 3.1.3), although cells of hypogeous species also are infested (Linsley et al. 1980). Mesostigmatid, laelapid, and cheyletid mites rove the nests of bees and sometimes consume other mites that feed on stored food or fungi (Eickwort 1979; DeJong et al. 1982; Reyes quoted in Delfinado-Baker et al. 1983).

Some mites apparently responsible for the death of immature and adult bees are listed in Table 3.3. The preponderance of these taxa receive attention from honeybee pathologists due to a destructive interaction with domesticated *Apis mellifera* (Bailey 1981; DeJong et al. 1982; Delfinado-Baker and Styer 1983). An aspect of honeybee biology that stands out in the analysis of varroids and laelapids is that

Table 3.3. Probable mite parasites of bees

Mite parasite	Host	References
Acaridae, *Horstia*	Xylocopinae (ANTH)	O'Connor 1988
Chaetodactylidae, *Chaetodactylus*	*Osmia, Hoplitis, Chelostoma, Megachile, Anthidium, Lithurge* (MEG); *Melitoma* (ANTH)	Linsley 1958, Krombein 1967, Linsley et al. 1980, O'Connor 1988
Chaetodactylidae, *Sennertia*	*Ceratina*, Xylocopinae (ANTH)	Daly, Stage, & Brown 1967, O'Connor 1988
Histiostomatidae, *Anoetus*	*Lasioglossum* (HAL)	Eickwort 1979
Histiostomatidae, *Histiostoma*	*Ptiloglossa* (COL)	Roberts 1971
Laelapidae, *Afrocypholaelaps*	*Hylaeus* (COL)	Daly & Coville 1982
Laelapidae, *Bisternalis*	*Melipona* (AP)	Baker, Delfinado-Baker, & Reyes 1983
Laelapidae, *Melittiphis*	*Apis mellifera*	Delfinado-Baker pers. commun.
Laelapidae, *Tropilaelaps*	*Apis dorsata, A. cerana, A. florea, A. mellifera, A. laboriosa* (AP)	DeJong et al. 1982, Koeninger et al. 1983, Delfinado-Baker pers. commun.
Macrochelidae, *Macrocheles*	*Apis mellifera, Bombus* (AP)	DeJong et al. 1982
Parasitidae, *Parasitus*	*Colletes*	Batra 1980a
Podapolipidae, *Locustacarus*	*Bombus* (AP)	Kistner 1982
Pyemotidae, *Pyemotes*	*Ceratina* (ANTH)	Daly, Stage, & Brown 1967
	Osmia (MEG); *Ceratina* (ANTH)	Krombein 1967, Daly, Stage, & Brown 1967
Pygmephoridae, *Pygmephorus*	*Bombus, Apis* (AP)	DeJong et al. 1982
Pygmephoridae, *Sicilipes*	Halictidae	Eickwort 1979
Pygmephoridae, *Siteroptes*	Halictidae	Eickwort 1979
Pygmephoridae, *Trochometridium*	*Nomia* (HAL)	Cross & Bohart 1969, Eickwort 1979
Saproglyphidae, *Vidia*	*Megachile* (MEG)	Krombein 1967
Scutacaridae, *Imparipes*	*Nomia, Lasioglossum*	Cross & Bohart 1969, Eickwort 1979
Scutacaridae, *Scutacarus*	*Bombus* (AP); *Lasioglossum* (HAL)	DeJong et al. 1982, Eickwort 1979
Suidasiidae, *Tortonia*	*Osmia, Megachile, Rhodanthidium* (MEG)	O'Connor 1988

Tarsonemidae, *Acarapis*	*Apis mellifera, Apis cerana, Apis dorsata* (AP)	DeJong et al. 1982
Tarsonemidae, *Tarsonemus*	*Apis cerana*	Delfinado-Baker, pers. commun.
Varroidae, *Euvarroa*	*Apis mellifera, Apis florea, Apis andreniformis?* (AP)	DeJong et al. 1982, Koeninger et al. 1983
Varroidae, *Varroa*	*Apis cerana, A. mellifera* (AP)	DeJong et al. 1982, Needham et al. 1988
Winterschmidtiidae	*Hylaeus, Amphylaeus* (COL)	O'Connor 1988

^aPossible parasitoid of eggs.
Abbreviations: ANTH, Anthophoridae; AP, Apidae; COL, Colletidae; HAL, Halictidae; MEG, Megachilidae.

such parasites are particularly damaging to *A. mellifera*. This bee lacks some behavioral counteradaptations of the other *Apis* species. The mites parasitic on native honeybees in the Old World tropics frequently attack mature larvae of introduced *A. mellifera*, appearing to prefer drones, and kill or deform them (Tangkanasing, Vongsamanode, and Wongsiri 1988). At least one large mite, *Varroa jacobsoni*, has been transmitted from *Apis cerana* to *Apis mellifera* and thus over much of the world within the past century (Bailey 1981; Needham et al. 1988). The drone cells of *A. cerana* are capped by an extremely tough cocoon, lacking in *A. mellifera*, which conceivably prevents dispersal of the mite (Koeniger, Koeniger, and Wijayagunasekera 1981). In addition, these authors show that damage to the colonies of *A. cerana* is minimized due to the completion of the mite life cycle only in the cells of drone bees. Both immature drones and workers of the novel host, *A. mellifera*, are attacked by this parasite. In contrast to workers of *A. mellifera*, workers of *A. cerana* use their mandibles to quickly remove and kill *Varroa* encountered within the nest (Peng 1988; Peng and Fang 1988). A related mite genus, *Euvarroa*, has recently expanded its host species from *A. florea* to include *A. mellifera*, and species of the laelapid *Tropilaelaps* have apparently transferred both from *A. florea* and giant honeybees to *A. mellifera*. It seems likely that these and other mites can be transmitted between bee species at nests or foraging sites, but perhaps only within the genus *Apis* do such transfers lead to mite establishment among novel hosts (Eickwort 1988b). About 160 mite species of 16 families are associated with *Apis* (Delfinado-Baker pers. commun.).

Phoretic mites, frequently adult females, depend on bees to carry them between flowers and foraging sites. The phoretic lifestyle is widespread among mites in general (Eickwort 1979; O'Connor 1982). I found an adult of *Neohypoaspis ampliseta* on the hindtibia of a *Trigona* that was gathering resin. This is an instance in which phoretic mites may transfer between foraging bees in the process of gathering nesting material. Also in Panama, anthophorid bees of the genus *Tetrapedia* have hundreds of mites *Chaetodactylus* living in their nests and feces, apparently only as scavengers (Baker et al. 1987; Roubik 1987b). I have observed the mites arrive on bees foraging for loose dirt, with which they fill spaces around nest cells. Several bees and usually more than one anthophorid genus gather soil within a small area. Seeing a mite walking freely within one such patch of dry dirt suggests to me that mites disperse between nests in this way, as do leiodid beetles (Roubik and Wheeler 1982).

Among occasionally benign but also damaging parasites of bees are eellike nematodes. Nematodes are abundant soil-dwelling invertebrates that infest adult bees, eventually occupying much of the gut and metasoma. Nematode eggs are ingested by adult bees; alternatively, adults or larvae burrow into the immature bee or are merely phoretic. Their natural hosts probably include most ground-nesting bees and other ground-dwelling invertebrates, and repeated occupation of a nest area seems to increase the probability of infestation (Alford 1975; Batra 1980a; Bailey 1981; Crowson 1981; Poinar 1983, p. 183). Tropical nematodes affecting bees are poorly known. Rozen and Michener (1968) state that cell contents of a South African soil-nesting bee, *Scrapter,* are consumed by nematodes. Rozen (1984b) gives an account of aphelenchoid nematodes infesting nests and provisions and crawling over the larvae of *Ptiloglossa;* they are shown passing through pores of the cocoon operculum but were not found in the bodies of bees or known to harm them. Bailey (1981) mentions that *Agamermes* parasitizes honeybees in Brazil. It is therefore likely to have native apoid hosts. Giblin, Kaya, and Brooks (1983) report that nematodes of *Huntaphelenchoides* infested the reproductive tracts of both sexes of an *Anthophora,* and the genus *Acrostichus* did the same to *Halictus*. These did not kill the bees but fed upon fungi or bacteria in their nest. Fungi and bacteria also are the food of rhabditoid and aphelenchoid nematodes, although both groups have been found in the Dufour's glands of tropical *Megalopta* and *Halictus* (Halictidae) in Brazil (Lello 1976) and in North American *Colletes, Halictus,* and *Anthophora* (Batra 1980a). Batra suggests that the larvae consume Dufour's gland lactones (triglycerides or free fatty acids). Nematode infestation of temperate zone bumblebees prior to winter hibernation prevents development of host ovaries, also preventing bees from nesting (Alford 1975; Poinar 1983).

3.2.8 Parasitic bees

Obligate parasitism of other bees has arisen in halictids, megachilids, anthophorids, apids, and perhaps a few colletids and ctenoplectrids. Known or reasonably certain tropical associations are listed in Table 3.4. Parasitic species are lacking in Melittidae, Andrenidae, Fideliidae, Stenotritidae, and Oxaeidae, though these families are parasitized by other bees (Table 3.4). It thus seems that the less species-rich or smaller families (Chapter 1) are unlikely to produce parasitic species, regardless of their often greater age, but the tiny family Ctenoplectridae has a probable cleptoparasite (Michener and Greenberg 1980).

Loss of structures associated with a parasitic life-style has been addressed in Section 2.2.4, and it is complemented here by a brief discussion of derived features of parasitic species that are lacking in related nonparasitic bees. A consistent anatomical feature of some parasitic bees is high ovary and ovariole number, as well as small egg size (Iwata and Sakagami 1966; Alexander and Rozen 1987). Nonparasitic apids and anthophorids possess four ovarioles per ovary, whereas the remainder of known bee groups possess three. In contrast, most cleptoparasitic anthophorids have five ovarioles per ovary, and *Psithyrus* (Bombinae) has 6–18. *Apis mellifera* has an estimated 180 ovarioles per ovary. A larger number of

Table 3.4. Parasitic bee genera known from the tropics

Parasite	Host	Locality	References
Abromelissa (ANTH)	*Centris* (ANTH)	neotropics	Snelling & Brooks 1986
Acanthopus (ANTH)	*Ptilotopus, Centris* (ANTH)	neotropics	Rozen 1969b, Snelling & Brooks 1986, Pickel 1928b
Agalomelissa (ANTH)	*Centris* (ANTH)	neotropics	Snelling & Brooks 1986
Aglae (AP)	*Eulaema* (AP)[a]	neotropics	Zucchi et al. 1969
Allodape (ANTH)[a]	*Allodape* (ANTH)[a]	Eq. Africa	Michener 1970a
Allodapula (ANTH)[a]	*Allodapula* (ANTH)[a]	S. Africa	Michener 1970a
Austrosphecodes (HAL)	Halictidae[a]	neotropics	Michener 1978
Braunsapis (ANTH)[a]	*Braunsapis* (ANTH)[a]	Australia	Michener 1961, 1970a
	Braunsapis (ANTH)[a]	S. Africa	Michener 1970a
	Braunsapis (ANTH)[a]	Malaya	Michener 1966, 1970a
Chelynia (MEG)	*Anthidium, Heriades* (MEG)	America	Linsley 1958
Cleptotrigona (AP)[a]	*Hypotrigona* (AP)[a]	Africa	Michener 1974a
Coelioxys (MEG)	*Megachile* & Megachilinae (MEG)	worldwide	Linsley 1958, Iwata & Sakagami 1966
	Anthophora, Centris (ANTH)	America, neotropics	Rozen 1967, Batra & Schuster 1977
Ctenioschelus (ANTH)	*Centris* (ANTH)	neotropics	Roubik unpub.
Ctenoplectrina (CTEN)	*Ctenoplectra* (CTEN)	Old World	Michener & Greenberg 1980
Doeringiella (ANTH)	*Capoulicana* (COL)	neotropics	Rozen 1984b
Echthralictus (HAL)	*Homalictus* (HAL)	Samoa	Michener 1980
Effractapis (ANTH)[a]	*Allodapinia*	Madagascar	Michener 1977b
Epeolus (ANTH)	*Colletes* & Colletidae (COL)	America, Asia	Linsley 1958, Iwata & Sakagami 1966
Ericrocis (ANTH)	*Centris, Anthophora* (ANTH)	America	Linsley 1958
Euaspis (MEG)	Megachilinae (MEG)	Asian	Iwata & Sakagami 1966
Eucondylops (ANTH)[a]	*Allodapula* (ANTH)[a]	S. Africa	Michener 1970a
Eupetersia (HAL)	Halictidae[a]	Africa	Michener 1978
Exaerete (AP)	*Eulaema, Eufriesea* (AP)[a]	neotropics	Bennett 1972, Dressler 1982, Kimsey 1982a, Zucchi et al. 1969

Parasite	Host	Distribution	Reference
Hesperonomada (ANTH)	Exomalopsis (ANTH)[a]	America	Linsley 1958
Hopliphora (ANTH)	Centris obsoleta (ANTH)	neotropics	Roubik unpub.
Hypochrotaenia (ANTH)	Exomalopsis (ANTH)[a]	neotropics	Zucchi 1973, Raw 1977
Inquilina (ANTH)[a]	Exoneura (ANTH)[a]	Australia	Michener 1965a, 1970a
Lestrimelitta (AP)[a]	Meliponinae (AP)[a]	neotropics	Sakagami & Laroca 1963, Sakagami, Roubik, & Zucchi, in press
Macrogalea (ANTH)[a]	Macrogalea (ANTH)[a]	Eq. Africa	Michener 1970a
Melecta (ANTH)	Anthophora, Habropoda (ANTH)	America, Orient	Linsley 1958, Lieftinck 1983
Mesonychium (ANTH)	Centris (ANTH)	neotropics	Snelling & Brooks 1986
Mesoplia (ANTH)	Centris, Epicharis (ANTH)	neotropics	Rozen 1969b, Snelling 1984
Microsphecodes (HAL)	Lasioglossum, Habralictus (HAL)[a]	neotropics	Michener 1978, Michener, Breed, & Bell 1979
Morgania (ANTH)	Nomia (HAL)	Africa	Bischoff 1923
Nasutapis (ANTH)[a]	Braunsapis (ANTH)[a]	S. Africa	Michener 1970a
Nesoeupetersia (HAL)	Halictidae[a]	Madag.-India	Michener 1978
Nesoprosopis (COL)	Nesoprosopis (COL)	Hawaii	Perkins & Forel 1899
Nomada (ANTH)	Nomia, Halictus, Agapostemon (HAL)[a]	America	Linsley 1958, Abrams & Eickwort 1980
	Andrena (AND); Exomalopsis (ANTH)[a]	America	Cane 1983b, Iwata & Sakagami 1966, Raw 1977
Odyneropsis (ANTH)	Ptiloglossa (COL)	America	Rozen 1966, Roberts 1971
Paracrocisa (ANTH)	Anthophora (ANTH)	N. Africa	Lieftinck 1977
Parathrincostoma (HAL)	Thrinchostoma (HAL)	Madagascar	Michener 1978
Paraepeolus (ANTH)	Tapinotaspis (ANTH)	neotropics	Rozen 1984a
Pseudodichroa (ANTH)	Scrapter (COL)	S. Africa	Rozen & Michener 1968
Ptilocleptis (HAL)	Halictidae[a]	neotropics	Michener 1978
Rathymus (ANTH)	Epicharis (ANTH)	S. America	Vesey-Fitzgerald 1939, Camargo, Zucchi, & Sakagami 1975
Sphecodes (HAL)	Lasioglossum, Corynura, Halictini (HAL)[a]	pantropics	Michener 1978
	Augochlorella, Agapostemon (HAL)[a]	pantropics	Michener 1978, Iwata & Sakagami 1966
	Dasypoda (MEL); Eucera (ANTH)	pantropics	Michener 1978
	Calliopsis, Andrena (AND)	pantropics	Michener 1978

Table 3.4 (cont.)

Parasite	Host	Locality	References
Stelis (MEG)	Euglossa (AP);[a] Ceratina (ANTH)[a]	neotropics	Bennett 1966, Zucchi et al. 1969, Daly et al. 1967, Linsley 1958
	Heriades, Megachile (MEG)	India	Kapil & Jain 1980
Temnosoma (HAL)	Augochlora, Augochloropsis (HAL)[a]	neotropics	Michener 1978
Tetralonioidella (ANTH)	Habropoda, Elaphropoda (ANTH)	Oriental	Lieftinck 1983
Thalestria (ANTH)	Oxaea (OX)	S. America	Hurd & Linsley 1976
Thyreus (ANTH)	Anthophora, Amegilla, Asaropoda (ANTH)	Old World	Rozen 1969a, Iwata & Sakagami 1966, Lieftinck 1972
Triepeolus (ANTH)	Melissodes, Anthophora (ANTH)	America	Linsley 1958, Hurd & Linsley 1959
	Ptiloglossa (COL); Protoxaea (OX)	America	Hurd & Linsley 1976, Rozen 1984b
Xeromelecta (ANTH)	Anthophora (ANTH)	America	Linsley 1958

[a] Social species.

Abbreviations: AND, Andrenidae; ANTH, Anthophoridae; AP, Apidae; COL, Colletidae; CTEN, Ctenoplectridae; HAL, Halictidae; MEG, Megachilidae; MEL, Melittidae; OX, Oxaeidae.

ovarioles allows more oocytes to mature simultaneously, and thus more eggs can be laid in a shorter time. Furthermore, cleptoparasitic bees carry a larger number of mature oocytes in each ovariole than do the nonparasitic solitary bees. Comparisons made by Alexander and Rozen (1987) show that mature oocyte number differs by a factor of three or four between related parasitic and nonparasitic species. The selective advantages include attaining higher oviposition rates and perhaps also allow compensation for high larval mortality.

Few completely parasitic species retain the capacity to handle pollen and nest materials. Extensions of the pygidial area, the terminal abdomenal segments used to shape and compact cells and burrows, have been reduced or completely lost in parasitic Halictidae, as have the area of hairs flanking this zone (Michener 1978). Simplification and reduction in bristles characterizes the labrum of such parasites. A penicillum, or small brush used for spreading the liquid of the cell lining, is present on the hindbasitarsus of nest-making halictids but largely absent among parasites (Batra 1964; Michener 1978). However, either perfectly normal structures or the complete loss of the same structures has been noted among parasitic halictids. The basitibial plate, which functions to support the nesting bee as it pushes its legs against burrow walls, displays this variability in parasites. Other structures of the legs are more developed in parasites, such as spikes on the apex of the tibia and spiny projections along its edge. Because the parasites entering burrows of other bees may encounter them head-on and engage in fighting, these probably afford better traction on the walls of the burrow, as a parasite pushes the host bee (Michener 1978). Mandibles of adults are sometimes slightly reduced in parasites and may serve no important aggressive role. Parasites' mandibles can also be inserted further apart in the head capsule and can be pincerlike structures, with a capacity to grasp the host (Michener 1978). Also, the cuticle is thicker in parasitic species, which provides defense from the host's sting. An elongated and disproportionately large gut is characteristic of the cleptobiotic stingless bees, which carry all pillaged food in the crop (Cruz-Landim and Rodrigues 1967).

a. Types of parasites. In the most general sense, an apoid parasite uses the provisions and nest of another bee to produce its own offspring. Larvae of a social parasite also receive food from host workers or the reproductive host females. Solitary bees may parasitize social species, but the reverse apparently does not occur. Some permanently eusocial bees are parasitic in the sense that they remove food or nest material from a host nest and also must usurp a host's nest in order to reproduce. Such species, termed *cleptobiotic* bees or robbers, and the *cleptoparasitic* solitary bees share certain traits, although most of the social species have lost the ability to forage. Facultative and obligate parasites, social parasites (bees that must have their offspring fed by other species), and the cleptobiotic species share comparable traits, although derived independently. Three of their variations are recognized in the bumblebees alone and indicate steps by which parasitism might evolve:

1. Temporary parasites facultatively usurp the nest of conspecifics. Such parasitic *Bombus* kill the host queen and then usurp its nest. There may be no initial interaction with workers, because if the first brood has yet to emerge, only the foundress queen and her immature progeny are present in the nest; the usurping queen forages, also feeding her worker offspring.
2. Obligately parasitic *Bombus*, two of which are known (Michener 1974a; Alford 1975; Itô and Sakagami 1985), do not produce workers, only more reproductives, after usurping a host nest and brood cells. They can have slightly reduced pollen-handling structures or appear normal, but neither forages pollen.
3. Obligate social parasites *Psithyrus* kill host eggs or larvae, and some kill the host queen *Bombus* (summary by Fisher 1987). *Psithyrus* has no corbicula, and neither it nor the obligately parasitic *Bombus* appears to produce the wax needed for nest construction. An evolutionary sequence could thus be that the parasite is first aggressive toward both the adult host and its offspring and then becomes more dependent on host workers to rear larvae; concomittantly, it loses the ability to gather food for larvae, and finally it loses the ability to build a nest.

The obligate social parasite *Inquilina* studied by Michener (1965a) was not aggressive toward adult host allodapines *Exoneura*, yet no eggs or larvae of the host were present, and there were no adult *Exoneura* in nests occupied by multiple *Inquilina*. In such species it seems that host larvae are either selectively eliminated by the adult parasites or that host oviposition is somehow inhibited.

Two types of parasitic bees are highly eusocial; one is a facultative parasite analogous to *Bombus*, and another is obligately cleptobiotic. The latter, *Lestrimelitta* in the neotropics and *Cleptotrigona* in Africa, rob food and nest material by raiding nests of other bees (Portugal-Araújo 1958; Nogueira-Neto 1970a; Michener 1974a; Sakagami, Roubik, and Zucchi in press). Food is carried in the crop, but the nest material is placed on hindtibiae. The known highly eusocial temporary parasites and nest usurpers include at least lowland African races of the honeybee, *Apis mellifera*, which occasionally invade the nests of conspecifics. In natural conditions, colonies of this bee often form "megaswarms" (Chapter 4), and they may locate and combine with other colonies in the wild (Kigatiira 1988). Another eusocial bee, *Tetragonisca angustula*, usurps nests of other stingless bees, *Plebeia, Nannotrigona, Scaptotrigona, Melipona,* and even one *Lestimelitta* in Panama and Brazil (Sakagami and Laroca 1963; Nogueira-Neto 1970a; Roubik pers. obs.). Both the facultatively parasitic *Apis* and *Tetragonisca* usually forage and build nests in the normal manner, yet their opportunism is extreme; they frequently nest and reproduce in disturbed habitats and within cities. Unlike *Psithyrus* or *Bombus*, the *Apis* enters the nest of a host colony as a swarm (Michener 1975; Silberrad 1976; Fletcher 1978; Kigatiira 1984), after which the queen presumably kills the resident queen, as does *Bombus*. A similar pattern may exist for *Lestrimelitta* and *Tetragonisca*, although the necessary observations have not been made (Sakagami, Zucchi, and Roubik in press). For the *Apis*, the host nest is usurped, as is the labor of the worker population not killed by the invading workers.

Food and building materials are usurped by both solitary and social parasites, but the social parasites alter the nest structure and work the building material. *Bombus* frequently usurps a nest and then modifies its architecture. *Psithyrus* modifies nest structure to a lesser degree by sometimes constructing its own brood cells. The stingless bees *Lestrimelitta* and *Cleptotrigona* can usurp a nest and modify nest architecture, or they steal food and building material, which is then transported to their nest (Sakagami and Laroca 1963; Nogueira-Neto 1970a; Michener 1974a; Sakagami 1976; Sakagami et al. in press). The *Lestrimelitta* make a large temporary entrance at the nest they are raiding, a modification likely to make the raided nest more visible to arriving nest mates and repellent to returning host foragers or secondary parasites (Sakagami et al. in press). No other major architectural modifications are known to be made by these bees when they establish a new nest within a raided nest of another colony.

The highly modified social parasites and robber bees are for naturalists the opposite of other bees. Their adults are unknown from flowers and are collected only in bee nests. At least the 7 meliponine cleptobiotic species of *Lestrimelitta* and *Cleptotrigona*, and probably all 13 social parasitic species among the allodapine genera or subgenera *Inquilina, Macrogalea, Allodape, Allodapula, Eucondylops, Effractapis, Nasutapis,* and *Braunsapis* (Michener 1965a, 1970a, 1974a, 1977a,b, 1983; Camargo and Moure in press; Sakagami et al. in press) rely completely on hosts for all of their resources. The genera *Eucondylops, Effractapis, Nasutapis,* and *Inquilina* found in nests consist solely of parasites. Those found in the nests of other species had unworn wings but had offspring of all sizes, fed by the host. In each of these parasites the mouthparts and feeding structures, eyes, pollen-handling structures (Section 2.2.2), and wing veins are reduced, and the bees have not been seen at flowers. This suggests that they never forage and seldom fly. The cleptoparasitic bees of other bee groups at least forage for nectar. Discovery of a stingless bee nest resembling that of *Frieseomelitta*, but containing a bee described as a new genus, *Trichotrigona,* which had no pollen comb (rastellum) on the hindlegs (Camargo and Moure 1983), suggests that it conceivably is a type of social parasite (Wcislo 1987a). It has never been collected at flowers but does have pollen grains on its body (Zucchi pers. commun.). The related bee groups *Tetragonisca* and *Frieseomelitta* have a weakly defined pollen comb and corbicula, although they do collect pollen.

As might be expected, the host of a parasitic bee is frequently a related species (Bohart 1970; Wilson 1971; Michener 1974a, 1978; Sakagami 1976; Pamilo,

Pekkarinen, and Varvio-Aho 1981; Rozen 1984a; Itô and Sakagami 1985; see also Table 3.4). Both host and parasite are of the same bee family in over 80% of 52 parasite genera having tropical representatives (Table 3.4). Most of the examples in the table are solitary bees, but obligate cleptoparasites (solitary bees) and cleptobiotic robbers and social parasites likely evolved from the facultative parasites or occasional inquilines of social bee colonies. Many of their resource-gathering traits are lost, as is wax production, and new features for defense often are added. Although gregarious nesting in some anthophorid, halictid, and megachilid groups seems to promote occasional intraspecific nest usurpation (parasitism) by a nesting female (Eickwort 1975; Rozen 1984a; Packer 1986), the mere fact that the same type of nesting site is used conceivably allows facultative parasitism in species with no aggregating tendencies. For example, host and parasite bees of the same subgenus account for over 90% of facultative parasitism in *Bombus* (Sakagami 1976). The nesting sites are generalized; often they are the abandoned burrows of rodents (Michener 1974a; Alford 1975). Although increasing specialization of nest usurpation likely led to the evolution of *Psithyrus* (Itô and Sakagami 1985; Williams 1985), a given species of *Psithyrus* apparently has phylogenetically more varied hosts than does a facultatively parasitic *Bombus* (Alford 1975).

The large anthophorid subfamily Nomadinae consists solely of cleptoparasites (Rozen 1966; Michener 1974a), and the Anthophorinae has three parasitic tribes, Ericrocini (= Ctenioschelini), Melectini, and Rhathymini, that parasitize bees in related tribes (Camargo, Zucchi, and Sakagami 1975; Rozen, Eickwort, and Eickwort 1978; Rozen 1984a; Snelling 1984; Snelling and Brooks 1986; Alexander and Rozen 1987). The anthophorid tribe Allodapini also has numerous parasitic species (Michener 1977a). Why has this bee family produced so many obligate parasites? It appears that the family as a whole has active larvae that crawl about the food mass during feeding. They may be preadapted to rapidly find a host larva and food mass, and able to reach the host provisions after emerging from a concealed egg placed in the wall of the host bee's cell or away from the host larva (Rozen et al. 1978; Rozen 1984a; Torchio and Trostle 1986). For contemporary allodapines no such larval behavior exists because food masses are very small and replenished often by adults that constantly care for the larvae and carry them about the open nesting cavity (Figs. 3.4 and 3.9). Because any adult bee in the nest has access to all immatures and food stores, larval mobility per se would not be a requirement for destroying competing host immatures.

Nomadini and other parasitic nomadines push their egg through the cell wall of the host, after making a notch on the inside of the host's cell (Rozen 1977c). Nomadine eggs often taper and have corrugations or annular rings about their middle, which probably allow the egg to bend in a U shape when inserted through the hole or to expand as it absorbs water (Alexander and Rozen 1987). Superparasitism by nomadines is common, which is remarkable for the following reasons. The adult parasite is constrained to a short time period in which it must enter the host cell, make an incision, usually at the base of the cell where pollen provisions will be placed, insert an egg and avoid host detection; if parasitizing a provisioned cell it may remove the host egg (Section 3.2.8d). Other bee families, such as the Megachilidae, have parasitic species with highly mobile larvae (e.g., *Coelioxys*), but this is not characteristic of the entire family. In such cases, the mobility of larvae may have arisen in conjunction with obligate parasitism.

The outside of the cell wall can be the oviposition site for both megachilids and anthophorids (Rozen and Michener 1968; Stephen et al. 1969; Rozen 1984a). Such varied egg placement schemes suggest that some host bees can detect the presence of parasite eggs and remove them (Rozen et al. 1978; Parker et al. 1987). A period of cell inspection precedes egg laying in all bees that have been studied, including the highly eusocial species that regularly eat eggs laid by workers (Section 3.3.4). Excluding the halictids, cleptoparasites usually conceal their egg in the host provisions or in the cell wall, but some parasitic anthophorids place the egg near the top of the cell, and some megachilids place their egg next to the host egg, on the top of larval provisions (Fig. 3.4).

Natural history correlates of an evolutionary tendency toward parasitism are

1. general similarity in life history traits (e.g., nesting site, seasonal activity, food sources) of the host, such that an association can persist long enough for temporary parasitism to become permanent (Wcislo 1987a); and
2. larval mobility, or crawling around the food mass while feeding (Rozen 1984a), because a sedentary larva is unlikely to be effective in locating and killing the host larva with which it competes for food.

Additional characteristics suggest that close relatives are sometimes most likely to evolve parasitic relationships because

3. for *Psithyrus,* the few probable parasitic colletids (Perkins and Forel 1899), and the allodapines (Michener 1974a) there is often close physical resemblance between host and parasite, and in the first group this includes coloration (Pamilo et al. 1981); and
4. there is a higher probability of nest usurpation, or accidental oviposition in the cell prepared by another female, if at least some degree of sociality exists or nests and cells are tightly aggregated; communal nesting or cooperation among sisters seem likely to be prerequisites (Michener 1974a; Eickwort 1975; Myers and Loveless 1976; Raw 1984; Rozen 1984a; Packer 1986).

Such social and nesting traits are relatively rare in bee families that have not repeatedly evolved parasitic species (see Section 1.2.3).

Whether the bees of a parasitic genus or group are monophyletic in the narrowest sense and represent divergence from a single ancestral parasite or are in fact amalgamations of more distantly related bees that became similar due to ecological convergence is not always clear. Michener (1970a) discusses evolution of the eleven known allodapine social parasites. In this case, seven closely resemble their host genera, which are not closely related to each other, and four species are classified in three exclusively parasitic genera. The former seven resemble hosts so closely that they are classified in the same genus; a parasitic lifestyle in the allodapines has clearly evolved several times. All, however, share the traits of reduced eyes, mouthparts, wing venation, and pollen-carrying structures. Parasitic Euglossini of the genus *Exaerete* bear greater resemblance to *Euglossa* than to their host *Eulaema* and *Eufriesea* (Moure 1964; Kimsey 1979, 1987). The other parasitic euglossine genus is *Aglae*, clearly related to host *Eulaema* and *Eufriesea* (Kimsey 1987). In the euglossines, it is conceivable that a temporary parasitic relationship between *Euglossa* sisters did not lead to permanent parasitism until the incipient parasite shifted to another host, probably ancestral *Eulaema* or *Eufriesea,* and then evolved to *Exaerete*. Morphological evidence summarized by Kimsey (1987) suggests that in parasitic euglossines obligate parasitism arose twice.

Several ancestors might be expected for Nomadini; at least five genera parasitize *Exomalopsis* (Rozen 1984a). Therefore, a different *Exomalopsis*, rather than a single ancestral nomadine, may have produced each of them, particularly since communal nesting is common in the Exomalopsini. On the other hand, some nomadine bees have more than one host species, and thus parasitization of an *Exomalopsis* need not imply evolution from this lineage. Some support exists for these proposed routes to permanent parasitism. Rozen (1984a) suggests that communal Exomalopsini oviposit in cells constructed and provisioned by their sisters, and he includes a field observation of two eggs of *Exomalopsis solani* placed in a single closed cell. Bees having more advanced social behavior, however, might not show such tendencies. For example, I have seen *Euglossa imperialis* tear down completed and provisioned cells of sisters. They reuse the resin nesting material rather than oviposit in or otherwise use the completed cell. Strong behavioral traits like these may favor or subdue intraspecific parasitism. The fact that several reproductive females live in a single nest in each group is perhaps the factor most closely related to development of parasitism.

Eickwort (1975) gives an account of facultative parasitism among megachilids that nest gregariously: Over 50% of the bees usurped nest cells made by conspecifics, and this resulted in rates of parasitism similar to that inflicted by other organisms. Pamilo et al. (1981) found good evidence for monophyly in the genus *Psithyrus*, each species of which usually parasitizes one or a few closely related *Bombus,* but they also suggest that certain host–parasite species pairs of bumblebees seem more closely related to each other than to any other taxon. If this were the case, then the independent evolution of parasitism could be responsible for some *Psithyrus*. Michener (1974a), summarizing the literature on bombines, further notes that the male genitalia are similar in the different species of *Psithyrus*, indicating that common ancestry rather than ecological convergence has likely produced the bees included in this genus. Owen (1983) found a higher chromosome number in one species of *Psithyrus* than in *Bombus*. Chromosome study of more *Psithyrus* would verify this sharp distinction.

Bohart (1970) estimated that a parasitic lifestyle evolved independently at least 15 times among lineages of bees and that roughly 19% of the known apoid taxa are parasitic, nearly 3,700 species. Hurd (1979) places the percentage of cleptoparasitic species in the American bee fauna north of Mexico at 20%. A lower figure (14%) can be extracted from the fauna of Panama listed by Michener (1954), when excluding 47 highly eusocial bees. A similar figure, 13%, was obtained from a survey of bees at the edge of the tropical zone in southern Brazil (Camargo and Muzcato 1986). A broad summary by Wcislo (1987b) shows similar ranges in the percentages of parasitic bees in many different habitats. For example, the ranges are 9–28% for 38 surveys in North America (28–45° N latitudes), compared with 8–27% among 12 surveys in the neotropics and subtropical South America (22° N–35° S latitude). A few surveys from paleotropical localities yield ranges of 12–22%. Because the apids, predominantly eusocial species, are largely aseasonal and compose a major portion of tropical bee fauna (Chapter 4), and their parasitic species usually have several hosts, they perhaps should not be included in comparisons with the temperate bee fauna. Their large representation in part of the tropics may be due to their ability both to thwart natural enemies and to store food (Section 4.3.5). If the above estimates are representative for solitary and nonapid eusocial bees, it implies that apids are considerably less likely to evolve permanent parasites. As perennial colonies became a fixed population trait of these bees, the seasonality seen as a central factor in the evolution of obligate parasitism in bees (Wcislo 1987b) was eliminated.

At least one variety of *Apis mellifera* steals food and may steal building material or usurp nests (Michener 1975). The stealing of building material (resin) is common and also reciprocal among stingless bees and euglossines. I have seen two *Euglossa* species remove cerumen and resin from outside the natural nests of two stingless bee species. Likewise, small *Plebeia* remove resin from the nest entrance area of *Euglossa*. Removal of cerumen building material from nest entrances has also been observed among conspecific and other stingless bees (Nogueira-Neto 1949; Sakagami and Laroca 1963) along with honey and pollen pillaging from broken nests, although the cerumen of some species is unattractive to others (Sakagami and Camargo 1964). Thus, although most apids are social and may opportunistically thieve food, building materials, and nest sites of other species, this seldom leads to the evolution of obligately parasitic species.

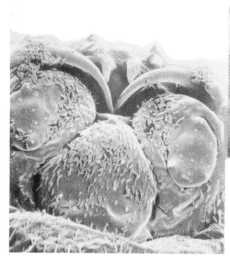

Figure 3.24. Structural adaptations of parasitic bees. Large hooklike mandibles of the parasitoid larva of *Protepeolus singularis* (left), which kills its host larva and then consumes its provisions after moulting to a more sedentary and less well-armed form (Rozen 1978). On the right is the larva of another parasitic anthophorid, *Triepeolus*, which possesses large hooked mandibles and lateral projections that serve to stabilize the larva on the surface of liquid provisions (Rozen 1984b). Copies of the photograph of *Triepeolus* by M. A. Cazier and the scanning electron micrograph of *Protepeolus* used courtesy of the American Museum of Natural History.

b. Larval parasite biology. The larvae of cleptoparasites have unusual morphological specializations that increase their chances of survival in the host cell. Rozen (1966) discusses larval mouthpart morphology of the parasitic anthidiine megachilid *Stelis*. The species with primitive, bidentate mandibles (the same type found in nonparasitic anthidiines) do not kill the host larva; rather, the adult bee removes the egg. Rozen's theory is that aggressive larval adaptations are necesary if the adult parasite does not remove potential enemies or competitors. An *Odontostelis* (= *Hoplostelis* in current usage, Griswold and Michener 1988) parasitizes *Euglossa* in this manner (Bennett 1966). On the other hand, *Stelis* larvae with simple, sharp mandibles do kill the host larva. Often the instars of a cleptoparasitic larva have long, pointed mandibles with which they kill the host larva (Rozen et al. 1978; Rozen 1984a; see also Fig. 3.24). Subsequent larval moults result in larvae having a simple apex and reduced cusp of the mandible. Parasitic halictid larvae lack these specialized traits, perhaps because the host larva is usually killed by the adult bee (Michener 1978). Mandible structure of the first and second instars varies with their feeding schedules and contact with host larvae. For example, larvae that tunnel through the host provisions feed first on pollen and have blunt and rather short mandibles. By the time they come into contact with the host larva, they have undergone a moult to a form with enlarged or pointed mandibles, used to crush or damage the host or merely disrupt its provisions and orientation on the food mass so that it dies. Larvae of parasitic anthophorids have projections from the body segments that allow locomotion and also may prevent them from drowning in liquid host provisions (Fabre 1921; Rozen 1977c). Eickwort (1975) interprets the behavior of specialized host-killing larvae as a preadaptive trait that leads to successful interspecific parasitism and thence to obligate parasitism. Combined adult and larval aggression should perhaps promote the greatest generalization of parasites. Such behavioral flexibility, displayed among species in the genera *Sphecodes* and *Stelis*, is effective in overcoming the defenses of different host species. When a strong chemical weapon is added to aggressive adult behavior, as is the case for *Lestrimelitta* (Section 3.2.8d), then the host species are even more varied.

One can speculate that a reason for superparasitism by a female is high larval mortality due to aggression by the host. For species that place eggs in open cells, the parasite egg risks attack from both the host adult and larva. Attack of a host by a related group of larvae could be an effective means of parasitizing a larger and more powerful adversary. On the other hand, there is often no certainty that multiple parasite eggs do not represent the activity of more than one parasite. The aggressive behavior of parasitic anthophorid larvae in the same host cell is striking, and only one parasitoid survives (Torchio and Trostle 1986). Among parasites, larval aggression might be selectively favored both by competition with other parasites and combat with the host, thus driving the repeated evolution of apoid cleptoparasites.

The larvae of bees that parasitize social bees are sometimes sedentary, not equipped with the sharp mandibles used by other apoid parasites to kill the host, and it is the adult female that destroys the host larva or is aggressive toward the host adult (Michener 1974a). Aggressive behavior of some *Sphecodes* (Halictidae) is subdued toward the adults of social halictids that it parasitizes, to the extent that one species apparently coexists for an extended period with host adults, more so than do bombines or allodapines (Eickwort and Eickwort 1972; Michener 1974a). *Sphecodes* shows considerable variety in its host species and behavior toward them (Table 3.4); some hosts are chased from nests or killed (Michener 1974a).

c. Adult parasite biology. The adult parasite tends to be nearly the same size as the host, but this is often not true of host and parasite egg (Iwata and Sakagami 1966). Eggs of cleptoparasites are usually small compared with all bees but apids, although parasitic Old World anthophorids lay eggs that are big for their adult size (Section 3.3.3). Young larvae have several means of successful predation on the host larva, which may be one factor regulating the size of eggs. Smaller parasite investment in an egg could be a means of compensating relatively larger numbers of eggs produced, particularly because more than one egg is often placed in a host cell. In addition, larval traits of host bees, perhaps including a larger egg size for nonapids or bees that encounter parasite larvae, suggest that a battle between larvae can be crucial. For example, Camargo et al. (1975) provide detailed drawings of larval mandibles through the various moults of *Epicharis rustica* and its neotropical anthophorid parasite, *Rathymus*. The first instars of both species have sharp, pointed mandibles – clearly not a requirement for pollen feeding. The later instars of *Epicharis* have rounded mandibles, which again suggests that in larval life, instars of this age have no need for defense. South African *Scrapter* (Colletidae) are parasitized by *Pseudodichroa* (Anthophoridae) (Rozen and Michener 1968). These anthophorids have pointed mandibles. The young *Scrapter* larvae abruptly turn their heads and open and close their mandibles to full extension. Such behavior is probably common in megachilid and anthophorid hosts and parasites. Rozen et al. (1978) found in detailed studies of another parasitic anthophorid that, although far smaller than its host, the larva attacked when the host was in a quiescent stage, immediately prior to moulting. A means of accomplishing this was through an extended egg incubation period, the result of which was a first-stage parasite larva preying upon a much larger host, in its third or fourth developmental stage. Thus, although the host had well-developed mandibles throughout its larval life, they could not be used in defense. Again, in this study the host (*Diadasia*) had strong, pointed mandibles, which were extended and retracted continuously during feeding. In contrast, the parasite *Protepeolus* kept the mandibles closed while feeding on the pollen stores, but when in its parasitoid stage, these have an elongate pointed hook.

Aggressive interactions between host and parasite should tend to increase the defensive traits of each, but the thickness of the cuticle and size of the sting are generally more developed in parasitic species (Michener 1974a). *Exaerete* kills the host egg of a *Eulaema* or *Eufriesea* after opening the cell of the host (Bennett 1972), whereas *Psithyrus* seldom kills the queen *Bombus* of the host nest (Alford 1975). In *Psithyrus* the sting is larger than that of the host, but this is not the case for *Exaerete*. However, the integument of *Exaerete* is very smooth and thick, which may give it considerable protection from hosts. Variation in mode of parasitism within a single genus may be linked to differing degree of defensive parasitic traits – for example, in *Sphecodes* (Halictidae), which can kill many host bees or cause no damage to the adults (reviewed by Michener 1978). A megachilid bee genus *Stelis* is a parasite of three bee families (Table 3.4). Bennett (1966) saw this cleptoparasite drive *Euglossa* hosts from their nests, and it sometimes kills them (Roubik pers. obs.), ostensibly either by biting or stinging, but the larva of *Stelis* is generally thought to be the aggressor, not the adult (Michener pers. commun.).

Like the larvae, adult female bees have traits corresponding to their roles as apoid cleptoparasites or hosts. Considering that a 50% parasitism rate is not un-

common in those relationships that have been studied (Rozen et al. 1978; Rozen 1984a; Vinson et al. 1987; see also Section 3.2.5), selection for host traits reducing such parasitism should be strong. The response of a host bee to a parasite varies from apparent indifference to pursuit or biting if the encounter takes place within or even near the nest (Rozen et al. 1978). These authors also suggest that entry of bees into nests of other individuals in an aggregation could decrease the rate of cleptoparasitism by disturbing oviposition by a parasite while the host is foraging. Even as an incidental result of tight nesting aggregations, this extension of ordinary cell inspection behavior might confer a selective advantage. In the same paper by Rozen and co-workers, three general traits of parasitic apoids were described:
1. perching near the nest entrance of the host, which allows visual surveillance of host activity and orientation to the exact location of the cell entrance;
2. limited pollen or provision feeding upon entering the host cell; and
3. frequent grooming and cleaning when outside host nests, potentially necessary to prevent contaminating cells with fungi or other microbes that could kill the parasite's larva (Section 3.2.5).

d. Host selection by parasites. Cleptoparasitism between families occurs most often by anthophorid attack on a melittid, colletid, halictid, andrenid, or oxaeid. Of these hosts, only Colletidae and Halictidae have extant cleptoparasitic species. A megachilid occasionally parasitizes an anthophorid, but the reverse apparently does not occur. The very widely distributed *Sphecodes* (Halictidae) parasitizes halictids, melittids, andrenids, colletids, and anthophorids. *Nomada* (Anthophoridae) and *Stelis* (Megachilidae) parasitize at least three bee families. *Triepeolus* (Anthophoridae) parasitizes bees of four families (Rozen 1969b).

Cleptoparasitic bees probably detect active nests by using chemical cues, including the odor of tumuli and the host burrow. This could certainly be facilitated in closely related host and parasite by shared sensory development and nest site discrimination. However, if a host burrow or nest site is in use for a longer period than one parasite generation, then the problem of host location is solved. Several authors contend that parasites follow hosts to their burrows from foraging or mating sites (Rozen 1977c; Tengö and Bergström 1977; Abrams and Eickwort 1980). Following from a shared resting aggregation site (Chapter 2) is also possible. Cane (1983b) reported olfactory host recognition in the anthophorid *Nomada pseudops* and two other species. Their host is an andrenid bee, *Andrena alleghaniensis*. Parasites in this case oviposit before the host and thus are restricted to a very short time interval between completion of the cell and cell closing. Based on Cane's experiments and observations, the *Nomada* did not attempt to locate host bees in flight and follow them to nests; rather, they located the host nest. During a period of a few hours in the afternoon, the *Nomada* could inspect up to 300 nests. Its reaction to various experimental stimuli showed high sensitivity to kairomones, pheromones, and the aromatics of pollen and possibly nectar. Females were more attracted to a pollen-laden host than to a host or pollen alone or to another potential host bee genus. Furthermore, they were highly attracted to a burrow from which another *Nomada* had departed. Four positive reactions of a parasite to the host burrow were defined: hovering, perching a few centimeters from the nest, inspecting the nest entrance, and entering the burrow. Competiton between parasites was evident when they hovered face to face (see Rozen et al. 1978; see also Section 3.3.3). The attractiveness of a recently entered and potentially parasitized cell emphasizes the likelihood of larval competition or occasional conspecific egg consumption among parasitic bees. The risk of encounter with the host is significant. The usual defensive behaviors can and do damage parasites (Michener 1974a; Cane 1983b).

Parasitism by opening the cell of the host has several associated traits. In most cell closures of anthophorines there remains a thinner point in the middle, the *micropyle,* resulting from a small hole in the cell cap through which the tongue of the nesting bee protrudes while finishing the closure. Melectine parasites chew open the cell in this area and then widen it so that the abdomen can curve within the cell and attach an egg to the wall. The cell is then repaired, probably with glandular secretions of the parasite, so that the cell is again properly closed (Torchio and Trostle 1986). Some parasitic *Sphecodes* (Halictidae) may also open and then reseal a host cell, removing the egg or larva (Eickwort and Eickwort 1972). Among the Megachilidae, facultative parasites open the cells of conspecifics, resealing them after oviposition and removal of the host egg (Eickwort 1975).

Further aspects of appearance and behavior of tropical parasites are conducive to speculation. Bright colors and light blue or green markings are common in cleptoparasites such as *Thyreus* and *Nomada* (Anthophoridae), and conspicuous metallic blues or greens are characteristic of *Aglae, Exaerete* (Apidae), *Acanthopus, Mesoplia,* and *Ctenioschelus* (Anthophoridae). Other traits apparently only related to extranidal behavior include frequent bursts of sound while in flight, both in *Aglae* and *Acanthopus*. The bright colors may perhaps stand out for bees. Preliminary study of the responses of male and female *Centris flavofasciata* to a nest parasite, *Mesoplia,* showed a pronounced aggressive response of males. When D. Yanega and I placed dead male and female *Centris* on the tips of long grass blades in a row with a female *Mesoplia,* the male *Centris* repeatedly flew rapidly at and even collided with the *Mesoplia*. No other interactions were recorded, and study of the same bees in Costa Rica demosntrated this cleptoparasite to be a serious natural enemy of *C. flavofasciata*. Because even relatively dull colored parasitic bees can be recognized and attacked by a host (Rozen et al. 1978), there seems to be little selective advantage to the parasite for having inconspicuous coloration, particularly when it only enters a vacant host burrow. The female *Mesoplia* did hover face-to-face in territorial display. Nonetheless, the bright colors of parasites may function as intraspecific signals. The bursts of sound made by some genera and bright colors may be a territory sign for other parasites, in a manner analogous to brightly reflecting colors of male tropical butterflies (Silberglied 1984). The function of bright color or of the other auditory and visual signals of parasitic bees, if this is true, is not to attract other parasites but to signal that an area has already been searched for potential hosts. Avoidance of conflicts with competitors could select for such a seemingly mutualistic signaling behavior.

The degree of specialization by parasitic apids is low compared with solitary bee species. A weak inference can be made that *taxonomic* degree of specialization increases with the proportion of parasitic species in the apid tribe or subfamily. This would indicate that the superior colony and nest defense systems of highly eusocial bees exclude many potential cleptoparasites, and those bees that successfully bypass these defenses have access to many species. Of the four apid groups, the percentages of obligately parasitic species, along with source of the figures on total taxa in groups, are as follows: Bombini, 16% (44/300, Williams 1985); Euglossini, 4% (6/166, Kimsey and Dressler 1986); Meliponinae 2% (7/450, Camargo and Moure in press); and Apinae has no obligate parasites. Both *Psithyrus* and *Bombus* usurp the nests of few species, but local *Psithyrus* in North America do sometimes parasitize all local species of *Bombus* (Michener 1974a and pers. commun.; Sakagami 1976; Pamilo et al. 1981), whereas the six parasitic euglossines may attack several of the 70 or so *Eulaema* and *Eufriesea* (Zucchi et al., 1969; Kimsey 1982a; Roubik and Ackerman 1987). The six neotropical *Lestrimelitta* have over 300 *potential* meliponine host species, and the one African *Cleptotrigona* may raid nests of many small African meliponine species. Tropical *Bombus* have few social parasites, as there are few *Psithyrus* in lowland tropical Asia and the neotropics (Williams 1985). Nest usurpation by other *Bombus* may nonetheless occur (Michener 1974a; Itô and Sakagami 1985). Known cases of nest usurpation in this genus were given for 26 of 70 species reviewed by Sakagami (1976), but the actual proportion that usurps nests might be much higher (Michener pers. commun.). In all of the above comparisons, the actual hosts in nature are unknown. Furthermore, the degree of preference or specialization, like that of a floral preference versus visitation records (Section 2.4.3), is unknown for any parasitic apid species (but see below for an account of apparent specialization by *Lestrimelitta limao*).

Chemical deterrence or damage of a host bee by its parasite is common in the stingless bees. Numerous accounts of attacks by *Lestrimelitta* show that they kill bees of far larger size. Many host workers abandon the nest, although the queen and a group of workers, mostly callows, remain within the nest but avoid the robbers and move little during the raid (Schwarz 1948; Kerr 1951; Sakagami and

Laroca 1963; Nogueira-Neto 1970a; Roubik 1981b). Very elaborate hovering defensive behavior, similiar to that constantly expressed by *Tetragonisca angulstula* (Wittmann 1985), is evoked in colonies of *Melipona fasciata* invaded by a few scouts of *Lestrimelitta* and may also involve the alarm pheromones of this *Melipona* (Smith and Roubik 1983; see also Fig. 3.25). The neotropical *Oxytrigona* attacks honeybee workers many times its size and kills the workers at the nest (Schwarz 1948; Nogueira-Neto 1970b). Both the citral and formic acid produced by these species could be toxic if introduced through a bite (Blum 1981). The latter group has no known stingless bee hosts but possibly engages in intraspecific nest robbing, as does *Lestrimelitta* (Section 3.1.5). It is worth noting that *Oxytrigona* apparently has a strong taxonomic affinity with *Lestrimelitta, Cleptotrigona, Hypotrigona,* and *Trigonisca,* and at least one species of the last group also produces citral in the mandibular glands. The African host–parasite association of *Hypotrigona* and *Cleptotrigona* and their similar nest structure (Portugal-Araújo 1958) suggest close relationship. All these groups share broad genal areas, and the pollen-collecting species have similar hindtibiae, useful characteristics in the taxonomy of stingless bees (Moure 1961; Wille 1983a). Another neotropical meliponine, *Melipona fuliginosa,* attacks and kills honeybees at the hive and sometimes steals honey (Schwarz 1948; Nates and Cepeda 1983). Although it has no obvious mandibular gland odors used in nest raids, it is larger than *Apis mellifera* and other stingless bees and attacks colonies of other *Melipona* (Roubik 1983a). The Kayapó Indian bee specialists of the eastern Amazon report that both *Scaptotrigona* and Africanized honeybees raid nests of stingless bees (Posey and Camargo 1985).

Host protection from parasites by allomones or what has been termed "masking pheromones" is unknown in solitary bees but seems possible in neotropical stingless bees. Most supporting data, however, come from studies of stingless bee colonies brought together in a small area by researchers. Subsequently, the apparent failure of *Lestrimelitta* to perceive a potential host may be an artifact caused by the presence of more attractive colonies. For example, Nogueira-Neto (1970a) saw repeated pillages at individual colonies in the course of 23 attacks by *L. limao,* while surrounding colonies were untouched, including some of the same species that were robbed. I have watched many attacks by *L. limao* of *Scaptotrigona, Tetragonisca, Scaura, Plebeia, Nannotrigona, Paratrigona,* and *Melipona fasciata,* while the same meliponary contained one to four other *Melipona* species and *Frieseomelitta, Trigonisca, Cephalotrigona, Tetragona, Partamona,* and *Trigona.* I have seen colonies of the last two attacked elsewhere. Kerr (1951) and Sakagami, Roubik, and Zucchi (in press) note that *Lestrimelitta limao* seems to avoid nests of some *Frieseomelitta.*

Although the colonies of many meliponine subgenera and species are raided by *Lestrimelitta limao,* long-term observations where most of the local stingless bee colonies are known reveal preferences for *Nannotrigona, Scaptotrigona,* and some *Plebeia* (Sakagami et al. in press). This was found over a large geographic area, and these authors suggested a probable basis for the behavior in the quality of larval food stolen by *Lestrimelitta.* Provision energy and protein values of the preferred hosts of *Lestrimelitta* are higher than those of other stingless bees. In addition, the host bees display behavior that is often effective in preventing raids but also minimizes escalated fighting within the nest after it is penetrated by the robbing bees. The complex interplay of host and robber attractants, allomones, kairomones, and pheromones during raiding and retreat of *Lestrimelitta,* and the initial defense followed by quiescence of the host, changing to renewed defensive behavior at the robber's retreat, will require sophisticated studies before the complete basis for this sequence of events is understood.

Lestrimelitta is a persistent yet prudent robber that seldom kills a colony on its first raid nor eliminates the queen, but returns several times (Michener 1946; Sakagami and Laroca 1963; Nogueira-Neto 1970a; Laroca and Orth 1984; Sakagami et al. in press). The presence of a particular nest of *L. limao* on Barro Colorado Island, Panama, for six years and the intraspecific raids already mentioned suggest to me that a colony establishes its foraging territory and periodically exploits colonies within foraging radius. Crypsis, lack of perceptible odor, and small size of stingless bee nest entrances are among the potential defenses against this colony predator. The most efficacious defense from an active raiding party, however, is possibly that reported by Portugal-Araújo (quoted in Kerr 1984a), who saw *Melipona seminigra* block the entrance of its nest with a hard resin/mud ball.

3.3 Mating and brood production

Combined visual and olfactory senses regulate the mating process of bees. In addition, the physical condition of potential mates is sometimes tested by both sexes, as the male often clings to the female while in flight, and in a few species it may also briefly fly with her (Eickwort and Ginsberg 1980; Thornhill and Alcock 1983). Precopulatory attractants can include pheromones released from mandibular, labial, or Dufour's glands (Smith, Carlson, and Frazier 1985), or both sexes may release pheromones before and during copulation. Furthermore, male bees sometimes spread pheromonal attractants over their bodies (Velthuis and Camargo 1975a,b).

3.3.1 Selecting a mate

Acoustic signaling is at least implicated by the loud and high-pitched buzzing flight of males, which is augmented when male *Xylocopa,* for example, vie for mating territory (Velthuis and Camargo 1975b; Eickwort and Ginsberg 1980; Louw and Nicholson 1983). These sounds are audible to humans from some 100 m, but their significance and mechanism of perception in bees remain obscure (but see Michaelsen, Kirchner, and Lindauer 1986; see also Section 2.3.6). In Java, Lieftinck (1955) noted similar loud sounds by *Xylocopa,* but Velthuis and Camargo also mention species that hover silently. Postcopulatory stimuli such as stridulatory vibration and stroking by the male may enhance the chances of fertilization as well as diminish female receptivity to other males in melittids and anthophorids (Rozen 1977b; Alcock and Buchmann 1984). Smith (Smith 1983; Smith et al. 1985) and Barrows (1975) suggest that odors permitting discrimination of nest mates from unrelated individuals (Section 3.2.2) also promote outbreeding. Smith shows that males display a preference for unfamiliar female pheromones.

Chemical features of the mating act still reveal new and surprising details. Recent discovery of male-to-female transfer of mandibular gland chemicals, which later conceivably permit the female of a cleptoparasitic species to gain entrance to a host nest (Tengö and Bergström 1977), and of odors that possibly reduce female attractiveness after mating (Frankie, Vinson, and Coville 1980; Kukuk 1985) underscore the subtle features that might influence the mating behavior of bees. There is no satisfactory evidence that bees produce what have been termed *antiaphrodesiacs,* to diminish male mating attempts (Wcislo 1987a). It is, however, certain that *Lasioglossum zephyrum* (Halictidae) can immediately recognize an unreceptive female that it has previously encountered (Smith 1983; Wcislo 1987a). Wcislo points out that extremely few female bees are available to male bees, which are thus unlikely to discriminate between females and will attempt to mate whenever given the opportunity. The ways in which females discriminate among the males are more relevant to bee mating behavior. Information as basic as the type and location of glands producing compounds employed in mating are still very incomplete. For example, Vinson, Frankie, and Williams (1986) described for the first time two masses of dorsal thoracic glands in male *Xylocopa* of two very common species (Fig. 3.25), which had previously been overlooked as a source of the fragrant terpenoids produced by territorial males.

Structural modifications related to mating abound in male bees, some of which suffice to indicate their mating behavior. Male bees have one more flagellar antennal segment than the females, corresponding to increased olfactory sensitivity. The compound eyes of some males are greatly enlarged and sometimes nearly cover the head (Appendix B), thereby providing the visual acuity necessary to pursue a female at a mating area. In these rather uncommon groups, males of related species that remain in one place to wait for females have eyes of normal size, probably because they do not require the complex visual information used by patrolling

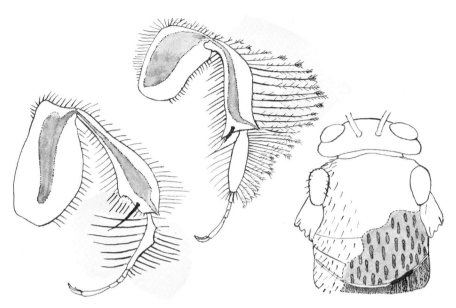

Figure 3.25. Scent glands in the thoracic terga and hindlegs of anthophorid bees. Volatile odors from the glands are released by males at their mating territories (Williams et al. 1984, 1987; Coville et al. 1986; Vinson et al. 1986). Shown on the left are the hind legs of two species of *Centris,* and to the right is a male of *Xylocopa fimbriata.* Glandular regions are darkened; ducts at the apex of the tibia of *Centris* release odors from the legs, and odors escape from an opening near the tip of the scutellum of *Xylocopa* on the metathorax.

males. These include some males of *Ptiloglossa, Centris, Xylocopa, Bombus, Protoxaea, Eulaema,* and *Euglossa* that perch at a designated territory. Males of other *Xylocopa, Centris,* and *Bombus*, as well as those of *Andrena, Nomada, Triepeolus, Hoplitis, Calliopsis, Megachile,* and *Callanthidium,* patrol areas without setting up territories (Eickwort and Ginsberg 1980; Kimsey 1980). That the former group has less conspicuously enlarged eyes is evident in *Xylocopa* and *Bombus* displaying these contrasting behaviors (Stiles 1976; Anzenberger 1977). In the tropical species *X. nigrita,* which perches while waiting for females to approach its presumably chemically marked territory, the eyes are smaller than in patrolling males of two sympatric African species (Fig. 3.26). Enlarged, hairy fore- or midtarsi are used for grasping the female and possibly in performing part of courtship. A femur spike projects from the forelegs of males of some Old World *Xylocopa;* the two spikes hold the female in position during mating (figs. 7 and 8 in Anzenberger 1977). The enlarged hairy foretarsi of such males in this genus resemble those of some male megachilids. Male *Acrosmia* (Megachilidae) have terminal metasomal projections and spines on their sterna that conceivably have a

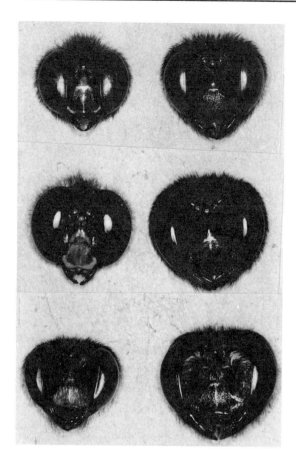

Figure 3.26. Mating behavior revealed in the sizes of compound eyes among male bees of three African *Xylocopa* species (Anzenberger 1977). All the males have relatively larger eyes than their respective females, but the males on the top and middle in the left column possess eyes that are extremely enlarged. This trait is associated with active patrol and flight during mating activity, while the male shown at the bottom primarily remains stationary in its mating territory. Original photographs provided by Gustl Anzenberger.

courtship function (Griswold 1983). Projecting spikes on the edges of terminal abdominal segments are actually rammed into other flying insects by male megachilids of *Anthidium manicatum* (Wirtz et al. 1988). A large, blunt spur at the base of the midfemur permits some males to grasp the female during copulatory flight. Some male halictid and andrenid bees have unusually large heads, which are probably used to fight for mates within communal burrows (Eickwort and Ginsberg 1980; but see Kukuk and Schwarz 1987). Horns are present on the heads of both males and females of some megachilid and anthophorid bees, but whether these are useful in mate competition at nesting sites, as is the case for beetles (Eberhard 1985), remains to be seen.

Solitary bees do interact aggressively when competing for mates (Wirtz et al. 1988), but this is usually a ritualized hovering face-to-face, occasionally followed by brief pursuit or grappling on the ground. The bright facial patterns of most male bees probably provide a stimulus that triggers male contests. Thus they may be the evolutionary product of sexual selection, or competition, strictly between the males. Usually, sexual selection by females occurs in each generation, producing rapid and at times exaggerated secondary sexual characteristics (West-Eberhard 1983; Eberhard 1985). Facial maculations of male bees likely have no significance to females, because males approach them from behind (Eickwort and Ginsberg 1980). The facial markings might enhance the tendency of males to establish territories and hierarchies. I suggest that this is accomplished with less cost if they are highly visible as potential competitors to other males. The size of the marks on the face are variable; if strictly related to size, they may aid a bee to judge the size of its competitor. However, more serious fighting takes place between males at nest sites. In milder cases, males of *Centris*, for example, grapple and bite as they detect and then attempt to excavate the nest cell of a virgin female, rarely causing mutual damage (but at times removing the female's wings, Alcock 1979). Large size differences correspond to this behavior; the larger males weigh three times more than the minor males. Such asymmetry is to the advantage of most unevenly matched rivals, which rapidly establish a hierarchy while avoiding damaging combat (Parker and Rubenstein 1981). In contrast, the huge nesting aggregations of tropical *Centris* observed by Camargo (pers. commun.) are littered with decapitated males. The males kill each other, and their size polymorphism is slight, a widespread correlate of escalated conflict.

Three types of encounter sites for mating are recognized for the bees (Alcock et al. 1978; Eickwort and Ginsberg 1980):

1. nest sites,
2. female resource collection sites, and
3. nonresource sites.

Males of a particular species may encounter females in more than one type of site (e.g., Cane et al. 1983), although it is unknown to what extent individual males switch between them. Clear cases of behavioral flexibility, at times dependent on displacement from mating areas by larger or more powerful males, are found in anthophorid, oxaeid, andrenid, megachilid, halictid, and colletid bees (Linsley, MacSwain, and Raven 1963; Alcock, Jones, and Buchmann 1976; Barrows 1976; Alcock 1979; Eickwort and Ginsberg 1980). Bees unsuccessful in securing mates at a nest site hover and wait for females to pass at another type of site, or hovering bees switch to patrolling large areas. On the other hand, the highly eusocial species are quite rigid in their mating sites. Males form a resting aggregation near the nest for meliponines (Engels and Engels 1984; see also below), and the drones of *Apis* fly within a mating congregation area about 30 m above the ground, in which mating queens release attractant pheromones (Seeley 1985; see later paragraphs). Raw

Figure 3.27. A hovering male of *Xylocopa hirsutissima* in its mating territory (Velthuis and Camargo 1975a). Copy of original drawing by J. M. F. Camargo reproduced with his permission.

(1975) was first to record that tropical male bees (*Centris*) deposit marking pheromones and thus maintain territories, just as some of the temperate bombines that patrol along trap lines (Free and Butler 1959; Schremmer 1972; Free 1987; see also Table 2.4). Chemical attraction of both sexes to the first two types of sites seems likely, but attraction to the last site is probably visual and also chemical at close range.

Searching the environment for mates is not a random process. Male aggregation sites or individual mating territories related neither to nests nor to resources are known in *Apis, Xylocopa, Bombus,* Euglossini, and *Centris,* which are among the few bees to be studied intensively. Such sites are frequently associated with landmarks (Thornhill and Alcock 1983, p. 192). In the tropics, these include prominent hilltops or rock outcroppings; the tops or canopies of vegetation, including trees with massive fruit crops; relatively open or well-illuminated areas in the forest understory; light-colored tree trunks; and depressions in the horizon. The last point is extensively discussed for drone congregation areas of *Apis* by Ruttner and Ruttner (1966, 1972), and their evidence points to nonchemical selection of the sites preferred by *A. mellifera*, some of which have continued in use for 12 years (Seeley 1985). Velthuis and Camargo (1975a,b) proposed that both the males and females of Brazilian *Xylocopa hirsutissima* fly to hilltops and establish territories by perching and hovering to encounter mates (Fig. 3.27), after which the females are attracted to mandibular gland odors of males.

Virgin queens of the four common *Apis* species produce the same mate-attracting pheromone, 9-oxo-2-decenoic acid (9-HDA) and 10-HDA, in the mandibular glands, and several more acids are present but have unknown functions (Free 1987). The former compound gradually predominates in mated queens. However, workers of one south African race, *A. mellifera capensis*, produce 9-HDA when laying eggs (Crewe and Velthuis 1980). Where three or four species of *Apis* are sympatric in southeast Asia, the males of each assemble for mating at different times of the day, and thus are able to use the same site and mating pheromones (Koeninger and Wijayagunasekera 1976; Koeniger et al. 1988). Although these species differ widely in size and interspecific mating is thus unlikely, other tropical bees of the same size and, indeed, genitalic structure do mate within a common territory (Raw 1975). In this case, males of three *Centris* species probably produce different mating pheromones in mandibular glands or hind femora (Stort and Cruz-Landim 1965; Raw 1975; Vinson, Williams, and Frankie 1982; Williams et al. 1984).

In a manner analogous to vertebrates, in which the quality of a territory defended by a male provides a criterion for mate selection (Fretwell and Lucas 1970; Wilson 1975), indirect discrimination based on territory quality takes place in bees. This enhances the selection operating on males to recognize the types of territories in which females are likely to be found or even to defend them. Severinghaus, Kurtak, and Eickwort (1981) saw that *Anthidium* males of larger size maintained larger mating territories at flowers than did smaller males, and they not only pursued but also attacked and damaged other foragers within these territories, as do other perching megachilids. These bees are polyandrous and unusual because the males are larger than the females. However, smaller nonterritorial males had reproductive success at times superior to that of large males, and females showed no tendency to reject smaller males during attempted copulations. Females were choosy about their feeding territories but not about the males within them. This observation also indicates that an ideal free distribution of males among mating territories may occur as a result of male competition (Fretwell and Lucas 1970). Males of differing competitive ability may segregate among mating sites, with a result that better territories are more intensively used, but male fitness approaches equality in different sites. The polymorphism of males might be permitted when

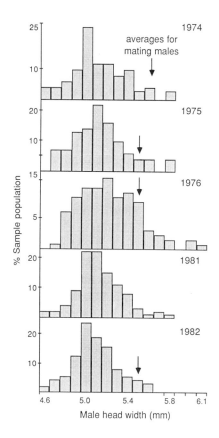

Figure 3.28. Mating success as related to size among males in populations of *Centris* (Alcock 1984).

female competition for flowers is keen, so that territories of all males are entered. A selective advantage also undoubtedly exists for females that can produce smaller and less costly male progeny in these settings (Severinghaus et al. 1981; see also Section 3.3.5). Janzen (1964) noticed that in Mexican *Xylocopa tabaniformis* not only did males pursue any insect within a territory by flowers, but a result of this behavior was protection of a food resource for potential mates. In contrast, other anthophorids seem to ignore all flying insects but the female of their species (Brooks 1983). In other bees that fight with rival males at the nests of females, as does *Centris pallida* (Alcock 1984), larger males are more successful in achieving copulation (Fig. 3.28).

Evidently, the mating system of a bee species can determine the chances for mating at a particular site. For females that mate several times, the mating occurs at pollen or oil sources that only the females collect, whereas females that mate only once are receptive at nectar sources also used by males (Eickwort and Ginsberg

3. Nesting and reproductive biology   275

1980). The frequent records of male anthophorids holding mating territories in tropical flowering trees providing only pollen (Frankie and Baker 1974; Frankie, Vinson, and Coville 1980; Arroyo 1981) suggest that these bees are polyandrous (Alcock and Buchmann 1984). The mating sites of *Centris* might be chosen carefully by males so that females are likely to pass but intense foraging traffic by other species is unlikely (Coville et al. 1986).

If mating site selection is adjustable, then the same should be true of territory size and placement. Either the mate-encounter sites of bees are used by many unaggressive individuals or a territorial individual attempts to prevent the usurption of his area by another male. Perching or hovering territorial males defend rather small areas, no larger than several square meters, whereas those of patrolling bees appear at least an order of magnitude larger (Eickwort and Ginsberg 1980). Territory placement and ownership often change each few days (Severinghaus et al. 1981). The drone congregation areas of *A. mellifera* are 2,000–31,000 m^2 and have been found in Austria at a stable density of roughly 1 per 3,000 km^2 (Ruttner 1985; Seeley 1985). More than 100 drones can be attracted from distances of 30 m downwind from the queen as she enters the congregation area (Gary 1962). Mate competition is intense, but males do not inflict damage upon rivals.

A hierarchy exists among aggressive patrolling bees. Younger and more vigorous males are able to displace older males, regardless of size, in the megachilid *Hoplitis* and possibly in most bees (Eickwort 1975; Eickwort and Ginsberg 1980). Raw (1975) noted that when hovering or perching male *Centris* in Jamaica were permanently removed from their territories, they were replaced by another male the following day. The duration of territory ownership may be for a few days or a few weeks, depending on the degree of competition between males. For a subordinate male, no territory may be owned, and he subsists by staying near the territory of another male until chased away (Alcock 1979; Eickwort and Ginsberg 1980). Sneak mating by insinuating males in the territories of others is likely, such as that documented by Severinghaus et al. (1981).

Precocious emergence of males can be expected in bees like colletids, andrenids, and anthophorids, in which males dig females out of their cells to be the first to mate with them. In these taxa, a nesting aggregation exists and the males locate females by pheromonal cues, possibly detecting the degree of relatedness in a female yet to emerge from her burrow (Alcock, Jones, and Buchmann 1976; Eickwort and Ginsberg 1980). Multiple matings are likely to occur at resource-based encounter sites for other taxa (Table 3.5) nesting in scattered and isolated localities. If the males emerge before females in a particular nest, a pattern common among anthophorid and megachilid bees that build cells in a linear arrangement (Section 3.1), then males are more likely to mate with nonsisters.

Bees that set up mating territories by food plants of the females are obliged to

Table 3.5. *Nest dispersion and mating site in polyandrous bees*

Bee taxa	References
Scattered nests, mating at resource	
Colletidae	
Ptiloglossa jonesi	Cazier & Linsley 1963
Caupolicana yarrowi	Hurd & Linsley 1975
Oxaeidae	
Protoxaea gloriosa	Cazier & Linsley 1963
Megachilidae	
Anthidium spp.	Haas 1960, Jaycox 1967, Alcock, Eickwort, & Eickwort 1977, Severinghaus, Kurtak, & Eickwort 1981
Anthidiellum spp.	Turrell 1976
Andrenidae	
Perdita texana	Barrows et al. 1976
Nomadopsis puellae	Rozen 1958
Anthophoridae	
Triepeolus	Alcock 1978
Peponapis pruinosa	Matthewson 1968
Halictidae	
Augochlorella endentata	Eickwort & Eickwort 1973
Lasioglossum spp.	Michener & Wille 1961, Brooks & Cane 1984
Melittidae	
Meganomia	Rozen 1977
Aggregated nests, mating at nest site	
Colletidae	
Ptiloglossa guinnae	Roberts 1971
Colletes cunicularis	Bergström & Tëngo 1978
Crawfordapis luctuosa	Otis et al. 1982, Roubik & Michener 1984
Andrenidae	
Calliopsis andreniformis	Shinn 1967
Andrena erythronii	Michener & Rettenmeyer
Melitturga clavicornis	Rozen 1965
Nomadopsis anthidius	Rozen 1958
Anthophoridae	
Centris spp.	Alcock, Jones, & Buchmann 1976, Frankie et al. 1980 Roubik pers. obs.
Tetralonia minuta	Rozen 1969
Emphoropsis pallida	Bohart et al. 1972
Ptilomelissa	Roubik pers. obs.
Halictidae	
Nomia melanderi	Bohart & Cross 1955
Melittidae	
Meganomia binghami	Rozen 1977

Source: Modified from Thornhill & Alcock 1983.

move in accordance with the daily changes in flower density and distribution (Severinghaus et al. 1981). Even patrolling males stop at several to a few dozen areas within their trap line, and their movement on the circuit is repeated many times during a day, usually in the same direction (Eickwort and Ginsberg 1980). They may therefore assess resource distribution and abundance, at times without actually feeding at flowers. The correspondence to foraging behavior is a close one (Section 2.3.6), although males of solitary bees can delay their foraging activity until the afternoon (e.g., Raw 1975) while exclusively pursuing females during the morning hours when the females are out of their burrows. Louw and Nicholson (1983) saw that territorial male African *Xylocopa* spent 34% of their time in flight, 65% emitting buzzes while stationary on vegetation, and only 1% of their time feeding at flowers.

Temporal partitioning of foraging and mating behavior by males makes it likely that they do not use some of the nectar sources of the females. The males of *Centris flavofasciata* that I have observed hovering in their mating territories near nests compete for females in the morning and then forage in the late afternoon. Their replenished nectar loads contain high concentrations of sugar, 70–75%, late in the day. No female bee was found carrying nectar containing more than 62% sugar. As predicted by nectar selection based upon metabolic activity (Section 2.3; Bertsch 1984), the more active sex should discriminate more against dilute nectar. Territory holders sometimes opt for new perching or hovering territories, and thus they are not found at a particular site on successive days (Velthuis and Camargo 1975a; Eickwort and Ginsberg 1980).

Nonaggressive males can participate in joint pheromonal marking and patrolling of extensive territories. The neotropical *Bombus pullatus,* for example, was found to maintain a group flight path extending 2.5 km, patrolled by several hundred males (Stiles 1976). Stiles suggests that male *Bombus* from the large and often continuously active colonies in the tropics (Sakagami 1976; Plowright and Laverty 1984) tends to search for established flight paths rather than to initiate new ones. The larger mating territory of this tropical species, compared with two temperate species by Stiles, suggests that in areas having dense populations and large numbers of related males, the mating territories are larger and more stable. This line of reasoning can be extended to the highly eusocial apids. Both the stingless bees and honeybees participate in forming drone congregations that are used repeatedly, associated with a nest site for the former (Kerr et al. 1962).

Odoriferous chemical compounds from vegetation are collected and sequestered by male bees of the Euglossini (Cruz-Landim et al. 1965; Dodson 1967; Dressler 1982; Williams 1982; Ackerman 1985) and *Anthophora abrupta* (Norden and Batra 1985). In the former group, the males have brushy pads on the forebasitarsus that adsorb a fragrance that is then passed to a slit on each hindtibia (Chapter 2). Chemicals sequestered within the glandular pouches in the hindlegs are modified and passed to other areas in the body (Roberts et al. 1982). Roberts and his coworkers performed chemical analyses on *Eufriesea purpurata,* widely known in the Amazon region to collect chemical pesticides such as DDT and lindane. They found that DDT was detoxified, and although largely confined to the hindlegs, it also appeared in other sections of the body. The anthophorid studied by Norden and Batra has a row of fine, flattened hairs on the tip of the labrum, used to adsorb and later apply the sap of an Umbelliferae to a mating territory, mixed with mandibular gland secretions. For the euglossine bees, the odor present at the hindtibiae is sometimes acrid and foul, closely resembling skatole or a musty odor, but for many the odor released after pinching the tibia is slightly sweet.

Application and use of the chemicals collected by the euglossine bees are still completely without a proven explanation. An early hypothesis was that the fragrances are modified to form mandibular gland substances that attract females (Dodson 1975; Dressler 1982). Williams and Whitten (1983) show that the mandibular gland contents of euglossine bees are very different from the chemicals collected at orchids, gesneriads, aroids, and other fragrance sources, but they are practically unique in each species (see Table 2.4). Moreover, some male bees that have never collected from sources of chemicals have most compounds found in mandibular glands of males collected in the field (Williams and Roubik pers. obs.). Hindlegs of euglossine bees have a strong smell, and when they are crushed, other males sometimes attempt to collect the odor, but females are not attracted to it. Attraction of females to crushed heads of males was witnessed by Williams (1982), and I have seen female *Euglossa* come to intact males that were resting on vegetation after I had forcibly extracted nectar from them. However, no evidence has been given that any odor normally presented by a euglossine male attracts females. Earlier suggestions that the odors attract other males to form *lek* areas (Vogel 1963; Dodson 1966; Kimsey 1980), in which passing female bees select a mate, have no convincing proof (Eickwort and Ginsberg 1980; Dressler 1982; Ackerman 1985).

Notwithstanding some disappointing results of chemical studies, male euglossines have elaborate display and precopulatory behavior, and several untested hypotheses have been advanced to explain their significance. Male *Eulaema* and *Euglossa* perch individually on tree trunks, make brief flights away from them, and then return. When they are stationary, some produce intermittent buzzes and elevate the wings (Kimsey 1980). This behavior may continue for hours, and it concludes when a female flies close to the male, whereupon she is seized in the air and a brief copulatory event occurs as the pair falls to the ground (Dodson 1966). Precopulatory visual signaling is likely in *Eulaema,* because the species studied has a conspicuous striped abdomen and bicolored wings (Appendix Fig. B.17). When another male approaches a perching male, aggressive pursuit or extended hovering in a restricted looping flight path occurs until one male departs (Roubik pers. obs.), and in other species the territory holder unhesitatingly chases an intruding male from his display site (Kimsey 1980; Dressler 1982; Schemske and Lande 1984). The last authors showed that the large, uniformly metallic green *Euglossa imperialis* increased the frequency of territorial display behavior after collecting an artificial fragrance. These controlled experiments were performed in two cages, each containing several male bees. Cineole was presented in one cage for 30 min, where it was collected by the males. Thus the propensity to establish a mating territory is related to the collection of chemicals. Kimsey (1982b), following a line of sexual selection theory developed by Thornhill (1979), presented a new hypothesis consistent with these observations. She proposed that the fragrances collected by males provide a means for a female to indirectly assess the quality of a potential mate. In this scenario, the female prefers males able to collect rare chemicals from the environment, presumably in addition to finding food sources and avoiding predation. I would like to add another hypothesis, developed during baiting studies of male bees in Panama (Chapter 4). The females of some *Eulaema,* such as *E. leucopyga,* intensively collect skatole from fecal sources. Their males do not collect this substance, in marked contrast to other *Eulaema.* Skatole may have a function in protecting bee nests from the raids of army ants and from microbes (Brown et al. 1979; see also Section 3.2.6). Male bees might pass the chemical in their hindtibiae to females during brief copulation events, which then permit the female to protect her offspring. No male euglossine is known to inhabit a nest (Zucchi et al. 1969). For the male bees themselves, some collected odors may repel predators or parasites at the exposed resting places that the males use each night. The striking variation in size, color, and behavior of euglossine species promises interesting explanations for mating behavior and mate selection in this group.

Female bees that mate several times usually do so in a short time period. The sperm of some 4–15 males is mixed, but only a fraction is stored in the spermatheca, from which it is ejected to fertilize ova that develop into females (Section 3.3.6). This is the reproductive pattern of *Apis* and some anthophorid bees (Eickwort and Ginsberg 1980; Seeley 1985; Torchio and Trostle 1986; reviews in Rinderer 1986). In contrast, some megachilid and andrenid bees are continuously receptive to new males and mate throughout their lives. Most bees, including the meliponines, appear to mate only once (Michener 1974a; Sakagami 1982; Wille 1983a). No information is available on the number of times an individual male can successfully mate, although some male halictines engage in multiple copulation (Michener 1974a, p. 252). Only apids are restricted to one copulation, because the male genitalia do not disengage. In *Apis* the male is killed, but the meliponine male does not die immediately after copulation.

Visual mating signals attributed to the light-colored male genitalia or exposed membranous area of the female reproductive tract are known in the queen *Apis* and anthophorids (Thornhill and Alcock 1983; Ruttner 1985; Seeley 1985; Torchio and Trostle 1986). In the honeybee, male genitalia remain briefly in the queen to ensure that the semen is kept in the reproductive tract; the genitalia are easily removed by the next male and are apparently designed to facilitate extraction by him (Koeniger 1983, 1986). Thus queens of *A. cerana* accomplish as many as 10 mating flights in a day (Woyke 1975; Moritz 1986). The relative costs and benefits of single or multiple matings have yet to be analyzed for bees; nor is there a clear indication of whether the stored sperm of several males compete, thereby providing furthur opportunity for mate selection. However, Martinho (quoted in Kerr et al. 1980) has purportedly shown that sperm of a single male may be used by multiply mated *Apis mellifera* for up to 20 days. Mixing of sperm from several males within the spermatheca of *A. mellifera* occurs, but there is evidence of clustering by semen of the same type (Koeniger 1986; Laidlaw and Page 1986; Page 1986).

3.3.2   Larval development and nutrition

The egg of a bee usually has minimal contact with food or cell substrates; only one or both tips of the egg touch the provisions, and the egg is gently arched. An embryo continues to develop within the egg, not hatching for 2–5 days following oviposition and occasionally not until after 12 days (Stephen et al. 1969; Sommeijer, Houtekamer, and Bos 1984). As Stephen and co-workers emphasize, more extensive incubation periods are caused by a temperature decrease, and they also identify 23 types of egg placement for various bee genera, which are plausibly related to embryo development rate. Embryo development reveals slight changes that are due to genetic differences among races. For instance, Harbo et al. (1981) showed that Africanized and European honeybee eggs differed by about four hours in the delay before hatching.

The eggs of solitary bees are large, both relative to the size of the adult bee and compared with the small eggs of some apids. Large *Xylocopa* produce eggs that are 15–20 mm in length (Krombein 1967; Iwata 1976), whereas the egg that produces a queen *Bombus* is a mere 3 mm long (Stephen et al. 1969). The eggs of apines and meliponines, as well as those of cleptoparasites, are generally less than 3 mm in length. A study considering 92 bee species of six families (Iwata and Sakagami 1966) demonstrated that most cleptoparasites lay eggs less than 3 mm in length, and *Apis* produces an egg 1.5 mm long, half the size produced by meliponine bees as large as *A. mellifera*. Some of the Old World *Thyreus, Euaspis,* and *Epeolus* (anthophorid and megachilid cleptoparasites) have a larger egg, 3–5 mm long. Solitary bees of widely varying sizes produce eggs 3–6 mm in length. Before hatching, the embryo rotates 90° (the Colletidae) or 180° within its egg and then emerges in the position in which the first instar feeds (Torchio 1984). It rapidly moults to a second instar which may be of almost the same size. Two other moults follow. Thus bees normally exhibit five larval instars. Most larval growth occurs in the last stadium, after connection of the midgut to the hindgut, rapid absorption of nitrogen and other tissue-building molecules, and subsequent defecation. Defecation by earlier instars is known in megachilids, *Xylocopa, Ceratina,* and allodapines, in which case feces are not liquid and are often attached by silk threads to the cell wall or otherwise prevented from contaminating the food of the larva. In the allodapines and also many *Ceratina,* the nesting female removes the feces of her progeny from the nest during their development (Michener 1974a). As stated earlier (Section 3.2.5), all larvae that build cocoons, except some highly eusocial apids, defecate before they spin the cocoon.

Bee provisions have an energetic value that seems to vary slightly among species. However, the amount of protein and free amino acids in provisions can vary substantially among related species (Hartfelder 1986). Bees achieve considerable control over the consistency of larval provisions by evaporating water from their crop nectar before it is regurgitated, combined with pollen, in the provision mass. The labor involved is very clear in large *Xylocopa* that spend many minutes at their nest entrance extruding and retracting a pollen-nectar droplet from their mouthparts (Camillo and Garófalo 1982; Roubik pers. obs.). Meliponine bees make provisions containing 0.4–1.3% free amino acids, 40–60% water, and 5–14% sugar (Hartfelder 1986). The protein content (mg/g fresh weight) of provisions varied 20-fold among the seven species studied by this author. Buchmann and I gathered the provisions from nests of 16 stingless bee species in Panama and found that their energy value varied by about a factor of two. Provisions of *Trigona silvestriana* produce 6.3 J/µl (joules per microliter), or 18,769 J/g dry weight, whereas those of *Nannotrigona testaceicornis* contain the energy of 13 J for the same amount (17,343 J/g dry weight). Such readings vary much with moisture content; when energy content per gram dry weight is considered, the values are more uniform. *Melipona fasciata* was highest with 20,322 J/g, and the *Nannotrigona* was lowest. These values are higher than pollen alone, which produces on the order of 14,000 J/g (Southwick and Pimentel 1981; but see Section 2.1.6). Closely comparable data exist for *Xylocopa capitata* in southern Africa. Its provisions averaged 18,300 J/g (Louw and Nicholson 1983). Slightly higher values were recorded for the provisions of *X. sulcatipes* (20,600 J/g) and *X. pubescens* (19,080 J/g) in Israel (Gerling et al. 1983). *Megachile rotundata,* studied in New Zealand by Wightman and Rogers (1978), had provisions of higher energetic value, about 23,500 J/g.

Measurement of gross production efficiency by bees, a good indication of ability to convert protein to tissue as well as to capture the energy released by chemical breakdown, shows that they are highly efficient. In *Xylocopa* and *Megachile,* just mentioned, during development 54–59% of all energy ingested was incorporated into tissues (as opposed to being lost in defecation and respiration products). In addition, among phytophagous insects, apparent digestibility of leaves has been measured at 30–60% (Waldbauer 1968). This value was considerably higher (78–89%) for *Apis mellifera* fed several types of nutritionally rich pollen; it assimilated 78% of all nitrogen consumed (Schmidt and Buchmann 1985). The *Xylocopa* and *Megachile* had higher efficiencies, utilizing about 90% of ingested nitrogen. Although this sample of species is small, it appears that less dietetically specialized species like *Apis mellifera* are less efficient in digestion of a particular resource. They prefer a mix of pollen to a single species (Schmidt 1984). This implies a lesser likelihood of evolution in response to a particular food species, although it does not suggest a particular cause-and-effect relationship, rather than chance preadaptation (Chapter 2.4).

The digestive process of larvae may be aided by adult bees through contribution of their glandular secretions. Larval salivary glands are active and change though larval life, producing increasing amounts of glycoproteins and glycosaminoglycans for silk production in the mature larva of cocoon-spinning bees (Mello and Vidal 1971, 1979). Comparative study of digestive enzymes produced by worker bees of *Apotrigona nebulata* in Africa, European *Apis mellifera,* and neotropical *Melipona beecheii* (Delage-Darchen, Talec and Darchen 1979; Delage-Darchen and Darchen 1982) showed that 15 of 18 enzymes were found in all three species, either in the mandibular glands, hypopharyngeal glands, salivary glands, or the gut. The relative quantities of these enzymes varied. Considering the supplemental digestive enzymes provided by *Bacillus* and the potentially large contribution of yeasts to larval nutrition (Section 3.2.4), as well as the antibiotics found in female mandibular glands or Dufour's glands, larval feeding can involve the direct contributions of four types of organisms. Dufour's gland secretions were found placed upon the pollen provisions in cells of *Megachile*, where they are possibly consumed by the larva and contribute fatty acids and triglycerides to its diet (Williams et al. 1986). Among eusocial bees, several to hundreds of different workers contribute their glandular products and crop contents to the food placed within brood cells (Michener 1974a; Sakagami 1982; Sommeijer, Beuvens, and Verbeek 1982).

The admixture of floral products and glandular secretions consumed by larvae is complex and incompletely known. Larval nutrition requires vitamins, lipids, minerals, sugars, and proteins (Hydak 1970), all of which are diluted in larval provisions. Pollen species vary substantially in chemical composition, containing from a few to nearly all of the amino acids essential for manufacturing structural proteins (Kauffeld 1980; see also Chapter 2). Therefore, provisioning bees may be expected to supplement or possibly compensate for floral food quality with their own glandular products. Relatively specalized bees that gather pollen from a single plant genus or related genera might evolve specializations that enhance production efficiency. Strickler (1979) compared both the efficiency of pollen harvest and brood production of specialist *Hoplitis* with three other megachilids and *Ceratina* collecting the pollen of *Echium vulgare*. She found that the specialist produced broods more efficiently, and although larval assimilation efficiency was not assessed, it was one component of the combined adult and larval efficiencies measured by this study. A later study (Strickler 1982) revealed that adult body size, throughout the active season, was consistent for the specialist bee but varied, becoming progressively smaller, for the generalist species (Torchio and Tepedino 1980). A potentially confounding element in studies of provisioning by bees is the fact that some portion of the food is occasionally left uneaten. Natural selection has not eliminated the error in matching food to a given larva, although the progressive feeding by allodapines and some apids certainly limits food wastage.

Larval development proceeds according to the quality of larval food, and the caste determination of highly eusocial species is mediated by the larval endocrine system (Hartfelder 1987). The phenomenon is amply demonstrated in *Apis mellifera,* in which queens result from feeding with worker glandular secretions past the stage of the second instar. These glandular secretions are complex mixtures containing approximately 15–20% protein and 15–30% simple sugars; but as both are present in nectar and pollen, the additional vitamins, sterols, and acids produced by the adult are significant in larval development, although still incompletely assessed. In *Apis mellifera*, emergence of the embryo from the egg is delayed for 3 days (up to 5 in the drone), followed by 5.5 to 6.5 days of feeding by workers (Michener 1974a; Dietz 1982). The queen larva is fed for the shortest period and the drone for the longest time, yet the total interval from the onset of larval feeding to completed adult development for the queen, worker, and drone is 12.5, 18, and 21 days, respectively. The same pattern appears in *Apis dorsata*, in which developmental periods were recorded as 13.5, 19, and 22 days (Qayyum and Ahmad 1967). Because queens and drones weigh more than twice as much as workers of *A. mellifera,* yet drones are only slightly larger than workers in *A. dorsata* (Appendix Fig. B.21), developmental rate in this genus seems to have little relation to the amount of food given to a larva.

In *Apis mellifera,* however, the *rate* of larval feeding is correlated with caste development. Queen larvae receive glandular food at a steady tempo, but the worker is fed at a diminishing rate (Brouwers, quoted in Beetsma 1985). Either a worker or queen can result from an individual egg, but the first two instars are fed a similar mixture of honey, pollen, and glandular secretions produced primarily in the hypopharyngeal glands (Michener 1974a). Hypopharyngeal gland products are phagostimulants by virtue of their high sugar content; moreover, the higher titer of glucose than fructose in such secretions fed to queens, compared with workers, suggests that the former encourages more rapid development (Ninjhout and Wheeler 1982; Beetsma 1985). These authors provide evidence that queen development is triggered by activation of the corpora allata of the third instar, which triggers production of juvenile hormone (JH) during feeding by the fourth and fifth instars. Immature *Apis* are sensitive to application of JH primarily as fourth instars (where they show caste differentiation), but in *Bombus* this occurs at the prepupal stage (Ninjhout and Wheeler 1982). Physiological determination of queens also appears late in the development of stingless bees; it can be induced by JH application to prepupae (Campos, Velthuis-Kluppell, and Velthuis 1975; see also Section 3.3.5).

3.3.3   Reproduction of solitary bees

The reproductive cycle of bees in the tropics can be approached by considering the correspondence between adult activity and flowering (Janzen 1967a; Frankie et al. 1983) and duration of the developmental sequence of egg to adult (the *EA period*), in addition to the ultimate general causes of seasonality (Tauber et al. 1986; Danks 1987). Adult bee phenology and the EA period determine the maximum number of broods appearing during a given year but do not necessarily indicate how many are produced by a single female. Generation length can vary and also far exceed the EA period (Garófalo 1985). Escape from parasites can also relate to seasonality, either through unpredictable or protracted appearances of the host (Price 1980).

Bees that nest in seasonally inundated areas, such as *Epicharis* and *Centris* (Section 3.1), spend most of their lives under water. The resting stage occurs for the prepupae (Roubik and Michener 1980), as appears to be true for most bees that do not pass inactive periods as adults (Sage 1968; Stephen et al. 1969; Michener 1974a; Kapil and Jain 1980; Garófalo et al. 1981; Camillo and Garófalo 1982; Camillo et al. 1983; Garófalo 1985; Maeta et al. 1985; Coville et al. 1986). Adult diapause occurs at least in halictids, megachilids, andrenids, and some anthophorids. This takes place either in the natal nest or in a shallow hibernaculum excavated by a mated female. Most male bees (*Xylocopa* is one exception) undergo no such diapause, although individual males and even groups of males of one or of several species form resting clusters during daily inactive periods, either in burrows or on vegetation (Schremmer 1955; Evans and Linsley 1960). Fertilized females of *Agapostemon* (Halictidae) are known to diapause in large numbers in their natal nest, like some other halictids (Michener pers. commun.).

Completion of only one or two broods or generations during a year is frequent among tropical bees that have been studied, which do not differ in this regard from bees in the temperate zone. However, most have been studied in decidedly subtropical areas or near the border of the temperate zone. Three to eight annual broods appear to emerge among solitary bee species having adults present throughout the year (Section 4.1). However, these data are often extrapolated from population trends rather than from direct observation of nesting females. Bonelli (1976) recorded up to four yearly brood cycles corresponding to two generations in the social Ethiopian carpenter bee, *Xylocopa combusta*. Similarly, Beeson (1938) saw four annual brood cycles in the Oriental *X. latipes*. Brood cycles did not directly correspond to generations in these bees because newly emerged adults assist their mother and contribute to the production of sisters before making their own nest. Multiple broods are produced during a single season by many halictids in temperate areas, and each may have proportionately more males, produce no males until the second brood (Stephen et al. 1969; Batra 1987), or produce progressively larger workers (Breed 1975). This trend also appears in tropical halictids (Packer 1985). Packer and Knerer (1987) find that caste differentiation decreases in more seasonal tropical climates of *Halictus ligatus* and that less pronounced seasonal changes in rainfall are correlated with continuous brood production.

Univoltine (single-brood) and multivoltine species are found in the same family and genus, although regional differences also appear in single species. Thus in the subtropical–tropical region of Ribeirão Preto, Brazil, one species of *Lithurge* is univoltine and another species is bivoltine (Garófalo et al. 1981; Camillo et al. 1983). In the dry forest of Pacific Costa Rica, for example, Heithaus (1979a) shows that *Centris analis* is active at flowers for about three months. In the wetter environment of central Panama, I find both the males and females of another subspecies of this bee active at nest sites throughout the year. Variation in many traits among races of a bee species is well documented by Ruttner (1988). *Eufriesea* is a euglossine genus made up largely of species active for only one or two months each year, presumably a single generation, although some of its species are probably bivoltine (Roubik and Ackerman 1987). Other species in the same tribe may often have three or more reproductive cycles within a year (Garófalo 1985; see also Chapter 4).

Little work has been done to determine if physical factors such as temperature, radiation, humidity, photoperiod, or some combination of meteorological events trigger cessation of diapause and emergence of adult tropical bees. Reproductive diapause of adults, as often occurs in Xylocopinae (Michener 1985), remains another subject for future study. Bees nesting in the soil of xeric temperate areas often emerge after spring rains and the major flowering period (Linsley 1958).

Soil-dwelling tropical bees often show the opposite trend, as they emerge after cessation of rains, when the soil dries and flowering of trees begins (Chapter 4.1). Because the roots of trees share the soil environment of bee cells, both being within a few meters of the soil surface, it is likely that some of the same physical changes that lead to flowering also promote pupation and adult bee emergence. Flowering peaks also occur during the wet season (Opler, Frankie, and Baker 1980a,b; Bawa 1983), and particular species are most active at that time (Sections 2.4 and 4.1).

An idea of how many generations of solitary bees can be produced during one tropical year can be gained from surveying adult life spans and the EA period. The two periods are often similar, and so their sum can be divided into a year's length to indicate the maximum number of yearly broods. Foraging success ultimately controls the number of broods or offspring of a female. Dietetic specialization and larval developmental rate interact to lengthen or shorten the EA period, and parasite or predation pressures may place bounds upon brood number and duration. But an example from the simplest type of feeding relationship illustrates the central role of food availability. Louw and Nicholson (1983) studied *Xylocopa capitata* where it forages almost exclusively at the papilionaceous legume *Virgilia divaricata*. By their calculations, a female bee must harvest pollen and nectar from about 1678 unvisited flowers to rear one offspring. The remainder of its adult activity, and perhaps seasonality, is undoubtedly influenced by this stark relationship.

Wood-nesting bees can vary markedly in their annual brood number in the tropics, but *Xylocopa* and *Ceratina* have been studied in some depth and provide comparative data. Many of their species are social during at least a brief period of the life cycle (Section 1.2.3). The EA period is usually 42–67 days (Bonelli 1976; Anzenberger 1977; Camillo and Garófalo 1982). These authors note that temperate xylocopines (Malyshev 1935) have been reported to spend some two weeks more as resting prepupae. Two southern Brazilian species studied by Camillo and Garófalo produce two yearly broods, but EA periods are variable. In equatorial Africa and the Serengeti and on the island of Rubondo in Lake Victoria, two broods per year are produced by four species of *Xylocopa* (Anzenberger 1977). The life span of an adult female is generally terminated when her brood emerges as adults, although the first bees to emerge are fed by their mother. Based on combined information on nest construction, provisioning of cells, and an estimated four cells made by a female during her life, the four species of *Xylocopa* would be expected to live for about 60 days. My computations assume that a few days are required for mating and that the female lives long enough for her first two progeny to emerge. Adding to this period her preadult life, the bee lives for 100 days, passing over 40% of her life as an immature. Bonelli (1976) mentions life spans of up to five months among more colonial African *Xylocopa*. Hurd and Moure

(1960) discuss *X. nogueirai*, which makes up to eight cells in a bamboo internode and thus does not need to invest as much time excavating a space for each cell. Large neotropical *X. frontalis* excavate cells but can also produce four to eight progeny (Camillo and Garófalo 1982). Sakagami and Laroca (1971) reviewed the bionomics of neotropical xylocopines, providing data that the modal number of cells completed by females of seven *Xylocopa* was two to four, with some producing up to eight (see also Camillo and Garófalo 1982). They show that *Ceratina* also are bivoltine. Female *Ceratina* in Punjab, India, were found to live for over 70 days (Kapil and Kumar 1969), and the EA period for tropical species may be near 30 days (Sakagami and Laroca 1971). Based on all the above information, it is likely that some *Xylocopa* and *Ceratina* complete three broods in a year, but Sakagami and Laroca suggest that two yearly broods are probable where there is a single pronounced dry season. Bonelli (1976) summarized all literature on *Xylocopa* and found a few examples of four broods, or possibly generations, in the paleotropics. In contrast, Iwata (1976) notes that some Southeast Asian *Xylocopa* complete a single generation in a year, as is also true of some African species (Watmough 1974.)

The Xylocopini reutilize parental nest substrates that are relatively free of seasonal changes in suitability or availability. They should be among the tropical bees most likely to produce more than one generation and brood in a year. Even these genera are univoltine in the temperate zone (Sakagami and Laroca 1971). In either case, the inactive period between reproductive episodes is passed in the adult stage, whereas the prepupae or other immature forms of most nonxylocopines are present at this time. To judge from the diverse rainfall and climatic regimes in the tropics it can be inferred that these tropical bees reproduce from one to four times in a year. More than one brood, or progeny that reach the adult state, are possible, however, during each generation. But these would have shorter EA periods than average adult longevity.

Some information can be added on the life span of euglossine bees reared in captivity (Ackerman 1985), which also indicate probable maximum generation length in this tropical bee group. Large *Eulaema* and *Exaerete* lived for up to 120 and 154 days, respectively. Females were found to be at least twice as long-lived as the males, in agreement with earlier work by Dodson (1966). Furthermore, male bees caught and tagged in the field were seen up to 56 days later, including species of *Euglossa*, *Eulaema*, and *Eufriesea*. Dodson (1966) reports the resighting of a tagged male *Eulaema cingulata* 165 days later. Bennett (1965) had the impression that *Eulaema* live for several months in normal circumstances, and Dodson and Frymire (1961) again saw a male *Eufriesea surinamensis* that they had marked at flowers four months earlier. Completion of up to at least two generations and several broods in a year is therefore likely and perhaps the rule for euglossines (Section 4.1). Most species are present throughout the year, and seasonal peaks provide the only means of assessing potential generation length unless observation nests are employed. Nesting females of *Euglossa cordata* in protected nests on a university campus in southern Brazil lived 39–191 days (Garófalo 1985).

3.3.4 Reproduction by queens and colonies

Queen bees begin reproductive activity shortly after mating, but their colonies may not reproduce, at least by colony fission, during their lives. The male contribution to colony fitness is potentially large but is also essentially unknown for any tropical social bee. Despite the fact that a wide variety of eusocial bees sustain colonial living for at least several years, their queens normally live less than a year (Malagodi et al. 1985), but may under unusual circumstances live for up to four years (Darchen 1977b), after which they are superseded (Rothenbuhler, Kulinčević, and Kerr 1968; Michener 1974a; Sakagami 1976, 1982; Plowright and Laverty 1984; Seeley 1985). Queen *Apis* are normally replaced after two or three years in the temperate zone, but sometimes after a single year, at least in apiaries, in the tropics (Smith 1966; Kerr 1974; Seeley 1985). Temperate-zone *Apis mellifera* queens fail to return from one-fifth of all mating flights (Ruttner 1988, p. 188). Malagodi and co-workers (1986) report average longevity of approximately 200 days for queens of *A. mellifera ligustica* and 150 days for Africanized honeybee queens that were artificially inseminated and kept in apiaries in southern Brazil. A few queens of the 35 that were studied lived much longer, but neither study followed longevity of queens departing with swarms, which normally occurs at intervals of less than a year (Winston 1979b; Boreham and Roubik 1987). The varied patterns of survival and potentially great longevity seem comparable in meliponines, but long-term studies of marked queens of either group in a natural environment are lacking.

In the Meliponinae and Apinae, colony age structure and worker survivorship are tied closely to food supply and foraging activity. Worker longevity varies inversely with colony activity and food acquisition (Winston 1979b, 1981; Roubik 1982d; see also Fig. 3.29). Reproductive cycles and brood rearing are determined by pollen abundance (for brood production) and general flower availability in the environment. The effect of different floral and weather conditions can be pronounced. Even across a transect of only 40 km, from the forested Atlantic slope to the less forested and drier side of the Panama Canal area, honeybee swarm movement and colony reproduction were at times asynchronous (Fig. 3.30). A general flowering episode in both areas lasts roughly half the year, from dry through early wet seasons (Croat 1978; Wolda and Roubik 1986), but a single pronounced

3. Nesting and reproductive biology   289

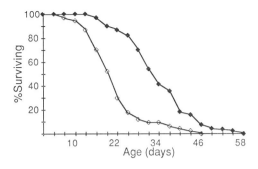

Figure 3.29. Seasonal differences in worker survivorship among cohorts of *Melipona fulva* (Apidae) in lowland forest of French Guiana. Workers survived an average of up to twice as long when foraging activity was less (after Roubik 1982d).

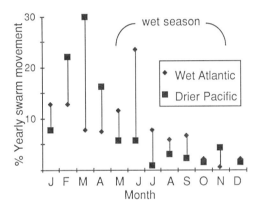

Figure 3.30. Monthly swarm movement by feral colonies of Africanized honeybees in two regions of central lowland Panama (after Boreham and Roubik 1987). The Atlantic side of Panama is more heavily forested and receives twice the rainfall as the Pacific side. Data provided courtesy of M. M. Boreham and the Panama Canal Commission.

reproductive peak is evident on the Pacific side of the canal whereas the reproductive peak and a simultaneous peak of nest abandonment occur during the wet season on the Atlantic side (Boreham and Roubik 1987). Peak seasonal pollen availability to foraging honeybee colonies occurs during or preceding each season (Roubik et al. 1984; Roubik 1988; see also Chapter 4).

Brood production often closely matches accumulation of surplus pollen within the nest (Stanley and Linskens 1974; McLellan 1978; Kauffeld 1980; Roubik 1982c,d; Fletcher and Ross 1985; Fig. 3.31). Worker survivorship, however, appears to change simultaneously and almost inversely with the acquisition and storage of food, reaching its shortest span when foraging activity is most intense (Fig. 3.29). In addition to the influence of such factors of intracolony ecology, the reproductive activity of females is a complicated process. For example, Bego (1982) found that seasonal change in temperature and precipitation had no correlation with the production of males in colonies of *Scaptotrigona*. Key questions concerning food acquisition and reproduction of such permanently social bee colonies remain unanswered. Their degree of independence from current resource

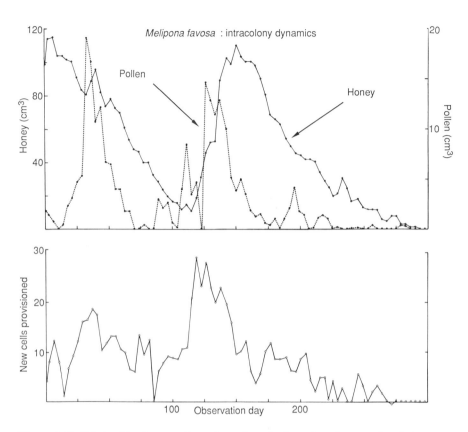

Figure 3.31. Relationships between fluctuating colony food supplies (honey and pollen) and brood production by *Melipona favosa* (Apidae) in lowland forest of French Guiana (after Roubik 1982d).

availability in the environment is implied by the mere existence of stored food in the nest, but colony activities such as food storage area construction are regulated by the amount of incoming food (Section 3.1.4). Whether such colonies or bees possess biological clocks or diapause dormancy (Section 3.1.2) that regulates the production of males or promotes general population synchrony, regardless of resource conditions, remains to be investigated. Within colonies, as will be described, the behavioral and chemical interactions between females are varied and at times quite elaborate, even in colonies with a single queen and uninseminated workers. Their relations to resource conditions (both within and outside of the nest) are still not well understood.

3. Nesting and reproductive biology 291

Extensive study of the oviposition process and its variation among higher taxonomic groups in Meliponinae helps to illustrate the complex manner in which brood production is orchestrated (review by Sakagami 1982; Sakagami, Yamane, and Inoue 1983; Sommeijer 1984, 1985; Sakagami and Yamane 1987). Diverse interactions take place among workers and queens in nests of the stingless bees. Basically, the queen receives most of her food either from the buccal discharges of workers or by consuming their eggs, and she stimulates them to discharge food in brood cells, seal the cells, and solicit more food from other workers.

Reproductive swarming depends not only on the supply of food and immature bees that the colony can afford to invest in new nest establishment but also on location and preparation of a nesting site; the latter process varies widely in its duration (summary by Inoue, Sakagami, Salmah, and Yamane 1984; see also Section 3.1.4). In contrast with the colonies of *Apis mellifera*, in which a healthy queen normally influences worker behavior in some way through at least a few dozen chemicals originating in tergal, tarsal, and primarily the mandibular glands (Seeley 1985; Velthuis 1985; Free 1987), stingless bees are portrayed as "less effective" in their mode of queen influence in colony processes (Fletcher and Ross 1985). Our scant understanding of compounds used by stingless bees in the nest limits such comparisons. The worker bees of *Scaptotrigona depilis* contained 70 chemical compounds in extracts of their heads (Engels et al. 1987). Might queen meliponines possess a comparable array of chemicals? From the above data, chemical signals within the nest or during foraging activity seem at least as complex in the Meliponinae as in the Apinae. Free (1987, p. 22) suggests that a number of the substances found in bee exocrine secretions may serve as keeper chemicals that delay volatilization of active pheromone components. Much work remains to be done on the chemical ecology of colonial bees before tabulated compounds can readily be interpreted in terms of behavior or ecology.

Queen or worker control of development and reproduction of colonies is sometimes due to degree or type of social evolution, but it is also an evolutionary product of life history traits and ecology. As a consequence of multiple mating, honeybee workers are not, on average, more related to their queen than to their own offspring (Hamilton 1972). Thus their benefit from producing sisters rather than personal offspring need not be considerable. Male production by some of the workers can be due to a lack of control by the queen but also may allow her to produce all female offspring (Michener 1974a, p. 253), which should lead to high fitness in genetically viscous populations. Inbreeding does increase relatedness among siblings but does not, however, promote the evolution of nonbreeding females that are found in all obligately eusocial insects (Craig 1982). If the original honeybees mated with only one male, as do contemporary stingless bees, then

Hamiltonian kin-group selection could provide the major force that established permanent eusociality in both groups.

However, the putative vestiges of parental manipulation, in addition to kin-group selection, can be seen in honeybee societies (Seeley 1985). Because selection has proceeded to an extreme degree in the honeybees, there is little to substantiate either a "queen control" or "worker control" theory of evolution for honeybee colonies. These females have evolved an obligate relationship, and its evolutionary development is difficult to retrace. Odor production by the queen, if it much resembles that of primitively eusocial halictid bees, probably arose from a direct correlation between the size and reproductive capacity of a female and the amount of odor she produced (Smith and Weller in press). The fact that a honeybee queen produces many different odors need not imply that these were ever selectively advantageous in manipulating conspecifics. It could also mean that conspecifics once used odor as an indication of potential fecundity and that in honeybees the odors simply indicate the presence of a queen (Seeley 1985). The queen of some stingless bees also produces odors that signal her presence to workers (Sommeijer 1985). These are dispersed within the brood area when a queen exposes her tergal abdominal glands, which are absent in workers (Cruz-Landim 1967; Cruz-Landim, Höfling, and Imperatriz-Fonseca 1980; Sommeijer 1985). If the queen did not disperse the odor, the functioning of the colony could be disrupted, but she does not have a dominant role in colony activity due to this behavior. Furthermore, a few stingless bees such as some *Trigona, Scaptotrigona, Partamona,* and *Ptilotrigona* have either colonies several times larger than those of any *Apis* or more than one brood area separated by as much as a meter. Stingless bee queens perform in a structurally more complex setting and should thus differ from *Apis*. Odor production by the queen nonetheless occurs in these two groups. Considering also that construction of a new nest cannot possibly be controlled by the queen stingless bee, because it involves diverse environmental variables that the queen cannot assess (Section 3.1.3), workers must control a major aspect of colony reproduction.

Sakagami and Zucchi (1974), Sakagami (1982), Sakagami et al. (1983a,b), and Sakagami and Yamane (1987) have gathered considerable information on the oviposition processes and associated behavior of stingless bees in Brazil, Africa, Sumatra, and Taiwan. Sommeijer, Houtekamer, and Bos (1984) performed related studies with two *Melipona* and *Frieseomelitta* in Trinidad. Consistent trends are found within supraspecific groups (subgenera), both in cell provisioning and queen oviposition. However, they vary widely among higher categories, suggesting independent evolution of reproductive behavior patterns. Four types of provisioning and cell construction patterns are recognized:

1. *Successive.* Provisioning of cells occurs singly; thus as soon as one cell is provisioned and the queen oviposits, provisioning of the next cell is initiated.

This pattern is almost entirely restricted to species with small colonies, about 1,000 workers or less; but surprisingly it also occurs in *Trigona* and *Tetragona,* both of which may have more than 10,000 workers in a nest with two to five times this many brood cells. Furthermore, the brood cells are either constructed simultaneously, such that all are the same size and only one oviposition cycle is completed each day (*Frieseomelitta* and *Trigonella*), or cells are under varying stages of construction at any given time. Successive provisioning by bees of relatively moderate colony size is displayed by the Asian *Tetragonula* (Sakagami, Yamane, and Inoue 1983).

2. *Synchronous.* Many cells – that is, a few dozen – receive provisions at about the same time. This pattern occurs in species with colonies of moderate size, having about 3,000 adult workers.
3. *Semisynchronous.* Successive cell provisioning is followed by synchronous provisioning during the growth of the colony, shown in *Scaptotrigona* and *Lepidotrigona*, the former of which ranges greatly in colony size but usually has colonies of less than 10,000 adults.
4. *Composite.* Synchronous provisioning occurs on all the cells of a given comb, but cells on two or three combs are provisioned in succession. This type of provisioning is evident in *Lestrimelitta,* which nonetheless builds cells synchronously on more than one comb in some populations (Roubik 1979a; 1983b). *Lestrimelitta* have colonies of several thousand workers.

A constant feature in the colonies of stingless bees is that the queen must be near a food cell to encourage provisioning workers to seek food from other workers, which is then brought back and discharged within the cell (Sommeijer, de Bruijn, and van de Guchte 1985). During this process, the queen often regurgitates a droplet of pollen, honey, and possibly a glandular secretion, holding this in place while moving the mandibles. Feeding of workers with this droplet has not been reported. Oviposition by the queen occurs in all of the patterns previously listed for cell provisioning and construction. She may, however, deviate from the cell construction scheme. For instance, cells that are constructed synchronously can receive provisions and queen ovipositions in batches (Sommeijer, Houtekamer, and Bos 1984).

Queen–worker interactions provide a wealth of comparative material with which to appreciate the varying social organization of stingless bee colonies. During the oviposition process, the queen often is closely attended by workers that discharge food provisions into the cell. They rush about in some groups, such as *Partamona* and *Geotrigona,* or move more sluggishly. In most of the species studied, the "postdischarge" worker rapidly leaves the site of cell construction. This behavior resembles escape from the queen and her domination, but to interpret it thus seems incorrect (Sakagami, Yamane, and Inoue 1983; Sommeijer, de Bruijn, and de Guchte 1985). The worker rushes to find another worker, one specialized in carrying provisions from colony storage pots to bees participating in brood production (Sommeijer et al. 1985; see also Fig. 3.32).

Other interactions between a queen and workers are highly ritualized and dis-

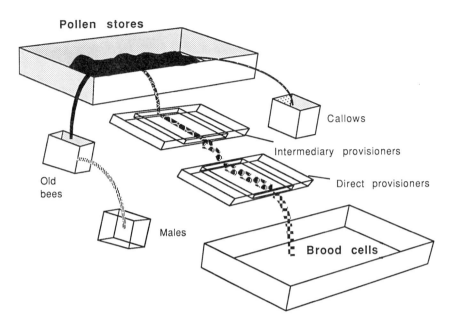

Figure 3.32. The restricted food flow within a colony of *Melipona favosa* (Apidae); adapted from Sommeijer et al. (1985).

tinctive. They provide striking examples of the vestiges of reproductive competition. Queens routinely inspect a cell before oviposition, eating some of the regurgitated provisions and any egg laid by a worker. Worker eggs are often placed on the tip or side of an open cell, not in a position where successful eclosion and feeding would take place (Fig.3.33). They are more spherical and smaller than the queen egg in *Melipona* and the African *Meliponula,* but far larger in *Scaptotrigona* and *Schwarziana,* or subequal in size in normal colonies of most species. The trophic or anucleate eggs not laid on provisions are eaten by the queen and, in some cases, also by workers (Sakagami and Zucchi 1963; Sakagami 1982). Large trophic eggs are laid by older workers. Neotropical *Cephalotrigona* even presents an egg directly to the queen, by flipping its abdomen over the dorsum so that the egg protrudes in front of the head. More unusual is the behavior of a worker *Lestrimelitta limao,* which eats its own egg while in frontal contact with the queen (Sakagami 1975). Furthermore, in normal colonies of *Melipona rufiventris,* Sommeijer, van Zeijl, and Dohmen (1984) showed that eggs laid by workers have no surface reticulations and easily topple into the provisions, making larval development less likely (Section 3.3.2). In a queenless colony, however, worker eggs laid on provisions are similar to those of a queen and develop into male bees.

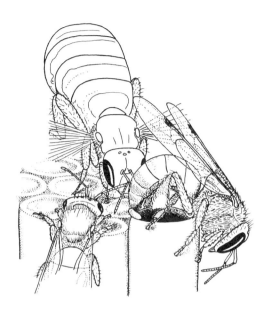

Figure 3.33. The queen of *Scaptotrigona* (Apidae: Meliponinae) preparing to eat a trophic (anucleate) egg being laid on the edge of a brood cell by a worker bee. Original drawing provided by S. F. Sakagami.

Sommeijer and his colleagues have thus given convincing proof that the queen influences egg development in workers, although the method is still unknown. Nonetheless, sneak oviposition is performed in some species after the queen has laid but before the cell is closed (Sakagami 1982). Cells containing the eggs of a queen are apparently not opened by workers. However, Sakagami (1982) reports that cells are opened and worker eggs and provisions are removed or eaten in queenless colonies. In queenright colonies, it is significant that worker-laid eggs are precocious and hatch rapidly, in the process consuming the fertilized egg laid by the queen (Beig 1972). As in honeybees, several worker eggs apparently may be laid in the cells of queenless colonies, and larvae compete within the cell (see also Section 3.3.2).

In contrast with some social halictines, bombines and apines, in which worker ovarian development is absent or largely inhibited, worker stingless bees of all the groups studied but *Duckeola, Frieseomelitta*, and some Asian *Tetragonula* have completely developed ovaries and are thought to produce most of the males in a given colony (Michener 1974a; Bego 1982; Sakagami 1982; Sakagami, Yamane, and Inoue 1983). Many stingless bee colonies have males during most of the year. Some species have drone populations equal to one-half the number of adult workers (Roubik 1979a, 1983a; summary by Sakagami 1982). I have seen colonies of *Melipona marginata* with relatively large drone populations that help ripen the honey. Imperatriz-Fonseca and da Silva (quoted in Sakagami 1982) discovered in work with meliponines in southern Brazil that males contribute to colony maintenance in other ways. The males may secrete wax, work with cerumen, and transfer to storage pots the nectar of incoming foragers. These attributes apply to a seemingly natural group, *Plebeia–Schwarziana–Melipona* (Camargo 1974). Males and virgin queens in most stingless bee colonies, however, are normally driven from the nest or killed.

Regulation of drone, worker, and queen production is so different in *Apis* from the stingless bees that their independent evolution of permanent sociality appears unquestionable (Winston and Michener 1977; Sakagami 1982; Wilson 1985b). A queenless colony of *Apis* utilizes young larvae to make "emergency" queens, whereas no such behavior occurs in meliponines (Michener 1974a; Ruttner 1988). Colony fission of *Apis* occurs when a group of very young workers departs with the mother queen. Although *Apis* and the meliponines differ in their swarming behavior, both groups display, to greater or lesser extent, dependence of a daughter colony on the mother colony (Ruttner 1988, p. 116). This author notes that gradual nest construction for a daughter colony can take place before the swarm departs in *Apis dorsata*. During a single reproductive cycle, two or three swarms normally emigrate, following the primary swarm, each with a virgin queen (Seeley 1985). The feedback system that regulates brood production, particularly queen production but also that of drones and workers, is pheromonal and influenced in at least three ways by the queen (Free 1987). One mode is purely physiological. The principal queen substance 9-ODA is acquired by workers that frequently lick the queen and touch her with their antennae and then disperse this chemical throughout the nest (Velthuis 1972; Seeley 1985). Queen supersedure occurs with failing pheromone production; worker ovarian development and construction of queen cells proceed. More than one queen pheromone must be present to inhibit queen cell construction and the single components are ineffective alone (Free 1987). Free emphasizes that while 9-ODA and its hydrated form 9-HDA can, when applied experimentally in large quantities, compensate for the lack of other components present in the mandibular gland secretion of a queen, further unidentified chemicals of the glands, including tergite glands, are necessary to regulate worker ovarian development. A second mode is also chemical, but difficult to understand without considering the physical influence of the nest. When the queen walks under the brood combs, she deposits an unidentified "footprint" pheromone from glands in the fifth tarsal segments (Lensky and Slabezki 1981; see also Fig. 3.13). If the queen infrequently passes through this area, queen cells may be produced there. However, queen cells are also made in uncrowded nests and are not made on the side combs of the nest, where queens seldom go (Free 1987). Volatile pheromones originating in immature bees have an effect nearly equal to that of the queens's glands in inhibition of worker ovarian development.

Drone production by *Apis* is often confined to swarming periods. The number of drones and drone cells produced is regulated by a negative feedback originating in part from the queen and in part from adult drones and drone cells already in the nest (summary by Seeley 1985; Rinderer, Hellmich, Danka, and Collins 1985). About 1% of the workers in colonies of *Apis mellifera* normally have developed ovaries, but the workers of tropical races more readily develop ovaries when not in direct contact with the queen and brood (Fletcher and Ross 1985). Workers with developed ovaries are harassed by other workers, but they are also more often fed by them (Velthuis 1985). Ovarian development lags several days behind disappearance of a queen and her larvae. Seeley (1985) postulates that the role of 9-ODA in signaling the queen's presence was perhaps secondarily derived from a function of attracting male bees. In fact, the queen's mandibular glands, which produce 9-ODA, are largest at the time she mates. The reverse appears true in stingless bees. Cruz-Landim, Höfling, and Imperatriz-Fonseca (1980) observed a mated queen of *Paratrigona* with far larger mandibular glands than a virgin. When the four paired tergal glands were examined in this species, virgin queens and laying queens displayed similar development.

Honeybee workers only lay unfertilized eggs in the absence of a queen and her larvae. However, a southern African race of *A. mellifera* is thelytokous and thus produces both workers and queens (Section 3.3.5). The drones produced by laying workers of *A. mellifera* are smaller than normal drones when they are reared in the smaller cells of workers. Even drones produced by worker ovipositions in drone cells are more variable in wing hook number (Lee 1974), and some are probably at a disadvantage during mating flights (Section 2.3.3). Virgin queens display behavioral dominance over workers, but the workers regulate the timing of queen departure with a swarm and, in preparation for swarming, cause a fecund queen to reduce her weight by 25% (summary by Seeley 1985).

In both the honeybees and stingless bees, the flow of food among colony members is typified by uneven distribution and discrimination based upon widely differing factors (Butler 1974; Korst and Velthuis 1982; et al. 1983; Inoue, Sakagami, Salmah, and Nukmal 1984; Sommeijer, de Bruijn, and de Guchte 1984). No trophallaxis between mature females, however, normally occurs in the social halictines, bombines, and some eusocial anthophorids, in which the dominance of one egg layer over other individuals derives from direct behavioral interaction as well as through chemical means (Michener 1974a, 1984). Food transmission in the nests of highly eusocial apids influences brood development, worker and male reproductive status, and queen oviposition potential (Fig. 3.32). In some of the stingless bees – *Melipona favosa,* for example – all food transfer between nonforaging adults is initiated by begging rather than by a spontaneous offering (Sommeijer et al. 1984). Queens and older workers are usually unsuccessful. Further, the quality of food (sugar concentration) exchanged between workers has correlates with phylogenetic group as well as with the degree of ovarian development (Martinho 1975; Korst and Velthuis 1982). These findings show that the "social stomach" of some colonial Hymenoptera (Wilson 1971) does not exist in colonies of some highly eusocial bees, for their food is not readily shared.

The question of how often highly eusocial bee colonies reproduce in normal forest or other tropical habitats is relatively obscure (Singh 1962; Reddy 1980; Delgado and Amo 1984; Boreham and Roubik 1987), although studies of African honeybees in Africa and their descendants in South America provide reasonably complete data, reviewed in Chapter 4. The dry season is universally appreciated as a major swarm production season for tropical *Apis* (Fig. 3.29). One to four swarm cycles in a year have been surmised for *Apis* (Ruttner 1988); in the tropics three of the swarm cycles may be completed following wet-season flowering peaks in Panama (Chapter 4). In marked contrast, stingless bees seem either to undergo colony reproduction only once during the dry season (Nogueira-Neto 1970b; Posey 1980; Bego 1982; Inoue et al. 1984) or to reproduce so infrequently that they may barely succeed in replacing their colony during the decades in which it exists. The transition period during which the daughter colony depends on the food and nesting materials of the parent colony may be protracted in stingless bees but nonetheless end in failure of the new colony and abandonment of the nest site (Nogueira-Neto 1953; Wille and Orozco 1975).

Although data are scant for the highly eusocial colonies of honeybees and stingless bees, it seems that both reproduce conservatively. That is, colony investment is biased toward the mother queen rather than the colonies of her daughter queens. Data to be reviewed suggest that new daughter colonies, as a rule, have lower expected fitness than the mother colonies. Supporting field data are lacking, but the following description of the process of reproductive swarming help to establish the hypothesis. Conservatism in colony reproduction might help to avoid competition for resources and mates, which occurs between an established colony and offspring colonies in the immediate vicinity. A given queen's output of queens and drones is uniformly promoted by workers so long as she has higher potential fitness than her daughter queens. Both honeybee and stingless bee queens may live for one to four years. Over this time the rate of egg laying and supply sperm stored in the queen's spermatheca decline (Page 1986). However, unless a queen has declined to nearly one-half of her original productivity, the workers will not replace her with their sister queens. Seeley (1985) offers this conclusion on the basis of his concept of genetic relatedness in honeybee colonies where workers control production of queens but cannot control the patrilines represented by those queens. According to Seeley's formula, the value (genetic similarity) to workers of female reproductives produced by the mother queen is approximately 1.8 times that of reproductives produced by queens that are sisters of the workers. The genetic relatedness of a worker to a sister queen, compared with that of the female offspring of a sister queen, is measured as a function of the number of males mating with queens and is given by the following formula:

$$\text{Genetic relatedness} = \frac{2(n^2 + 2n + 2)}{(n + 1)(n + 2)}$$

where n is the number of males that fertilize the sister queen (the numerator in the above equation) or the queen offspring of a sister queen (the denominator of the equation). Here the payoff to workers that have foregone their personal reproduction is expressed solely as the production of new queens and colonies by their colony. Thus for n from 1 to 20, the ratio remains between 1.67 and 1.91. Therefore, although the queen *Apis* (*mellifera* and *cerana*) mates with several to more than 20 drones (Ruttner, Woyke, and Koeniger 1973; Woyke 1975; Adams et al. 1977; Seeley 1985; Koeniger 1986), and the stingless bee with apparently only one (Sakagami 1982; Wille 1983a), workers are expected to show a bias toward supporting the fitness of their mother at the expense of the fitness of their sister queens. This applies to replacement of the mother queen in a given nest, but may also hold back the production of queens that establish new colonies. The relatedness of a worker to new queens produced by her colony is always higher when those queens are sisters rather than nieces.

An intriguing correlation between worker behavior and genetic relatedness to reproductive females comes from studies of *Bombus hypnorum* (summary by Fletcher and Ross 1985). In this case, queens were not produced unless there were 1.8 times more larvae (potential reproductives at the end of the colony cycle of this primitively eusocial bee as at that time only queens and males are produced) than workers. If there were fewer than this number, then the workers would perhaps have not derived their full potential benefit from helping their mother queen. If this ratio is not mere coincidence, then the reproductive behavior of *Bombus* may provide confirmation of Seeley's rule, according to which queens should produce 1.8 times as many potential reproducers as could workers by helping sisters, or via personal offspring. Therefore, bumblebees should not contribute to queen production and subsequent dissolution of the colony until this ratio has been reached.

Perhaps the colony fission that results in new colonies occurs only when both mother and daughter colonies have even chances for survival. However, queen replacement in a given nest is much more conservative, and a bias toward the mother colony seems to continue even when a colony undergoes reproduction. This bias is expressed in completely different ways and has a contrasting explanation in the stingless bees and honeybees. The bias can be seen in the manner of nest construction and proportion of workers that leave the mother nest to form new colonies.

Underlying the variation between meliponines and apines is the fact that a laying meliponine queen cannot fly and thus cannot leave the nest to initiate a new colony. Although the new nest is initiated and provided with stored food before meliponine swarming, the old nest, along with its greater worker population and food reserves, is maintained by the old queen. In *Apis,* the old nest is occupied by a new queen, but she still requires more time than the departing colony to produce her first reproductive swarm. In making 19 observations in French Guiana of reproduction by prime swarms and the colonies remaining in the nests that produced them, Otis (1982) recorded more rapid reproduction by the swarms. This reflects two biases in colony fission. First, the prime swarm that leaves with the queen comprises about two-thirds of the workers (Seeley 1985), all of which are completely engorged with honey, and many carry six to eight wax scales produced with the nest's resources (see Table 2.1). Second, nearly all of the youngest bees in the nest depart at this time, leaving most of the older bees behind; thus greater longevity in members of the swarm further increases the difference between mother and daughter colonies (Winston and Otis 1978; Seeley 1985).

Inoue et al. (1984) compare an Asian stingless bee to temperate and tropical *Apis mellifera,* showing that over twice the portion of the mother colony leaves with a swarm of *Apis* compared with *Tetragonula*. However, the completed nest left by *Apis* almost certainly represents a larger investment in the fitness of an offspring colony than does the incipient nest provided by the meliponines. The disparity is reduced considering that two or three additional swarms are produced in the same swarming cycle by *Apis*, with the final colony to occupy a nest being far smaller than the swarm of the mother queen and her afterswarms (Winston 1981; Otis 1982; Seeley 1985; see also later paragraphs). In the temperate zone, one set of observations showed that only 25% of new colonies nesting during the summer survived the winter, while 80% of the colonies in established nests did so (Seeley 1985). This did not even include the portion of swarms that failed to establish nests. Presumably, many if not most of the swarms that failed were those headed by new queens, rather than the mother queen and the larger primary swarm that accompanies her. Thus continuation of the colony remaining in the parent nest is important to the original queen's fitness among both stingless bees and honeybees.

Studies of the survival of dispersing honeybee swarms in tropical areas have not been made, but one might predict greater nesting success in the absence of resource scarcity and extreme low temperatures. Of the few daughter swarms produced in a tropical swarming cycle, the mother queen and her large primary swarm seem to have the best chance of survival. In addition, she will be the first to lay eggs and take advantage of the favorable resource conditions that brought about the swarming cycle, despite the need for construction of the new nest (Otis 1982).

Neither honeybees nor stingless bees share their peculiar reproductive pattern with Bombini or most eusocial Halictidae, in which lone mated queens disperse from the mother colony and, sometimes after a period of dormancy, establish a new colony completely on their own. Many social Euglossini, Xylocopinae, and a few Halictinae reuse the parental nest, occupying it sometimes for only a brief period when the mother is still present (Michener 1974a). The nest provides large stores of resin in the first bee tribe and principal burrows in the last two subfamilies, each of which must enhance the fitness of the new reproductives due to time and energy thus conserved.

The reason that a mother queen *Apis* leaves the nest rather than remain there, as do meliponine queens, includes several components. One of these relates to the fact that honeybees can produce more than one swarm in a reproductive cycle, whereas the stingless bees apparently never do this. First, the presence of more than one inseminated queen in a honeybee nest provokes a battle to the death between queens. They have a large sting that is not barbed and thus can be used repeatedly. Since no mechanism exists for protection of one queen from another by workers, a queen cannot without considerable risk occupy the nest in which her own reproductive daughters will emerge. Some races of *A. mellifera* – the Mediterranean *syriaca, lamarckii, intermissa,* and *sicula* – maintain multiple virgin queens in nests, at least until one has successfully mated (Ruttner 1988, p. 171). In contrast, stingless bee workers regularly kill virgin queens or, in some groups, imprison them within the nest; and the mother queen loses her flight ability when she begins to lay eggs (Sakagami 1982). Second, the multiple mating of honeybee queens can lead to sibling rivalry (Getz, Brückner, and Parisian 1982; Getz and Smith 1983). Uncooperative interactions among workers are normally suppressed by the chemical influence of a laying queen and her larvae (Free 1987), but when the prime swarm and mother queen depart, groups of workers can recognize and form "subcolonies" with the emerging virgin queens to which they are most related (Getz et al. 1982; Breed 1983; Page and Erickson 1986; summary by Gadagkar 1985). These groups arise due to the uneven mixing of sperm stored in the queen's spermatheca, which although representing several paternal lines, tend to be used in batches (Page 1986). Part of one male's sperm will fertilize a number of ova during a short time, and the sperm of each male is not used equally (Taber 1955; Kerr et al. 1980; Page and Metcalf 1982; Moritz 1983). One can postulate that if the genotypic distinctness of these lineages is extreme, then swarms continue to depart until the remaining colony is below a survivable size, despite inheritance of the parental nest.

3.3.5   Genetic variability of bees

In bees (and the order Hymenoptera in general), all normal males are haploid and therefore represent mature gametes, having only one set of maternal chromosomes. All female Hymenoptera, whether solitary or social, are diploid. In bee colonies, the females having the same parents have been termed "supersisters" (Laidlaw 1974), because their genetic parents are their mother and grandmother. This is so because recombination between chromosomes does not occur during spermatogenesis by the haploid males; thus the maternal gamete that gave rise to any male bee is passed directly to the offspring of the females that it sires.

Accumulating evidence attests to infrequent expression of more than one allele at the genetic loci of bees; they tend to be relatively monomorphic. The hypothesis most frequently invoked is that haplodiploidy reduces the amount of genetic variation in a population (summary by Sylvester 1986). Deleterious recessive genes are eliminated rapidly because they are always exposed to selection when present in males. Other hypotheses involve the expected environmental stability in the nests or cells of bees. Nesting conditions may be so stable that genetic variability does not promote survival during larval growth and development. Alternatively, strong selection on females may have diminished the variability in genes regulating nesting behavior, or produced a small range of genetic variants adapted to varied conditions. The first hypothesis is favored by comparative evidence from termites. These insects, though colonial and able to regulate the nest environment, are diplodiploid. Their genetic variability seems to be much higher than that of Hymenoptera (Sylvester 1986). No tests have been made refuting the other hypotheses.

Environmental variability experienced by a population tends to be strongly related to phenotypic variation (Bryant 1974; Hedrick et al. 1976; Hedrick 1986). Lack of genetic polymorphism was striking in bee studies using the technique of gel electrophoresis (Mestriner and Contel 1972; Snyder 1974; Wagner and Briscoe 1983). The 45 enzyme systems studied for *Apis mellifera*, using various types of gel electrophoresis, have shown that the loci controlling them (which usually include more than one locus per enzyme system) are less than approximately 10% polymorphic (Sylvester 1986). This technique has enjoyed wide application since the 1960s because it is relatively easy to use (Lewontin 1974). The macromolecules studied in electrophoresis have roughly 80% correspondence to underlying differences in DNA sequence, for more than one DNA sequence can produce a similar macromolecule (Lewontin 1985). Amino acid substitutions in a gene are revealed when gene products are forced to migrate through a gel in an electrical field. Differential migration of such gene products are made visible by enzyme–substrate binding reactions; thus separate bands appear for different phenotypes. These are genetic if shown to obey the rules of Mendelian segregation at a locus, but they are usually regarded as equivalent to morphological or other phenotypic variants under genetic control.

Current population genetic appraisal of animals shows a general tendency for 33% polymorphic loci and 10% heterozygosity at loci in an average population (Lewontin 1974, 1985; Nevo et al. 1984). Genetic study of insects such as *Drosophila*, the most intensively researched animal genus, indicates 40% polymorphism and 14-15% heterozygosity. Even excluding perhaps excessively polymorphic insect species such as *Drosophila*, Hymenoptera show low polymorphism (Sylvester 1986).

Summarizing the available information on bees, Kerr (1974) wrote that 3 out of 27 (11%) of the loci studied in tropical and temperate *Apis mellifera* were polymorphic, and this proportion has not changed significantly since more sensitive techniques, such as isoelectric focusing, have come into use during the past decade (Sylvester 1986). Snyder (1974) found that none of 25 electrophoretic loci was variable in *Augochlora* and *Lasioglossum* (Halictidae) as well as in one species of *Bombus* in Kansas. These were studied using a variety chemical conditions, but further research on the same populations of *Lasioglossum* have revealed polymorphism in 5 out of 29 loci (Crozier, Smith, and Crozier 1987). Several *Bombus* studied in northern Europe showed low heterozygosity and polymorphism: 11–14% of the loci studied were polymorphic, and 0–4.8% of loci in an average genome were heterozygous (Pamilo, Varvio-Aho, and Pekkarinen 1978a; Pekkarinen 1979; Sylvester 1986). Pamilo and co-workers also obtained estimates of 0–6% heterozygosity in some Andrenidae, a colletid, and a melittid. Mean heterozygosity of 1–4% was found for one *Megachile* and one *Nomia* (Lester and Selander 1979). Similar levels of heterozygosity have been found in one anthophorid bee and several genera of wasps (Sylvester 1986). Wagner and Briscoe (1983) examined 20 loci of two Australian meliponines, *Tetragonula* and *Austroplebeia*, encountering a complete lack of enzyme polymorphism within species. However, Yong (1986) found allelic variability in two of nine electrophoretic loci among 20 colonies of one Malaysian stingless bee and also some degree of variation in other stingless bees from peninsular Malaysia.

The largest survey study has been that of Snyder (1977), who analyzed a wide range of social and solitary bees. In his study, an average of between four and eight individuals were assayed at 48 loci, in each of 49 species representing 41 genera and 8 families. This study applied sodium dodecyl sulfate electrophoresis, a comparatively conservative technique in which molecular weights rather than molecular charges are assayed. Although insufficient samples were taken to determine the degree of polymorphism in populations of these bees, in 20 of the 48 electrophoretic loci, 90% of the bee species shared the same form of the gene product.

The above results perhaps indicate remarkable genetic convergence, but they more likely reveal the inherent biochemical conservatism within the Apoidea. Making several assays of an individual using different chemical systems (usually employing sequential acrylamide gel electrophoresis) reveals considerable differences in gene products, often undetected in a single assay in one set of chemical conditions (Hung and Vinson 1977; Lewontin 1985). Gene products failing to appear in one chemical system may nonetheless be shown in another. Furthermore, like other taxonomic phenotypes, band patterns can display false convergence. Similarity in gene product chemistry does not necessarily show that the underlying DNA sequences are the same. Geographic variability also increases the difficulty of characterizing the genetics of bee populations. For example, in India three *Apis* species – *florea, dorsata,* and *cerana* – showed no variation in 12–20 colonies at two electrophoretic loci (Nunnamaker et al. 1984). The same species were studied from 25 colonies in Sri Lanka, and 12–25% polymorphism was found at 16 loci; studies of *A. mellifera* using the same sequential technique also reveal polymorphism that previously was undetected (Sheppard and Berlocher 1984; Sheppard 1985).

Monomorphism at a genetic locus need not correlate with morphological or behavioral uniformity, as conspicuous evolutionary modifications have been found in populations expressing very slight biochemical divergence (Lewontin 1974). However, all such electrophoretic surveys deal with only a few dozen of a species' genes. They are therefore susceptible to sampling errors, but the remarkable lack of polymorphism in two very distinct stingless bees led Wagner and Briscoe (1983) to develop an explanation that may have general significance for bees. It should be clear from Sections 3.1 and 3.3.2 that bees carefully regulate the environment and food of immatures, despite the temperature variability in nests of both eusocial and solitary species. If the bee nest cell environment is taken to be quite stable, then genetic variability with a function in development and physiology may be attenuated in bees due to selection operating upon the nesting behavior of females. Furthermore, selection operating on larval metabolic functions that are geared to certain pollen or food sources, both in free-living and parasitic bees, conceivably reduces the amount of genetic variability maintained within populations. On the other hand, as suggested by Sylvester (1986), selection for a generalized genotype that would allow similar larval or adult performance under varied conditions might account for monomorphism in bee populations. Comparative studies of related species displaying varied control of the nest environment or diets should help to test this hypothesis.

Biochemical genetic study of bees has arrived at the threshold of understanding genetic variation as a function of environment. A recent discussion of the general phenomenon of low genetic variability in Hymenoptera by Gruar (1985) concluded that selectively "neutral" alleles, thought to account for high genetic variability shown in electrophoretic work, are rare in bees. However, work recently performed to study variation in honeybee mitochondrial DNA, which is only passed on from female to female, shows that these genes also display relatively low polymorphism (Moritz et al. 1986). The cause of such reduced variation can therefore not be explained on the basis of haplodiploidy. Clearly, experimental work using mitochondrial DNA in female bees can progress toward allowing an understanding of how nesting behavior and bee metabolism affect the genome. Sequential gel electrophoresis and intensive work at the DNA level should be useful to reveal which types of genes belong to polymorphic or monomorphic groups (Lewontin 1985), and the probable causes of these patterns.

Preservation of more than one allele at a locus in haplodiploid populations is promoted by sex-limited genes, which are only expressed in the diploids; deleterious recessive genes can persist because they are never exposed in males. Kerr (1974, 1976) and Rothenbuhler et al. (1968) state that 20–40% of genes in *A. mellifera* and about 17% in *Melipona marginata* are sex-limited. These estimates might imply that genetic variability is preserved in bee populations; but a low level of electrophoretic variability is common among bees. In all Hymenoptera, alleles that produce mortality or sterility in the haploid males are eliminated each generation. This selection also reduces the frequency of recessive genes that would have similar deleterious effects if expressed in females (Crozier 1977). Thus genes with detrimental effects are eliminated more rapidly than would be possible in a diplo-diploid population given a particular selection intensity. This applies only to non-sex-limited genes. Strong selection on females to regulate their nesting behavior is necessary and perhaps sufficient to rapidly eliminate mutant alleles affecting nesting and reproduction. As suggested above, experimental work would be useful to establish such strict regulation, which also might apply to metabolic pathways associated with certain resources. Dominance might tend to evolve for such genes, which could in turn mask the effects of detrimental recessive genes and then eventually produce monomorphic loci (Lewontin 1974; Spiess 1977). Because outbreeding is promoted in many different ways (Section 3.3.1), some portion of the female genome retains genotypic and probably phenotypic variability. The traits related to pheromonal cues used in mating are likely to be quantitative, and they should display additivity. In summary, selection for female reproductive success has possibly led to fixation of single genes in bee populations. In contrast, the fitness costs of inbreeding, a direct result of the genetics of sex determination (Section 3.3.6), have promoted some genetic variability, which may be preserved among sex-limited genes.

Deleterious recessive genes are increasingly exposed to selection by inbreeding; the mortality arising from expression of recessive alleles is termed *genetic load*

(Kerr 1967; Spiess 1977; Page 1980; Kerr and Almeida 1981; Schemske and Lande 1985). Genetic load has been quantified in terms of *lethal equivalents* for tropical and temperate species of *Drosophila*, for which it was found to be 1.43–1.46 lethal equivalents per gamete (Spiess 1977, p. 289). Recent calculations for several tropical bees were obtained by Kerr and Almeida (1981). They ranged from 1.10 to 1.57 for *Apis* and four *Melipona* and were 1.08 for *Scaptotrigona* and *Lestrimelitta*. These estimates considered diploid males to be either eliminated by workers or otherwise unviable. A conclusion from these studies is that, in the relatively viscous populations maintained by bees, particularly the highly eusocial species that do not disperse over a great distance, a sizable portion of the larvae of each generation may die, in effect, genetic deaths. In haplodiploid populations, these deaths may often involve sex-limited female traits (Kerr 1975, 1976).

3.3.6 Chromosomal variation, sex determination, and sex ratios

The haploid chromosome number of bees has been measured in a few dozen taxa and ranges from 6 (potentially a mistaken karyotype; Ramberg, Kukuk, and Brown 1984) to 25 (Kerr and da Silveira 1972; Kerr 1974; Owen 1983; Ramberg et al. 1984; Moritz 1986). A haploid chromosome number of 16 is widespread, known in Megachilidae, Anthophoridae, Apidae, and Halictidae. Speciation driven by ployploidy and chromosome fusion is evident in the three bee families for which several species have been studied. Halictid females putatively display diploid ($2N$) chromosome numbers of 12, 16, 24, 26, 32, and 34. Female anthophorids possess 16, 24, and 32 chromosomes; the Bombinae (Euglossini and Bombini) display $2N = 32, 36, 38, 40$, and 50 (Owen 1983), and meliponines display varied $2N$ of 16, 18, 28, 30, 34, and 36. In contrast, the four common species of *Apis* and presumably all members of this genus are $2N = 32$ (Kumbkarni 1964; Fahrenhorst 1977). As shown by Kerr (1972), low-temperature shock can lead to polyploid sperm production in *Melipona,* and this author also believes (Kerr 1974) that the survival and fertility of diploid males could lead to the production of triploid workers or queens.

The sex ratio of bees is a subject that spans ecology and genetics, including kin selection, the mechanisms of sex determination, genetic load, and investment in reproductives. More importantly, it involves the timing of reproduction and the mortality patterns of larvae and adults of both sexes. Basic to the understanding of these phenomena is the fact that females sequester the sperm in a spermatheca, from which some sperm is excluded after mating, and egg fertilization is subsequently controlled (Gerber and Klostermeyer 1970; Michener 1974a; Page 1986). Furthermore, an ovipositing female occasionally fails to fertilize eggs that she deposits in cells intended for females, usually those having larger size or more food (Raw and O'Toole 1979; Knerer 1980). The first authors reported a 7% surplus of male *Osmia* due to such errors.

Immature bee mortality, particularly in highly eusocial colonies, can vary substantially as a consequence of the genetics of bee sex determination. One group of polymorphic loci in bees includes those related to sex and caste expression. Inbreeding in any bee population reduces the number of alleles at the sex locus, so that when the population is monomorphic, half of all offspring are presumably not viable; in *Apis mellifera,* usually 12 sex alleles are maintained and declining brood survival accompanies reduction of allelic variants within the population (Woyke 1986). Kerr (1974) reviews four sex-determination schemes that have been postulated for bees, relating them to two more envisioned for the wasps. The basic mechanism in bees was thought to involve loci distributed over a few to several chromosomes, mostly autosomes (chromosomes present in both sexes). More recent work with honeybees favors a model involving a single, multiallelic locus (Laidlaw and Page 1986). It seems that male-determining alleles function in a weakly additive or nonadditive manner, and female-determining alleles are additive. For example, if an individual carries an equal number of male- and female-determining genes, it is a female. In theory, 8–16 alleles at a single locus determine the sex of *Apis* (Woyke 1979, 1986). If both sex-determining alleles at a locus carried by a diploid zygote bee are the same (homozygous), then it becomes an unviable male, which is usually eaten while it is a larva in two species of *Apis* that have been scrutinized for this trait (Woyke 1963, 1980). The fate of diploid male *A. cerana* is less clear, and some may reach maturity (Woyke 1980; Hoshiba, Okada, and Kusanagi 1981); tremendous variability in drone size within single nests could be due to this postulated survival, at least for colonies I have seen in southern Japan.

Diploid eggs occasionally result from unfertilized female eggs, despite a sex-determination system that does not allow homozygosity at certain loci in females. Normal males only have one allele at each of the sex loci or the single sex locus, because they are haploid. A heterozygote is a normal female. Unfertilized eggs of workers occasionally (1–2% of eggs layed by workers) give rise to diploid workers in European *Apis mellifera* (Mackensen 1943). The figure is near 50% for the unfertilized but fully viable diploid eggs produced by workers of the southern African *A. mellifera capensis*. In this thelytokous species (uninseminated diploids can produce fertile diploid offspring), new queens can be produced by workers in queenless colonies (review by Kerr 1974). Thelytoky has further been detected by Daly (1966, 1983) in the North African and Mediterranean *Ceratina dallatorreana* and by Maeta, Kubota, and Sakagami (1987) in the cleptoparasitic anthophorid *Nomada japonica.* Both these anthophorids and *A. m. capensis* appear to undergo automixis, thereby allowing recombination between chromosomes before parthenogenic production of eggs. Genetic variability may still be generated, suggested in the *Ceratina* from varied coloration patterns among offspring (Daly 1966). Furthermore, the males of *C. dallatorreana* and *N. japonica* are unknown, making them appear to be obligately parthenogenic bees.

In the stingless bee subgenus *Tetragona* and neotropical *Bombus atratus,* Tarelho (unpublished work cited by Kerr 1974) and Garófalo (1973) have found evidence that two loci with at least two alleles, located on different chromosomes, are responsible for sex determination. In the former, heterozygotes are female and a double homozygote is a diploid male; and in the latter, a double heterozygote is a female and homozygotes at one or both loci are diploid males. Plowright and Pallett (1979) modify this hypothesis to account for the inordinately high proportion of diploid, and therefore subfertile, males that would be produced through inbreeding by *Bombus*. The proportion would be 75% if only sib–sib matings are considered, but 94% if aunt–nephew matings also occur. As these authors show, inbreeding has very negative consequences for colony growth and reproduction, and they suggest that worker–male conflict should be particularly intense in a nest of *Bombus* to ensure that males (usually produced by workers) do not mate with the virgin females.

Caste and sex determination in *Melipona* follows another pattern (Kerr 1974; Camargo et al. 1976; Kerr and Almeida 1981; Kerr 1987). In contrast with *Apis,* in which the development of a queen from a diploid egg is determined by the amount and quality of larval food (Section 3.3.2), *Melipona* display a system whereby the potential for development to a queen has a genetic component. A recent version of this hypothesis is given by Camargo et al. (1976) in studies of four *Melipona*. They suggest that two loci are involved in sex determination. Workers are produced by homozygosity at one locus, double homozygotes are diploid males (for *M. quadrifasciata,* C. A. Camargo 1979) or sterile workers (Campos, Kerr, and da Silva 1979), and queens are double heterozygotes. As noted in the last two papers and originally alluded to by Kerr, Stort, and Montenegro (1966), and then demonstrated by Campos et al. (1979) and Kerr, Akahira, and Camargo (1975), the endocrine system and JH production of the prepupa triggers development of a female *Melipona* into a queen. Prior to this stage, it seems that heterozygosity at a single genetic locus determines the sex or JH activity of the larva, as it may in *Tetragona, Apis*, and *Bombus*. Whether either food quality or quantity is primarily responsible for the switch from worker to queen is unknown.

Following the discovery of genetic influence in meliponine caste determination, apparently confined to *Melipona*, Darchen and Delage-Darchen (1971, 1977) offered the view that caste determination in *Melipona beecheii* could be influenced by massive overfeeding of larvae. At that time, experimental data from several studies were incomplete, but they gave rise to the notion that queens are reared when food is plentiful, and as a consequence of the sex-determination mechanism just outlined, fully one-fourth of the colony population potentially consists of queens (Michener 1974a). The consequence to a colony would be disastrous, because the queens would be killed by workers or ejected from the nest, resulting in substantial wastage of colony resources. Now it is clear that no such event occurs in normal colonies of stingless bees. First, queens are slightly smaller than workers, and their provisions do not differ quantitatively. Second, the JH production induced by feeding and its effect on the corpora allata is controlled by the types of sugars in the provision mixture and rate of larval feeding (Section 3.3.2). As in the bumblebees and honeybees, production of reproductives should be carefully controlled by workers and the queen. In all such bees, the quality of larval food now seems to be of greater importance than the amount provided. The interaction of kin-group selection and queen presence/dominance prohibits maladaptive queen production and the squandering of colony resources on queens that serve no purpose.

However, many virgin queens are killed by workers of a queenless meliponine colony before one of them is accepted (Kerr 1987). Kerr observed that the workers of *Melipona compressipes fasciculata* kill all virgin queens for five days after the mother queen is removed, thus killing 50 queens before allowing one of them to mate and resume female production. Because colony production of queen *Melipona* is said to vary seasonally from a few percent to 25% maximum, approximately 10% of the females reared during one year might be killed (Kerr 1987). These data imply strong selection processes by workers as they may evaluate their sister queens, or perhaps a high mortality of virgin queens outside of the nest during attempted mating flights. If Kerr (1987) is correct, then the apparently excessive queen production by this stingless bee genus is necessary not just to ensure that a queen is available to supersede the mother queen or to form a new colony, but that a *particular* queen genotype may be selected by the workers or other agents. Worker–queen conflicts that give rise to the variable sex ratios of a colony are reviewed below, and it is now conceivable that these also apply to worker interactions with sister meliponine queens. The removal of queen cells and their developing immatures by the workers of *Apis* (Seeley 1985), and the ejection of larvae from nests by *Bombus* (Pomeroy 1979) may prevent such conflict from taking place among adults in these social bees.

A population property of all organisms (and therefore not necessarily a property of an individual bee colony or group of nests) is that female investment in males and females should be equal in a population at equilibrium, at least in large, outbreeding populations (Fisher 1958). Equilibrium theory provides a model from which to judge departures from it, as the implicit assumption for such an equilibrium is that the forces of natural selection, immigration and emigration, mutation, genetic recombination, and nonrandom mating are insignificant. Predictions corresponding to a particular population are thus seldom correct, because "equilibrium" is a device for exploring predictions rather than for describing conditions most likely to be found.

In bees, maternal investment includes food, eggs, and nest construction. In the perennial eusocial species, maternal investment can be thought to include the efforts of workers in nest construction, foraging and colony maintenance. Seeley (1985) discusses some different methods that have been applied in studies of *Apis mellifera* to predict the expected investment in drones and queens. If the nest material and the size of reproductives are included as colony investments, then a comparable sex investment in males and females would occur when one male is produced for each 16 virgin queens (Macevicz 1979). In nature, the ratio is nearly a thousand times skewed toward males. During one reproductive cycle of a large, normal colony, about 3,000–10,000 males and 4–12 queens are produced. Here the population of colonies of all sizes that contribute to the local breeding population may show quite a different pattern, perhaps conforming to the ratios given by Macevicz, but no adequate study has been made of this difficult field problem. For the honeybee, colony investment ratios are indeed flexible, evident in drone production by weak colonies, even when no queens are reared (Seeley 1985).

Solitary bees often display skewed sex ratios. The most complete study to date considered the sex ratios of a megachilid, *Osmia,* in the temperate zone (Frolich and Tepedino 1986). These authors' work demonstrated that despite a sex ratio that changed through the active season of the adults, beginning with a female bias and ending with a ratio favoring males, the net weight of offspring of each sex was nearly equivalent. Frolich and Tepedino maintain that the mortality rate of these solitary adults would be similar throughout their life (in normal field conditions) and that a decline in foraging efficiency would affect both the size and sex ratio of progeny during the month or so that a reproducing female bee would survive. These authors also showed that new adult bees emerging from these linear nests delayed their emergence by up to four days to avoid killing bees in the outer cells, but that when their emergence was artificially delayed by 17 days, emerging males and females either circumvented the outer cells in their nest burrow or destroyed them as they left (Tepedino and Frolich 1984).

Jamaican *Megachile zaptlana* studied by Raw (1984) nested in aggregations in old beetle burrows; a sample of nests from approximately 50 females showed a 9:1 sex ratio, favoring males. Torchio and Tepedino (1980), Torchio and Trostle (1986), and Jayasingh and Freeman (1980) described male-biased sex ratios of megachilid and anthophorid bees. The ratios varied between 1.5:1 and 2.5:1 among the nests of individual females of a nesting aggregation. Torchio and his co-workers present convincing evidence, based on female investment as determined by the fresh adult weights of both sexes, that a skewed sex ratio matching the one they recorded could be predicted based on bees' weight and, quite possibly, maternal investment. Further, they showed that the level of parasitism on male bees, which were those outermost in the nest and thus most vulnerable, reduced the male/female ratio. When the higher male mortality was considered, expected and observed sex ratios were in conformity. An interesting question arises when considering these data in light of evolutionary theory. Are males smaller and therefore less costly to produce because more are lost to parasites, thus reflecting balancing selection? In a general critique of this balanced mortality hypothesis, Williams (1966) proposed that increased production of offspring was more likely due to genetic load mortality than the action of parasites or predators. If the males express hemizygous lethal alleles, then this theory may also account for certain skewed sex ratios in bees.

Parasitism and developmental mortality are clearly separated in one study. *Megachile* and *Chalicodoma* nesting in Jamaica (Jayasingh and Freeman 1980) sustained developmental mortalities of 49–58%. However, mortality of *Chalicodoma* was due solely to intrinsic factors (possibly physical conditions or genetic load?) and not parasitism. All three bees make cells in a linear arrangement with the slightly smaller males outermost in the series. *Megachile* displayed a male bias in sex ratio, and immature males were intensively attacked by the chalcidoid *Melittobia*. The resinous cell cap of *Chalicodoma* prevented entry by this parasite, and after immature mortality its sex ratios approached equality. Similar studies of *Megachile nana* in India (Kapil and Jain 1980) showed a lower immature mortality (9–44%). Furthermore, variation in sex ratio of adults emerging during three generations produced during the year and over three different years was pronounced in India. Kapil and Jain found females more abundant than males during much of the year, but the male/female ratio varied markedly, from 1:0.33 to 1:1.69. Their data presumably reflect actual brood ratios and not merely those of surviving adults. An opposite trend during a single season has been found for other megachilids: More males are produced in greater proportion as the season progresses (Torchio and Tepedino 1980). The differences could be due to changes in mortality and food supply and shifts in female reproductive investment.

Social bees display variable investment ratios in which genetic relatedness among females plays an important part (e.g., Fisher 1987; Frumhoff and Schneider 1987). Studies in South and equatorial Africa on the ratios of male to all female pupae in nests of 25 social allodapine bee species led Michener (1971) to conclude that sex ratio did not change seasonally. Moreover, a bias toward females was very pronounced. All but three species showed ratios favoring females 1.4–14 times over males (grand mean = 1.75 more females than males). Michener goes on to suggest that mortality of young females (workers) may be particularly high. Because size differences between males and females are slight in species that I have seen, this information at least suggests that differential mortality may be offset by queen regulation of the sexes of broods, rather than as the result of less "costly" females.

As summarized by Fisher (1987), sex ratios in the genus *Bombus* vary from a preponderance of males to ratios favoring workers in the colonies of different species. If workers are in complete control of colony investment in males and females, more investment in females is expected. Trivers and Hare (1976) formulated an explanation for observed sex ratios in haplodiploid colonies. Because workers may in theory share only the paternal half of their inheritance with a sister (0.5 genetic relatedness) or both maternal and paternal genes (1.0 relatedness), the average relatedness between workers is 0.75. This is higher than the expected relatedness to brothers, which is 0.5 when the queen mates with a single drone. The ratio of investment in bee sexes varies both with the proportion of males produced by workers and the queen and also with the degree of influence by a queen and workers in directing colony brood production. These and other variables have been explored by many authors in proposing routes to the evolution of insect colonies; theoretical development thus far outstrips the empirical data (Danks 1983). Fisher (1987) suggests that an even investment in males and females is likely when both workers and the queen control brood rearing; some type of collective colony control seems likely during most times. However, when colonies are temporarily queenless or undergoing colony reproduction, that control may shift dramatically. Parasitic social bees such as *Psithyrus* favor female production, although they influence the colonies they invade to produce a preponderance of males, and normal host colonies display equal investment in males and females (Fisher 1987). Colonies composed of hosts and parasites thus display sex ratios influenced by not only host queens and workers but also by the parasite queen.

A further possible source of selection acting to regulate differential production of sexes by colonies is that of adult mortality during mating and foraging. One data point is available for *Apis mellifera*, in which queen mortality during mating flights was measured at 15–30% (Ruttner 1980). It is difficult to envision a way in which female investment might be regulated precisely, lacking suitable feedback. However, the variation among females within a population in the sex ratios of their progeny might result from differences in mortality of the sexes if these are consistent over many generations, and if sex-ratio biases are controlled genetically.

4 Community ecology

Community ecology is a relatively new and challenging area of biology, as it attempts to explain broad patterns and correlations that can offer penetrating insights. A confounding part of this discipline is that phenomena under study have multiple explanations. To evaluate them, natural history and long-term population data are critical. Case studies of species interactions invite discussion of the rules of viable biological communities (reviews edited by Price, Slobodchikoff, and Gaud 1984; Strong, Lawton, and Southwood 1984; Shorrocks 1985; Diamond and Case 1986). Often, either competition and resources or predation and natural enemies constitute the primary set of interactions used to explain the number and type of organisms that can be found in a habitat at a given time. Yet these factors are not mutually exclusive. In a statistical sense, one must be more important than the other. In any given year or season, however, varied natural phenomena could have pervasive influence on the individuals of a population. Gathering the necessary natural history information, usually far from complete in either the fossil record or in the records compiled by biologists, forms a basis for sorting out the alternatives. At the same time, the study of mortality and fecundity patterns and their causes provides general and applicable insight. Collecting and evaluating the data stimulates appreciation of both population biology and natural history; their distinctions may gradually blur as both phases of research are more widely applied. Real organisms and their population processes might be neatly converted into abstract numbers and mathematically consistent concepts, and vice versa (Caughley 1977).

One emerging generalization from the synthesis required by community ecology is that biological communities include many ad hoc species associations. According to this view, community members often have not significantly evolved within the community they occupy. Organisms have entered the community, found sustaining interactions there, and reproduced. This process is a cornerstone of island biogeography (MacArthur and Wilson 1967), but the events it proposes that structure communities – colonization, extinction, and competition – may never attain an equilibrium (Price 1984; Strong et al. 1984). As Price maintains, species-rich communities are generally too complex to permit the evolution of closely specialized relationships, and there may be a shortage of appropriate colonizers to fill vacant niches. The implications at the community level are interesting, because they suggest that the match between a consumer and its resources is not likely to be precise, regardless of competition. Communities are dynamic; while certain taxa dwindle in number or, for varied reasons, become locally extinct, they are replaced by other organisms (Hubbell and Foster 1986). On the other hand, there may exist species that evolved in the communities they occupy and that, in theory, have maintained pivotal interactions with other such organisms over long periods of time. Thus groups of plants and animals may thrive and perish together and are often thought to be mutualists (Beattie 1985; May and Seger 1986; McIntosh 1987). The former type of association is plausibly more common where extinction and immigration have been frequent or climatic changes on a large scale have occurred. A clear contrast between temperate and tropical communities does not, however, necessarily follow from this observation. Whether tropical conditions are inherently more or less stable than those of temperate areas is probably a moot point. Local species tend to adjust to the range of environmental circumstances they encounter and are thus, on average, probably neither more nor less vulnerable to change than species in another type of environment (Janzen 1967c; Tauber, Tauber, and Masaki 1986). Among insect populations, which have been monitored in many tropical and temperate habitats, no striking differences were found in variation of their annual abundances and relative stability (Wolda 1978).

The organizing principles of a bee–flower community sometimes depend upon contrasting feeding structures and preferences of bees useful in the avoidance of competition (Pyke 1982; Ranta 1982; Pleasants 1983). Although this approach has some appeal in relatively simple assemblages of species, such as the bumblebees of an Alpine meadow, it has yet to be applied to larger species assemblages. Studies of more complex communities, such as the avifauna of tropical islands or lizards inhabiting desert regions, have gone a few steps further by characterizing each potential colonizer or resident species on the basis of vagility, reproductive rate, or feeding niches (Diamond 1978; Pianka 1986; Case and Cody 1987). For tropical bees, too little is known of natural history or ecology to carry out this type of analysis for local faunas. Perhaps at this stage, groups of ecologically (or phylogenetically) similar species, sometimes called *guilds,* provide an adequate starting point for discussing the ways in which apoid communities are formed. When much more information is available, individual species can be discussed rather than genera or guilds. To wait for those data before attempting to approach the subject at all might insure fewer mistakes in early analyses; however, it would also be far less stimulating or self-correcting. Furthermore, this approach might require the collection of much more information than could be gathered by the end of this century (approximately the amount of time remaining before the forest and natural habitats of the tropics are largely degraded). Full evaluations of trophic interrelations based on purely empirical methods rapidly become so complex that each

4. Community ecology 315

study requires a different set of explanations, and a model is needed to accommodate them all. This research problem has often been noted and frequently leads to discussion of the relative merits of deterministic versus robust or stochastic models. One is not useful without very precise information, and the other depends most upon selected central tendencies of populations. The population characteristics should perhaps be meaningful discriminators of different types of communities and thus cannot be chosen without a broad information base.

This chapter presents a wide sampling of the types of information obtained in study of bee communities. One theory suggested from these preliminary data is that competition may have structured some tropical bee communities in which the highly eusocial bees play a central role. The generalized foraging abilities of the highly eusocial species have permitted them to preempt feeding niches that lack specialized consumers. The type of competition that occurs between such bees can also help to explain differences in the way that tropical and temperate *Apis mellifera* forage and store their food, and it may also explain why virtually no highly eusocial bee colonies exist outside of the tropics. However, as will be shown for several experiments with Africanized honeybees in the neotropics, relentless food competition does not rapidly produce major results in complex tropical communities.

The discussion that follows is intended to illustrate bee ecology in terms of populations and communities, neither of which can be seen in perspective without natural history studies like those referenced in preceding chapters. There are population studies, interpretations of fossil and other historic data, description of species richness and composition, quantitative estimates of energy and nutrients that cycle through bees and their nests, a broad overview of an essentially tropical bee group – the highly eusocial species – and their evolution, and discussion of a large-scale experiment now occurring in the neotropics with the spread of honeybees introduced from Africa. Economic benefits from tropical bees are considered chiefly in Section 4.3.

4.1 Bee seasonality, abundance, and flower preference

Competition for unpredictable resource flushes of the type that typify the Southeast Asian dipterocarp forests (Appanah 1985) corresponds to bee community patterns documented in Section 4.2. The most common bees in this habitat are the apids and social Xylocopinae. These are found as adults throughout the year, but their reproduction and life cycles fit very well with boom-or-bust resource conditions. Allodapini, Ceratinini, and Xylocopini can be singularly long-lived as adults and, in the last group, as immatures (Maeta et al. 1984; Michener 1985; see also Section 4.3). Their two- or three-year lifespans and presumably slow metabolic rates dur-

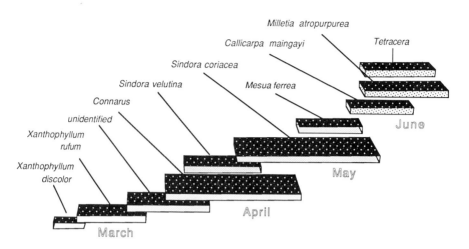

Figure 4.1. Sequential flowering of major floral resources, forest trees, used by *Xylocopa* (Anthophoridae) in lowland forest of Malaya (after Appanah 1985).

ing adult diapause (a very rare trait in bees) allow endurance of dearth periods. Moreover, adults can rapidly discover and harvest floral resources during a mast flowering year, such that reproduction can occur during the sequential flowering of forest tree species (Appanah 1985; see also Fig. 4.1). Appanah also notes that *Xylocopa* are commonly found dead on the forest floor after a community-wide flowering episode. The relatively unrestricted reproductive and foraging activity of bees during such periods likely shortens their life spans (Chapter 3), as broods are produced rapidly. Contrasting life histories among Southeast Asian *Xylocopa* mentioned in Section 3.3, notably production of one to four generations in a year, are possibly one result of adjustment to extreme fluctuations in resource availability. Reproductive variability in individual species might also be expected.

A variety of resource-use studies have been repeated in different tropical habitats, thereby establishing a basis for comparative study, but other types of comparative work are lacking. Some effects of natural disasters have been studied and provide substantial insight, but studies of more ordinary regulators of bee populations are still undeveloped. In particular, there are no quantitative studies of the impact that predation or parasitism has on natural bee populations over a period of years. However, accumulation of microbial natural enemies at nesting sites has been strongly implicated for some solitary bees (Batra, Batra, and Bohart 1973). Insecticide applications and the devastating effects they can have, followed by recovery of a temperate bee fauna, has been discussed by Plowright, Pendrel, and McLaren (1978) and Thomson, Plowright, and Thaler (1985); it is also mentioned for tropical countries by Crane and Walker (1985). However, the tropical references have not included comparable study, nor have there been satisfactory assessments of whether population depression by pesticides has an effect similar to, for example, landslides or unusually high mortality due to natural enemies or the weather.

Volcanic eruptions and other natural, large-scale disasters have caused large changes in bee populations. Yamane (1983) has documented recolonization by both ground- and twig-nesting bees on Krakatau, an island decimated by volcanic activity during the past century. Wille and Fuentes (1975) analyzed how certain bees and wasps might have been affected by ashes deposited by eruption of the Irazu volcano in Costa Rica. Populations evidently recovered rapidly. Of 16 bee species censused periodically at flowers between 1962 and 1966, only three seemed less abundant at flowers a year after the erruption than they had before it took place in 1963–5. These were the only three native highly eusocial bees in the study. In contrast, a small colletid (*Chilicola*) that nests in dead stems was perceptibly more abundant the year following the erruption. Its population possibly responded to the increased availability of nest sites, apparently associated with a planted hedgerow tree, *Erythrina*, and other vegetation killed by volcanic ash. In contrast, the temperate studies of Plowright and co-workers (1978; Thomson et al. 1985) showed that bumblebees were more resistant to insecticide application than were several native solitary bees in the same habitats.

4.1.1 Community studies of floral choice

Methods for determining the seasonal activity and abundance of bees have been devised relatively recently. In contrast with the vast majority of insect monitoring programs (Wolda 1978, 1983; Itô 1980; Williamson 1984), most long-term data on native bee populations come from the tropics. Long-term studies include those made for six or more successive years, while shorter-term data are more common and often involve two or three years (Wolda 1983). Surveys lasting for a complete year can indicate the seasonal activity of some bees and the identity of some of their resources, but they do not assess the variability that surely determines the course of many interactions. Independent surveys of a single year can, however, bring into focus general trends. The longer studies at single sites are useful in additional ways. An adequately designed monitoring program can

1. confirm or disprove trends that seem to hold during one-year surveys,
2. provide insight on yearly brood or generation number,
3. show variance in abundance and thus allow statistical comparison with other populations, and
4. provide indirect data on flowering phenology and selection pressures that may have been important during the evolution of flowering plants.

Table 4.1. *Extensive tropical bee and flower surveys*

Survey location	Bee species	Plant families	Plant genera	Plant species	Prominent bee resources	Refs.
Guanacaste, Costa Rica	192	52	126	168	legumes, borages, composites	Heithaus 1979a,b
Ribeirão Preto, S. Brazil	212	35	86	123	composites, mallows, amaranths, bignones, legumes, mints	Camargo & Muzucato 1986
Belém, Amazonia	239	54	95	119	legumes, Rubiaceae, mints, composites, Sterculiaceae	Ducke 1901, 1902, 1906, 1925
Kourou, French Guiana	210	53	108	198	legumes, Rubiaceae, composites, Sterculiaceae	Roubik 1979a,b

This type of information can be combined to identify some of the processes and relationships that give rise to biological communities, in which species or groups of species have characteristic functions.

Bee phenology data and quantitative impressions of their relative abundance have been reported for a full year or more at the flowering plants near Belém and Ribeirão Preto, Brazil; Kourou, French Guiana; and lowland Guanacaste Province, Costa Rica (Table 4.1). These habitats are in the seasonal lowland tropics, with a long dry season lasting at least four months and annual rainfall of 1,500–3,000 mm. The southern Brazilian site, Ribeirão Preto, is at the edge of the tropics and is largely second-growth and altered habitat; the Guanacaste sites are deciduous dry forest, oak forest, and savanna; the French Guiana site is moist forest and savanna; and that of Belém is moist–wet forest. The 10 highly eusocial bees native to the area were excluded in data from Ribeirão Preto, whereas Meliponinae comprise over 20% of the bee species found in the Belém and Kourou studies and almost 50% of all individual bees recorded at flowers in Guanacaste. Sampling with an insect net led to identification of 7,000–8,000 bees and their floral hosts in the Guanacaste and Ribeirão Preto studies. The other two studies also relied upon

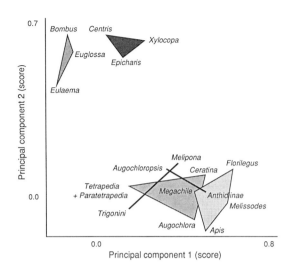

Figure 4.2. A simplified rendering of flower species associations using a principal-components analysis (after Roubik 1982c). Qualitative data on flower visitation by common bee genera in lowland forest and savanna were compiled for one year; lines in the diagram connect groups that are similar, and proximity of genera on the graph indicates similar taxonomic floral choice. The data show nothing about the relative importance of plant species to bees or vice versa, and the analysis assumes that species of a given genus tend to be similar. Large and small bees tend to cluster in separate portions of the graph, indicating dissimilar floral hosts.

observations away from flowers and made no attempt to quantify relative bee numbers.

Intensive destructive sampling has the drawback of producing hundreds of meliponines or other social bee species within a short time; so species presence or absence is more practical in some surveys. In fact, single flowering trees are said to draw up to 15,000 bees at once (Frankie et al. 1983). Sampling a small part of one such tree, during a short period within the total visitation time of bees, is thus likely to produce biased data. This is true because the type and position of foraging bees within a tree crown, or any other type of flower patch, change throughout the day and, to some extent, through the flowering period (Section 2.3). Nonetheless, each survey concluded with recorded floral visitation by about 200 bee species. From 120 to 200 flowering plant species corresponded to the bee visitation data.

Surveys provide samples of local flora and fauna, making necessary some qualification of the extent to which they may represent community-wide relationships. Large bees tend to visit a different group of flowers than smaller bees (Roubik 1982c; see Fig. 4.2). The relative contribution of pollination by bees compared to other animals can also gauged from this perspective. For instance, Frankie et al. (1983) describe flowers of Guanacaste that fit a general pattern of primary visitation and pollination by "large" bees (bees the size of worker *Apis mellifera* or larger). The information was gathered over several years and serves as a comprehensive treatment of tropical bees and the local flowers they visit. Large bees are the principal or exclusive pollinators of 22% of the tree species and 30% of woody vines and lianas. Direct comparison of total angiosperm species and those in the Guanacaste survey is possible, since a reasonably complete checklist of the lowland flora has been prepared by Janzen and Liesner (1980). This list includes about 900 species of 120 families in 546 angiosperm genera. By comparing the species lists, it is possible to estimate that 40% of total plant families and 20% of genera and species were visited by bees. Questions remain as to whether many plants that were not seen in flower have bee visitors and what proportion of them rely on pollination from other types of animals. It should be true that many small, relatively rare forest understory plants with brief flowering periods were missed. These seem likely to have small and at least temporally specialized bee visitors, such as halictids and stingless bees, but no detailed analysis has attempted to reveal such trends. The studies of both Ducke and Roubik (Table 4.1), which listed bees not recorded at particular flower species, included many meliponines and halictids that may be candidates for visitation of flowers less easily surveyed.

The above estimates from the large-bee flowers and one general survey give a very conservative impression of the number or proportion of a neotropical flora pollinated or visited by bees. Furthermore, the surveys probably underestimate total flower species used by bee species and total bees visiting flowers. The latter can be staggering. In Guanacaste, the papilionaceous tree *Andira inermis* has at least 70 species of bee visitors (Frankie, Opler, and Bawa 1976), and 47 appeared in the single-year survey (Heithaus 1979a). Another tree, the borage *Cordia inermis,* had 43 bee visitors in the survey of Heithaus, whereas roughly 300 insect species, likely including many more bees, have been associated with these flowers (Opler, Baker, and Frankie 1975). One might estimate that at least 50% more bee species visit a particular flower than is indicated by a year of transect sampling. Correction factors might be derived in other ways for numbers of plant species used by bees. Considering nectar, the maximum plant species utilized by a bee is probably close to that of total flowers having roughly the same dimensions as the proboscis; but this ignores all destructive visitation or use of existing perforations to extract nectar (Section 2.4.4). Other significant restrictions to nectar feeding may result from toxic substances in nectars (Chapter 2). To estimate true total pollen sources, pollen from bee nests could be compared with flower visitation data. Studies of this depth have not been realized.

A number of additional analyses were made of bee seasonality in Guanacaste (Heithaus 1979b,c). Of 74 common bee species in five families (excluding Apidae and therefore many continuously active species), 35% were found in dry and wet seasons, and twice as many were found only during the dry season as only during the wet season, which are of similar length. The Megachilidae and Anthophoridae had a preponderance of species active during the dry season; halictids were active in both dry and wet seasons, and the few andrenids and colletids were most often seen in the wet season or during both seasons. A surprising conclusion was that the tropical bees were similar in seasonality to temperate species. First, compared with 192 bee species found in southern Illinois (Pearson 1933), the two months of peak activity in each habitat showed foraging by 80% of the total bee fauna. Second, by dividing the total active season for bees into equal quarters at each habitat, we see that exactly the same proportion of bee species were active throughout the entire season, during one-quarter of the season, or during half to three-quarters of the season. These percentages were 21%, 26%, and 51%, respectively (Heithaus 1979a).

Similar seasonal appearances of bee families were determined for Panama (after Michener 1954) and for a southern extremity of the Brazilian tropical biota in São José dos Pinhais by Sakagami et al. (1967). The latter authors considered bee records on the rainy Atlantic side of Panama, including Barro Colorado Island. This showed a scarcity of Megachilidae and Halictidae during mid to late wet season and a preponderance of these two families and Anthophoridae during dry season. The same authors defined two patterns, which they called temperate and tropical, for the bees of a second-growth habitat in southern Brazil. Seasonal activity was apparent for Megachilidae and the Anthophoridae primarily nesting in the ground. These spent the cold season (a resource dearth period, hence similar to mid to late wet season at more tropical latitudes) as immatures. Halictidae, Xylocopinae (often the social, wood-nesting Anthophoridae), and many apids showed no such seasonal restrictions, presumably because they overwinter as adults and will fly and visit flowers as the weather permits. The same groups having greater persistence as adults during the year were also social, which was also shown by the above studies of Heithaus and likewise implied by studies in the paleotropics (Yoshikawa et al. 1969; Appanah 1985).

What plant groups appear most important to tropical bees? The general surveys help to answer this question, even though they are all based on neotropical observations and should to some extent fail to produce correct predictions for a chosen species. Furthermore, as will be shown, at least the highly eusocial bees seem to rely most upon a distinctive set of flowering plants, not visited by many of the large bees. A small number of plant families, however, repeatedly dominate the community surveys. The legumes (Caesalpinioideae, Papilionoideae, Mimosoideae), Rubiaceae, Sterculiaceae, Boraginaceae, Lamiaceae (Labiatae), and Asteraceae (Compositae) had many species visited by bees. In addition, the plants were visited by many bee genera. Heithaus (1979a) points out that among Caesalpinioideae, Mimosoideae, Papilionoideae, Asteraceae, and Boraginaceae in his organized and quantitative survey, the flowers were visited by bees out of proportion to their abundance in the flowering community. That is, bees seemed to display a preference for these taxa. For the larger bees in the dry forest community, Frankie et al. (1983) note prominent bee visitation at Bignoniaceae and Passifloraceae, particularly among the climber species. The oil-producing Malpighiaceae were visited heavily by bees. This would be expected because only bees collect the oil and it is apparently essential for nesting success (Section 2.2.3). More weedy plant species were included in the survey made in southern Brazil (Camargo and Muzucato 1986), where Malvaceae, Verbenaceae, Solanaceae, Convolvulaceae, Amaranthaceae, Onagraceae, and Euphorbiaceae were important bee flowers; the flower species visited by the most species was a small, yellow flower, *Sida* (Malvaceae). Raw (1976) reported that roughly 90% of all pollen species harvested by two Jamaican species of *Exomalopsis,* studied throughout the year at a few dozen localities, came from legumes, composites, and Rubiaceae. These included only 4 of the 24 recorded species brought to nests by returning foragers.

Characterized by pollen diet, many of the Apinae and Meliponinae can perhaps be labeled commensals of the tropical flora – they remove resources left by other pollinators, frequently bats, birds, moths and beetles (Roubik et al. 1984, 1986, 1988; Villanueva 1984; see also Table 4.2). These data are gathered in the forest, rather than in mixed or altered habitats, where some extensive pollen data have also been taken (Section 2.4). The bat resources are large flowers that leave a large residual pollen and nectar crop on the morning following anthesis. Pollination consequences of the mixed utilization syndrome are discussed briefly in Section 4.3. Plants pollinated frequently by bees, flies, and curculionid and dynastine beetles are the palms (summaries by Dobat and Peikert-Holle 1985; Gottsberger 1985, 1986; Henderson 1986), and some of their species are visited by huge numbers of stingless bees and honeybees. Henderson (1986) notes that olfactory and visual deceit probably occur often at pistillate flowers that attract bees but provide neither pollen nor nectar.

Not only do dioecious and monoecious plants receive many visits from colonies of highly eusocial bees (Bawa and Opler 1975; Bawa 1980), but such bees appear to obtain most of their food from these plants. The study by Roubik et al. (1986) showed that all of the principal pollen sources used for an interval in the rainy season by honeybees and stingless bees in a lowland Panamanian forest were dioecious or monoecious. Principal resources in this study and one earlier broad survey (Roubik et al. 1984) were for the most part nocturnally dehiscent. In upland areas, preferred resources of the bees were often wind-pollinated or offered no nectar (see Severson and Parry 1981 for a temperate honeybee comparison). These flowers included cultivated corn. It is not surprising that the highly eusocial bees could be preadapted to utilize the flowers offering only pollen, because they are the only tropical bees that store large amounts of carbohydrate in the nest and are thus not forced to seek energy sources each day.

However, many if not most plants visited heavily for pollen by the highly eusocial bees also provide nectar. This was true for *Spondias* (Anacardiaceae), a tree

Table 4.2. *Major pollen sources of honeybees and stingless bees in the seasonal lowland neotropics, characterized by a dry season lasting three to five months*

	Pollen source	
Site, bee	Dry season	Wet season
Mexico		
Apis mellifera (European)	*Ceiba asculifolia* (Bombacaceae)	*Bidens, Viguiera, Heliantheae* (Compositae)
	Heliotropium (Boraginaceae)	*Ipomoea trifida* (Convolvulaceae)
	Tillandsia[a] (Bromeliaceae)	*Caesalpinia, Leucaena,* Mimosoideae (Fabaceae)
	Bursera simaruba[a] (Burseraceae)	*Zea mays* (Gramineae)
	Croton niveaus (Eurphorbiaceae)	*Cardiospermum halicacabum* (Sapindaceae)
	Acacia, Psicidia (Fabaceae) Rubiaceae, Sapotaceae spp.	*Trema micrantha*[a] (Ulmaceae)
	Heliocarpus (Tiliaceae)	
Panama		
Apis mellifera (Africanized)	*Cavanillesia, Pseudobombax* (Bombacaceae)	*Spondias*[a] (Anacardiaceae)
	Cordia (Boraginaceae)	*Bursera*[a] (Burseraceae)
	Erythrina, Pithecellobium (Fabaceae)	*Baltimora, Clibadium, Melampodium* (Compositae)
	Cedrela (Meliaceae)	*Mimosa, Leucaena* (Fabaceae)
	Coussapoa[a] (Moraceae)	*Zea, Panicum* (Gramineae)
	Oenocarpus (Palmae)	*Bactris,*[a] *Chamaedorea,*[a] *Elaeis,*[a] *Oenocarpus* (Palmae)
	Zanthoxyllum[a] (Rutaceae)	*Genipa* (Rubiaceae)
Apis mellifera (Africanized) 3 *Melipona*, 7 Trigonini		*Spondias*[a] (Anacardiaceae)
		Protium tenuifolium (Burseraceae)
		Cecropia, Pourouma[a] (Moraceae)
		Socratea, Elaeis[a] (Palmae)
		Pouteria stipitata[a] (Sapotaceae)

[a]Dioecious, monoecious, or androdioecious plant taxa.
Sources: The data are summarized from quantitative palynological studies of pollen harvested by foraging bee colonies in Panama (Roubik et al. 1984, 1986; Roubik 1988; Roubik & Moreno unpub.) and from Veracruz, Mexico (Villanueva 1984).

genus reported to be especially sought by highly eusocial bees in three neotropical nest-pollen surveys (Sommeijer et al. 1983; Roubik et al. 1984, 1986). Absy and Kerr (1977) showed that *Miconia* (Melastomataceae), another species that sometimes provides both nectar (Section 2.1.6) and pollen, was visited throughout the year by *Melipona*. It is presumably one of its most important resources. This study took place in the altered floristic habitat of suburban Manaus, Amazonia; and *Miconia* is to large extent found in second-growth forest and at forest edges. However, Snow (1965) found similar year-long flowering and fruiting among the species of *Miconia* in Trinidad's forests; the study showed close dependence of bird populations on the fruit of these plants. The distribution of bees, or birds feeding on fruit, could well be influenced by distribution of a plant genus such as *Miconia*.

The implications of studies of a few dozen highly eusocial bee taxa in lowland neotropical forest (Roubik et al. 1984, 1986) are that the foraging specializations of these bees can be expected to include flowering trees having large inflorescences or panicles of many small flowers. Alternatively, the bees focus on enormous flowers having many hundreds of stamens, which include, for example, a number of bombacaceous genera (Eguiarte, del Rio, and Arita 1987; Roubik 1988). During the wet season they seem to forage mostly at palm trees, grasses, herbaceous flowers, and shrubs. It is worth mentioning here that similar shifts from trees (dry-season flowering) to flowers present in lower forest strata have been noted for other types of tropical bees (Heithaus 1979a,b; Ackerman 1985). Gentry (1983, 1985) suggests that 53% of the flora of Panama is pollinated by bats and hummingbirds. If this is true, then it is not surprising that varied foraging behavior of meliponine bees (Section 2.4) has allowed them to incorporate many of these resources in their diet and foraging repertoires. Some of their preferences undoubtedly evolved in concert with their ability to recruit nest mates, as a result of intense selection pressure to arrive rapidly before other competitors at rich resources (Section 4.3).

The instructive studies of several stingless bee species in subtropical São Paulo, Brazil, show some basic community patterns for this group in altered habitat (Kleinert-Giovannini 1984; Ramalho et al. 1985). They also give two valuable methods for analyzing data generated by counting pollen grains on microscope slides of samples taken from bee nests. Relative representation of pollen species in standard slide transects were summarized both as proportional frequency of plant families and proportional frequency of species belonging to given families. For instance, *Melipona marginata,* a small bee for its genus, utilized far more species of composites, legumes, and euphorbs than species of the other 75 plant families in its diet. But in terms of total pollen grains, which differ substantially in size and nutritional quality, it primarily used Myrtaceae (introduced *Eucalyptus* species), Melastomataceae, and Solanaceae. Another bee, a small Trigonini in the *Plebeia* group, used far more species of palms, Myrtaceae, Moraceae, and legumes than it used species of 25 other families, but most of its pollen came from composites, euphorbs, legumes, and moracs. It is intriguing that proportional pollen use by these bees is similar to general floral visitation preferences delimited in the four community-wide surveys previously discussed.

Keeping in mind this sample of the types of flowering plants tropical bees often prefer, we may turn to paleobotanical natural history for another dimension. A number of broad questions can be posed. How did bee–plant communities arise? Did bees diversify along with the flowering plants or, as suggested earlier, did their feeding and foraging behaviors often adjust to the floras they encountered, such as flowers designed to attract and feed bats, moths, hummingbirds, or beetles? Some correlations have been found between pollen exine type and pollinator group (summaries in Blackmore and Ferguson 1986), but they do not address how such correlations arose.

Fossil angiosperm pollen corresponding to extant families has been placed in chronological order of appearance by Muller (Muller 1981; also see Table 4.3). Almost all the contemporary pollen types first appear at the Cretaceous–Tertiary boundary, when traits corresponding to specialized bee pollination were already evident (Crepet and Taylor 1985; Taylor and Crepet 1987). These authors give an example in which derived flower morphology apparently preceded major change in pollen morphology by about 50 million years, the gap in time between the earliest fossilized papilionoid-like flower and pollen of modern Papilionoideae. Thus the appearance of recognizable fossil pollen might be unlikely to corroborate evolution of floral preferences by bees. However, because the papilionaceous legumes are certainly one of the most highly specialized bee flowers (Faegri and van der Pijl 1979), they may be a special case.

Among major bee-visited families given in Table 4.2, most used heavily by bees were apparently present before the Oligocene and even the Eocene. Thus they predate many of the earliest known fossil bees but do not in fact predate the appearance of meliponines (Chapter 1). Fossil evidence indicates the *minimum* age of taxa and thus cannot be used to test hypotheses regarding their initial appearance. Biogeographic evidence indicates an earlier origin of bees, but the anemophilous, nectarless Chloranthaceae, Betulaceae, Fagaceae, and Ulmaceae were well represented during the mid-Cretaceous period. Pollen of these plants is used by eusocial apids such as honeybees (Table 4.3) but, according to available data, by little else. Their appearance in the fossil record very likely corresponds to the early presence of catkinlike inflorescences and nectarless flowers typical for such plant families. Floral structure and fossil pollen morphology probably are in concordance. Plants pollinated by animals (rather than by wind) appeared relatively soon afterward, including Myrtaceae, Sapindaceae, Amaranthaceae, Sapotaceae, Bombacaceae, Caesalpinioideae, Onagraceae, and the palms. It is remarkable that pollen of composites, borages, and labiates, groups visited by many bees, appeared much later. Perhaps as with the papilionoids, radical change in their pollen morphology occurred long after bees were primary visitors.

Pollen rain is thought to contribute to fossilized pollen in much the same way

Table 4.3. *Pollen spectra: fossil pollen, modern pollen rain, modern honey bee diets*

Million years before present					% Total pollen			
120	65	20	5	Fossil pollen	33	20	10	Modern pollen
						% Pollen rain		
————————————————				Chloranthaceae			———————	Euphorbiaceae
————————————				Ulmaceae		———————————		Leguminosae
————————————————				Aquifoliaceae			———————	Meliaceae
	———————————			Sapindaceae			———————	Burseraceae
	———————————			Fagaceae			———————	Julianiaceae
	———————————			Betulaceae			—————	Boraginaceae
	———————————			Myrtaceae			—————	Bignoniaceae
	———————————			Annonaceae			————	Compositae
	———————————			Amaranthaceae			————	Palmae
	—————————			Ericaceae			———	Gramineae
	———————————			Sapotaceae			——	Tiliaceae
	———————————			Bombacaceae			——	Ulmaceae
	———————————			Caesalpinioideae			—	Cactaceae
	———————————			Onagraceae			—	Urticaceae
————————————————				Proteaceae			—	Malpighiaceae
	———————————			Palmae		% Annual honey bee pollen diet – highland		
	———————————			Lauraceae		—————————————		Fagaceae
	———————————			Polygonaceae			———————	Compositae
	———————————			Tiliaceae			———————	Betulaceae
	———————————			Sterculiaceae			——————	Chloranthaceae
	———————————			Euphorbiaceae			——	Palmae
	———————————			Anacardiaceae			——	Rubiaceae
	———————————			Apocynaceae			—	Onagraceae
————————————————				Clusiaceae			—	Lamiaceae
	———————————			Lecythidaceae		% Annual honey bee diet – lowland, dry		
	———————————			Umbelliferae			—————————	Bombacaceae
	———————————			Convolvulaceae			—————————	Anacardiaceae
	———————————			Moraceae			———————	Poaceae
	———————			Mimosoideae			———————	Leguminosae
	———————————			Malpighiaceae		—————————		Compositae
	———————————			Bignoniaceae			—————	Meliaceae
	———————			Cyperaceae			—————	Simaroubaceae
	——————			Malvaceae			—————	Boraginaceae
	———————			Rubiaceae			—————	Sterculiaceae
	——————			Agavaceae			————	Combretaceae
		———————		Dipterocarpaceae			———	Flacourtiaceae
		———————		Flacourtiaceae			——	Moraceae
		———————		Cucurbitaceae			——	Rutaceae
		———————		Chrysobalanaceae			—	Euphorbiaceae
		———————		Meliaceae		% Annual honey bee diet – lowland, wet		
		———————		Rhamnaceae			———————	Palmae
		—————		Vitaceae			———————————	Anacardiaceae
		—————		Boraginaceae			———————————	Bombacaceae
		————		Compositae			———————————	Sterculiaceae
		————		Verbenaceae			—————	Boraginaceae
		———		Dilleniaceae			—————	Bignoniaceae
		——		Brassicaceae			———	Burseraceae
		——		Labiatae			———	Compositae
		—		Papilionoideae			——	Moraceae
		—		Rutaceae			——	Meliaceae

Sources: Muller 1981; Roubik et al. 1984; Palacios Chavez 1985.

that it contributes to the pollen found on natural substrates in forest. A study of contemporary pollen rain in a tropical caducifolious forest in Mexico shows that although legumes were by far the predominant plant family in the habitat, combined legume pollen was only about one-third as common as that of Euphorbiaceae (Palacios 1985; see also Table 4.2). These data, taken in a habitat that has many similarities to Guanacaste, Costa Rica, and a similar bee fauna, provide another type of information. They show that bees use some of the most abundant pollens. It is still difficult to deduce why legumes are relatively poor dispersers of their pollen. It is often small and light in the Caesalpinioideae, but at the same time it is often held within poricidal anthers (Section 2.1.6). Whether of buzz-pollinated caesalpiniaceous legumes, of mimosoid legumes having large polyads, or of the intermediate-size pollen grains of papilionoids, legume pollen may well be largely removed by foraging bees, leaving little record of pollen abundance per se in the habitat. Such pollen data, of course, might be expected to show no consistent trend aligned with the number of flowers of a particular kind. Pollen per anther and pollen per flower available to bees are still imponderable quantities in the current ecology, let alone past evolution, of tropical habitats. A preliminary analysis of the quantity of floral oils and the biomass of specialized oil-collecting bees that they may support in desert regions has been provided by Buchmann (1987). Schmidt and Buchmann (1986) give a similar analysis for a cactus species providing rich resources and the number of honeybee colonies that their local populations could in theory support. The correspondence between resource biomass and the proportion of it incorporated by local bee populations is a logical and very desirable step toward building a general picture of bee community ecology.

Some of the preliminary results on pollen availability in the Amazon basin indicate predominance of two families utilized by bees, Palmae and Myrtaceae. In the Quaternary pollen taken from core samples in the central Amazon Basin, Absy (1985) shows that nearly one-half of the pollen taxa belong to these families. Extension and interpretation of such data could fill large gaps in understanding food utilization, foraging, and evolution of tropical bees.

4.1.2   Long-term monitoring studies

One result of monitoring programs has been the discovery of an unexpected variety in number of yearly bee generations, broods or population abundance peaks. Such peaks usually indicate the appearance of newly emerging adults in a habitat. Over evolutionary time, they can be viewed as points from which seasonal populations could diverge by selection favoring specialization and by taking advantage of new resources. They can also be seen as the result of past population trends that have become entrenched as a result of, or original causation by, such factors as physiological processes of development, resources, natural enemies, or climatic conditions. Resources include not only food and nesting material, but co-mimics and other types of mutualists (Chapter 2). The climatic conditions and pressure from natural enemies can further influence foraging, mating, and nesting success (Chapters 2 and 3). While experimental studies that adequately address the relative importance of these variables have yet to be conducted, the studies that follow indicate that from one to eight broods a year are produced by tropical bees, all in the forests of central Panama. Therefore, the selection pressures that influence adult emergence and activity are likely to vary. To understand some of these dynamics we must take into consideration that these habitats are largely monsoon (tropical moist) forests with a single dry season of two to six months, most usually lasting from December to April. With each abundance peak of a species, there is likely to be a different degree of specialization on the resource taxa available at a given time. Evolution of generalization or specialization is perhaps best appreciated in light of this potential variability. The variance and stability of bee abundance will be discussed in the following section, and other patterns found in studies covering one to seven years are interpreted here.

Six techniques have been employed in long-term studies that can reveal population processes of bees. The efficacy of such work depends upon sampling replicability. Four types of studies have been made in the tropics:

1. *Chemical baiting studies.* These have been performed in the neotropics with euglossine bees, the males of which are attracted to chemicals that mimic natural odor compounds that they seek at floral and nonfloral sources (Pearson and Dressler 1985; Roubik and Ackerman 1987).
2. *Light trap studies.* These have been made by placing ultraviolet fluorescent lights at varying heights above the tropical forest floor; continuous 24-hr operation provides data on noctural, crepuscular, and diurnal bees (Wolda and Roubik 1986).
3. *Trap nest studies.* Bees that nest in holes in wood are given a more than adequate quantity of nesting sites, and their adult numbers are monitored in the evening when bees remain within the nest holes.
4. *Inclusive sightings.* Reproductive swarms of honeybees have been monitored at consistent intensity in a large but defined area (Boreham and Roubik 1987).

Additional monitoring techniques are the following:

5. *Malaise trap studies.* Bees are captured as they fly into a net spread horizontally along the ground (Owen 1978).
6. *Inclusive flower monitoring.* Studies are made that monitor all the flower types in a given area (preferably one that exceeds the flight ranges of bees) or that focus upon flowers visited when few or no other resources are being used (Sakagami et al. 1967; Heithaus 1979a; Frankie et al. 1983; Camargo et al. 1984; Camargo and Muzucato 1986; see also Section 4.4.2).

Other insect sampling programs (Banaszak 1980; Southwood 1981) could be adapted for bees, but the preceding methods are practical and effective. All of these methods, particularly 1, 2, and 4–6, depend on bee flight activity, which changes due to meteorological conditions, the phase of the moon and cloudiness (for nocturnal species), and other factors. This implies that unless data are gathered extensively and consistently, allowing the spurious positive and negative influences of extrinsic factors to approach equality, the numerical results are suspect. Inclusive sampling programs, unless intensive or on a large scale, should probably be standardized by dividing total bees sighted by the number of observation hours, the number of plants or flowers surveyed, or the like. An implicit requirement of all studies is that the baits or sampling sites are located near, or themselves constitute, relatively attractive resources. If neither assumption can be made, then it at least must be true that the sampling takes place where bees fly between their nests and resources. Otherwise, meager quantitative data may represent the activity of the least biologically fit subpopulation rather than the average individual.

Light trap studies of bees in lowland forest of Barro Colorado Island revealed that the nocturnal halictid bees, *Megalopta genalis* and *Megalopta ecuadoria,* had only one consistent seasonality pattern (Wolda and Roubik 1986). More important, their populations behaved similarly within a given year. They were found in low numbers during the first three months of the year, corresponding to the dry season; abundance peaks usually occurred on and after the fourth month (Fig 4.3). Seven years' data were taken by continuous operation of two ultraviolet fluorescent light traps, one at 3 m and one at 27 m above the forest floor. During this time the two species were consistently correlated with each other in weekly abundance in each year, but slightly correlated with other bee groups or with their own abundance patterns in other years (Wolda and Roubik 1986; see also Fig. 4.4). Their abundance peaks occurred once during the dry season, 8 times during the dry-to-wet transition, 10 times in mid wet season, and 10 times during the late wet season. Any one of these peaks was an annual peak during a given year (Fig. 4.3). Each species had exactly 18 monthly abundance peaks during six years, although as many as five and as few as two appeared in a 12-month period. If abundance peaks coincided with adult emergence and activity, there were on average three yearly broods or generations. More than one brood could be produced by females of these colonies (Chapter 3). A major abundance peak occurred most years during the transition between dry and wet seasons, late March to mid-May. However, in one of the six years depicted in Fig. 4.3, peak abundance occurred during the late wet season (November), and in another year it occurred in the middle of the dry season (February). Minor abundance peaks were still evident during the dry-to-wet transition period.

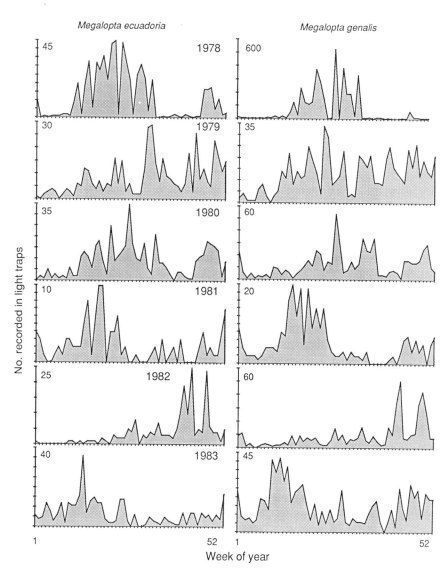

Figure 4.3. Weekly abundances of two nocturnal species of *Megalopta* (Halictidae) monitored for six years at light traps in lowland Panamanian forest (after Wolda and Roubik 1986).

A one-year study by Camargo et al. (1984) at flowers in the southern neotropics showed that *Oxaea flavescens* had three yearly abundance peaks (Fig. 4.5). Their survey included most of the local flowering plants; thus it likely produced an accurate picture of bee phenology and abundance in the habitat. Repetition of such

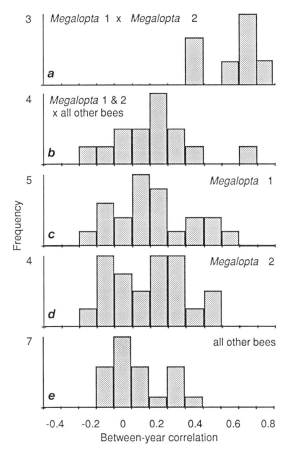

Figure 4.4. Seasonality in bee population dynamics: correlation coefficients of weekly abundances of two nocturnal bees (*Megalopta*) and 48 other species monitored for six years at light traps (after Wolda and Roubik 1986). All possible year-to-year correlations were calculated. The figure shows the frequency distributions of correlations. Starting from the top these are for the following: (a) both *Megalopta* in each of the seven years (7 values); (b) each *Megalopta* compared with all other bees (14 values); (c) the first *Megalopta* for all possible pairs of years (21 values); (d) the second *Megalopta* for all possible pairs of years (21 values); (e) all non-*Megalopta* for all possible pairs of years (21 values). Positive values signify some degree of regular seasonality, and negative values imply unpredictability.

studies, in addition to recording the flowering intensity of host plants, gives population profiles sometimes labeled *biocoenotics* (Sakagami et al. 1967; Laroca 1983). If these studies are done on a scale large enough so that some heavily utilized plant species do not escape detection (by occurrence outside the study area, for example), then the timing of bee emergence and adult reproduction can be tested for goodness-of-fit to the flowering phenology of plant species. Describing their yearly variance seems to be necessary before such patterns can be interpreted.

Although the above data imply that there were usually three yearly generations among some tropical and subtropical bees, they do not explain why bee numbers fluctuated. Supplementary data can be found that suggest probable causes. The magnitude of the peaks obtained from light trap studies should depend on local resources and the flight activity of bees as well as on their absolute abundance and whether they nest near the light. During the 1978 census, almost 10 times as many

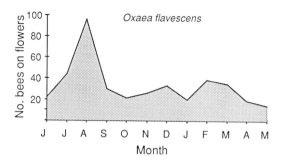

Figure 4.5. Monthly abundance of *Oxaea flavescens* (Oxaeidae) monitored at flowers for one year (adapted from Camargo et al. 1984).

M. genalis were collected in the trap as in other years. At the time of maximum catches, trees of the large caesalpiniaceous legume *Tachigalia versicolor* were flowering throughout the forest. Light traps were suspended from this large tree, which flowers only once before dying (Foster 1977). However, the species of *Megalopta* was the only bee of 50 species monitored that appeared in large numbers during flowering by *T. versicolor*. Because its abundance pattern generally corresponded to that of the other *Megalopta* species, a conclusion from this study was that the abundance of these bees reflected general factors that influenced absolute population size, not merely resource conditions in the immediate vicinity of the light trap. This conclusion is reinforced by the fact that similar bees forage nocturnally during several hours when there is sufficient light (Kerfoot 1967b); thus they could be attracted to the light on their way to and from nests and between flowers scattered over a large area. Both local flowering and bee numbers fluctuated from year to year, and their seasonal patterns varied. The central tendency of the bees was to emerge as adults during the wet season, and they probably responded to the same rain cues or other conditions that trigger flowering and bee emergence in seasonally dry habitats (Linsley 1958; Augspurger 1980). Multiple broods during the year certainly improve the chance of adult activity during flowering peaks of suitable plants. They would, of course, prevent narrow resource dependence. However, the remarkably large numbers of *M. genalis* during a flowering year of *Tachigalia* suggest that there may be specific adjustments in timing and synchrony of flower and bee appearance.

Crepuscular (early-morning) bees were also monitored during the light trap study; *Ptiloglossa* (Colletidae) and *Rhinetula* (Halictidae) were found throughout the year, but in numbers too low to detect seasonal trends or generation intervals. Seasonality was pronounced in the centridine anthophorids *Centris* and *Epicharis*, which were found almost exclusively in early wet season. This tendency was associated with the large size and/or dark color of those species, relative to local centridines as a whole. Whereas many *Centris* are active primarily in the dry season,

Epicharis and some *Centris* span both wet and dry seasons or are found only during the wet season (Heithaus 1979a; Frankie et al. 1983). These may tend to be larger or darker species (Wolda and Roubik 1986; see also Appendix Fig. B.1 for *Centris inermis*), which, due to thermoregulatory characteristics, should encounter more favorable foraging conditions during the wet season (Section 2.3). Specializations associated with foraging at relatively low temperature or light intensity could also be promoted by avoidance of competition from bees seldom foraging in these conditions.

Some bee species flew at different heights above the forest floor. Bee abundance in light traps at 27 m and 4 m above the same spot captured bees of either xylophilous or hypogeous nesting habit, respectively, although this difference in nesting placed no evident seasonal constraints on bee activity. The ground-nesting *Ptiloglossa* and *Rhinetula* were most abundant in the lower trap, and wood-nesting *Megalopta* were consistently more abundant in the higher trap. The degree to which these flight patterns were due to nesting or foraging localities alone was unknown, but nest locations were probably more relevant.

Trap nest studies of solitary bees have been successful in temperate areas (Parker and Bohart 1966, 1968; Krombein 1967), but have seldom been employed in the tropics (Parker et al. 1987). After some trouble with rain and colonization of trap nests by ants in tropical moist forest, sheltered nest sites that I placed near forest have produced much data for a few bee species in Panama. *Centris analis* nesting among a few thousand smooth-sided holes drilled in a hardwood board has shown almost constant adult activity during 3.5 years (Fig. 4.6). This species was recorded only during six months of the year at flowers in dry forest of Costa Rica (Heithaus 1979a), but was later recognized to have slight activity during the wet season (Frankie et al. 1983). In Panama, *C. analis* was regularly attacked by parasitoid *Anthrax* flies, chalcidoid wasps of the genus *Leucospis*, and sometimes by cleptoparasitic megachilids *Coelioxys*. Combined male and female bee abundance showed four major peaks during the dry and dry-to-wet seasons and four minor peaks during the wet season (Fig. 4.6). These occurred near intervals of 1.5 months, matching duration of the egg-to-adult period. A suggested pattern, without confirmation, is that the majority of these bees diapause as pupae during most of the wet season (nondiapause dormancy?; see Section 3.1.2), but a few emerge at the same intervals seen during the dry season. It may be significant that the trap nests used in this study were sheltered from rain. They may be representative of few natural nest sites.

Continuous monitoring of a 50-km^2 area along the Panama Canal made possible a study of honeybee reproduction and seasonal activity. This was initiated with arrival of the colonizing Africanized honeybee, strongly resembling *Apis mellifera scutellata,* in Panama (Boreham and Roubik 1987; see also Section 4.3.3). Be-

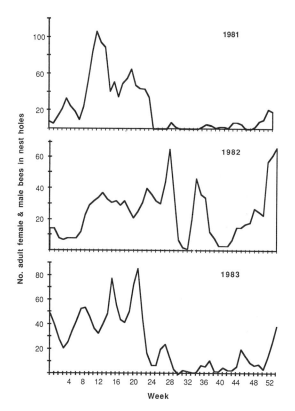

Figure 4.6. Seasonal abundance of *Centris analis* (Anthophoridae) in nesting holes in lowland forest of Pacific Panama (pers. obs.). Both male and female bees were surveyed each evening, when they could be seen in holes in a hardwood nesting board.

ginning in the second year of its residence, the number of swarms produced by this bee followed a pattern of three or four yearly peaks, the largest being in February and March, the middle of the dry season (Fig. 4.7). General floral richness in the habitat corresponds closely to the appearance of honeybee swarms (Figs. 4.7 and 3.30). A contrasting quality of swarms in the wet season was that extremely small and large swarms predominated in the more rainy Atlantic coastal area during the second annual swarm abundance peak. The peak was largest on the wetter Atlantic side of the canal transect and coincided with flowering of palms, herbaceous plants, and many forest trees (Croat 1978; Roubik et al. 1984, 1986). Colony absconding or fusion or the amalgamation of small swarms (Kigatiira 1988) into megaswarms may all have contributed to this second annual peak in swarm movement (Boreham and Roubik 1987). The two smaller swarm movement peaks seen in the wet season were probably related to small flowering episodes, but the major period of swarm movement coincided with flowering by trees in the dry season and early wet season.

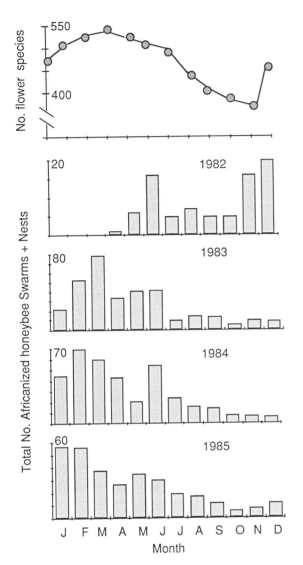

Figure 4.7. Monthly abundance of colonies of the Africanized honeybee recorded from continuous reporting in a defined region within the Panama canal area during four successive years (Boreham and Roubik 1987) and monthy numbers of flowering plant species on Barro Colorado Island, of the Panama Canal area (Croat 1978). The proportional representation of flowering trees is highest during approximately the first half of the year, and that of herbs is highest during the second half (Croat 1978).

Long-term studies of euglossine bee males arriving at chemical baits have provided diverse information on bee populations and flower visitation in three forests of central Panama (Roubik and Ackerman 1987). Artificial components of attractive orchid compounds were employed singly (Section 2.1), and three of the chemicals were sufficient to attract almost all local euglossines. The baiting studies were performed by placing blotter pads with the same amount of bait on the same tree each month and making counts each 15 min of all individuals and species at three baits during four hours. Four euglossine genera were attracted, *Eulaema, Euglossa, Eufriesea,* and the cleptoparasite *Exaerete*. Most individual bees stayed for less than five minutes; thus few were counted more than once. Over 40 of the 51 bee species arrived in adequate numbers to suggest their generation lengths. They usually showed three yearly abundance peaks, although most were found all year (Figs. 4.8 and 4.9). In contrast, *Eufriesea* usually were univoltine, appearing during the first month of the wet season and surviving one to two months. These bees are relatively large, and where several species were present, their equally large potential euglossine competitors *Eulaema* displayed no overlapping abundance peaks. Forty-three of the species were found in all three census sites, which differed substantially in their climate and biogeographic affiliations. Nonetheless, most bees showed two or three abundance peaks each year. Additional aspects of euglossine studies are given in Section 4.1.4.

4.1.3 How stable are bee populations?

In light of data gathered over a period of three to six years in two European localities – a suburban garden (Anasiewicz 1975) and three alfalfa fields (Owen 1978) – and over a period of three to seven years in Panama, there are clearly some remarkable examples of stability among temperate and tropical bee populations (Table 4.4). Comparison of the Panamanian light trap bees to those studied in Europe, and to insect populations in general, shows that population stabilities were roughly equal. They were the least variable of almost 3,000 insect species studied in 80 surveys (Wolda 1983; Wolda and Roubik 1986; Roubik and Ackerman 1987). Data used here were transformed to correct for differing numbers of yearly observations made in alfalfa fields (Anasiewicz 1975). It should also be pointed out that all the preceding studies differed in mode and presentation of censusing. For example, the study just mentioned included sightings and sweep nettings of 100 solitary bee species, 21 bombines, and *Apis mellifera,* presumably most from hived colonies. Fifty-one of these bee species were common, including some of each family that appeared – anthophorids, andrenids, halictids, melittids, apids, megachilids, and colletids. However, the data as originally presented considered bees in only three categories: honeybees, *Bombus,* and the remaining taxa. Thus it is necessary to pool data in the other studies to allow comparisons.

The annual variance (AV) statistic of Wolda (1978) and Williamson (1984) is one method used to quantify annual abundance changes of bees; the other is the stability index (SI) of Wolda (1983), and both are given where applicable in Table 4.4. Each measures statistical variance of transformed raw census data. The latter

4. Community ecology   337

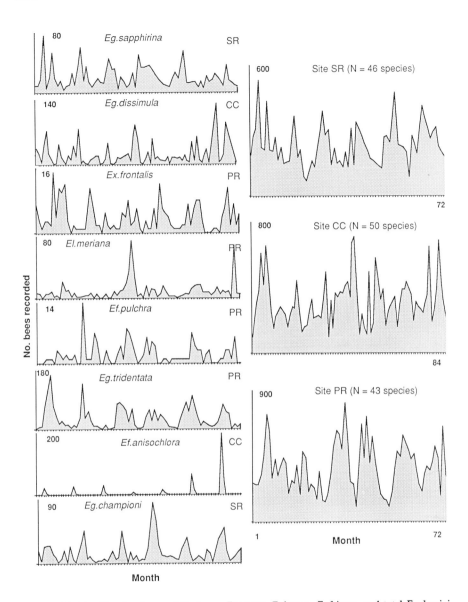

Figure 4.8. Monthly abundance of *Euglossa, Exaerete, Eulaema, Eufriesea,* and total Euglossini monitored at chemical baits during six or seven years in Panama (after Roubik and Ackerman 1987). The left column provides samples of local species dynamics at each of the three sites (designated in the upper right corner of the graph). Graphs on the right side show the abundance of combined species at each site.

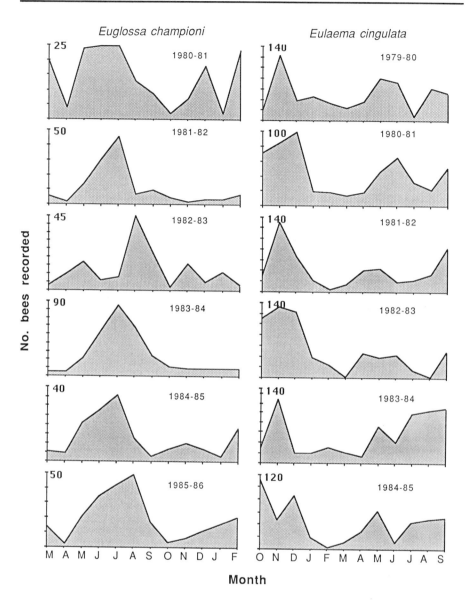

Figure 4.9. Monthly abundance patterns of the euglossine bees *Euglossa championi* and *Eulaema cingulata* from year to year at two sites in Panama (after Roubik and Ackerman 1987).

index is refined to apply to long-term census data that are compiled for at least six years. Admittedly, such data are very short-term compared with the life of a biological population, but they do constitute a separate category for census data taken by biologists (Wolda 1983). When data are taken for three or more years, abun-

Table 4.4. *Sample of stability and seasonality characteristics of bee populations*

Bee population	Yearly abundance peaks	Annual variance	Stability index	Study habitat
Euglossini (43 spp.)		0.004	−3.33	tropical forest
Euglossa sapphirina	3	0.008	−4.12	tropical forest
Euglossa imperialis	3	0.008	−3.87	tropical forest
Eulaema cingulata	3	0.009	−4.01	tropical forest
Euglossini (40 spp.)		0.009	−4.12	tropical forest
Apis mellifera	4	0.014		tropical, mixed habitat
Euglossini (50 spp.)		0.017	−3.4	tropical forest
Centris analis	8	0.02		tropical forest, mixed habitat
Euglossa deceptrix	4	0.021	−3.17	tropical forest
Apis mellifera		0.022		temperate, alfalfa fields
Bombini (12 spp.)		0.026	−2.502	temperate suburban garden, mixed habitat
Euglossa tridentata	2	0.029	−2.91	tropical forest
Eulaema meriana	3	0.031	−2.55	tropical forest
Non-*Megalopta* (48 spp.)		0.034	−2.99	tropical forest
Bombus terrestris		0.05	−2.23	temperate suburban garden, mixed habitat
Eulaema nigrita	3	0.052	−2.72	tropical forest
Euglossa gorgonensis	2	0.058	−2.69	tropical forest
Bombus pratorum		0.062	−1.11	temperate suburban garden, mixed habitat
Solitary bees (100 spp.)		0.067		temperate, alfalfa fields
Bombus ruderatus		0.077	−1.85	temperate suburban garden, mixed habitat
Bombus lucorum		0.082	−1.27	temperate suburban garden, mixed habitat
Bombus hortorum		0.089	−1.33	temperate suburban garden, mixed habitat
Bombus agrorum		0.094	−1.56	temperate suburban garden, mixed habitat
Megalopta genalis	3	0.12	−0.42	tropical forest
Megalopta ecuadoria	3	0.139	−0.67	tropical forest
Bombus lapidarus		0.146	−0.7	temperate suburban garden, mixed habitat
Bombus (17 spp.)		0.32		temperate, alfalfa fields
Bombus ruderatus		0.341	0.53	temperate suburban garden, mixed habitat

Note: Species ranked according to stability determined by three- to seven-year surveys by Anasiewicz (1975), Owen (1978), Wolda & Roubik (1986), Roubik & Ackerman (1987), Boreham & Roubik (1987), and Roubik (unpub.).

dance in successive years can be expressed as a ratio R that is equal to net replacement rate from one year to the next, regardless of the number of generations during a year. Statistical variance of the logarithm (base 10) of this ratio is the AV statistic. The smaller the value, the less the population variability. Most insect populations display AVs of 0.018–0.642 (Wolda 1978). As can be seen in Table 4.4, several tropical bees and combined species of several surveys have an AV less than 0.018. These are very stable, because the individual species are not rare. Rare species, by virtue of low abundance (less than an average of two or three in a year) have low variances, and thus their low AV does not indicate stability (Wolda 1983). Wolda has further determined that an insect population with mean annual ln N (natural logarithm) of less than 0.6 (where N is the total number censused during one year) is best omitted in analyses of long-term population studies. For long-term censuses, linear trends in the data make population variance appear more drastic than it really is (Roubik 1983b; Wolda 1983). After such time-series trends are filtered out, the variability of one population compared with another is usually about the same as shown by the SI statistics (Wolda 1983). This statistic is a log-transformed variance of abundance, given by the following formula:

$$SI = \ln(\text{Var}(\ln(N + 1)))$$

where N is as defined above. Other published long-term surveys of insects have produced SI values averaging –1.6 for the most stable groups in particular surveys (sometimes of closely related species, sometimes of insects not even in the same taxonomic order). These indices are for the mean averages; therefore, comparable SI values for the three euglossine surveys in Table 4.4 are –1.34, –1.52, and –1.71 (Roubik and Ackerman 1987). For the SI of an individual species, no insect was known to have a value more stable than –3.24. A few euglossines listed in Table 4.4 are more stable. Furthermore, the rough correspondence in rank between SI and AV indicates that Africanized *Apis mellifera* and *Centris analis* in Panama have also maintained relatively high stability.

For the euglossine bees studied by Roubik and Ackerman (1987), the most abundant species had more stable populations than did the less common bees. Moreover, assemblages of higher "diversity" (α-diversity or species-richness) were less stable). Yearly abundance peaks or annual generation number did not differ for stable as compared with unstable species, although most of the *Eufriesea* were univoltine and relatively unstable. It is worth noting that during a four-hour census, up to 800 bees were counted, which included only a fraction of those that came to the fragrance baits (bees were counted for less than 5 min during each hour). Short-term studies of euglossines, in which every individual bee was counted, give a maximum 1,300 bees arriving at baits in 1.5 hours at an island, where attractive bee resources presumably were scarce (Janzen 1981). In the Panamanian studies, there could easily have been thousands of euglossine bees within a few kilometers of any point within the forest. Their abundance was, however, far lower in other neotropical forests, ranging from about 15 to 30 bees arriving at baits of most of the same chemicals during one hour (Pearson and Dressler 1985). Such different estimates seem to result primarily from the way chemical baits are presented.

4.1.4   In-depth studies of euglossine bees

One of the most intensively studied groups of tropical bees, the euglossines, have increasingly been used to characterize plant–pollinator coevolution, chemical ecology, population dynamics, mimicry, parasite–host relationships, forest fragmentation, relationships with flowers, physiology, competition, mating behavior, and ecological aspects of body size. Independent studies of euglossine bees' relationships with plants – fragrances for males, resin and pollen sources for females, and nectar sources for both sexes – make their ecology and possible coevolution with plants subjects that can be discussed in uncommon detail. Since the discovery of male attraction to fragrance compounds (Dodson and Frymire 1961; Dodson et al. 1969), field studies have uncovered most of the extant species (except their females!) and many details on their floral visitation, despite the general paucity of data on pollen composition of nest provisions. Furthermore, the possible mutual dependence between orchids and their male euglossine pollinators can be assessed at the level of species, genus, and local community assemblage in a way still impracticable for most other tropical bees and the plants they visit for food.

The euglossines are common throughout the range of moist–wet tropical and subtropical forest in the Americas, from 32° S latitude to 25° N (Moure 1967; Kimsey 1982; Wittmann pers. commun.). They range in size from about 8 mm in length and 70 mg weight, for the small *Euglossa sapphirina,* to about 28 mm length and 1.13 g for *Eulaema meriana* (Casey et al. 1985). The bee assemblages consist of as few as 5 species and as many as 50. In the richest communities that have been studied, in central Panama, there are 15–30 euglossine species active at different times of the year (Ricklefs, Adams, and Dressler 1969; Roubik and Ackerman 1987). It is believed that the total numbers of species in most habitats is somewhat lower, which seems to be reflected by recorded totals of 30–40 species (Roubik 1979a; Janzen et al. 1982; Ackerman 1983a; Pearson and Dressler 1985). Altitudinal limits for the euglossines are close to those for the stingless bees, which reach at least 2,700 m at the equator (Vergara and Pinto 1980) and have progressively lower limits away from it, at the upper level of cloud forest formations where heavy epiphyte loads are found.

Euglossine bees as a whole visit at least 23 plant families for nectar, 9 for pollen, and 3 for resin. Nectar sources are often, but not always, flowers with long, fused corollas or complex floral structure. Known nectar sources are the Acanthaceae, Annonaceae, Apocynaceae, Bignoniaceae, Convolvulaceae, Dilleniaceae, Euphorbiaceae, Fabaceae, Flacourtiaceae, Gentianaceae, Gesneriaceae, Malvaceae, Marantaceae, Musaceae, Orchidaceae, Passifloraceae, Polygalaceae, Rubiaceae, Sterculiaceae, Turneraceae, Verbenaceae, Violaceae, and Zingiberaceae (Ducke 1901, 1902, 1906; Kimsey 1982a; Frankie et al. 1983; Ackerman 1985). The studies of Frankie et al. (1983) were focused on large anthophorid bees that visited almost all the families noted in the euglossine literature. Roubik (1979a, 1982d) found a close similarity in the floral resources used by the large anthophroids and bombines of lowland French Guiana. Therefore, some additional plant families noted by Frankie and co-workers may possibly also be used by euglossines. Pollen sources of euglossines are more restricted, constituted almost exclusively of nectarless, large, and often buzz-pollinated flowers, usually presented in small numbers at each plant and spaced widely in the forest (see Section 2.3.6). These include a few of the nectar sources, such as *Sabicea* (Rubiaceae), papilionaceous legumes, and bignones. For the most part, however, they are nectarless Solanaceae, Bixaceae, Caesalpinioideae (notably *Cassia* and *Senna*), Tiliaceae, Melastomataceae, Cochlospermaceae, Araceae, and Malpighiaceae.

The resin sources used by females for nest construction include floral resins of *Clusia* and *Dalechampia* [Clusiaceae and Euphorbiaceae (Armbruster and Webster 1979)] as well as fruit of Anacardiaceae [*Spondias* (Williams and Dressler 1976)]. Many other resin sources from wounds in trees are also used but have received little attention. Each of the above local surveys of euglossine food sources listed about 13 plant families. Because twice this amount has been documented, the flowers visited by even such a large group of widespread, perennially active, and relatively conspicuous bees may be poorly known.

Euglossine males carrying orchid pollinaria have been increasingly used to assess their pollination relationships with orchids (Janzen 1981a; Dressler 1982; Williams 1982; Ackerman 1983b; Pearson and Dressler 1985; Roubik and Ackerman 1987). The nature of dependency by euglossine species on orchid species is unknown (Section 3.3), but considering that species like *Euglossa imperialis* have been observed more than 20,000 times at chemical baits (Roubik and Ackerman 1987), there are perhaps more individual observations on these bees than any others. Nonetheless, a number of euglossines are not known to visit any species of orchid. Part of this impression may be sampling artifact. How do we know whether or not orchid bees, such as very abundant *E. imperialis,* might fail to visit chemical baits because they have already collected all the phytochemicals they require at sap, orchids, aroids, or fungi within the hundreds of square kilometers they traverse, or because there are hundreds of times more bees than orchid pollinaria? Might it also be true that the bees visit orchids lacking pollinaria or are otherwise nonpollinating visitors?

Arguments against the above possibilities for *Euglossa imperialis* are that observations at the orchids of which they occasionally carry pollinaria have yet to reveal visitation by them, despite the yearlong high abundance of the bees (Roubik and Ackerman 1987). Furthermore, the flowers of orchids are specialized in a variety of ways that would prevent nonpollinating visits. Orchid flowers cease fragrance production after pollination; bees that successfully gather fragrances are generally obliged to take the correct position for pollinia transfer (Dressler 1981, 1982; Janzen 1981a; Ackerman 1983b). That some bees attempt to remove pollinaria from their bodies or avoid visiting male flowers that forcibly apply their pollinaria to the bees has not prevented orchids from obtaining their pollination services (Romero and Nelson 1986).

The chances that sampling error or insufficient study create the impression that bees fail to visit orchids are notably diminished through intensive and extensive sampling. During one year, Ackerman (1983a) used 16 chemical baits that attracted 44 of the 57 euglossine species living in central Panama. Weekly censuses failed to produce a single orchid pollinarium on male bees of 15 species. These studies were carried out on the 15-km^2 island of Barro Colorado, to which few but resident species may be expected to arrive (see Powell and Powell 1987). On adjacent mainland, but during a seven-year study, 13 of the same species carried no pollinia (Roubik and Ackerman 1987). However, across a 75-km transect through both Atlantic and Pacific forests in central Panama, just 3 of the 15 species lacking known orchid hosts on Barro Colorado were not recorded with pollinaria (Roubik and Ackerman 1987). As of this writing, just one species of these euglossines has not been found with a pollinarium along the 75-km transect. This figure is the same if other records from South and Central America are considered (Kimsey 1982a; Ackerman 1983b). Nonetheless, nearly 40% of the species at each of the three Panama sites were never seen with orchid pollinaria (Roubik and Ackerman 1987). A principal conclusion from these long-term data is that, with each new season, there were few surprises in observed bee associations with orchids at a given locality. At the same time, bees appearing to lack orchid associations often visit orchids in nearby areas, and these shifts almost surely depend on the availability of other resources providing the same chemicals gathered at orchids.

Euglossine genera may show different tendencies to visit orchids due to their contrasting seasonality patterns and relative abundance in a given habitat. Summarizing data for the genus *Eufriesea,* medium-size to large euglossines of over 50 species, Kimsey (1982a) listed orchid associations for only 20 bees. These had an average of 2.6 orchid hosts (range 0–6), and the figure was 1.8 orchids/bee on average in central Panama (range 0–3) (Kimsey 1982; Roubik and Ackerman 1987). The *Exaerete* and *Eufriesea* surveyed by Roubik and Ackerman (1987) visited between zero and three species, averaging only one. The *Euglossa* and *Eulaema* in their study averaged slightly over four orchids per bee species (range 0–13). In marked contrast with the preceding two genera, these two genera were present throughout the year, as were most of their species. In general, for the local euglossine populations studied in Panama, the majority were associated with either one or no orchid species. Yet for the orchid species, combined bee visitation data for all three sites showed a preponderance of orchid species associated with just one species of euglossine bee (Roubik and Ackerman 1987).

The observation that a large portion of a local euglossine fauna displays no visits to orchids is perhaps to be expected in studies of a single year, but this figure remained near 25% with several years of data taken at multiple sites in Panama. The figure shrinks if all neotropical observations are combined, and such trends suggest that most if not all of the euglossine bees visit orchids, but this relationship varies widely with location. Ackerman (1983b) has shown large seasonal changes in the number of orchid species visited by a bee, along with varying proportions of species that carry pollinaria.

How can there have been close coevolution between orchids and euglossine bees – probably necessary to account for their highly specialized interactions – if many of the local species seldom or never visit orchids? Ackerman (1983b) suggested that bees do not visit orchids when their chemical needs are supplied by other sources. However, occasional visits to orchids still are likely to occur, unless strong preferences have evolved for orchids absent in some of the bee's geographic range (Roubik and Ackerman 1987). Coevolution could have produced a part of this relationship, but the restrictions for a one-to-one reciprocal evolution are so great that, although the process is conceivable, the terminology is perhaps inherently misleading. The usual caveat is that coevolution, if it has occurred, has been *diffuse* rather than singular and reciprocal between species (Feinsinger 1983; Schemske 1983; Howe 1984; Janzen 1985). Many different interactions with different populations or guilds might be responsible for speciation or specialization (Armbruster and Webster 1979; Frankie et al. 1983; Armbruster 1986; see also Section 2.4.3). Thus flowering plants may respond to their visitors as do plants to other kinds of herbivores: Heritable traits are selected when effective for a variety of interacting organisms, but perhaps for none of them in particular (Futuyma 1983; Janzen 1985; Strong 1985). This is a logical explanation given the general lack of historical or contemporary population data. What seems different for the euglossine bees (and other bees) is that they can include the most stable known insect populations (Table 4.4).

The evidence against close coevolution of euglossine bees and their host orchids is that a given euglossine bee may carry pollinaria of two or three orchid species, and even more than this number can be found among the bees of one species at a given site in a day (Dressler 1982; Roubik and Ackerman 1987). These are carried on different parts of the bee's body (Dressler 1981; Ackerman 1983b), which may indicate avoidance of competition for pollinarium placement sites by orchid species and hence a generalized foraging choice by bees. However, a behavioral variable not revealed in qualitative pollinaria data (presence or absence of a pollinarium species) is the proportion of bee visits to orchids due to floral deception, mimicry, or mistakes (Section 2.4). In a study by Roubik and Ackerman (1987), 9 of 30 pollinaria species carried by bees were produced by flowers that appear to offer no reward of any kind to their visitors. In conclusion, highly specialized plant-pollinator interactions might well be favored by euglossine population stability, but the degree to which their genetic basis reflects coevolution remains enigmatic.

No one knows the type of variation in resource availability experienced by bees. Phenology of plant and pollinator is often precisely matched but has been shown to be out of phase by six months for *Catasetum* orchids and their sole euglossine bee pollinator in Panama (Zimmerman, Roubik, and Ackerman in press). There are thus constraints more important than pollinator availability that determine plant phenology. If net floral resource availability of a given plant species is far more variable than are bee populations, bees seem less likely to evolve specialized floral preferences than are plants to respond to the type and availability of pollinators. Implications of work carried out on euglossine bees and orchids may apply only slightly for other flora having less specialized structures and attractants. The possibility of close coevolution with particular pollinators seems remote, but there is perhaps a better chance of floral manipulation of bees when food rewards also are provided, some of which cannot be collected or used by most bees (Section 2.4.2).

Subtle advantages of large body size appeared to explain some aspects of euglossine seasonality. When euglossine bees of two size classes (large and small) were analyzed for a number of years at three sites in Panama, it was apparent that small bees predominated in both wet and dry seasons (Roubik and Ackerman 1987). Bees of both sizes peaked in abundance at the same time, in the mid dry to early wet season. However, comparisons across the range of sites showed that proportionately more large bee *species* make up the euglossines of the coolest habitat, and the larger species in two of the three sites displayed their annual abundance peaks when conditions were coolest. Seasonal maxima in the activity of large euglossines during the cool season was also observed in a seasonally cooler habitat in southern Peru (Pearson and Dressler 1985). Ackerman (1985) suggested that larger bees were favored during the wet season in the warmer lowland forests of central Panama due to predominance of widely spaced food plants at this time.

The larger bees would have large flight ranges necessary to exploit these food sources (Sections 2.3.6 and 2.3.7). However, at all three sites studied for several years in Panama, the size characteristics of the general euglossine bee assemblage shifted during wet and dry seasons only one year out of three. The most noticeable change came about at the highland site (900-m elevation) during a year of extended drought. In this year the relative numbers of small bees showed no change during the year, but the numbers of large bees more than doubled after arrival of the wet season (Roubik and Ackerman pers. obs.). This tendency was dissimilar to all other years and may reflect extended pupal dormancy of the large bees in response to prolonged dry weather. Large body size thus seems most tightly coordinated with relatively cool temperatures and thermoregulatory ability (Section 2.3.4).

Significant positive correlations in monthly abundance were found between two *Eulaema* and their *Exaerete* parasite, at least where the former were relatively stable and abundant (Roubik and Ackerman pers. obs.). Each *Eulaema* was present throughout the year, but their combined annual abundances were far less stable at one site where no correlation was found with the abundance of *Exaerete*. The two other local species of *Exaerete* were very scarce, and their potential *Eufriesea* hosts (Kimsey 1982a) were highly seasonal. Pronounced seasonality can be viewed as a potential factor that allows escape from parasites in this bee genus, although the evolution of parasitism in bees as a whole might be favored by seasonality (Wcislo 1987a). Parasitism of their species by the three small *Exaerete* does not differ substantially from the probable attacks of the 18 *Eulaema* by the two other cleptoparasitic euglossines. However, the influence of meloid and mutillid parasitoids (Chapter 3) is unknown.

Müllerian mimicry systems in the Euglossini appear to be maintained by either the continuous presence or co-occurrence among the mimetic species. The three largest and most conspicuous euglossines of all species in central Panama are two *Eulaema* and their co-mimic *Eufriesea ornata;* all three are black with orange and yellow bands on the metasomal segments and have bicolored wings (Appendix Fig. B. 17). Two of the species were common through the year at one site in three, and only there were their monthly abundances positively correlated (Roubik and Ackerman pers. obs.). However, the *Eufriesea* was seasonal (but far less so than its congeners), and its marked divergence in color and size appears to have been promoted by year-round presence of the *Eulaema*. Seasonal appearance of *Eufriesea ornata* did not coincide with seasonal peaks of the *Eulaema* but displayed the opposite tendency. It appeared 11 of 18 times before or well after either of the *Eulaema* showed peaks in seasonal abundance. Automimicry of all three species (Section 2.3.8) probably involves their stinging females (which often collect mud or feces at ground level for nest construction and are likely to be encountered by predators). Furthermore, the male *Eufriesea* has a very striking patch of golden yellow hairs on the hindtibia, making it resemble a female bee carrying pollen.

4.2 Composition of bee assemblages

If we could easily observe all bees within the tropical landscape, their differing size and activity would seem remarkable. The rapid movements and flight of some would contrast with the slow, deliberate activity of others. Bees of a similar size and behavior might be seen gathering very different amounts of food from the same source. In addition, their principal nesting and foraging activity might be segregated along gradients of height, temperature, humidity, and other factors. Over time, we would also notice the presence of some bees throughout the year, while others appeared for one month and then vanished for eleven. Notwithstanding such variance, a simple species inventory allows certain comparisons between different regions. Such lists may not specify behavior or ecology, but they do reflect a type of evolutionary channeling that occurs over a large expanse of time, as immigration, extinction, and evolution change communities. Relatively few apoid faunal lists have been developed for the tropics (Michener 1979a), but there is perhaps sufficient variety among them to construct a global perspective on the composition of tropical bee assemblages.

Species lists of local bees have value in predicting which taxa might be found in survey of a particular habitat, but perhaps only by placing the emphasis on higher taxonomic groups. For example, Schwarz (1939) and Rothchild (1963) listed the bees of Sarawak found in museum collections; most were apids (Meliponinae and Apinae) and anthophorids (Xylocopinae). Yoshikawa et al. (1969) made a brief survey of bees on this island and at other Malesian localities in 1966 and reported the predominance of these two groups, both in species and in numbers of bees collected at flowers. The authors illustrate results from a single hour of collecting at flowers in Touaran, Sabah. Seven anthophorids and five Trigonini were found, in accordance with expectations that may have been founded upon the larger surveys. Like surveys of flower utilization (Section 4.1), numerous small collections can be used to build an appreciation of an entire fauna. The relative abundance or biomass of particular bee species, like the degree of generalization or specialization in their choice of flowers, are population statistics that are likely to vary and can best be described by combining several seasons' data. The surveys to be presented encompass one or more years of intensive field collecting at a variety of latitudes where bees are active through the year. Museum voucher specimens are not available for all of them, and some inaccuracies could surely be found. However, possible misidentifications should not obscure the major family-level characteristics.

348   Ecology and natural history of tropical bees

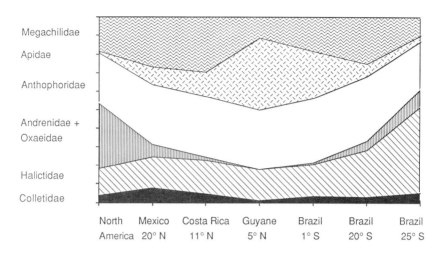

Figure 4.10. Bee family representation in the neotropics. Proportions of total species are based upon surveys in the literature. From left to right, these are Hurd (1979), S. Bullock and R. Ayala (pers. commun.), Heithaus (1979a), Roubik (1979b), Ducke (in Roubik 1979b), Camargo and Muzucato (1986), and Sakagami, Laroca, and Moure (1967).

4.2.1   The neotropics

Survey data from North America to southern Brazil are summarized in Figure 4.10. Five of the six tropical localities are in the lowlands with pronounced wet and dry seasons. Yearly precipitation is near 1,500 mm at Chamela, Guanacaste, and São José dos Pinhais (the highland site at 900 m); it varies between 2,000 and 3,000 mm yearly at the other sites. Second-growth and agricultural lands intervened strongly at the two southernmost localities. The earliest data represented here are those from collections made by Ducke in the Belém area during the first two decades of the 1900s. The remaining studies have been made since 1962; their respective literature references are given in the figure caption.

Anthophorids and colletids make up a stable portion of these apoid faunas. The first family accounts for 30–36% and the second for 2–7% of bee species. Halictid species number increases abruptly in southern Brazil, while megachilids there show a proportional decrease. A similar reciprocal trend between megachilids and apids is apparent in French Guiana. In this forest area separated from the coast by 15 km, I have found 69 species of Meliponinae, although only 45 were included among the apids in the study depicted in Figure 4.10. Because the other apids recorded from French Guiana have no more species than the number found in other neotropical areas, it appears that megachilids may have been replaced by stingless bees in equatorial forests and by halictids at the edge of the southern Araucarian subtropics. Oxaeidae and Andrenidae increase in species number toward the extreme latitudes of the neotropics, whereas a slight but opposite trend is evident among the anthophorids.

Whether the shifts in species number tend to reflect favorable nesting or other survival conditions, the results of food competition, or the ancient distributional history of these bees is a question too complex to be resolved here. The different floral regions represented even in lowlands at either side of Panama, one being Laurasian (northern) and the other Gondwanan (Gentry 1985), do not seem to lead to differences in the bee families inhabiting them. One possible explanation for similar bee family representation in the northern and southern hemispheres might be that substantial faunal exchanges took place during the Eocene and Quaternary (Michener 1979a; Simpson and Neff 1985; Camargo et al. 1988). During the slightly cooler, drier glacial periods in the Pleistocene, many bees of the southern and northern latitudes attained wider distributions than those they now occupy. Tropical bees, like many other animals, were likely confined to continental islands of moist forest during these intervals (Haffer 1969). Their reinvasion of expanding tropical forest during the Holocene should have coincided with the retreat of some subtropical bee fauna to latitudes where they are currently found.

4.2.2 The paleotropics

To reveal aspects of bee distribution that imply broad similarity through the Old World (Chapter 1), areas ranging from Nigeria to the South Pacific are presented in Figure 4.11. Some of the surveys date from the late 1800s, made in Egypt and Arabia near the southern Red Sea and in central India. A major regional survey based partly on museum material was undertaken by Michener (1965b), which includes the diverse Indopacific and Australian region, east of Weber's line to the Tasman Sea. Another survey of Hymenoptera of the Philippines was completed by Baltazar (1966). Under such varied meteorological and biogeographic conditions, some tentative patterns in bee family representation in the Old World tropics can be found.

Like the neotropical region, more paleotropical species of Anthophoridae are known than of any other family. Excluding the Australian bees and associated fauna, 33–42% of local species belong to this group, varying slightly by latitude or longitude. The pattern of reciprocal species representation appears again in apids and megachilids, as seen in the neotropics, especially near the equator. Mirroring the pattern in the neotropics, apids are the largest portion of the local bee community in the equatorial forests. On the other hand, the halictids do not seem to increase at southern latitudes as they do in the neotropics. Increasing species representation of Colletidae in the Australian region corresponds to an abrupt decrease

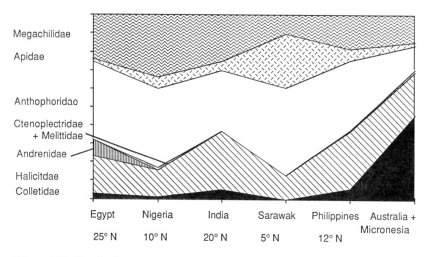

Figure 4.11. Bee family representation in the paleotropics. Proportions of total species are based upon surveys in the literature. From left to right, they are Walker (1871), Medler (1980), Rothney (1903), Rothchild (1963), and Schwarz (1939), Baltazar (1966), and Michener (1967).

in numbers of Anthophoridae. As we move from North Africa to India, a relatively large component of Andrenidae seems to be replaced by Halictidae. In addition, at low latitudes anthophorid dominance increases where that of halictids is least.

Judging by the similarity of Figs 4.10 and 4.11, some of the trends in tropical bee assemblages traverse continental barriers. In the paleotropics and neotropics, apids stand out as the bee group that expands most in the low latitudes. The majority are highly eusocial, but roughly 50–100% of the local species richness found in Meliponinae is matched by Euglossini in the American tropics, and many if not most Euglossinae are not eusocial.

4.2.3 Are bees less diverse in the tropics?

Tropical bees do not display the extremely cryptic and specialized habits of other tropical invertebrates, which for some surveys allows estimates of their actual species richness to eclipse the number of known species many times over (Erwin 1982; see also Stork 1987). New species and genera are common in most surveys of tropical bees, but the total field effort required to discover local taxa is greatly reduced because bees can be found at flowers. If we accept the premise that at least some parts of the tropics are relatively well known, then the bees of the world display singularly rich faunal regions outside the lowland tropics (Michener 1979). In particular, they are richest in warm, seasonally dry or xeric habitats, near the limit of the subtropics.

In Illinois, even during the 1920s and well after agriculture had greatly transformed this part of the world, more bee species could be found than in Costa Rica, Suriname, the region of Belém at the mouth of the Amazon, or the Philippines (reviewed by Michener 1979). Regional and local lists of palearctic bees, many above the 40th parallel, contain at least as many species as the tropical places summarized by Michener and in the added references in Figures 4.10 and 4.11. Roughly twice as many bee species have been found within areas of a few dozen kilometers in warm temperate areas of southern Europe and the southwestern United States than in any tropical area of comparable size. In both the Old World and New World tropics, within intact habitat and in more disturbed areas, there is good reason to believe that bee species assemblages are not particularly rich.

The taxonomic question of species richness is adequately resolved with available information, but qualifying this temperate–tropical comparison leads to a different picture of bee diversity. Diversity in *ecology* – spanning behavior and the modes by which bees influence travel of nutrients and energy through communities – is much greater in the tropics. To point out but one instance, some neotropical apids extensively utilize animal feces, extrafloral nectar, fruit, and dead animals (Chapter 2). Such phenomena seem rare in the temperate zone (Sections 2.1.1 and 2.1.8). Because many apids are active throughout the year, they potentially come into contact with many more resources, natural enemies, and mutualists than could conceivably fit into the active period of a temperate-zone bee. Some tangible consequences for other organisms are suggested in the following sections.

4.3   Roles of bees in communities

4.3.1   *General considerations*

The elements of bee ecology outlined in Chapters 1–3 are here given emphasis as part of larger community processes. Bees undoubtedly intervene in food webs of intricate and surpassing variety – for example, by utilizing pollen ordinarily dispersed by wind and harvested by small invertebrates. Honeybees may instead circulate this resource as excreta that support prolific fungal growth. The scope of bee ecology in tropical biotas very pointedly concerns the primary producers, the plants at the base of the food chain. Bees interact with plants as commensals, mutualists, and parasites (Section 2.4.1). As I suggest in the preceding chapters, the first and third categories are sometimes more apparent in the tropics because of the stingless bees and honeybees. Bees compete for plant products with other flower visitors as well as with ants and wasps for extrafloral nectar and the exudates of homopteran bugs. Competition of the latter types seems to involve only certain groups of stingless bees that forage diverse extrafloral nectars and tend

Homoptera. Flesh-feeding stingless bees also compete with ants, flies and wasps for carrion (Cornaby 1974; Roubik 1982a; Baumgartner and Roubik in press). Standing crops of fruit and seeds, the sole foods of diverse animals, are influenced by both pollination and competition generated by bee foraging activity. Furthermore, while bees provide domiciles for their own set of mutualists, commensals, and parasites, they also compete for some of the cavity domiciles of ants, wasps, birds, bats, and other vertebrates. Lodger bees utilize abandoned tunnels of beetles, wasps, or other bees that burrow in soil and wood as well as nests made by wasps, ants, and termites that construct free-standing nests or occupy cavities, and they also use the abandoned nests of birds and rodents. Bee larvae, provisions, and adults all are consumed by larger animals, with varied yet scarcely documented consequences for these consumers and their alternative prey. One case has been documented in which the aggressive behavior of *Trigona corvina,* which builds large, exposed nests on branches, can even improve the chances for survival of birds or wasps nesting nearby (Smith 1980). Quantitative studies of these events are very few, and their eventual elaboration should repay the effort.

The pollinating roles of tropical bees sometimes arise by default. Two brief observations serve to illustrate the concept until more complete pollination studies are forthcoming. The tropical flowers that dehisce at night or late in the day often produce copious quantities of nectar, along with the pungent smell that attracts and guides wide-ranging nocturnal foragers. Flowers displaying all such traits, normally associated with pollination by vertebrates or nocturnal animals (Faegri and van der Pijl 1979; Baker 1983), may nonetheless be pollinated primarily by diurnal bees. In Panama, Snow, Buchmann, and I experimented by placing muslin bags over inflorescences of trees (*Luehea seemanii,* Tiliaceae). The flowers opened in early evening, and we ran two sets of controlled open-pollination treatments. Some inflorescences were bagged during the day, after being open to visitors during the night, and others were bagged during the night and exposed only by day. The results were clear: Many more fruits and seeds were formed on the inflorescences open to diurnal pollinators. Moreover, similar work done independently in Costa Rica showed the same pattern (Haber and Frankie 1982), and these authors demonstrated the tendency of this tree genus to receive nocturnal pollinating visits from bats and sphingid moths. During the day, we recorded visits by halictid, meliponine, and anthophorid bees that collect pollen and nectar. Another bat-pollinated tropical plant was studied by Steiner (1984), who recorded nocturnal visits to *Mabea occidentalis* (Euphorbiaceae) by pollinating opossums and bats. During the next day, the residual nectar was gathered chiefly by *Trigona pallens*. Experimental bagging studies with this flower showed that the stingless bee may have been responsible for 30% of seed production (Steiner pers. commun.). In light of these

4. Community ecology

observations, it seems that flowers of many types not thought to be pollinated by bees might be highly dependent on them.

4.3.2 Community structure

A theme developed thus far is that the highly eusocial bees are active all year and during most daylight hours (see Figs. 2.35–2.39). This section deals with some neglected aspects of their general ecology before examining processes, involving such bees in particular, that might make a large contribution to the structure of tropical bee communities.

Bee colonies, for the duration of their lives, are recycling centers. Honeybee colonies are used for environmental monitoring of contaminant trace metals, pesticides, and soil elements (Tong et al. 1975; Bromenshenk et al. 1985). Their colony output of fecal material has received some recent attention in Southeast Asia due to aerial defecations of giant honeybees (Seeley et al. 1985). This "throughput" should be no different for other bees, although somewhat less spectacular.

The amount of protein and energy harvested by all resident bee colonies in a square kilometer of tropical forest is calculated below. Colony trash later ejected from nests could each year constitute nearly 0.01% of the aboveground forest nitrogen biomass, as given by Vitousek and Sanford (1986). Some conservative calculations based on the number of stingless bee colonies found in five hectares of cleared forest in eastern Panama, as well as in surveys near ground level elsewhere in Central America (Section 3.1), suggest that approximately 600 highly eusocial bee colonies occur in each square kilometer. The biomass of honey and pollen in nests and the rate of return by incoming foragers (and total quality and quantity of their foraging loads) are known for a few stingless bees (Chapter 2). Either nests or foragers can be used to calculate approximately how much material is sequestered by bees on a yearly basis. A method adequate for this discussion is to compute the fraction of a year that honey and pollen stores found in a nest at any one time would support brood production. Effectively portraying a local group as diverse as the stingless bees, with colony and bee size differing by two or three orders of magnitude, is no simple task. Their individual foraging, food storage, metabolism, and reproductive rates normally vary widely during the year and between species, and so the following computations may be indicative but imprecise. Ratios of average volumes of stored food to average brood provisions, taking into consideration dilution of the resources placed in brood cells, suggest that stingless bee colonies could continue to rear brood, and survive for 80–280 days with no incoming food (Roubik et al. 1986). I had added an expected adult longevity of 40 days to these figures, which now can be modified to roughly one to eight months. Though this is a considerable range, I will place an average value of four months' foraging as the equivalent of such resource stores. The costs of wax production, foraging, and so forth are ignored altogether, which should lead to an underestimate of colony resource harvest.

Nutritional and energetic values of colony stores can be multiplied by 3 (yielding the average yearly consumption of resources from the preceding discussion) for each of 189 forest stingless bee colonies (of 37 species) surveyed during three years (Roubik 1983b). These nests were studied as they were encountered, including nests high in trees, so that there should be little bias toward colonies of a particular species or nesting preference. In this study the percent sugar composition of the honey was determined, and the energy and pollen protein values can be derived. If honey is 70% sugar and pollen 3.5% nitrogen (Stanley and Linskens 1974; Southwick and Pimentel 1981; Roubik 1983b), colony recycling of these materials can be estimated. Taking the 22 species of stingless bees for which average values of honey storage are given in Roubik (1983b), 1.436×10^7 kJ would be harvested on an annual basis by 600 colonies in each square kilometer. Total pollen volume is a straightforward computation, equal to 146 l/yr for the 600 colonies. Corrresponding nutrient content would equal about five liters of nitrogen. The energy value, at 60 mg pollen (dry weight) per milliliter, would equal roughly 2.2×10^5 kJ. Thus the stingless bees inhabiting a square kilometer of lowland forest harvest about 1 billion kJ and five liters of nitrogen during a year. Considering that this is *more* than the total annual energy from primary producers given for a square kilometer of tropical forest (Odum 1971), the energetics of flowering and flower production are clearly underestimated (Heithaus 1974; Southwick 1984).

The harvest of the bee colony is later scattered over the forest as wet feces, moist parcels of colony trash (for meliponines), and dead bees. Dead bees are taken from the nest soon after they die; Gary (1960) found that 100 workers of European apiary *A. mellifera* were disposed of each day. Complete turnover of stingless bee colony worker populations, and probably those of tropical honeybees (but see Dyer and Seeley 1987), occur each 40–50 days. Therefore, approximately eight times the local adult bee population is scattered along the ground within a few dozen meters of each nest. For 600 colonies, and average colony size of 2,000–3,000 bees, this would amount to a few hundred kilos of bees per square kilometer each year and an energy content approaching 10^7 kJ (Roubik and Buchmann pers. obs.).

Trash and feces of the highly eusocial bees are distributed in such a way that their accumulations are probably a significant fertilizer. The scattering of meliponine trash pellets, which are cast pupal exuviae and feces, would over time form a track away from the bee nest in line with the entrance. A few stingless bees, like *Lestrimelitta* and *Hypotrigona*, drop their trash directly from the nest entrance, and I have weighed the deposits by a species of the former. It expels about 3 kg/yr and is of intermediate size among the meliponines. This suggests that 600 colonies distribute on the order of 1,800 kg/yr over every square kilometer of forest. For the *Lestrimelitta,* the trash pellets are 55% carbon, 4% nitrogen, and 8% hydrogen (analysis by Buchmann). The *Apis* distribute their waste material differently, during "cleansing flights" in which a group of workers collectively void liquid feces in the air (Seeley 1985). This probably happens once in the life of a bee, after it has been engorged with pollen as a young adult. The 30 mg or more voided by *Apis dorsata* during a flight, multiplied by a colony size of 40,000, would amount to about 10 kg/yr. With one report from India of 156 colonies in a single tree (Singh 1962), it is perhaps no coincidence that the bees are sometimes thought to bring good luck to farms.

Regarding community structure, efforts to examine the number and type of species that coexist are immediately hampered by large information gaps. Although experimental data show competition occurs among bees, their varied foraging behavior and foraging specializations (Chapter 2) can be interpreted as

1. the result of present, transient competition;
2. the result of past competition and some coevolution; and
3. the result of chance associations during ecological time.

The critiques of simple competition theory as a basis for community structure are legion (Wiens 1977; Schoener 1983; Andrewartha and Birch 1984; Strong 1985). Short-term competition has uncertain consequences for bee community structure, and long-term competition may bring about lasting patterns, but the most conclusive type of evidence involves variables that are difficult to measure (Roubik 1983c; Roubik et al. 1986; Wolda and Roubik 1986). For example, Hubbell and Johnson (1977) report uniform nest dispersion for some stingless bees of the subgenera *Trigona* and *Scaptotrigona*. The cleptobiotic stingless bee *Lestrimelitta limao* exists in neotropical forest at colony densities very similar to those theoretically permitted by exploitation of its host stingless bee colonies (Sakagami, Roubik, and Zucchi in press).These are indirect methods, but a persuasive argument for intraspecific competition as a prime determinant of population size can be built upon them. An implicit assumption is that equilibrium has been reached by consumers and resources in the habitat, and this tenet is openly questioned for many tropical forests (e.g., Hamburg and Stanford 1986). In terms of clear biological parameters – nest establishment, colony biomass, reproduction, and mortality – taking overdispersion of nests to show lasting, significant competition implies that all four factors adjust to one another. The rate at which fluctuating competition pressures affect gross community patterns seems to be slow, judging by study of competition between Africanized honeybees and native neotropical bees (Section 4.3). Such seemingly slow and weak interactions might nonetheless affect community structure over time.

MacArthur (1972) stresses the value of biogeographic studies in efforts to understand processes that build communities. The composition of island and continental faunas can be examined from this standpoint. Island biogeography of bees, as a specialized area of inquiry, is currently fragmentary (Janzen 1981b; Rust, Menke, and Miller 1985; Eickwort 1988a). Fewer species exist on islands than on adjacent mainland, and the island species often arrive there within floating logs but also disperse over open ocean (Michener 1979; Yamane 1983; Eickwort 1988a; see also Chapter 1). The highly eusocial bee species inhabiting forested islands along the Pacific coast of Panama are few in number (representing about 15–20% of the adjacent mainland species), none nests in the ground, and all use small nesting cavities. One endemic *Melipona* exists in what appears to me to be extraordinary abundance on the large forested island of Coiba. It may be an example of the competitive release enjoyed by a generalist forager when the number of competing species is reduced. The history of highly eusocial bees on islands can involve introduction by indigenous peoples (Chapter 1), but no mainland population of this particular *Melipona* is known. Janzen (1981b) found large differences in the abundance of euglossine species on a small island 17 km from the coast of Costa Rica, compared with a nearby mainland forest that had several times as many euglossine species.

Broad biogeographic patterns of tropical bees suggest that highly eusocial species play a part in maintaining relatively low bee species richness (Michener 1979; Roubik 1979a). Trends that stand out in the foregoing descriptions of tropical bee assemblages are the expansion of apids or halictids (many of which are social) where megachilid or andrenid species representation is reduced. Further, although bee species richness is highest in warm temperate habitats, these are also the habitats where few apids are found (Michener 1979). Nesting site availability is likely greater for apids than other bee groups in tropical forests, because most species use trees and resin to build their nests. The relative scarcity or absence of apids in the warm temperate habitats where bee species richness is highest also suggests a lack of nesting sites for apids. This nest site availability hypothesis is perhaps sufficient to account for contrasting species richness in the bee families of tropical and temperate areas. The seasonality of floral resources in the temperate zone, even in the warmest areas, might also tend to diminish the numbers of social bee species that can be supported throughout the year (Section 4.5). However, one cannot conclude that the moist or wet tropical habitats are, in general, less favorable than warm temperate areas for bees. Where the apids exist in large number, their perennial colonies are likely to preempt solitary bee species at many flowers. Over evolutionary time, the constant intervention of opportunistic apids may prevent some of the sustained interactions between solitary bees and flowers that can lead to specialization, coevolution, or speciation.

Niche-preemption by stingless bees and apids is potentially the central determinant of the moderate species richness seen in tropical communities. Detailed studies comparing the actual degree of diet specialization, and hence potential to dominate resources and influence the success of other flower visitors, have not been made of many Meliponinae and Apinae. However, the inverse diversity patterns of highly eusocial apids and other bees are noteworthy. Exceptions to the proposed negative interaction between these bee groups would help to clarify processes contributing to what appears to be a widely established pattern. Perennial apid colonies are the primary determinants of tropical apoid community structure.The concept is a simple one and has several lines of support:

1. High species diversity of flowering plants is characteristic of the tropics, where at least 30–40% of all flowering plants provide food for bees; the small number of tropical bee species is therefore unexpected (Chapter 2; Sections 4.1 and 4.2).
2. The local faunal makeup of bee species can include nearly 70 Meliponinae or 4 Apinae, and these bees, unlike any temperate bee, are ubiquitous and aseasonal. Furthermore, the paleotropics (Malesia in particular) has up to 20 Meliponinae and 4 Apinae in a single habitat, compared with 16 Meliponinae in dry deciduous forest at comparable neotropical latitudes (Chapter 2; Section 4.2).
3. The Meliponinae and Apinae are, as a group, unquestionably more generalized foragers than more seasonal tropical bee species (Section 4.1).
4. The highly eusocial apids of the neotropics share almost all of the floral resources used by bees of similar or smaller size, which include all but the large anthophorids and Bombinae (Heithaus 1979a,b; Roubik 1979a, 1982d).
5. As demonstrated by Michener (1979), global patterns of bee diversity are inversely correlated with regional numbers of the highly eusocial species. He listed a total of 89 bee genera and subgenera for the Oriental faunal region, 175 were included in Africa south of the Sahara, and 315 were given for the neotropics. Corresponding species of Meliponinae and Apinae number approximately 77, 46, and 300, respectively. This pattern could be explained as the result of an ecological hierarchy, whereby one species of *Apis* is the equivalent of several Meliponinae, which in turn take the place of many solitary and less social bees.

4.3.3   Ecology of Africanized honeybees in tropical America

The honeybee has recently become a successfully introduced species in the American tropics. Introduction of plants and animals from one continent to another is a common practice, and the neotropics has many that appear to be well established: mangos, sugarcane, bananas, coffee, citrus, fire ants (*Solenopsis invicta*), and cattle. With the possible exception of the ant species, none of these organisms could maintain long-lived feral populations in neotropical forest. Occasionally, an introduced species initiates colonizing populations that spread, and rarely, these invade natural habitats (Elton 1958; Parsons 1983; Case and Cody 1987). In 1956, some 26 mated queen *Apis mellifera scutellata* were transported from southern Africa to southern Brazil and were the most desirable lineages available for breeding stock; the African honeybee colonies, thousands of drones, and some African/European hybrid queens and workers were released in southern Brazil (Kerr 1957, 1984a; Nogueira-Neto 1964; Michener 1975, 1982b). As a result, there are now approximately one trillion honeybees of African descent living in cities, agricultural areas, and forests from northern Argentina to southern Mexico. They have naturally colonized some 16 million square kilometers and may consume two billion kilograms of pollen and 20 billion kilograms of nectar annually (Roubik 1987a). The honeybee's spread and colonization of tropical America is a vast experiment, as most of this territory, in particular the Amazon Basin, never before contained *any Apis mellifera*.

Seldom has there been continuous study of what happens to Africanized honeybee populations after they arrive, and it is still too early to appreciate many changes that their arrival may promote. A survey of the impact of introduced species in native species assemblages led Simberloff (1981) to conclude that most introductions had resulted in no changes in local species abundance or in community structure. The time scale over which a significant impact becomes evident following any species introduction, however, seems to lack substantive treatment. The invasion by Africanized honeybees shows why such changes are expected to be slow or quite difficult to detect. In contrast, the impact on beekeeping is swift and dramatic (Section 4.3.4), and this feral honeybee population has also stimulated considerable original research. Now it is beginning to reveal some characteristics of tropical plant–pollinator communities.

What has allowed this phenomenal honeybee spread and what are its apparent limits? Small introductions rarely culminate in establishment of a species unless the organism in question is capable of self-fertilization or disperses very slowly (summaries by Baker and Stebbins 1965; Parsons 1983; Strong 1985). *Apis mellifera* produces 50% unviable offspring from continued sib–sib matings (Section 3.3) and its reproductive or absconding swarms in Africa traverse large areas [>200 km by some accounts (Kigatiira 1988)], so neither condition applies to this bee. In fact, long-distance dispersal of swarms may in itself constitute a powerful means of adjusting to local habitats. The pooling of colony resources may arise due to coalescence of many swarms and the elimination of all but one queen before a long-distance dispersal event occurs (Kigatiira 1988). Kigatiira notes that such amalgamation even occurs among apiary colonies, and I believe this accounts for frequent claims that colonies are larger than those of temperate-zone *A. mellifera* (Kerr 1984a). Further, the presence of many queens in large swarms (Silberrad 1976) is likely restricted to composite swarms that have yet to select among their queens.

The small population introduced into South America from Africa was by no means isolated; it was surrounded by thousands of other colonies of temperate-zone origin. The initial success of Africanized honeybees was permitted by interbreeding with these local bees or invading established nests. Swarms and their queens and lone drones are able to invade hives of other colonies (Michener 1975; Rinderer et al. 1985). This at least preserved the African genotype, but it was likely to become gradually dispersed and diluted. Natural selection appears to have consolidated the African genome, and since then the Africanized honeybee has likely overwhelmed the European contribution to the gene pool by its sheer numbers (Taylor 1985; Boreham and Roubik 1987; Roubik 1987a). Evidence for persistence of the African genome in colonizing populations derives mainly from phenotypic traits. The bees that invaded Panama from Colombia in 1982 were, according to morphometric analysis (Daly and Balling 1978), almost indistinguishable from the feral, highly African honeybees that existed in South America during the 1970s (Roubik 1982d). The same was true for bees invading Honduras (Gary et al. 1985). Feral honeybees encountered in many areas in Brazil now show only slight differentiation from honeybees in South Africa (Stort and Bueno 1985). Furthermore, four successive years of study in Panama revealed that the colonizing honeybees became *more* like African honeybees over subsequent generations (Boreham and Roubik 1987). Wax and cuticular chemistry of the invading Africanized honeybees also indicate little change from the African source populations (Carlson 1988). Recently, behavioral assays have shown that although the Africanized honeybees of Venezuela fall within the range of variation found in African populations, they represent an extreme in their rapid stinging behavior (Rinderer 1988). The stinging behavior of Africanized honeybees, and other behavioral traits, may have changed in response to selection pressures in the neotropics, to the extent that these bees no longer closely resemble lowland African honeybees from which they are derived (Ruttner 1988).

The apparent limits to the spread of Africanized honeybees in the neotropics are related to explanations that can be given for their colonizing success. Some of the same traits making them successful tropical honeybees should severely limit their progress in temperate areas, likely in response to the same conditions that have isolated and diversified populations of *A. mellifera* in Europe and Africa. One factor is seasonal temperature; this single aspect of climate has formed the *only* basis of forecasting the eventual dispersion of Africanized honeybees in the Americas (Taylor 1977; Taylor and Spivak 1984; Dietz, Krell, and Eischen 1985; Krell, Dietz, and Eischen 1985; Taylor 1985). As one test of this hypothesis, I compared the known distribution of tropical apids north and south of the equator in the neotropics (Roubik 1987a). The euglossine genera *Eufriesea* and *Eulaema* and the meliponine genus *Melipona* each inhabit the American tropics from approximately 32° S latitude to 23–25° N. Their altitudinal distributions reach 2,400–3,000 m (Schwarz 1932; Moure 1950; Nogueira-Neto 1970b; Dressler 1982; Kimsey 1982a). Relevance to Africanized honeybees is implied by a shared southern and altitudinal distribution (Vergara and Pinto 1981; Kerr, del Rio, and Barrionuevo 1982; Roubik 1987a; Villa 1987).

Winter temperature regimes in which Africanized colonies can survive are not completely distinct from those of European honeybees. Each survived through one season in relatively cool winter conditions in southern Argentina, where the average highest temperature during the coldest weather is above 16° C (Taylor and Spivak 1984; Krell et al. 1985). As shown by the former authors, this climatic trait extends to much of the southern United States. Cold and prolonged winters decisively limit Africanized honeybee colony survival, because this tropical race does not efficiently maintain nest temperature or store the food to do so for more than a short period (Southwick and Roubik pers. obs.). It does, however, form a relatively compact cluster of workers around the queen during cold weather in southern Argentina, and it also survives periodically harsh winter weather in the highlands of southern Africa (Fletcher 1978; Dietz, Krell, and Pettis 1987). Villa (1987) studied occupancy of artificial nesting sites along an altitudinal transect ranging from 2,000 m to 3,000 m in Colombia and found that Africanized honeybee colonies diminished gradually in abundance with increasing elevation and wet conditions. Further, Villa noted that European honeybees had never occupied these nesting sites despite their presence in nearby apiaries for many years.

Rainfall, predators, parasites, and vegetation that provides honeybee resources may be decisive in determining rates at which Africanized honeybees penetrate North America, but if their numbers and spread decline as they approach their natural limit, extensive Africanization of resident European honeybees may not occur (Roubik 1987a). Slow colonization has taken place in southern Brazil and northern Argentina, where the Africanized traits of the hybrid population are not so apparent (Taylor 1985). Taylor maintains that colonization rate, likely determined by swarm movement from dense populations to less densely colonized areas, decreased from nearly 600 km/yr to 100 km/yr near the southern limit of their range. However, Dietz and his co-workers (1987) found no evidence in northern Argentina to corroborate this report. Both central Argentina and southern Texas have feral populations of European honeybees. This source of competition for mates, food, and nesting sites may severely limit the spread of Africanized honeybees (Roubik in press c). The degree to which the Africanized honeybee will be adapted to the habitat of the southern United States is not necessarily less than that of resident feral honeybees that were introduced from Europe.

Less rapid colonization could be either a cause or an effect of reduced Africanized honeybee nest density in habitats already occupied by apiaries of the European honeybee. In the tropics, the size of feral populations alone seems sufficient to confer a tremendous genetic and reproductive advantage upon Africanized *Apis mellifera*. The number of Africanized honeybee nests might be as great as 108 per

4. Community ecology

kilometer (Kerr 1984b; see also Section 3.1), or it might be less than 10 in this area (Roubik 1983b, 1987a; Taylor 1985). Such estimates ignore continuous emigration and immigration of colonies, and none is the product of careful surveys to uncover all the nests within one square kilometer. Even a number as low as six colonies per square kilometer gives the drones of feral colonies a spectacular reproductive advantage over those of apiaries. Ruttner (1985) suggests 7 km as a conservative flight range of drones. Therefore, the drones of a given European honeybee colony (or apiary) compete with the drones within a 14-km radius. This will include males of 3695 feral Africanized honeybee colonies, which greatly exceeds local or even regional colonies of European *A. mellifera* in most of tropical America. The work of Rinderer et al. (1985) also suggests that drones of Africanized honeybee colonies (from other apiary hives) readily enter hives of European honeybees, after which they are fed and drone production by the latter is diminished. Furthermore, the drones and virgin queens of the Africanized honeybee fly somewhat earlier in the day than do the European races, and thus a partial reproductive isolation may exist favoring perpetuation of African genotypes (Taylor 1985, 1988).

In a lowland tropical area of mixed primary forest, agricultural land, and second growth flanking the Panama Canal, the records of Boreham and Roubik (1987) suggested that Africanized honeybees literally "swamped" other varieties of *A. mellifera*. This occurred despite an established local apiculture with European honeybees. A list of traits favoring such *initial* colonizing success of the bees can be compiled from various overviews of their biology by Smith (1960), Michener (1975), Silberrad (1976), Taylor (1977, 1985, 1988), Fletcher (1978), Roubik (1978, 1980a, 1983b, 1987a, and in press a,c), Rinderer et al. (1982), Anderson et al. (1983), Winston, Taylor, and Otis (1983), Seeley (1985), Rinderer (1988) and others who summarize the biology of this bee on two continents (Needham et al. 1988; Fletcher and Breed in press; see also Section 4.5). Of principal interest are the following:

1. *opportunistic selection of nesting sites,* ranging from abandoned armadillo burrows and hollows in arboreal termite nests to exposed habits on tree branches and foliage, buildings, bridges, and refuse;
2. *dispersal ability favoring habitat selection* (distances of 32 km have been traversed to uninhabited oceanic islands, and longer distances are possible if all honey sources carried from the nest are directed to flight expenditures or if foraging occurs en route);
3. *heightened and persistent defensive capability* of nesting colonies and escape behavior of queens following nest disturbance;
4. *facultative swarm amalgamation,* including megaswarm formation, permitting queenless and exceedingly small colonies to survive or reproduce;
5. *superior competitive ability* allowed by foraging range, orientation, and recruitment abilities far exceeding those of native bees; and

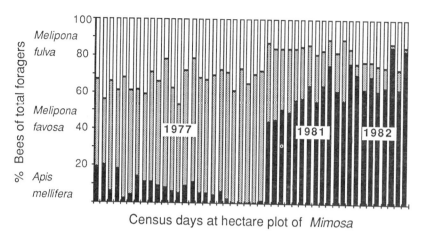

Figure 4.12. Changing relative bee abundance following invasion by Africanized honeybees in lowland forest of French Guiana. The two native bees visiting flowers of *Mimosa* are *Melipona* (Apidae) censused in a hectare plot (after Roubik 1987a).

6. *swarm production* during at least the major wet- and dry-season flowering periods and production of up to four swarms during each cycle (perhaps more as a consequence than a cause of the above).

Long-term studies of the immigrant Africanized honeybees show that their populations may wane some years after colonization, which would be in accord with the performance of most introduced species that establish feral elements (Elton 1958). Mechanisms to explain the phenomenon have not been documented, although Kerr (pers. commun.) believes that local predators learn to locate the colonies of honeybees and eventually drive down their numbers. Quantitative data on Africanized honeybee swarm and nest abundance across a broad transect come from the vicinity of the Panama Canal (Fig. 4.7). The feral colonies were intensively monitored for 48 consecutive months within an area of 50 km^2 where colonies comprising 1,175 swarms and nests were examined and identified (Boreham and Roubik 1987). Although more than a metric ton of the bees had been removed from the area, colony number increased markedly over two years, but then it gradually declined. Comparable qualitative data have been taken in a part of the Amazon forest. The keen attention of the Kayapó tribal bee experts for nearly two decades suggests that Africanized honeybees no longer diminish honey production of their coveted stingless bee colonies (Posey and Camargo 1985; see also Section 4.3.4), which I interpret to mean that a decline in population has taken place. A third study in French Guiana showed that Africanized honeybee representation on flowers slowly increased over five years (Fig. 4.12), which would mean a gradual increase in this forested lowland area over the first eight years of honeybee occupancy. Before 1974 there were almost no *Apis mellifera* in this part of the world (Taylor 1977).

The gradual increase in Africanized honeybee visitation to the single-hectare flower patch of *Mimosa pudica* shown in Figure 4.12, apparently at the expense of native meliponine visitors in a forest-savanna area of French Guiana, portends general and widespread replacement of tropical pollinators (Roubik 1978; Roubik et al. 1986). The figure shows that 1 in 14 pollinators was an Africanized bee in 1977, compared with 2 of 3 in 1981 and 3 of 4 in 1982. Studies at the same patch of flowers in 1977 (Roubik 1978) provided three indices of foraging profitability and population size among these bees. First, when the hives of honeybees were removed, the time spent on individual flowers increased significantly for each *Melipona,* but their relative numbers per flower did not change. Second, when more honeybee hives were added, the relative numbers of honeybees and *Melipona* in flower patches was not altered. And third, none of the bees interacted aggressively at resources or reacted to my surveillance, a behavior corroborated by many other observations at other types of flowers (Section 2.3). Therefore, the change recorded in the five-year period probably reflects the changing numbers (not necessarily colonies) of the bees in the immediate vicinity, and it implies a reciprocal trend. At the level of the flower patch, it implies that when bee numbers attain an equilibrium, there can also be additional adjustment due to flexible foraging behavior, such as altered foraging tempo at flowers (see Section 2.3.7). These conclusions are permitted because surrounding habitat had not been grazed, cut, or otherwise altered and the relative abundances of bees on flowers scarcely changed over successive days.

Field studies using Africanized honeybee colonies have indicated that competition for flowers is common in neotropical forests, but a reshuffling of foraging sites caused by competition seems slow to affect the reproductive fitness of local species. A series of experiments has shown that the Africanized honeybees unaggressively displace native bees from food sources and continue to forage despite harassment attacks by stingless bees that usually drive other foragers away from flowers (Roubik 1978, 1980a, 1981b, 1982c). Another experiment in a forest habitat having few or no local honeybee colonies tested whether such competition was sufficient to reduce colony food storage and brood production by two species of *Melipona* (Roubik 1983b). The introduction and removal of 10–15 honeybee colonies affected neither variable for Africanized or native bee colonies. A final experiment showed that although almost exactly the same pollen and nectar resources were used heavily by stingless bees and honeybees, and although a balance between colony foraging and floral resources existed, most stingless bee colonies escaped competition by the spatial partitioning of resources (Roubik et al.

1986). This study indicated that under conditions of Africanized honeybee colony density of one per square kilometer, the 25% reduction in colony resource harvest observed by stingless bees in competition with the honeybees would continue for 10 years before bringing about a change in the numbers of local stingless bee colonies.

Dynamics of changing bee populations can be modeled, allowing tests on the influence of perturbations (Roubik 1983b). Baseline data were taken for a number of years on approximately 100 species of native bees in Panama before the arrival of Africanized honeybees, and their continued monitoring should allow testing the long-term impact of the Africanized honeybee on native bees (Wolda and Roubik 1986; Roubik and Ackerman 1987). Fortunately, the year-to-year stability of these populations has been very high (Section 4.1). This means that changes due to the arrival of the immigrant bees might be detected if they occur, rather than lost among inherently chaotic population trends. One preliminary conclusion from these studies is that the tropical bee–plant assemblages in question do not exist in a delicate balance that is easily upset by introduction of bee colonies, despite the fact that such colonies may consume 40 kg of pollen and a few hundred kilograms of nectar on a yearly basis! The large foraging ranges of bees and their extreme flexibility in locating and harvesting resources may allow the system to be pushed very far from the norm before clear signs of the competitive process become evident at population or community levels.

Nest site competition between Africanized honeybees and native stingless bees seems minimal, because the honeybees accept cavities with large openings (Fig. 3.18). Although these are unsuitable for meliponines, they are used by nesting polybiine wasps, such as *Stelopolybia*, as well as by a number of larger animals (Roubik 1979a, in press a). The extent to which the Africanized honeybee deprives these native animals – among them opossums, birds, and bats – of nesting sites or refuges is unknown. Both nesting birds and *Nasutitermes* may help the honeybee to establish its populations. Parakeets excavate nesting cavities in the nests of these arboreal termites, which are used by the birds only during breeding season. At other times the cavities are occupied by small swarms of Africanized honeybees, and their combs are later expanded beyond the confines of the cavity, after which the outside of the termites' nest becomes a nesting substrate. Terrestrial termite nests are also used by the bees in forest environments (Kerr 1984b). However, nests built in the open, either under large arboreal termite nests or under tree branches, seem to permit longer colony residence than do nests built in cavities (Roubik in press a). This is perhaps due to high humidity and moisture in such closed spaces. It may be significant that over 75% of all nests found in the Panama Canal area were in buildings or other manmade structures (Boreham and Roubik 1987), where moisture may not pose similar problems. Furthermore, colony movement was greatly stimulated by the onset of rains on the Atlantic side of Panama, which potentially results from the flooding of dry-season nest sites (Boreham and Roubik 1987).

The impact of the introduced honeybee on plants is likely to be extensive, because *Apis mellifera* uses 20–25% of the local floral species that it encounters (Villanueva 1984; Roubik 1988, in press a). Such a range of interrelationships is difficult to document and to characterize. The reproductive output of plants and their success as seed and pollen parents can vary greatly from year to year (e.g., Foster 1982; Stephenson, Winsor, and Davis 1986; see also Chapter 2). One research tactic might be to identify the kinds of plants that compose a large portion of the honeybee's diet and then set up experiments to determine whether the honeybee improves, damages, or does not change the fitness of these species. The honeybee's predilection for wind-pollinated and dioecious species possibly hampers reproduction of the former and enhances that of the latter if certain conditions are met. For example, pollen dispersed diurnally by wind would be removed by the honeybee before it arrived at conspecific stigmas, thus reducing potential plant fitness. Pollen transfer between dioecious plants might be enhanced by the honeybee's large flight range, intrabee pollen transfer in the nest (Section 2.4.2), or the larger nectar load capacity (and hence number of flowers visited per foraging trip) of honeybees compared with neotropical bees visiting the same flower species (Roubik and Buchmann 1984).

A uniformly negative impact might be expected when honeybees dominate the crowns of flowering trees that previously supported diverse pollinators. Some of the native pollinators may have served for short-range pollen dispersal, and geitonogamy, whereas others might have transferred pollen over longer distances and increased outcrossing. Eliminating the services of certain types of pollinators would, over time, alter the genetic structure of the plant populations, with possibly drastic consequences.

The colonizing honeybees have the potential to upset resource utilization patterns and pollinator populations; they may therefore influence the reproductive success of plants with which they never have direct contact. One example is the orchids. Euglossine bees that pollinate orchids also visit some of the nectar and pollen sources used by honeybees. If the resources are substantially reduced for the euglossines, then orchids will receive less pollinating visits. However, the euglossine orchid-bees utilize many nectar and pollen sorces that ordinarily are not used by honeybees; these are primarily the flowers presenting nectar in long, tubular corollas and the buzz-pollinated flowers (Section 4.1.4). Another outcome is possible for such poricidally dehiscent plants. The *Melipona* visit many buzz-pollinated species as well as many nectar and pollen sources of the Africanized honeybee (Absy and Kerr 1977; Roubik 1979a; Roubik et al. 1986). The honeybee does not pollinate the flowers of poricidally dehiscent taxa (Chapter 2). Foraging generalists like *Melipona* preclude the possibility that Africanized honeybees can provide pollination services identical to that of competitors they may locally outcompete and displace. All such speculations on the indirect influence of honeybees would presumably come to light only if competition pressures were persistent and intense. Complex food webs and variety in mutualistic associations are a prominent characteristic of tropical ecosystems. Patterns in their organization will be exposed to closer analysis by the spread of the Africanized honeybee.

Africanized honeybees seem preadapted to invade and persist in mosaic tropical habitats, which gives them advantages over the native bees. Recently cleared agricultural plots allow establishment of Africanized honeybee nests in fallen, hollow trees. Within the plots there are few competing bees, because meliponine colonies live on, in, or near trees and consequently disappear as soon as the forest is cut. In contrast, a few Africanized honeybee nests in several hectares are common in such clearings in Panama. Particularly at forest edges, the honeybees take advantage of the newly created nest sites, forage both in the forest and on weedy vegetation that springs up in cleared habitats, and later abandon the nest sites when disturbed by predators, failing resources, or climatic conditions. Native bees do not show this combination of traits; they differ most by lacking the generalized nesting habits of the Africanized honeybee and the ability to abscond as colonies.

Total competition for nest sites and food should become more intense between honeybees than between the honeybees and meliponine bees, even though the nesting behavior of the honeybee allows it to increase the degree of food competition among all native species by depleting food in varied habitats. The eventual fitness reduction due to intraspecific competition among Africanized honeybees may curb their rate of population growth and the production of swarms (Fig. 4.7). Whether predators, parasites, or other factors reduce their numbers to a degree such that breeding success is diminished will be uncertain until their population dynamics and the effects of natural enemies become better known.

4.3.4   Tropical beekeeping

Beekeeping is a time-honored profession that not long ago provided almost the only source of concentrated sugar, although not the only sweetener (Caufeld 1985), for indigenous tropical people. Colonization of the tropics by Europeans led to establishment of European honeybees in every continent but Africa, where the native tropical honeybees predominated (Fletcher 1978). There are approximately 800 kinds of stingless bees (including the many geographic races), at least 7 species (at least 34 races) of *Apis* (Ruttner 1988; Tingek et al. 1988), and some 30 forms of *Bombus* that produce honey in the tropical latitudes. On a regional basis, the neotropics contains 70% of the varieties of stingless bees, many of which are widespread. Africa has some 15%, and most of the remaining 15% are found in Southeast Asia and Malesia. Three of the *Apis* extend from eastern Africa or the Middle East to China and the Philippines (Ruttner 1988). In addition, eight or nine distinctive races of *Apis mellifera* inhabit tropical Africa, and the *Apis cerana* group might contain additional species distinct from *A. cerana* and *A. vechti* (Maa 1953; Dutton et al. 1981; Ruttner, Pourasghar, and Kauhausen 1985; Williams 1985; Wu and Kuang 1986; Ruttner 1988; see also Section 1.2).

Artificial propagation of these honey-making bees would mean a boon for inhabitants of diverse regions. Unfortunately, colonies of less than a few dozen of the highly eusocial bees appear to be successfully produced by beekeepers (Nogueira-Neto 1970b; Nogueira-Neto et al. 1986). Thus no more than 5% of the honey-making bees are cultivated. Most of these are species that can persist in manmade habitats, which are a minority among local highly eusocial bee fauna (Michener 1946; Roubik 1983a). All such bees are to some extent utilized as nonrenewable resources. Their honey, cerumen, resin stores, wax, broods, and pollen are consumed directly or find other uses throughout the tropics (Bingham 1897; Sakagami and Yoshiakawa 1961; Crane 1975; Batra 1977; González 1983; Nightingale 1983; Posey and Camargo 1985). A good example is the little honeybee, *Apis florea*. Its entire wax nests, containing the brood, honey, and pollen, are harvested to be sold in the marketplace (Fig. 4.13).

The two little species of honeybees with only mildly effective stings and the stingless bees displaying little or no defensive behavior are in danger of extinction. In Zambia, for example, the cultural demand for honey has resulted in the virtual elimination of stingless bee colonies (Pierre pers. commun.). The conservation of woodland to provide bee forage and honey has gained local support, but the only highly eusocial bee conserved with the melliferous flora is *Apis mellifera*.

Colonies that readily sting or bite when molested are often exploited at night and killed or forced to abscond before honey is harvested. Among the *Apis,* if the queen is not killed, the colony emigrates following extreme disturbance, yet it may nest elsewhere and survive. For stingless bees, the egg-laying queen cannot fly; and unless a colony rapidly repairs its nest, the colony is killed by predators or opportunistic flies almost immediately (Chapter 3). Aggressive tropical species are subdued by using smoke from dry leaves or cattle dung. Even entire trees are burned by tribes that make a traditional living only by harvesting honey from African honeybees in Kenya (Morris pers. commun.). Such antagonistic interaction between humans and bees conceivably contributed to the evolution of honeybee defense (Seeley 1985), and it certainly has influenced human attitudes. In the Tsavo region of Kenya, the indigenous people fear certain snakes and bees above all other animals (Morris pers. commun.). Other African races of honeybees, in-

Figure 4.13. Nests of *Apis florea* sold in the central market of Bangkok, Thailand (Sakagami and Yoshikawa 1973). Original photograph provided by S. F. Sakagami.

cluding those of the western equatorial lowlands, do not have this reputation among local peoples.

Attempts to thwart the defenses of bees as their nests are pillaged have led to discovery of tranquilizing and repellent plants. Posey and Camargo (1985) report one such species from Mato Grosso, used originally for aggressive stingless bees and now also for Africanized honeybees. The Kayapó deposit wood shavings from the toxic vine *Tanaecium nocturnum* (Euphorbiaceae) at the bee nest entrance; defending bees are quickly stunned by its volatiles. Attacks of *Apis dorsata* are averted by repellent leaves of *Orophaea katschalica* (Annonaceae) and *Amonum aculeatum* (Zingiberaceae) in Malesia (Dutta, Ahmed, and Abbas 1983). Honey hunters scaling a large tree or rocky cliff to bring down nests of giant honeybees use smoke and protection, but they take tremendous risks (Fig. 4.14). Morse and Laigo (1969) state that inhabitants of the Philippines are primarily concerned with bee eradication and that honey harvest is incidental. However, giant honeybees of the Philippines often nest relatively close to the ground and thus are potentially more dangerous to humans (Starr et al. 1987). Bees of the *Apis dorsata* group are most effectively repelled at night by smoke and fire (Morse and Laigo 1969).

A consideration of increasing importance is whether colonies of *Apis mellifera* should be maintained in forest reserves or other protected natural habitat (Pyke and Blazer 1982; Kulinčević 1986; Rinderer 1986; Ruttner 1988). The last three authors signal the importance of such considerations in maintaining the richness of restricted geographic races of *Apis mellifera,* some of which are, for example, more resistant to disease or otherwise highly adapted to survive in certain settings. Regarding beekeeping with introduced honeybees, there are two general ways of viewing the question:

1. Is beekeeping damaging to native wildlife?
2. Is it likely to be profitable?

By answering the second question, the former is often resolved. For the humid or wet tropics, particularly in the lowlands but also including the cloud forests, European honeybees usually fail to produce surplus honey. Colonies are debilitated by fungal and microbial diseases, robbing and attack by ants, bees, and wasps, or nest destruction by vertebrate colony predators (Kerr 1978; Delgado and del Amo 1984; Villa 1987; see also Chapter 3). At another extreme, the dry tropics having annual rainfall below 20 cm and temperatures frequently above 45° C exceed the environmental tolerance of *A. mellifera,* although not of *A. florea* (Singh 1962; Ruttner, Pourasghar, and Kauhausen 1985; Ruttner 1988). Such habitats may be intact or disturbed, but they are far from ideal for apiculture. In contrast, more equable tropical climates have been generally transformed by human activity. Their bee fauna often is not depauperate, but native highly eusocial bees are less diverse or numerous than when the regions were forested. Floral resources are perhaps often unused due to absence of native tree-nesting colonial bees, or they are taken by bees adapted to disturbed habitats. As apparent with the descendants of African *Apis mellifera* invading the neotropics (Section 4.3 and

Figure 4.14. Nocturnal honey harvest from a hanging nest of *Apis dorsata* in southeastern Asia. Original drawing by Sally Bensusen.

Figure 4.15. A natural nest of Africanized honeybees in the mangrove forest of French Guiana. Photo by author.

Fig. 4.15), exploitation of this resource base is often profitable for beekeepers and not in conflict with conservation practices. In addition, reforestation with carefully selected leguminous tree species can satisfy demands for soil conservation, livestock forage, and fuel wood, while at the same time providing food for honeybees brought to areas where native bees have become scarce. Introduction of *Leucaena, Prosopis,* and *Calliandra* from Central America to various tropical localities typify this type of synergism (Anonymous 1983a,b, 1984; Clemson 1985). The caution applying to plant introduction should perhaps be the same for bees: "Such potentially invasive plants should be introduced only with great care and with serious consideration for the threat posed by their weediness" (Anonymous 1983b).

Genetic considerations often preclude the prolonged use of imported varieties of bees on a small scale, even in a manmade habitat that promotes their survival. Regardless of the type of bee, if total colony numbers are small and feral populations are not initiated, genetic inbreeding eventually leads to high brood mortality and subvital workers; small population size is responsible for poorly inseminated queens (summary edited by Rinderer 1986; Kerr 1987). Furthermore, exotic honey-producing bees, unless competitive with native bees, will fail to survive unless beekeepers intervene. Supplemental feeding and replacement of queens with new genetic stock are, however, infrequent in most tropical beekeeping. Notwithstanding queen longevity that may exceed two years, such bees do not permit a sustained-yield system. Instead, they are used for what could be called intensive cropping and short-term gain. Without continued attention and steady input of material and capital, they soon become unprofitable. Migratory beekeeping, whereby hives of *Apis mellifera* are moved into productive areas only for brief intervals, allows an escape from the problems of a particular habitat. Although some depredations of natural enemies and resource dearth may be avoided, genetic inbreeding is still a problem, and larger management and transportation costs are inevitable.

Seasonal beekeeping is attractive as a small-scale enterprise as practiced in Africa (Smith 1966; Kigatiira 1984) and more recently in much of the neotropics by trapping honeybee swarms during the dry season and then harvesting a single crop of honey and wax from the colonies. The year-round presence of swarms where none before existed has provided a source of revenue for many in the American tropics. Harvested honey is shared among members of the community, often as an appropriate compensation for complicity in allowing bee colonies to be maintained near residential areas. The Africanized honeybee, particularly in disturbed habitat, has developed into this type of perpetual resource (Michener 1975; Kerr 1984b; Boreham and Roubik 1987).

Components that distinguish the exploitation of native honey-making tropical bees are extensive practice on a small scale and firm ties to habitat and local culture. Honey hunting is pursued while hunting for other game and engages much of the men's time among equatorial forest peoples observed in the neotropics, Africa, Asia, and Australia (Hockings 1884; Nogueira-Neto 1970b; Crane 1975; González 1983; Nightingale 1983). Traditional beekeeping appears to be integrated into local economics and ecology. Describing the Kayapó Indians of the eastern Amazon, Posey (1983) mentions that smaller forest trees are felled to procure stingless bee nests, and the resulting opening in the forest attracts game and is used for planting vegetables and medicinal plants. Colonies nesting in hardwood trees are kept in a semidomesticated state; honey is harvested through a small hole that is later patched. Colonies that have been spotted in the forest have designated owners, who decide when and how to harvest from them. Nesting sites are prepared to attract some species that nest in the ground. Empty log hives, various wood or fiber containers smeared with honey, beeswax, or the scent of "lemon grass," which resembles the Nasanov pheromone of *Apis mellifera* (Free et al. 1984; Kigatiira et al. 1986; Free 1987), have been used through recorded history to attract reproductive swarms of tropical honeybees and stingless bees.

Honey is the major crop harvested from honey-making bees, but its quality and amount vary greatly, and certain stingless bees are exploited not primarily for honey but for resin and cerumen. Medicinal properties attributed to the honey of native stingless bees (probably associated with resin chemicals leached from the storage pots) make it preferred over the honey of *Apis mellifera*. It is kept for family consumption, whereas that robbed from nests of honeybees is sold (Posey and Camargo 1985). The honey of *Melipona* and some Trigonini is generally sweet, rather clear, and sometimes viscous. The honey of most Trigonini is strongly acidic, more watery, and darker (Nogueira-Neto 1970b; Wille and Michener 1973; Brand in press; Roubik pers. obs.). Trigonines and some Meliponini collect juices of rotten fruit or carrion in addition to mildly toxic nectars, and their honeys reflect this variety (Schwarz 1948; Gonnet et al. 1964; see also Section 2.1.5). Stingless bees nesting normally within the forest rarely store more than a liter of honey or pollen (Roubik 1979a, 1983a). Their ripened honey seldom exceeds 70% sugar and is therefore more watery than the honey of *Apis*. It does not ferment in the nests, presumably due to the resin chemicals and hydrogen peroxide (Section 2.1.2). Large amounts of cerumen are taken from the same stingless bees that store considerable food (Fig. 4.16). In addition, very large resin deposits (nearly a kilogram per nest) are maintained by a few species like *Ptilotrigona lurida occidentalis* of western Colombia (Parra pers. commun.). The Chocó Indians utilize this material to patch and waterproof their boats. Melted cerumen serves a similar purpose but due to its wax content has many more potential uses, which are similar in the New World and Old World tropics (Schwarz 1945).

No other apids store as much honey in single, natural nests as the giant honeybees. Their colonies have about the same number of workers as those of *Apis mellifera,* the only other honeybee frequently maintaining populations of over 20,000; individual bee size is considerably larger (Morse and Laigo 1969; Seeley 1985). Colonies of the *A. dorsata* group may store several dozen kilograms of honey on their single comb (summary by Morse and Laigo 1969). If India, possessing four native *Apis* (*cerana, dorsata, florea,* and *laboriosa*), is representative for the Indopacific region, *Apis dorsata* is the most remunerative honeybee. Its wax had at one time accounted for 80% (Phadke 1961) and its honey 75% (Ghatge 1951) of the commercial trade.

When honey-making bees are released from the usual pressures of competitors and predators, then certain species can excel in colony honey storage. Under these conditions, the species of *Apis* that make multiple combs (*mellifera, cerana, vechti*) and a few species of *Melipona* are potentially highly profitable. Their nests can be given more than adequate space for food storage, and honey-storage containers (or empty wax comb) can be provided by the beekeeper. Colony populations are augmented by artificial feeding and other manipulations that produce larger colonies than permitted by forage conditions. It is much to the beekeeper's advantage that this be done immediately preceding a major flowering period. The result is that surplus nectar is harvested and stored. Comparable apiculturists' manipulation of most Trigonini is impossible, because colonies use sticky cerumen to wall off the

Figure 4.16. An opened nest of *Tetragona clavipes* (Apidae) from the lowland forest of French Guiana. The brood comb and large numbers of honey storage pots are shown on half of the tree section containing the nest; also shown is a 1-m scale (Roubik 1979b).

nest area or fill excess nest space with pillars, even though the bees accept supplemental food. Tropical bees like *Melipona compressipes fasciculata, M. beechei, M. seminigra, M. scutellaris,* and *M. quadrifasciata* produce 5 kg or more honey in a year (Schwarz 1948; Bennett 1964; Murillo 1984; Nogueira-Neto et al. 1986; Kerr pers. commun.). *Apis mellifera* can yield more than 200 kg, and *A. cerana* might produce over 30 kg (Crane 1975; Inoue pers. commun.). If empty comb could be given to giant honeybees, their colonies might also display exaggerated hoarding.

Certain temperate versus tropical differences in foraging suggest that heavy honey storage is maladaptive for some kinds of tropical bees. If the bees are manipulated to store more honey, as often are the Africanized honeybees in parts of South America (Michener 1975; DeJong 1984), then colony reproduction by fission is encouraged, after which colonies dwindle. Artificial breeding and selection techniques would likely fail to uncouple these two traits, particularly when there is no colony advantage to be gained from delaying reproduction. In all studies to date, brood production by tropical bee colonies is closely tied to colony food harvest (Section 3.3.4), and colonies should be unlikely to maintain and feed unnecessarily large populations of workers. In contrast, honeybee colonies in cold temperate areas initiate brood rearing in midwinter, anticipating the arrival of a favorable but short foraging season in which food must again be intensively gathered in order to preserve the colony through winter (Seeley and Visscher 1985; Ruttner 1988). The means by which food hoarding is controlled involve multiple chemical stimuli, both of the wax comb and the bees (summary by Free 1987). However, the environmental conditions favoring surplus food storage more than brood production are generally lacking in the tropics. There is no extended dearth season comparable to a temperate-zone winter of several months. When resource conditions are poor or colonies perceive climatic or floral stimuli that signal better flowering conditions, many tropical *Apis* usually abandon their nests (Ruttner 1988). Furthermore, all *Apis* rob honey from the nests of other *Apis* species (Dyer and Seeley 1987). Such behavior is evidence of flexible means of coping with changing food distribution and abundance. The storage of food, rather than utilization of comb space for brood, can be a wasteful liability for tropical honeybees in their natural habitat.

Foraging and brood-rearing patterns observed in *Apis mellifera* suggest a relaxed threat of food scarcity in the tropics. In the temperate latitudes, colonies of this honeybee metabolize a minimum of 20 kg of honey during the winter (Seeley 1985; Seeley and Visscher 1985). If they do not have this assurance of stable nest temperature, chances are very high that they will perish. Contrasting behavior between tropical and temperate varieties of the same bee can be explained by extending this theme to include the means by which a colony assesses "profitability." Its currency (Cheverton et al. 1985; see also Chapter 2) appears to be quite different for bees in some tropical and temperate environments.

In a shared apiary site in agricultural, lowland Venezuela, European and Africanized honeybees of similar colony size were compared in foraging activity and rate of hoarding (Rinderer, Collins, and Tucker 1985). European honeybees responded more markedly to a period of high resource abundance, and they produced more honey. On the other hand, ordinary floral abundance seemed to create conditions in which the Africanized honeybees stored more honey. The profitability of foraging (determined by rate of increasing hive weight) seldom exceeds moderate levels in southern Africa, the original habitat of the honeybees introduced in Brazil (Silberrad 1976; Anderson, Buys, and Johannsmeier 1983, p. 149; Nightingale 1983; Ruttner 1988; see also Section 4.3). This at least suggests that the Africanized honeybee now of the neotropics had during its previous history usually exploited resources that were not abundant. Furthermore, it should be clear that resource abundance is not equivalent to the abundance of flowers. Foraging profitability is also determined by abundance of competing bees.

Competition affects colony foraging decisions and can certainly influence the evolution of foraging style and food storage patterns (Chapter 2). Flowering characteristics make another distinctive contribution. The flowering season of the temperate zone is relatively short, and as stressed in preceding sections, natural floral resources there are not so concentrated nor rare in space and time as those of the tropics. There are seldom large trees or inflorescences that offer both pollen and nectar, and there are almost never the number of different flower species blooming at one time (Fig. 4.7). Spatially, the species of temperate flora are widespread and the representation of individual species is often high in the floral community, whereas tropical communities present more scattered flowers of a species, or extensive blooming by single species only for short times (Section 2.3.7). Foraging and food storage by bees should be molded by these contrasts, but how far can comparisons be drawn? Fortunately, the natural distribution of *Apis mellifera* allows an intraspecies comparison of variants that have long histories either in tropical or temperate environments.

It has be suggested that European honeybee colonies are naturally selected to maximize cost efficiency [(gain − cost)/cost], which maximizes the energy harvest efficiency of colonies and also prolongs worker longevity (Seeley 1986). This tendency conserves nest space and the forager population, but both are incorporated fully when nectar harvest conditions are highly favorable. In this view the rate of harvest is not the primary consideration of the colony. Rather, it is the final accumulation of resources and the status of the colony and nest at the time resources cease to be available with the onset of winter. An implicit assumption, and one that a correct evolutionary interpretation of this behavior depends upon, is that the "best" resources are normally found and that it is better for the colony to postpone intensive scouting and foraging until they appear. Colonies are in effect selected to respond to a restricted range of resource quality and quantity. If the response is inherently different among races, then the local resource conditions can become a primary evolutionary factor. A great deal of supporting data concerning variation in honeybee food storage and brood production have been given by Ruttner (1988), but there are currently no ecological studies testing their relative merits. It is thus a seeming contradiction that the honeybees from areas where the flowering season is short and where winter conditions are decisive to colony survival might behave in the tropics as though operating on an unlimited time-and-energy budget. This would, however, be the predicted behavior if the stimulus mobilizing a temperate colony's foragers was lacking in the tropical setting.

In contrast to the temperate honeybee, certain lowland tropical *A. mellifera* have less seasonal time restrictions but a more complex foraging environment and possibly more risk of predation. Their ability to avoid competition from the myriad types of native bee colonies, and thereby to occasionally locate an exceptional resource that can be monopolized, may select for increased individual scouting. It is the scouting activity that determines rate of resource patch discovery (Section 2.3.5). Maximizing the rate of patch discovery is one means available to *Apis mellifera* to avoid competition. In addition, as resource conditions become less predictable the colony should rely more upon scouting and less upon efficient resource harvest. Even more basic, the tendency of honeybee colonies to forage primarily by recruitment (Seeley 1985) rather than by individual scouting is probably maladaptive in the tropics. An explanation is that a recruit must try to find a resource several times before it is successful, and this trail-and-error recruitment should be more prolonged, or become completely ineffective, as the physical complexity of the habitat increases. Massive and directed colony foraging, as opposed to individual or multigroup scouting and foraging, should often be less profitable in the mature lowland forest of the tropics.

As already discussed in Section 2.2.7, the most efficient nectar harvest behavior for an individual honeybee is imbibement of highly concentrated nectar, but the colony very rarely shows this behavior because its rate of energy intake is higher when poorer resources are utilized. Evolutionary adjustments to prevailing resource conditions could establish the thresholds that determine when a honeybee colony responds by massive recruitment to its major food sources. The comparative studies of Rinderer and co-workers and those of Seeley provide preliminary support for these hypotheses. Continued research on the role of the vibration dance (Schneider et al. 1986) and its potential for regulating colony response to local foraging conditions is one of several areas that seem to promise more clues to the ways in which honeybee species and races have adjusted to their habitats. Such work can provide glimpses of the evolution of foraging behavior (Gould and Towne 1987). These may eventually allow explanations of different foraging responses among the dozens of temperate and tropical honeybees and their geographic types, in the varied environments and habitats occupied by each.

Foraging behavior of lowland Africanized honeybees indicates that temperate honeybees have reversed their mode of colony assessment of foraging profitability, probably as they colonized the temperate zone. Tropical bee colonies may have less time to find resources before they are located by competitors. One known correlate is that colony recruitment to an artificial feeder requires less time for Africanized than for European honeybees (Nuñez 1979). It has also been suggested that tropical bees have less time to reproduce before the attack of a significant natural enemy (Winston et al. 1983; Seeley 1985). Because the Africanized honeybee stores honey under less favorable conditions than the European honeybee (Rinderer et al. 1985), it is correct to state that its mode of foraging behavior is better adapted to consistently low resource availability, whether due to competitors or the flora. Furthermore, the rate at which decisive selective agents act upon colony survival should be greater in the tropics, given the aseasonal nature of colony activity. These comparisons lead to a concept that modifies the fundamental currency of patch quality that was proposed for temperate *A. mellifera* (Seeley 1986). An overriding determinant of colony foraging behavior by lowland tropical *A. mellifera* might not be to achieve cost efficiency but instead to achieve a maximized net rate of energy intake [(gain – cost)/ time]. Reduction of resource discovery time by the colony (as opposed to time devoted to foraging by workers) can best be accomplished by increased colony allocation to scouting. The success of a temperate bee colony that minimizes colony foraging cost is produced by specialization upon superior resources. Success of a tropical bee colony that minimizes absolute time in relation to patch discovery and harvest may be achieved by continuous harvest of more varied resource types, punctuated by brief, intense harvest within relatively small patches. To reinforce the dichotomy, when superior resources are difficult to locate, scouting time necessarily increases, and as foraging conditions become more homogeneous, less scouting is necessary to survey the habitat. As discussed in Chapter 2, the complexity of resource distribution in the tropics is unlikely to resemble that of temperate areas. A few species flowering widely may be the norm for temperate habitats, and more patchy distribution of rewarding flowers might typify tropical areas. When plantations or agriculture has created the resource landscape such distinctions may disappear.

Honeybee management strategies are not easily predicted from an adaptationist approach to bee behavior, and yet they may come across clearly when the beekeeper considers how different types of colonies perform. The harvest of honey, and most important the proportion harvested by a beekeeper, will strongly determine the long-term performance of colonies. For example, from the above discussion some differences between temperate and lowland tropical honeybees can be paraphrased as follows: When an exceptionally rewarding resource is encountered, some tropical bees hedge their bets and continue to scout for new patches; but temperate bee colonies abandon most other options and concentrate on getting the most from a few flower types. If bees are rather specific in the type of forage they will harvest heavily, then the removal of their entire honey stores jeopardizes the colony. However, over the active period of the bee colony, averaging the rate of resource harvest and storage (or conversion to broods) may reduce temporary differences observed between different honeybee races. One four-month study utilized pollen-collecting traps to compare net colony pollen harvest by 26 Africanized and European honeybee colonies in lowland Panama (Roubik et al. 1984). The average daily trapped pollen per colony was approximately 55 g dry weight for each type of honeybee. The uncertainties in making direct comparisons using pollen trap data stem from the fact that these two honeybee ecotypes are of different size [and thus identical pollen traps may have slightly different trap efficiency (Anderson et al. 1983)] and the rate of pollen trapping reflects the colony's response to pollen shortage as well as to its normal foraging activity (Waller, Caron, and Loper 1981). A similar study that considered the rates at which pollen or other foragers returned to colonies of the two ecotypes (Danka et al. 1987; see also Section 2.3.9) showed that the Africanized honeybees tended to forage more intensively for pollen, at least during a fairly early portion of the day. The clear preferences that this bee has for nocturnally dehiscent flowers makes such comparisons even more intriguing (Roubik et al. 1986; Roubik 1988). A conclusion from this type of work is that, for a beekeeper, either type of honeybee seems productive. Nonetheless, the processes that determine productivity among different ecotypes can be quite different. Productivity will decline if the bee variety is not suited to the management technique, because at a particular level, say pollen harvest or honey harvest, productivity has differing costs and benefits among honeybee ecotypes. Differences in colony behavior most likely rest upon the thresholds for profitable foraging and food storage in the habitats where races of honeybees evolved (Ruttner 1988). In the case of pollen harvested from colonies, the Africanized and European honeybees might have yielded similar amounts of pollen due to the greater trapping efficiency from the latter, and a greater rate of pollen harvest by the former.

Successful bee management requires another major consideration: Which type of bee can be least productive? All highly eusocial bees apparently cannibalize their young broods when food is particularly scarce (Winston et al. 1979; Roubik 1982c). Colony relocation and relatively rapid swarm production are characteristic of tropical honeybees, but direct comparisons with temperate honeybees are often inconclusive for lack of ecological background data (Winston 1981). Furthermore, Africanized *Apis* cannot be induced to stay in hives even by intensive feeding if no source of natural floral pollen is available (Silberrad 1976; Roubik pers. obs.). The need to examine relative costs and benefits of acquiring, exploiting, and maintaining bee colonies promises continued research and development in tropical apiculture. Eventually, honeybee or stingless bee colonies might be matched to their most suited environment and habitat. Experimental work with isolated populations underlines the difficulty of maintaining sufficient genetic diversity; when it is lacking, brood populations are half diploid drone bees (Kerr 1987; summary by Rinderer 1986). The substantial growth in tropical apiculture during recent decades using *Apis mellifera* has been indicated by Crane (1980), and management of stingless bees is reviewed by Nogueira-Neto et al. (1986). Contributions to developing productive and ecologically sound tropical beekeeping practices will require learning the lessons offered by these bee species and experimentation with many more.

4.3.5 *Why are so many highly eusocial bees tropical?*

It is not accidental that only one highly eusocial bee lives entirely outside the tropics, a small minority exist in the subtropics, and only two span both tropical and temperate areas (Sections 1.2.2 and 3.1.2). These two bees, *Apis cerana* and *Apis mellifera,* are the only social bees, besides the insular Saban honeybee *A. vechti,* that make several vertical nest combs of wax (Section 3.1.4). Wax combs give profoundly effective insulation to layers of honeybees sandwiched between them (Southwick 1985a,b, 1987). Among the remaining types of highly eusocial bees, lacking such extraordinary architectural adaptations, none can survive several months of winter conditions. We can reason that the costs of maintaining permanent colonies are too high outside of tropical environments, because there are no geographic barriers to prevent colonization of the temperate zone. Further, many conditions favoring evolution of permanently social bee colonies are likely to be uniquely tropical. Perennial bee colonies do respond in varied ways to the thermal environment, but I believe that their evolution and biogeography is chiefly determined by a broad range of ecological interactions.

To become highly eusocial, a bee lineage must give up the reproductive options for some female bees (Section 1.2.3). Rather than temporarily postpone reproductive activity or occasionally serve as a helper to increase inclusive fitness, a highly eusocial female is programmed for a social role within a colony. Unlike other types of social bees, she cannot live solitarily, and her colony can be initiated only in conjunction with a swarm of workers. Colony foundation by swarming is a trait uniquely established among highly eusocial bees. Evolution of permanent sociality must depend on tropical habitat characteristics that relate to swarming. Additional selective pressures that helped to establish obligate sterility for workers and the inability of queens to nest and forage include correlates of colony defense, longevity, and the resources needed for reproduction and survival.

Permanent eusociality has been attained by many tropical bees. The thesis that I will pursue here is that such a radical change is adaptive only because the tropical environment permits it. It is inconceivable that permanently eusocial colonies evolved from any but the temporarily or primitively eusocial colonies (Section 1.2.3). Selection pressures that promoted evolution of permanent eusociality can thus be viewed as extensions of those favoring eusociality. Each of the three proposed routes to eusocial colonies (Fig. 1.4), the subsocial, semisocial, and quasisocial, will be examined. Highly eusocial colonies are also perennial, meaning they may continue indefinitely (Michener 1974a). The transition to permanence and rigid social roles therefore must be understood as tropical phenomena. Formation of a highly eusocial bee colony could favor (or be favored by) the following:

1. augmented resource gathering and hoarding capacity,
2. improved buffering from environmental shocks, and
3. increased protection from natural enemies.

A tendency toward delayed reproduction and advanced competitive ability might attend such developments in some social insects (Jeanne and Davidson 1984), but this may occur only after obligate colonies have evolved. Some observations concerning such traits will be tautological. The outcome of natural selection seems inextricably woven to its causes, although comparative analyses have developed satisfactorily in spite of this handicap (Michener 1985; Eickwort 1986). The primary question to be addressed is whether any of these proposed benefits seems more likely to lead to fixation of highly eusocial traits in the tropics.

One of the potential limitations to permanent colonies is cold weather. Because stingless bees do not regulate their nest temperatures as closely as *Apis* (Chapter 3) and nearly one-third of the 34 or so varieties of *Apis* build exposed nests, a mere three bee species, including nine geographic races, can maintain permanent colonies in the temperate zone. Two of the cavity-nesting *Apis* (*cerana* and *mellifera*) and the largest and hairiest exposed-nest species (*A. laboriosa*) are found in the temperate zone. However, some stingless bees nest in areas that experience intermittent cold weather, such as in the Colombian Andes at 3000-m elevation, or in Borneo, New Guinea, Taiwan, and southern Brazil at 1,000–3,000 m (Schwarz 1939; Nogueira-Neto 1970a; Vergara and Pinto 1981; Sakagami and Yamane 1984; see also Chapter 3). Even tropical honeybees in South America and Africa exist where there is snow on the ground during part of the year (Fletcher 1978; Dietz et al. 1987; Ruttner 1988). Freezing temperatures thus seem not to limit the distribution of tropical bee colonies. Rather, it may be the duration of the cold period and the availability of food, both in the habitat and within the bees' nests, that constrain the existence of apines and meliponines. A number of other correlates – for example, humidity, precipitation, mutualists, natural enemies, and vegetation – might prove to be singularly important, but there have been few analyses of such factors (Roubik 1987a; Villa 1987). Once permanent colonies have evolved, their resistance to prolonged cold weather and the storage of honey to endure it are likely to limit their geographical distribution. Nonetheless, as emphasized in Section 4.3, temperature alone is sometimes a poor indicator of the zones in which colonies of highly eusocial bees can survive. Other ecological limitations are likely to be more important.

Climate and flowering seasons also determine whether more than one generation of bees can be produced during the year. The temperate-zone winter may be too long and the favorable season too short, for the formation of colonies consisting of mother and offspring. Eickwort (1975) made this observation for a gregarious megachilid bee that has but one yearly generation. He extended the generality to halictids and anthophorids having less than two yearly generations and to the longevity of a foundress below one year (Eickwort 1981). However, as is true of many such correlates, the dependent variable is not always apparent. Overlapping generations are certainly favored by equable climate, but the complete overlap of generations might be a consequence rather than a cause of sociality, and eusociality is not necessarily favored by extending suitable foraging and flight conditions. The halictid *Halictus* (*Halictus*) *ligatus* in fact lost a eusocial life style as it colonized the tropics from the temperate zone (Michener and Bennett 1977). Its multivoltinism in subtropical localities such as southern Florida seems to permit cleptoparasitic behavior within the species (Packer 1986), and this may be a reason that colonies are at a selective disadvantage in the tropics. Eickwort (pers. commun.) also observes cleptoparasitic behavior among temperate halictines. In addition, eusocial behavior is often fixed in another *Halictus,* subgenus *Seladonia,* and it is displayed by the mature colony of all 10 species reviewed by Sakagami and Okazawa (1985), and a temperate, subalpine member of this subgenus is solitary (Eickwort pers. commun.). Many of its taxa are temperate and palearctic, but they are also abundant in tropical Africa (Michener pers. commun.). The subgenus probably arrived in tropical America since the Pleistocene (Brooks and Roubik 1983). One eusocial halictid species, *Lasioglossum marginatum,* is known to have developed long-lived colonies in temperate Europe (Michener pers. commun.). The reasons that eusocial colonies of temperate or subtropical areas have generally not become long-lived – for example, by diapause of adults or immatures during part of the year – are not readily explained by temperature regimes. Furthermore, subsocial colonies that persist for more than a year are common in the allodapines in subtropical, some temperate, and tropical habitats, but these also are initiated by single females and have not made the transition to permanent eusociality (Maeta et al. 1985). Only single females initiate the primitively eusocial colonies of *Bombus*, which are found predominantly in temperate climates (Michener 1974a; Williams 1985). One conclusion to be drawn from these examples is that eusocial colonies can persist and apparently evolve in temperate climates, but those that do are initiated by single females rather than by swarms.

Division of labor in bee colonies seems neither more nor less advantageous in tropical areas than in temperate areas. Guarding, foraging, and egg production roles are mutually reinforcing and should evolve jointly to produce perennial colonies. Auxiliary females that guard the nest or forage may be necessary to enable the egg-laying female to live long enough to share the nest with her offspring. Strict guarding and foraging roles would have slight selective advantages when the egg-laying female also occasionally forages or guards (thereby risking predation), because her role could suddenly be open to another female. A biological quandary is evident when guards also are reproductives; bees must cease to defend their brood cells from other bees, such that cooperation in brood rearing can occur (Sakagami and Zucchi 1978; Velthuis 1987). The small multifemale associations of *Ceratina* and other xylocopines tend to consist of two females; the guard is also the principal egg layer (Sakagami and Maeta 1984; Michener 1985). In colonies of more than two females, there is the need for increased nest defense to protect the investment in brood that is made by nonreproductive females. Thus the colony of three bees will usually consist of a small forager, a medium-size guard, and a larger queen, and the guarding behavior of the queen diminishes. Perhaps for this reason, the combination of cooperative brood rearing and full reproductive potential of all females (quasisocial behavior) is very rare, and purely quasisocial bee species may not exist in nature (Michener 1974a). As several of the above authors have made clear, usurpation of resources by nest mates is likely to be more frequent when a female does not closely guard the brood cell she is preparing, and this condition is common during the quasisocial colonial phase (Garófalo 1985).

In the tropics, Eickwort (1981) maintains that the "instant societies" produced by cooperation between females of the same generation would have particular value in protecting the nest from natural enemies and quickly producing offspring. Such semisocial colonies tend to be small and inconspicuous, yet their nests and the guarding behavior of females prevent depredations by small natural enemies. Facultative association of females for the purposes of increased fecundity or survival typifies semisocial colonies, and these seem to have flexibility that constitutes a selective advantage over eusocial colonies. Semisocial colonies have evolved in many species of tropical halictids, but there is only one known nontropical semisocial halictid bee (Sakagami, Hoshikawa, and Fukuda 1984). Because quasisociality promotes frequent reproductive competition and conflict between females and eusociality entails a delay in acquiring the colony labor force, semisocial colonies could have special selective advantages. The obvious inefficiency of frequent conflicts between nest mates is avoided, and flexible social roles can be tailored to current ecological and group characteristics. An adaptive paradigm of semisocial colonies has apparently worked well for some halictids. Colonies can still be founded by individuals, and auxiliary females are not necessary for successful reproduction.

Does subsociality lead more readily to highly eusocial colonies in the tropics, and could it explain how both apines and meliponines evolved there? Although a subsocial route to permanent eusocial colonies was likely that taken by xylocopine–apine ancestors of modern highly eusocial bees (Sakagami and Michener 1987), eusocial life is rare in all social Xylocopinae (Michener 1985). As this author puts it, "evolution of castes in Xylocopinae has occurred in situations in which cooperative nesting must not be a major advantage." However, temporarily eusocial colonies exist within some populations of allodapines, *Xylocopa,* and *Ceratina*. Persistent guarding behavior by an individual in the nest of social xylocopines may be advantageous (Sakagmai and Maeta 1977 and pers. commun.). Even male *Xylocopa* are known to guard the nest from ants (Camargo and Velthuis 1975), which might otherwise eliminate most of the immatures (Gerling et al. 1983).

Conditions favoring subsociality apparently do not enhance the likelihood that the fixation of eusociality will follow. In order to live in subsocial colonies, xylocopine adults already possess many attributes of permanently eusocial bees. Reproductive females are reportedly long-lived, surviving for two to four years, and prereproductive females remain in the nest (Michener 1985). Mutual feeding by adults occurs within the nest. Pollen is stored on the walls of the burrow by a few *Ceratina* and all allodapines, where other females can use it, and nectar is also deposited in small quantities on the walls or on larvae (Daly 1966; Michener 1968b, 1985; Kislow 1976; Maeta et al. 1985). These traits so closely parallel features of contemporary highly eusocial colonies that it appears that the selective advantages giving rise to their permanent eusociality must have been of degree and not of kind. Further, it is likely that the same circumstances allowed colonies to persist for an extended time and to swarm.

From the standpoint of social behavior alone, the preceding outline suggests that permanent associations of female bees have no inherent selective advantage, so that traits uniquely associated with highly eusocial colonies can be examined largely to the exclusion of complex social interactions and the genealogies among bees. The conditions that favored the evolution of sterility in workers of obligately eusocial bees seem to have been largely ecological, and they may not have required novel social interactions or mechanisms. Rather, traits that already existed in the subsocial bees were placed in a different context, and that context also favored the extreme specialization of a queen that led to her inability to nest or forage and to the extreme specialization of workers, preventing them from mating or becoming primarily egg layers.

A novel trait, the evolution of multiple wax glands (Section 3.1.4), may explain the tremendous expansion of one element of presocial (Wilson 1975) colonial life that allowed the formation of highly eusocial bee colonies. Wax allows bees to make containers in which food can be stored for later use. In natural habitats, meliponines store up to a few liters of honey and pollen. Tropical honeybees also hoard both honey and pollen, although not to the degree that temperate *A. mellifera* stores honey in order to survive the cold season (Section 4.5). Duffield et al. (1984) suggested that free-standing brood cells built from wax allowed concentration of brood in a small area and minimum cell-building effort. This is probably an added benefit, although many of the highly eusocial bees build relatively small nests and have colonies maintaining less than a few hundred brood (Chapter 3). However, wax is a costly building material. Its production would not necessarily lead to larger brood populations but certainly would allow them to be grouped together more readily. Apines produce the most complex type of wax, but meliponine, apine, and bombine waxes are broadly similar (Blomquist et al. 1985; Hepburn 1986). An added feature of wax utilization by meliponines is that it is almost never used in pure form but is mixed with a variety of plant resins (Chapter 2). This makes it pliable, so that it can be reworked for some period (a property lacking in most resins and waxes).

Honeybees and stingless bees produce wax when stimulated by increased colony nectar intake (Michener 1974a; Seeley 1985; Hepburn 1986). In honeybees, more comb may be added to store food or increase brood production when food is plentiful. Among stingless bees, new storage pots can be built rapidly by increased wax production, utilization of small wax and resin deposits in the nest, and the reutilization of cerumen. In both groups a large new volume of storage area can be created overnight. In addition, the honeybees may convert comb area occupied by eggs or small larvae to food storage space by consuming some of the brood. No primitively eusocial bees, including bumblebees, can in this manner devote nest space and resources to food storage. For the bumblebees, food storage pots are rarely created de novo and most often require empty cocoons as their bases (Dias 1958; Alford 1975).

The potential to harvest and store a large amount of food in a short time is unique to the honeybees and stingless bees. There is perhaps no other characteristic so tightly linked to the tropical environment and the evolution of bees that can only exist in colonies, although the extensive use of resin was permitted and perhaps required of the permanent colonies that evolved in the humid tropics. Resource hoarding seems highly advantageous in the tropics due to the particular types of floral resources found there. Most extremely rewarding flowers, like the large pollen- and nectar-bearing inflorescences pollinated by vertebrates (bats, birds, marsupials) or by very inefficient pollinators such as beetles and flies, are tropical (Janzen 1975; Gottsberger 1986; Henderson 1986). So are many of the dioecious flower species that are preferred pollen and nectar sources of highly eusocial bees (Chapter 2 and Section 4.2). To a surprising extent, many types of resources permitting large-scale hoarding of pollen and honey, in natural habitat, do not exist in the temperate zone. These resources are characteristically patchy rather than spread over large areas, and this usually pertains *both* to unexploited flowers and plant species dispersion. Bees able to rapidly locate and exploit this type of food source in the tropics and successfully store the food must have had a marked selective advantage over other flower-visiting Hymenoptera.

Rich and widespread resources are of course available in temperate forests during part of the year. One can only speculate that their variety, duration, and nectar reward (as most of the trees are wind-pollinated) were insufficient to support the evolution of the type of patch specialists that eventually became the highly eusocial apids. Notably, extremely large inflorescences visited by bees are lacking in temperate areas, unless they are introduced tropical species such as corn.

Food storage allows colonies to persist in spite of seasonal resource scarcity, and it also is a prerequisite for reproduction by swarming. Other traits probably enhanced by food storage include

1. evolution of highly organized foraging activity, such that queens specialize in producing a foraging force, while other bees forage, receive and store food, and guard the stored food and brood that attract a larger group of predators; and
2. initiation of new nests provisioned through the resources of the mother colony (Section 3.3.4).

If the nesting materials, pollen, and nectar or honey resources were not readily accessible, then the survivability of a group of workers and a queen, rather than an individual female, would be extremely low. A surplus of food in the mother nest and the constant availability of resin and pollen permit the establishment of provisioned meliponine nests prior to swarming. Had there been no method of storage of these resources, then reproduction by swarms could not have evolved. Thus wax, the availability of food in large but sparse parcels, and the presence of resin seem required for the evolution of these social bee colonies.

Workers that accompany the queen in a swarm serve for protection and carrying the resources of the mother colony, but they must also forage immediately upon colony initiation. Nest-founding workers not only replace the nesting abilities that the queen has lost, but may be decoys for flying predators that could easily eliminate an individual queen in flight (Section 3.3.4). In addition, worker honeybees carry wax and honey derived from the mother colonies; the worker stingless bees carry building material from the mother colony to build the new nest as well as transfer honey there. If pollen were not available, neither type of bee could initiate brood provisioning in a new nest, regardless of pollen stores in the mother nest.

Because the general division of labor between foragers, guards, and egg layers seems to provide mutually reinforcing advantages and selective pressures favoring semisocial and eusocial colonies, the storage of food, communication between foragers, and evolution of greater defensive capability seem mutually reinforcing in the evolution of highly eusocial colonies. Division of labor has been augmented by additional refinements, most of which are chemical (Chapters 2 and 3). The barbed sting, potent venom, and alarm pheromones of the honeybee, and the alarm pheromones, sticky resin depositing, and biting behavior of stingless bees, are examples of orchestrated defensive specializations found only among the highly eusocial bees. Recruitment of nest mates to resources is another among their unique traits. In contrast, bumblebees (which store very little food) include some highly defensive species among the tropical bees, but none is known to recruit to resources, and none uses alarm pheromones to promote massive colony defense (Sakagami 1976; Morse 1982; Plowright and Laverty 1984). The defensive behavior of bumblebees is probably a general correlate of nests that are not protected within rock or wood cavities and thus are exposed to predators. Broods of a bumblebee colony, irrespective of honey or pollen stores, are susceptible to attack, and perhaps most of their vulnerability is due to small colony size.

The early ancestors of the Meliponinae and Apinae were probably exposed to relatively unspecialized predators of bee nests, and the rate of attack probably increased over evolutionary time as colonies increased their food storage and brood population. Predation pressure may have served to augment the defensive specializations and reproductive roles within colonies. Colonies of meliponines and apines most exposed to attack are the hypogeous stingless bees and the honeybees nesting on tree branches or rock ledges. They perhaps share this nesting habit with their respective ancestral groups. Ancestral apines were almost certainly like contemporary *A. dorsata* and *A. florea* that build open nests (Michener 1974a; Ruttner 1988). This type of honeybee later evolved into forms building multiple nest combs, which brought about a "quantum jump" in cold tolerance (Section 3.1.2). The existence of varied ground-nesting stingless bees in Africa and the neotropics, and their general absence in Asia except for the highly derived *Tetragonula collina* (Wille and Michener 1973; Sakagami 1978; Wille 1983a; see also Chapters 1 and 3), suggests antiquity of hypogeous species, especially those found only in Africa. Wille (1979b) suggests that the hypogeous *Meliplebeia* and *Axestotrigona* of Africa and neotropical *Nogueirapis* are among the most primitive groups of meliponines.

Honeybees, stingless bees, and bumblebees probably experienced strong selection for rapid fixation of the basic social roles of guarding, foraging, and egg laying, which I have postulated to have evolved toward fixation in concert. These roles are present in facultatively social bees and are apparently established by social interaction, whereas in the permanently social colonies caste determination is largely trophic, and the different social roles are to some extent programmed genetically. Apid defensive behavior or highly protected nesting sites might be taken as a general indication of the strength of the selection process and resulting genetic change, and a parallel situation is found among less-social bee families. Michener (1985) argued that the reason for the lack of fixed morphological differentiation among female twig-nesting *Ceratina* is that parasite and predator pressures are only moderate; the Halictini, in contrast, nest on the ground, where they are more exposed to natural enemies. The importance of guards for these social halictids may have quickly led to the behavioral and morphological divergence expressed in colonial females, whereby some are always guards. For the eusocial apids, massive food storage seems to have been a factor that in particular allowed meliponines and apines to diverge from their ancestors, proceeding to extreme specialization in nesting, defense, and foraging.

Once food storage was possible, selection pressures for protection from microbes were intensified, but the specific traits relating to this problem were probably not novel in the highly eusocial bees. As outlined in Chapter 2, microbes that retard spoilage by other microbes are found in the food provisions of both solitary and eusocial bees. Furthermore, the microbes that frequently coexist in social bee nests may conceivably have a role in detoxification of nectar or pollen, as postulated in general for herbivores by Jones (1985). Glandular secretions of the bees serve to supplement or replace some of these defenses (Hefetz 1987), as do the substances contained in resin manipulated by Euglossini, Apinae, and Meliponinae.

If any of the varied germicides used by bees has particular relevance to social evolution in the tropics, it is probably resin. Resin is not only a germicide but also a defensive tool against invertebrates and, for the euglossines and meliponines, the most important building material. Defense of the nest entrance from ants is presumably due to resin and its chemical constituents, which still lack corresponding bioassays. Many naturalists have noted the avoidance of stingless bee nest entrances by raiding ant columns when a sticky resin barrier is present (Khoo and Yong 1987). In response to harassment by even a few ants, stingless bees remove resin from a deposit inside the nest entrance and quickly build an external protective ring. The little honeybee species also protect their nests with a similar resin barrier (Chapter 3). Thick resin placed on cells could also protect them from predators, parasitoids, or competing (egg-eating) conspecific females, particularly in the parasocial and small eusocial nests of the euglossines (e.g., Garófalo 1985).

Because resin is not used by Bombini but constitutes almost all construction material used for nesting by the sister group, the Euglossini, a basic tropical–temperate schism within this apid subfamily is apparent. Both groups produce wax, but the more extensive development of this trait in Bombini suggests that resin at least retained important values for lowland tropical euglossines. A further hypothesis is that resin utilization was needed for successful food storage by the first tropical, permanently eusocial apids. Later, perennially social bees either used resin facultatively, as modern Apinae appear to do, or they replaced the antimicrobial and general defensive properties of resin with their own behavior, glandular products, and mutualistic microbes or mites.

Appendix A
Extant families, subfamilies, tribes, genera and subgenera of the Apoidea: a partial checklist

Compiled by R. B. Roberts with additional input from H. V. Daly (Ceratinini), T. L. Griswold (Megachilidae), W. E. LaBerge (Eucerini), D. W. Roubik (Apidae), and R. R. Snelling (Hylaeinae, Centridini, Ericrocini). Entries have been updated to include Moure and Hurd (1987), Roberts and Brooks (1987), Sakagami and Michener (1987), Daly (1988), Griswold and Michener (1988), Ruttner (1988), and Sakagami and Khoo (1987).

Note: Taxa marked with an asterisk (*) are pictured in Appendix B.

COLLETIDAE
  Euryglossinae
    Argohesma
      Brachyhesma*
      Brachyhesma
      Henicohesma
      Microhesma
    Dasyhesma
    Euryglossa*
      Callohesma
      Dermatohesma
      Euhesma
      Euryglossa
      Parahesma
      Xenohesma
    Euryglossina
      Euryglossina
      Microdontura
      Turnerella
    Euryglossella
    Euryglossula
    Heterohesma
    Hyphesma
    Melittosmithia
    Pachyprosopis
      Pachyprosopis
      Pachysoprosopina
      Pachyprosopula
    Parapachyprosopis
    Quasihesma
    Sericogaster
    Stenohesma
    Stilpnosoma
    Xanthesma
  Hylaeinae

Amphylaeus*
Agogenohylaeus
Amphylaeus
Analasteroides
Calloprosopis
Eupalaeorhiza
Geophyrohylaeus
Hemirhiza
Heterapoides
Hylaeoides*
Hylaeus*
  Alfkenylaeus
  Cephalylaeus
  Cornylaeus
  Edriohylaeus
  Euprosopellus
  Euprosopis
  Euprosopoides
  Gnathoprosopis
  Gnathoprosopoides
  Hoploprosopis
  Hylaeteron
  Hylaeus
  Macrohylaeus
  Meghylaeus
  Metziella
  Nesoprosopis
  Paraprosopis
  Prosopella
  Prosopis
  Prosopisteron
  Pseudhylaeus
  Rhodohylaeus
  Sphaerhylaeus
  Xenohylaeus

Hylaeorhiza
Meroglossa*
Nothylaeus
Palaeorhiza
  Anchirhiza
  Heterorhiza
  Palaeorhiza
  Xenorhiza
Pharohylaeus
Psilhylaeus
Xeromelissinae
  Chilicola*
    Anoediscelis
    Chilicola
    Chilioediscelis
    Heteroediscelis
    Idioprosopis
    Oediscelis
    Stenoediscelis
  Chilimelissa
  Xenochilicola
  Xeromelissa
Colletinae
  Paracolletini
    Anthoglossa
    Bicolletes
    Callomelitta
    Dasycolletes
    Eulonchopria
    Goniocolletes
    Hesperocolletes
    Hexantheda
    Lamprocolletes
    Leioproctus*
      Aeganopria

Anacolletes
Andrenopsis
Baeocolletes
Belopria
Biglossa
Biglossidia
Brachyglossula
Ceratocolletes
Chrysocolletes
Caldocerapis
Colletellus
Colletopsis
Ctenosibyne
Edwyniana
Euryglossidia
Exocolletes
Filiglossa
Glossurocolletes
Goniocolletes
Halictanthrena
Hoplocolletes
Leioproctus
Lonchopria
Microcolletes
Nesocolletes
Nodocolletes
Nomiocolletes
Notocolletes
Perditomorpha
Phenacolletes
Protodiscelis
Protomorpha
Spinolapis
Stenocolletes
Urocolletes
Neopasiphae
Niltonia
Paracolletes
  Anthoglossa
  Paracolletes
Parapolyglossa
Polyglossa
Scrapter*
Tetraglossula*
Trichocolletes*
  Callocolletes
  Trichocolletes
Colletini
  Colletes*
  Hemicotelles
  Mourecotelles
  Xanthocotelles
Diphaglossinae
Caupolicanini
  Caupolicana*

Alayoapis
Caupolicana
Zikanapis
Crawfordapis*
Ptiloglossa*
Diphaglossini
  Cadegula
  Diphaglossa
  Policana
Mydrosomini
  Bicornelia*
Ptiloglossidiini
  Ptiloglossidia
STENOTRITIDAE
  Ctenocolletes*
  Stenotritus*
OXAEIDAE
  Notoxaea
  Oxaea*
  Protoxaea*
  Mesoxaea*
HALICTIDAE
Dufoureinae
  Conanthalictus
    Conanthalictus
    Phaceliapis
  Dufourea
    Dufourea
    Halictoides
    Michenerula
    Micralictoides
    Penapis
    Protodufourea
  Rophites
    Rophites
    Rhophitoides
  Sphecodosoma
  Systropha
  Xeralictus
Halictinae
Nomiodini
  Ceyalictus
  Nomioides
Halictini
  Agapostemon*
  Agapostemonoides*
  Caenohalictus
  Dinagapostemon
  Echthralictus
  Eupetersia
  Habralictus*
  Halictus
    Halictus
    Seladonia
    Vestitohalictus

Homalictus
Lasioglossum
  Afrodialictus
  Australictus
  Austrevylaeus
  Callalictus
  Chilialictus
  Ctenomia
  Dialictus
  Evylaeus
  Glossalictus
  Habralictellus
  Hemihalictus
  Lasioglossum
  Nesohalictus
  Oxyhalictus
  Paradialictus
  Paralictus
  Parasphecodes
  Sphecodogastra
Mexalictus
Pseudochilalictus
Microsphecodes
Pachyhalictus*
  Dictyohalictus
  Pachyhalictus
Paragapostemon
Parathrincostoma
Patellapis*
  Chaetalictus
  Lomatalictus
  Patellapis
Pseudagapostemon
Ptilocleptis
Rhinetula
Ruizantheda*
Thrincohlaictus
Thrinchostoma*
  Diagonozus
  Eothrinchostoma
  Thrinchostoma
Sphecodes*
  Callosphecodes
  Sphecodes
Zonalictus*
Augochlorini
  Andinaugochlora
  Ariphanarthra
  Augochlora
    Aethechlora
    Augochlora
    Mycterochlora
    Oxytoglossella
  Augochlorella*
  Augochlorodes

*Augochloropsis**
Auochloropsis
Paraugochloropsis
*Caenaugochlora**
  Caenaugochlora
  Ctenaugochlora
Ceratilictus
Chlerogas
Chlerogella
Corynogaster
Corynura
  Corynura
  Callochlora
Corynurella
Halictillus
Megalopta
Megaloptidia
Megommation
  Emgaloptilla
  Megommation
  Megaloptina
*Neocorynura**
  Neocorynura
  Neocorynuroides
*Paraoxystoglossa**
Pereirapis
Pseudaugochloropsis
Rhectomia
*Rhinocorynura**
*Temnosoma**
Thectochlora
Nomiinae
Austronomia
Crinoglossa
Crocisaspidia
*Hoplonomia**
Mellitidia
*Nomia**
  Acunomia
  Curvinomia
  Dieunomia
  Epinomia
  Nomia
Ptilonomia
Reepenia
*Rhopalomelissa**
Steganomus
ANDRENIDAE
*Alocandrena**
Ancylandrena
*Andrena**
  Andrena
  Aporandrena
  Belandrena
  Callandrena

Chaulandrena
Cnemidandrena
Conandrena
Dactylandrena
Derandrena
Diandrena
Eremandrena
Euandrena
Geissandrena
Gonandrena
Hesperandrena
Iomelissa
Larandrena
Leucandrena
Melandrena
Micrandrena
Nemandrena
Oligandrena
Onagandrena
Opandrena
Parandrena
Pelicandrena
Plastandrena
Ptilandrena
Rhaphandrena
Scaphandrena
Scoliandrena
Scrapteropsis
Simandrena
Taeniandrena
Thysandrena
Trachandrena
Tylandrena
Xiphandrena
Euherbstia
Megandrena
  Erythrandrena
  Megandrena
Orphana
Panurginae
*Acamptopoeum**
Anthemurgus
Austropanurgus
*Calliopsis**
  Calliopsima
  Calliopsis
  Perissander
  Verbenapis
Callonychium
  Callonychium
  Paranychium
Camptopoeum
Cephalurgus
Epimethia
*Heterosaurus**

Hypomacrotera
Liopoeum
Liphanthus
  Leptophanthus
  Liphanthus
  Melaliphanthus
  Neoliphanthus
  Pseudoliphanthus
  Tricholiphanthus
  Xenoliphanthus
Melitturga
Melitturgula
Metapsaenythia
Nomadopsis
  Macronomadopsis
  Micronomadopsis
  Nomadopsis
Panurginus
Panurgus
Parafriesia
*Parapsaenythia**
Perdita
  Allomacrotera
  Alloperdita
  Coekerelliia
  Cockerellula
  Epimacrotera
  Glossoperdita
  Hesperoperdita
  Heteroperdita
  Hexaperdita
  Macrotera
  Macroterella
  Macroteropsis
  Pentaperdita
  Perdita
  Perditella
  Procockerellia
  Pseudomacrotera
  Pygoperdita
  Xeromacrotera
  Xerophasma
Poecilomelitta
Protoandrena
*Psaenythia**
Pseudopanurgus
Pterosaurus
Rhophitulus
Spinoliella
  Peniella
  Spinoliella
Xenopanurgus
MELITTIDAE
Meganomiinae
  *Agemmonia**

A. Apoidea: a partial checklist

Ceratomonia
*Meganomia**
*Uromonia**
Melittinae
  Dolichochile
  Macropis
    Macropis
    Paramacropis
    Sinomacropis
  Melitta
  Rediviva
  *Redivivoides**
Dasypodinae
  Dasypodini
    Capicola
      Capicola
      Capicoloides
    Dasypoda
    Eremaphanta
      Eremphanta
      Popovapis
    Hesperapis
      Amblyapis
      Carinapis
      Disparapis
      Hesperapis
      Panurgomia
      Xeralictoides
      Zacesta
  Promelittini
    Promelitta
  Sambini
    Haplomelitta
      Astrosamba
      Haplomelitta
      Haplosamba
      Metasamba
      Prosamba
    Samba
CTENOPLECTRIDAE
  *Ctenoplectra**
  Ctenoplectrina
FIDELIIDAE
  *Fidelia**
  Neofidelia
    *Parafidelia**
MEGACHILIDAE
  Lithurginae
    *Lithurge**
      Lithurge
      Lithurgopsis
    Microthurge
    Trichothurgus
  Megachilinae
    Megachilini

*Chalicodoma**
  *Austrochile**
  *Callomegachile**
  *Carinella**
  Cestella
  Chalicodoma
  Chalicodomoides
  *Chelostomoides**
  Cuspidella
  Cigronoceras
  Dinavis
  *Eumegachilana**
  *Gronoceras**
  Hackeriapis
  Largella
  *Maximegachile**
  Morphella
  Neglectella
  Psedomegachile
  *Rhodomegachile**
  Sarogaster
  Schizomegachile
  Stenomegachile
  *Thaumatosoma**
*Chrysosaurus**
  Chrysosaurus
  Dactylomegachile
  Steloides
  Zonomegachile
*Coelioxys**
  Acrocoelioxys
  Allocoelioxys
  Boreocoelioxys
  Coelioxys
  Cyrtocoelioxys
  Dasycoelioxys
  Glyptocoelioxys
  Haplocoelioxys
  Hemicoelioxys
  *Liothyrapis**
  Melanocoelioxys
  Neocoelioxys
  Platycoelioxys
  Rhinocoelixoys
  Schizocoelioxys
  Syncoelioxys
  Xerocoelioxys
*Creightonella**
Cressoniella
  *Austromegachile**
  Chaetochile
  Cressoniella
  Dasymegachile
  Holcomegachile
  Neomegachile

  Ptilosarus
  Rhyssomegachile
  Trichurochile
  Tylomegachile
Eumegachile
  Eumegachile
  Grosapis
  Mitchellapis
  Sayapis
  Schrottkyapis
*Megachile**
  Amegachile
  Callochile
  Delomegachile
  Digitella
  Eurymella
  Eutricharaea
  Litomegachile
  Megella
  Megachile
  Paracella
  Platusta
  Xanthosarus
Megachiloides
  Argyropile
  Derotropis
  Megachiloides
  Phaenosarus
  Xeromegachile
*Pseudocentron**
  Acentron
  Grafella
  Leptorachina
  Leptorachis
  Melanosarus
  Moureana
  Pseudocentron
Osmiini
  Archeriades
  Ashmeadiella
    Arogochila
    Ashmeadiella
    Chilosmia
    Cubitognatha
  Bytinskia
  Chelostoma
  *Heriades**
    Heriades
    Michenerella
    Neotrypetes
    Physostetha
  Hofferia
  Hoplitis
    Acanthosmia
    Alcidamea

Allosmia
Anthocopa
Andronicus
Arctosmia
Atoposmia
Cyrtosmia
Dasyosmia
Eremosmia
Erythrosmia
Formicapis
Glossosmia
Haetosmia
Hexosmia
Hoplitis
Hoplosmia
Ulsosmia
Liosmia
Megalosmia
Micreriades
Monumetha
Odontanthocopa
Odonterythrosmia
Paranthocopa
Pseudosmia
Robertsonella
Tridentosmia
Jaxaretinula
Kumobia
Megaloheriades
Metallinella
Noteriades
Ochreriades
Osmia
  Acanthosmioides
  Aceratosmia
  Centrosmia
  Cephalosmia
  Chalcosmia
  Chenosmia
  Cryptosmia
  Diceratosmia
  Euthosmia
  Exosmia
  Helicosmia
  Hemiosmia
  Melanosmia
  Monilosmia
  Monosmia
  Mystacosmia
  Neosmia
  Nothosmia
  Orientosmia
  Osmia
  Pyrosmia
  Trichinosmia

Othinosmia
Prochelostoma
Proteriades
  Acrosmia
  Cephalapis
  Hoplitina
  Penteriades
  Proteriades
  Xerosmia
Protosmia
  Chelostomopsis
  Protosmia
Pseudoheriades
Stenoheriades
Stenosmia
Wainia
Anthidiini
  Adanthidium
  Afranthidium
  Allanthidium
    Allanthidium
    Anthidianum
  Anthidiellum*
    Anthidiellum
    Chloranthidiellum
    Eoanthidium
    Pygnanthidiellum*
    Trianthidiellum
  Anthidioma
  Anthidium*
    Anthidium
    Callanthidium
    Melanthidium
    Nivanthidium
    Proanthidium
    Severanthidium
    Stenanthidium
    Tetranthidium
  Anthodioctes*
    Nananthidium*
  Apanthidium
  Apianthidium
  Archianthidium
  Asianthidium
  Aspidosmia
  Atropium
  Ausanthidium
  Axillanthidium
  Aztecanthidium*
  Bathanthidium
  Borhranthidium
  Capanthidium
  Carinanthidium
  Cyphanthidium
  Dianthidium

  Adanthidium
  Deranchanthidium
  Dichanthidium
  Dianthidium
  Protoanthidium
  Dolichostelis*
  Doxanthidium
  Epanthidium
  Euaspis*
  Gnathanthidium
  Heteranthidium
  Heterostelis
  Hoplostelis
  Hypanthidium*
  Hypanthidiodes
    Anthidulum
    Hypanthidium
    Icteranthidium
    Immanthidium
    Melanthidium
    Manthidium
    Meganthidium
    Mesanthidiellum
    Mesanthidium
    Neanthidium
    Nigranthidium
    Notanthidium
    Oranthidium
    Oxyanthidium
  Pachyanthidium*
    Micranthidium
    Pachyanthidium
    Trichanthidium
  Paraanthidium
  Paranthidiellum
  Paranthidium
    Mecanthidium*
    Paranthidium
    Rapanthidium
  Parevaspis
  Plesianthidium
  Pseudoanthidium
  Pycnanthidium
  Reanthidium
  Royanthidium
  Rhodanthidium
  Saranthidium
  Serapista
  Stelis*
    Afrostelis
    Chelynia
    Heterostelis
    Melanostelis
    Microstelis
    Pavostelis

A. Apoidea: a partial checklist

  Protostelis
  Stelidina
  Stelidium
  Stelis
  *Trachusa**
  Congotrachusa
  Heteranthidium
  Leganthidium
  Massanthidium
  Trachusa
  Trachusomimus
  Ulanthidium
  Trianthidium
  Tuberanthidium
  Xenanthidium
 Dioxini
  Dioxys
ANTHOPHORIDAE
 Nomadinae
  Ammobatini
   Ammobates
   Caesarea
   Melanempis
   *Morgania**
   Omachthes
   Oreopasities
   Parammobatodes
   Pasites
   Pasitomachthes
   *Pseudodichroa**
   Pseudopasites
   Sphecodopsis
  Ammobatoidini
   Ammobatoides
  Biastini
   Biastes
   Neopasites
    Micropasites
    Neopasites
  Caenoprosopidini
   Caenoprosopidis
  Epeolini
   Doeringiella
   Epeolus
   *Odyneropsis**
   *Thalestria**
   *Triepeolus**
   *Trophocleptria**
  Epeoloidini
   Epeoloides
  Holcopasitini
   Holcopasites
    Holcopasites
    Trichopasites
    Schmiedeknechtia

  Isepeolini
   Isepeolus
  Neolarini
   Neolarra
    Neolarra
    Phileremulus
  Nomadini
   Brachynomada
   Centrias
   Hesperonomada
   Hexepeolus
   Hypochrotaenia
    Hypochrotaenia
    Micronomada
   Kelita
   Melanomada
   Nomada
    Holonomada
    Laminomada
    Micronomada
    Nomada
    Nomadita
    Pachynomada
   *Osiris**
   Paranomada
   Triopasites
  Protoepeolini
   *Leiopodus**
   Protoepeolus
  Townsendiellini
   Townsendiella
    Eremapasites
    Townsendiella
    Xeropasites
 Anthophorinae
  Anthophorini
   Amegilla
    *Amegilla**
    *Asaropoda**
    *Glossamegilla**
    *Zonamegilla**
   Anthophora
    Anthophora
    *Clisodon**
    Micanthophora
    Deltoptila
    *Elaphropoda**
    *Emphoropsis**
    Habrophorula
    *Habropoda**
    Heliophila
    *Pachymelus**
  Ancylini
   Ancyla
   Tarsalia

  Canephorulini
   Canephorula
  Centridini
   Centris
    Acritocentris
    *Centris**
    Exallocentris
    *Hemisiella**
    Heterocentris
    *Melanocentris**
    Paracentris
    Ptilocentris
    Trachina
    Wagenknechtia
    Xanthemisia
    Xerocentris
   Epicharis
    Anepicharis
    Cyphepicharis
    Epicharana
    Epicharis
    Epicharitides
    *Epicharoides**
    Hoplepicharis
    Parepicharis
    Triepicharis
    *Ptilotopus**
  Eucerini
   Alloscirtetica
    Alloscirtetica
    Ascirtetica
    Dasyscirtetica
    Scirteticops
   Agapanthinus
   Anthedonia
   Cemolobus
   Dithygater
   *Eucara**
   Eucera
   *Florilegus**
    Euflorilegus
    Florilegus
    Florirapter
   *Gaesischia**
    Agaesischia
    Gaesischia
    Gassischiana
    Gaesischiopsis
   Gaesochira
   Idiomelissodes
   Lophothygater
   *Loxoptilus**
   Martinapis
    Martinapis
    Svastropis

Melissina
Melissodes*
  Apomelissodes
  Callimelissodes
  Dasyhalonia
  Ecplectica
  Eumelissodes
  Jeliomelissodes
  Melissodes
  Pachyhalonia
  Psilomelissodes
  Tachymelissodes
  Zonolonia
Melissoptila
  Comeptila
  Melissoptila
  Ptilomelissa
Megascirtetica
Micronychapis
Pachysvastra
Pectinapis
Peponapis*
  Colocynthophila
  Eopeponapis
  Peponapis
  Xenopeponapis
  Xeropeponapis
Simanthedon
Svastra
  Brachymelissodes
  Epimelissodes*
  Svastra
  Svastrides
  Svastrina
Syntrichalonia*
Tetralonia*
Tetraloniella
Thygater*
  Nectarodiaeta
  Thygater
  Thygatina
Trichocerapis
Xenoglossa
  Eoxenoglossa
  Xenoglossa*
  Xenoglossodes
Ericrocini
  Abromelissa*
  Acanthopus*
  Agalomelissa*
  Ctenioschelus*
  Ericrocis
  Hopliphora*
  Mesocheira*
  Mesonychium*

Mesoplia*
  Eumelissa
  Mesoplia
Eucerinodini
  Eucerinoda
Exomalopsini
  Ancyloscelis*
  Caenonomada*
  Chalepogenus
  Chilimalopsis
  Eremapis
  Exomalopsis*
    Anthophorisca
    Anthophorula
    Diomalopsis
    Exomalopsis
    Isomalopsis
    Megomalopsis
    Phanomalopsis
  Lanthanomelissa
    Lanthanella
    Lanthanomelissa
  Monoeca*
  Paratetrapedia*
    Amphipedia
    Arhysoceble
    Lophopedia
    Paratetrapedia
    Trigonopedia
    Tropidopedia
    Xanthopedia
  Tapinotaspis*
    Tapinorhina
    Tapinotaspis
    Tapinotaspoides
    Teratognatha
Melectini
  Brachymelecta
  Melecta
    Melecta
    Melectomiimus
  Nesomelecta
  Thyreus*
  Xeromelecta
    Xeromelecta
    Melectomorpha
  Zacosmia
Melitomini
  Diadasia*
    Dasiapis
    Diadasia
    Diadasiana
  Energoponus
  Leptometria
  Melitoma*

  Ptilothrix
  Teleutemnesta
Pararhophitini
  Pararhophites
Rathymini
  Rathymus*
Tetrapediini
  Tetrapedia*
    Lagobata
    Tetrapedia
Xylocopinae
  Allodapini
    Allodape*
    Allodapula
      Allodapula
      Allodapulodes
      Dalloapula
    Braunsapis*
    Compsomelissa
    Effractapis
    Eucondylops
    Exoneura*
      Brevineura
      Exoneura
      Exoneurella
    Halterapis*
    Inquilina
    Macrogalea*
    Nasutapis
  Ceratinini
    Ceratina*
      Calloceratina
      Catoceratina
      Ceratina
      Ceratinidia
      Ceratinula
      Chloroceratina
      Crewella
      Euceratina
      Lioceratina
      Xanthoceratina
      Zadontomerus
      Comsomelissa
      Simioceratina
      Ctenoceratina*
      Megaceratina
      Pithitis*
  Manueliini
    Manuelia*
  Xylocopini
    Lestis
    Proxylocopa
      Ancylocopa
      Proxylocopa
    Xylocopa

Acroxylocopa
Afroxylocopa
Alloxylocopa
Apoxylocopa
*Biluna**
Bombioxylocopa
Calloxylocopa
Cirroxylocopa
Copoxyla
Ctenoxylocopa
*Cyanoderes**
Cyphoxylocopa
Dasyxylocopa
Diaxylocopa
Dinoxylocopa
Epixylocopa
Euxylocopa
*Gnathoxylocopa**
Hoplitocopa
Hoploxylocopa
Ioxylocopa
*Koptortosoma**
Lieftinckella
Megaxylocopa
Mesotrichia
Mimoxylocopa
Monoxylocopa
Nanoxylocopa
*Neoxylocopa**
Nodula
*Notoxylocopa**
*Nyctomelitta**
Oxyxylocopa
Perixylocopa
*Platynopoda**
Prosopoxylocopa
Rhysoxylocopa
*Schoenherria**
Stenoxylocopa
Xenoxylocopa
Xylocopa
Xylocopina
Xylocopoda
Xylocopoides
Xylocopsis
Xylocospila
Xylomelissa
Zonohirsuta
APIDAE
  Bombinae
    Bombini
      *Bombus**
Alpigenobombus
Alpinobombus
Bombias
Bombus
Brachycephalibombus
Coccineobombus
Confusibombus
Crotchiibombus
Cullumanobombus
Diversobombus
Eversmannibombus
Fervidobombus
Festivobombus
Fraternopbombus
Funebribombus
Kallobombus
Laesobombus
Megabombus
Melanobombus
Mendacibombus
Mucidobombus
Orientalibombus
Pressibombus
Pyrobombus
Rhodobombus
Robustobombus
Rufipedibombus
Senexbombus
Separatobombus
Sibiricobombus
Subterraneobombus
Thoracobombus
Tricornibombus
Psithyrus
Euglossini
  *Aglae**
  *Eufriesea**
  *Euglossa**
    Dasystilbe
    Euglossa
    Euglossella
    *Glossura**
    *Glossurella**
  Eulaema
    *Apeulaema**
    *Eulaema**
  *Exaerete**
Apinae
  *Apis**
Meliponinae
  Meliponini
    *Melipona**
Trigonini
  *Cleptotrigona**
  *Dactylurina**
  Lestrimelitta
  *Meliponula**
  Trichotrigona
  Trigona
    Aparatrigona
    Apotrigona
    Austroplebeia
    *Axestotrigona**
    Celetrigona
    *Cephalotrigona**
    Dolichotrigona
    Duckeola
    Friesella
    *Frieseomelitta**
    Gentiotrigona
    Geotrigona
    Heterotrigona
    *Homotrigona**
    Hypotrigona
    Lepidotrigona
    Leurotrigona
    Liotrigona
    Lophotrigona
    *Meliplebeia**
    Mourella
    *Nannotrigona**
    Nogueirapis
    Odontotrigona
    *Oxytrigona**
    Parapartamona
    Paratrigona
    Pariotrigona
    *Partamona**
    Platytrigona
    *Plebeia**
    *Plebeilla**
    *Ptilotrigona**
    *Scaptotrigona**
    *Scaura**
    Schwarziana
    Schwarzula
    *Tetragona**
    *Tetragonisca**
    *Tetrigona**
    *Tetragonula**
    *Trigona**
    Trigonella
    Trigonisca

Appendix B

Illustrated tropical bee genera, arranged by family and geographic area

Note: Individuals are females or workers unless otherwise indicated.

Figure B.1 *(facing)*. ANTHOPHORIDAE: **1,2** ♂ *Centris (Centris) flavofasciata* 1×, Panama; **3** *Epicharis (Epicharoides) albofasciata* 1.2×, Panama; **4** *Centris (Hemisiella) trigonoides* 1.3×, Panama; **5,6** ♂ *Xylocopa (Notoxylocopa) tabaniformis sylvicola* 1.3×, Guatemala; **7** *Xylocopa (Neoxylocopa) brasilianorum* 1×, Brazil; **8** *Centris (Melanocentris) conspersa* 1×, Brazil; **9** *Ptilotopus americanus* 1×, French Guiana; **10** *Centris* 1×, French Guiana; **11,12** ♂ *Xylocopa (Neoxylocopa) grisescens* 1×, Brazil; **13,14** *Centris (Centris) inermis* dark and light forms 1.3×, Costa Rica; **15** *Acanthopus palmatus* 1×, French Guiana; **16,17** ♂ *Xylocopa (Schoenherria) muscaria* 1.2×, Brazil; **18** ♂ of 7, 1×.

B. Illustrated tropical bee genera 399

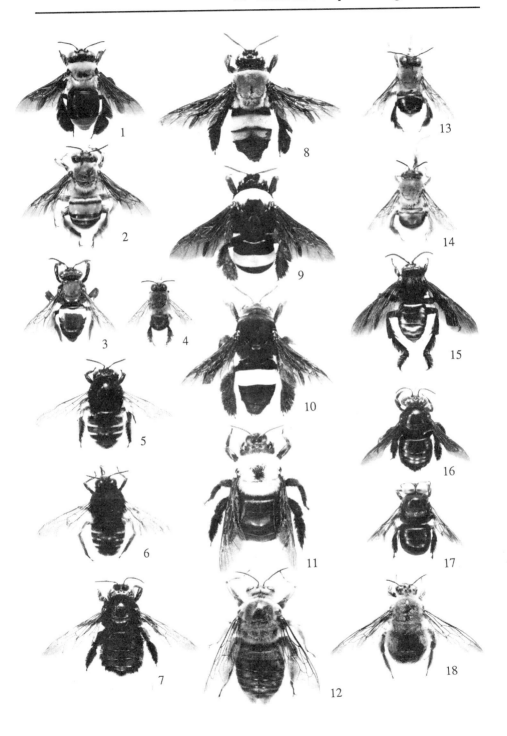

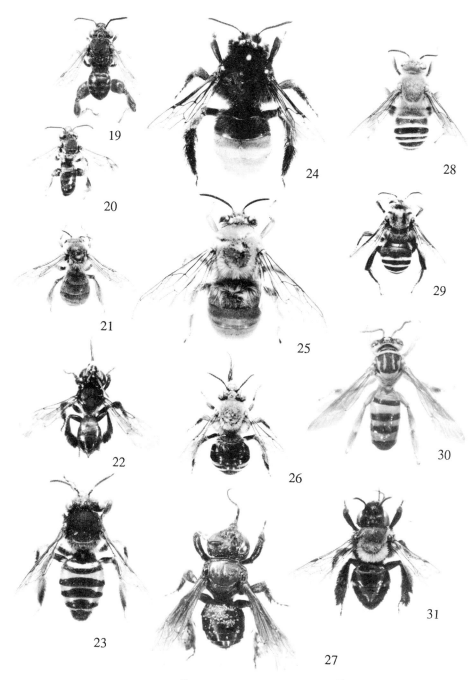

Figure B.2. ANTHOPHORIDAE: 19,20 ♂ *Ancyloscelis* 3×, Colombia; 21 ♂ *Diadasia* 1.5×, Brazil; 22 ♂ *Paratetrapedia* 3×, Mexico; 23 *Caenonomada* 3×, Brazil; 24 *Xenoglossa gabbii* 1.5×, Nicaragua; 25 *Xenoglossa fulva* 1.5×, Mexico; 26 *Peponapis fervens* 2×, Brazil; 27 *Tetrapedia* carrying mites on metasoma, 3×, Brazil; 28 *Diadasia* 2×, Mexico; 29 ♂ *Melitoma taurea* 2×, Mexico; 30 *Paratetrapedia* 3×, Mexico; 31 *Tapinotaspis caerula* 3×, Brazil.

B. Illustrated tropical bee genera 401

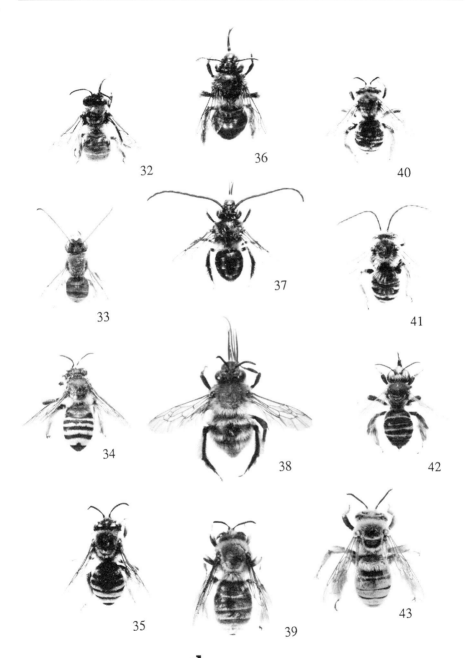

Figure B.3. ANTHOPHORIDAE: 32,33 ♂ *Gaesischia exul* 2×, Costa Rica; 34 *Florilegus* 1.5×, French Guiana; 35 *Loxoptilus longifellator* 1.5×, Mexico; 36,37 ♂ *Thygater analis* 1.5×, Costa Rica; 38 *Peponapis* 1.5×, Panama; 39 *Syntrichalonia exquisita* 1.5×, Mexico; 40,41 ♂ *Melissodes thelypodii* 1.5×, Costa Rica; 42 *Exomalopsis mellipes* 1.5×, Mexico; 43 *Svastra (Epimelissodes) petulca* 1.5×, Mexico.

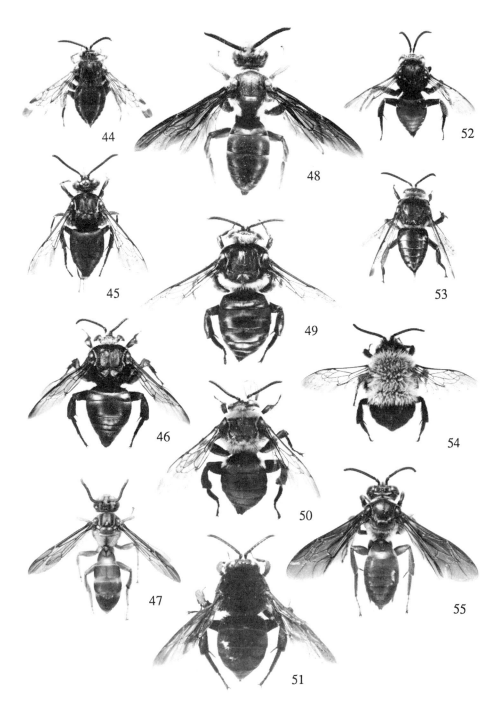

Figure B.4. ANTHOPHORIDAE: **44** *Mesocheira* 2×, Mexico; **45** *Thalestria* 2×, Brazil; **46** *Mesonychium* 2×, Panama; **47** *Odyneropsis* 2×, Panama; **48** *Rathymus unicolor* 2×, Bolivia; **49** *Ctenioschelus* 2×, French Guiana; **50** *Mesoplia insignis* 2×, Mexico; **51** *Hopliphora* 2×, French Guiana; **52** *Mesonychium littorium* 2×, French Guiana; **53** *Agalomelissa duckei* 2×, Costa Rica; **54** *Abromelissa lendliana* 2×, Chile; **55** *Rathymus* 2×, Panama.

B. Illustrated tropical bee genera 403

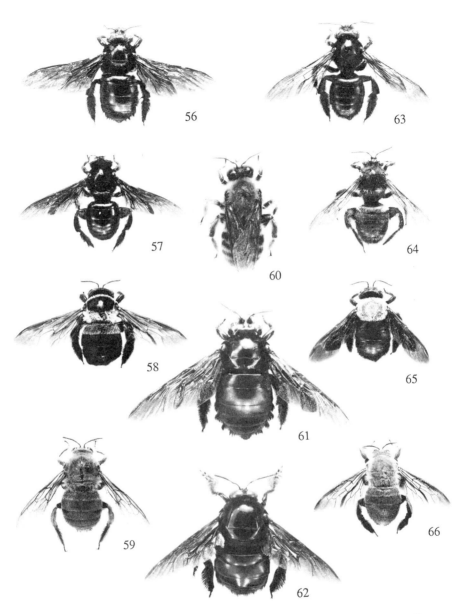

Figure B.5. ANTHOPHORIDAE: 56,57 ♂ *Xylocopa (Biluna) auripennis* 1×, India; 58,59 ♂ *Xylocopa (Koptortosoma) ghilianii* 1×, Philippines; 60 *Xylocopa (Nyctomelitta) tranquebarica* 1×, Thailand; 61,62 ♂ *Xylocopa (Platynopoda) latipes* 1×, Sumatra; 63,64 ♂ *Xylocopa (Biluna) tranquebarorum* 1×, Taiwan; 65,66 ♂ *Xylocopa (Koptortosoma) confusa* 1×, Java.

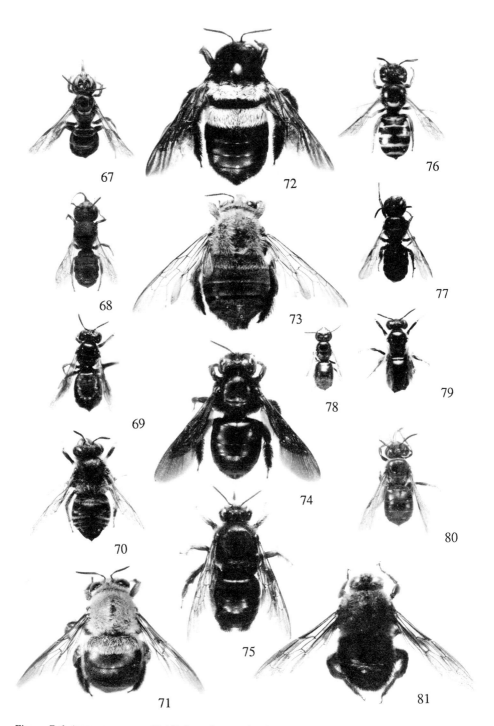

Figure B.6. ANTHOPHORIDAE: **67** *Allodape derutata* 3×, Ghana; **68** *Pithitus* 3×, Tanganyika; **69** *Exoneura bicolor* 3×, Australia; **70** *Macrogalea candida* 3×, Tanzania; **71** *Xylocopa (Cyanoderes) caerulea* 2×, Sarawak; **72,73** ♂ *Xylocopa (Koptortosoma) mossambica* 2×, Africa; **74,75** ♂ *Xylocopa (Gnathoxylocopa) kobrowi* 2×, Kalahari; **76** *Ceratina compacta* 3×, Philippines;

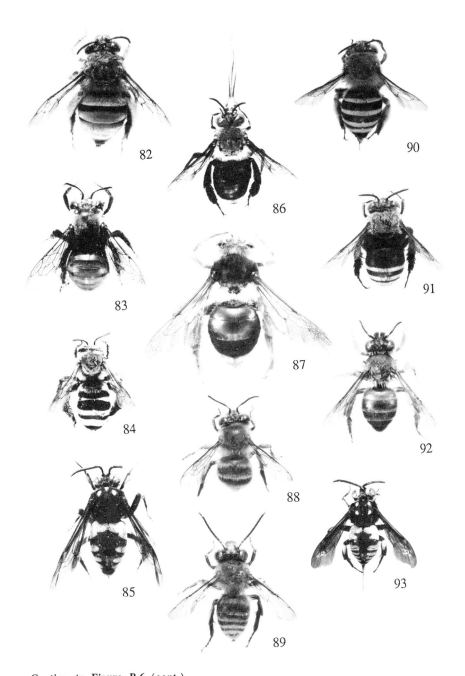

Caption to Figure B.6 (cont.)
77 *Ctenoceratina tanganyicensis* 3×, Kenya; 78 *Braunsapis plebeia* 3×, New Guinea; 79 *Halterapis nigrinervis* 3×, S. Africa; 80 *Braunsapis calidula* 3×, Tanzania; 81 ♂ of 71, 2×.

Figure B.7. ANTHOPHORIDAE: 82 *Amegilla (Asaropoda) bombiformis* 2×, New Guinea; 83 *Eucara* 2×, Kenya; 84 *Tetralonia phryne* 2×, India; 85 *Thyreus nitidulus* 2×, Indonesia; 86 *Amegilla (Glossamegilla) fimbriata* 2×, Thailand; 87 *Pachymelus grandidieri* 1.5×, Madagascar; 88,89 ♂ *Habropoda* 2×, Thailand; 90 *Amegilla (Zonamegilla) andrewsi* 2×, New Guinea; 91 *Amegilla* 2×, Thailand; 92 *Elaphropoda impatiens* 2×, Thailand; 93 *Thyreus* 2×, New Guinea.

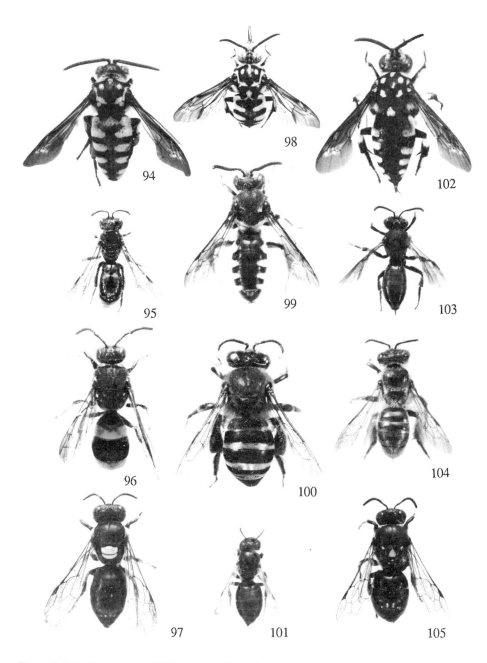

Figure B.8. ANTHOPHORIDAE: **94** *Thyreus* 3×, Thailand; **95** *Scrapter sphecodoides* 3×, S. Africa; **98** *Thyreus insignis* 3×, Sri Lanka; **99** *Morgania* 3×, Kenya; **102** *Thyreus historio* 3×, Sri Lanka; **103** *Pseudodichroa capensis* 3×, S. Africa. COLLETIDAE: **96** *Hylaeoides* 3×, N. Australia; **97** *Amphylaeus morosus* 3×, N. Australia; **100** *Trichocolletes hackeri* 3×, N. Australia; **101** *Euryglossa adeleidae* 3×, N. Australia; **104** *Leioproctus caerulescens* 3×, N. Australia; **105** *Meroglossa impressifrons* 3×, N. Australia.

Figure B.9 *(facing).* MEGACHILIDAE: **106** *Megachile (Austromegachile) microsoma* 3×, Bolivia; **107** *Chrysosaurus* 1.5×, Panama; **110** *Coelioxys* 3×, Panama; **111** *Pseudocentron* 3×, Venezuela;

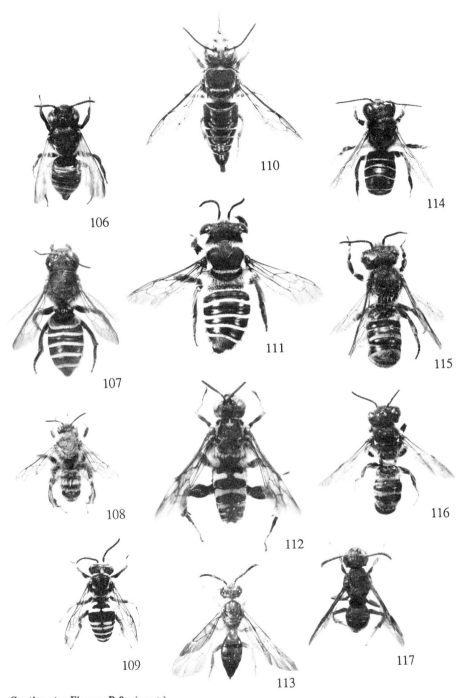

Caption to Figure B.9 (cont.)
114 *Chrysosaurus* 1.5×, French Guiana; **115,116** ♂ *Chalicodoma (Chelostomoides) otomita* 3×, Mexico. FIDELIIDAE: **108** *Neofidelia profuga* 2×, Chile. ANTHOPHORIDAE: **109** *Triepeolus* 2×, Costa Rica; **112** *Leiopodus* 3×, Brazil; **113** *Osiris* 3×, Panama; **117** *Trophocleptria schrederi* 3×, Colombia.

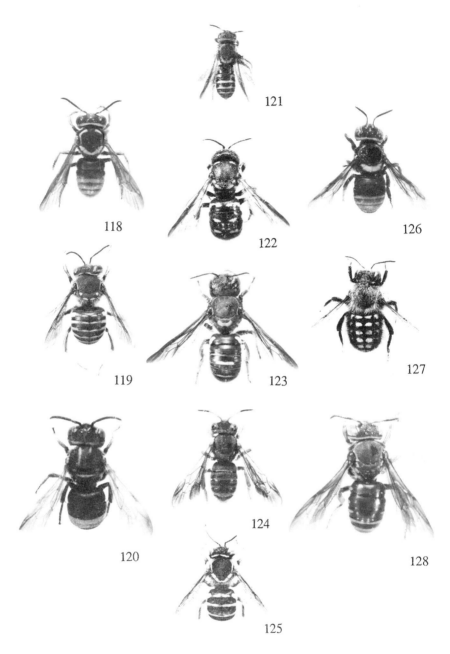

Figure B.10. MEGACHILIDAE: 118 Anthidiinae 3×, Panama; 119 ♂ *Anthidium* 2×, Brazil; 120 *Odontostelis bivittata* 3×, Panama; 121 *Nananthidium willineri* 4×, Bolivia; 122 *Dolichostelis perpulchra* 3×, Mexico; 123 *Aztecanthidium* 2×, Mexico; 124 *Paranthidium (Mecanthidium) macrorum* 2×, Mexico; 125 Anthidiinae 2×, Panama; 126 *Anthodioctes* 3×, Costa Rica; 127 *Anthidium funereum* 2×, Argentina; 128 *Hypanthidium flavomarginatum* 3×, Brazil.

Figure B.11 (*facing*). FIDELIIDAE: 129 *Parafidelia pallidula* 2×, S. Africa. MEGACHILIDAE: 130 *Chalicodoma (Callomegachile) rufipes* 2×, Uganda; 131 *Trachusa* 2×, Cameroon; 132 *Chalicodoma*

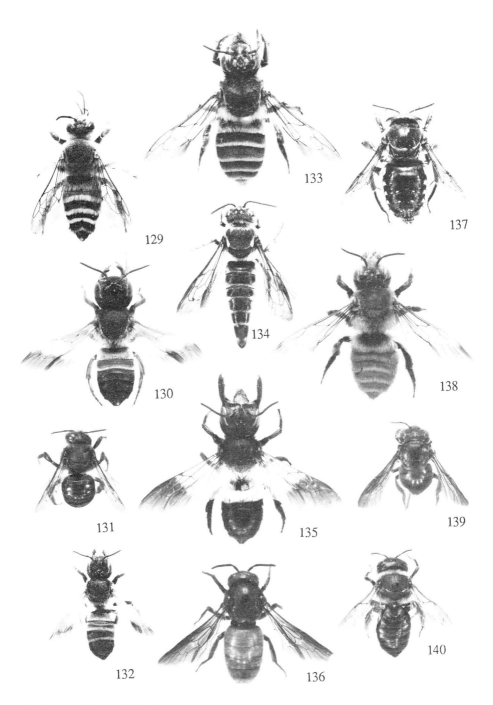

Caption to Figure B.11 (*cont.*)
(Carinella) torrida 2×, Cameroon; **133** *Cahlicodoma (Gronoceras) cincta* 2×, Cameroon; **134** *Coelioxys (Liothryapis) scocioensis* 2×, Cameroon; **135** *Chalicodoma (Maximegachile) maxillosa* 2×, S. Africa; **136** *Euaspis abdominalis* 2×, S. Africa; **137** *Anthidium (Nivanthidium) niveocinctum* 2×, Malawi; **138** *Creightonella rufa* 2×, Uganda; **139** *Pachyanthidium bouyssoni* 2×, Uganda; **140** *Lithurge* 2×, Africa.

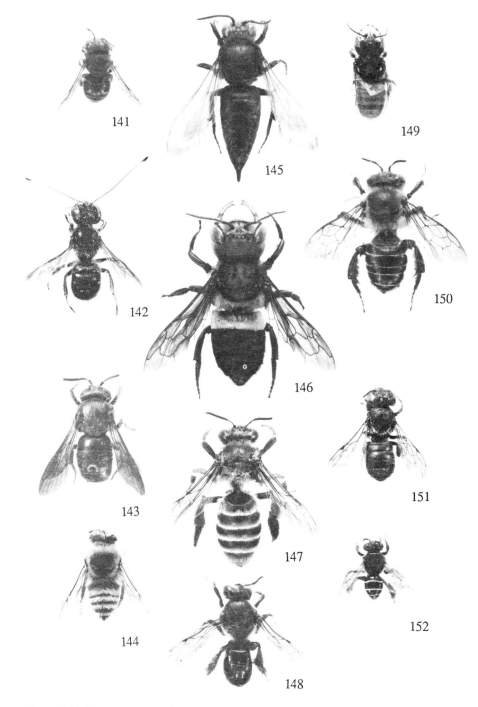

Figure B.12. MEGACHILIDAE: **141** *Anthidiellum frontorecticulatum* 3×, Kenya; **142** *Chalicodoma (Thaumatostoma) duboulaii* 3×, Australia; **143** *Euaspis carbonaria* 3×, India; **145** *Coelioxys ducalis* 1.5×, Indonesia; **146** *Chalicodoma (Eumegachilana) godeffroyi* 1.5×, Indonesia; **149** *Chalicodoma*

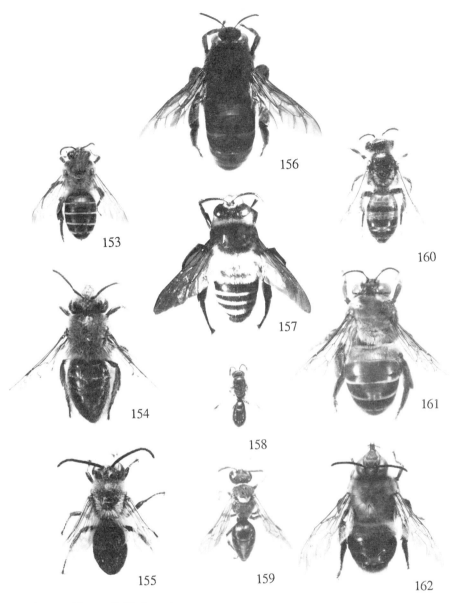

Caption to Figure B.12 (*cont.*)
(*Rhodomegachile*) *deanii* 4×, N. Australia; **150** *Lithurge* 3×, Bali; **151** *Heriades othonis* 4×, Java. STENOTRITIDAE: **144** *Stenotritus pubescens* 2×, W. Australia; **147** *Ctenocolletes nicholsoni* 2×, W. Australia. CTENOPLECTRIDAE: **148** *Ctenoplectra fuscipes* 3×, Malawi; **152** *Ctenoplectra albolimbata* 3×, Kenya.

Figure B.13. COLLETIDAE: **153** *Bicornelia inusitata* 2×, Panama; **154,155** ♂ *Colletes* 3×, Brazil; **156** *Crawfordapis luctuosa* 2×, Panama; **157** *Caupolicana* 2×, Ecuador; **158** *Chilicola* 4×, Costa Rica; **159** *Hylaeus* 4×, Panama; **160** *Tetraglossula* 3×, Brazil; **161,162** ♂ *Ptiloglossa* 2×, Panama.

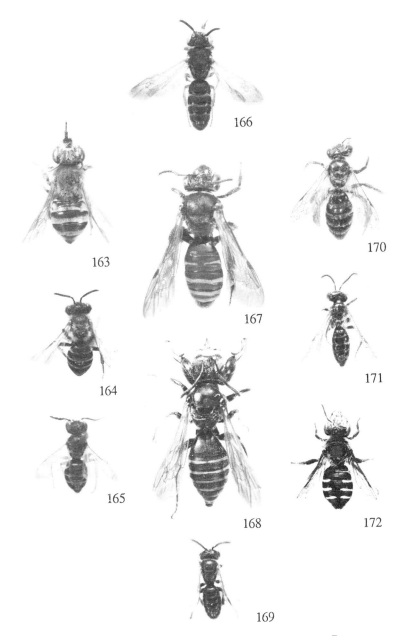

Figure B.14. ANDRENIDAE: 163 *Acamptopoeum* 3×, Brazil; 164,165 ♂ *Calliopsis hondurascica* 3×, Brazil; 166 *Parapsaenythia paspali* 3×, Argentina; 167,168 ♂ *Psaenythia* 3×, Brazil; 169 *Heterosaurus* 4×, Argentina; 170,171 ♂ *Andrena vidalesi* 3×, Costa Rica; 172 *Alocandrena porteri* 2×, Peru.

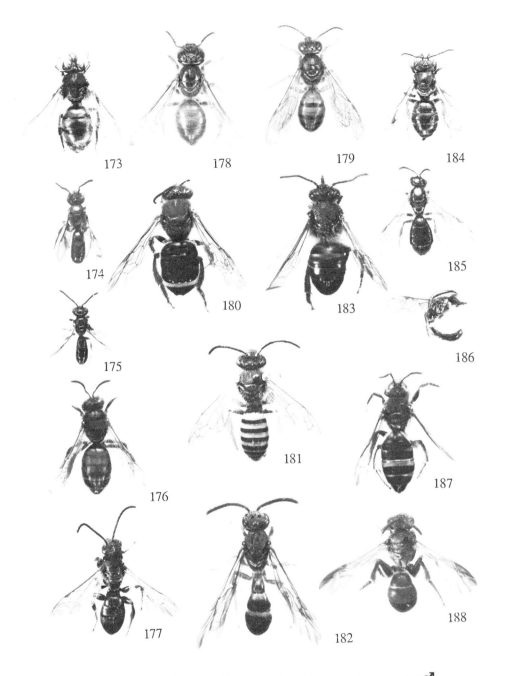

Figure B.15. HALICTIDAE: **173** *Caenaugochlora costaricensis* 2×, Costa Rica; **174,175** ♂ *Augochlorella bidentata* 3×, Costa Rica; **176** *Ruizantheda mutabilis* 3×, Argentina; **177** *Paraoxytoglossa transversa* 3×, Brazil; **178,179** ♂ *Megalopta ecuadoria* 2×, Panama; **180** *Agapostemon nasutus* 3×, Costa Rica; **181** *Agapostemon* 3×, Mexico; **182** ♂ *Agapostemon* 3×, Panama; **183** ♂ *Rhinetula dentricus* 3×, Panama; **184** *Augochloropsis electra* 2×, Brazil; **185,186** ♂ *Habralictus canialiculatus* 4×, Brazil; **187** *Agapostemonoides* 3×, Panama; **188** *Neocorynura* 3×, Panama.

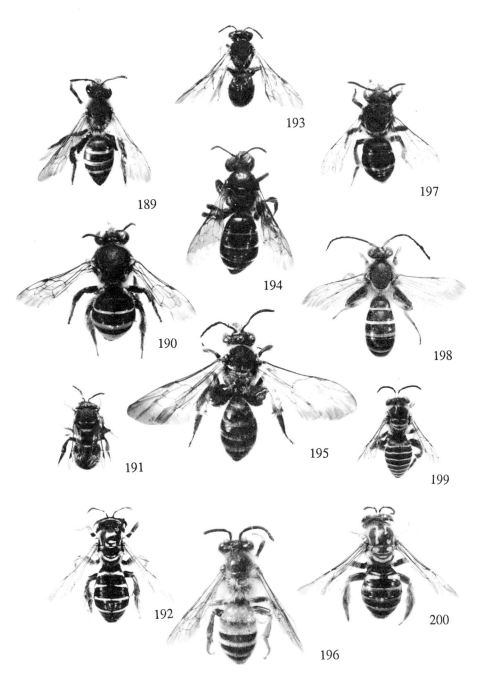

Figure B.16. HALICTIDAE: 189 *Zonalictus* 3×, Kenya; 190 *Nomia* 2×, New Guinea; 191 *Pachyhalictus* 2×, Malaya; 193 *Nomia (Rhopalomelissa)* 3×, Thailand; 194 *Patellapis montagui* 3×, S. Africa; 195 *Thrinchostoma producta* 3×, S. Africa; 197 *Nomia (Hoplonomia) pulchribalteata* 3×, New Guinea; 198 ♂ *Nomia* 3×, Thailand. MELITTIDAE: 192 *Agemmonia tsavoensis* 2×, Kenya; 196 *Redivivoides simulans* 3×, S. Africa; 199 *Uromonia stagei* 2×, Kenya; 200 *Meganomia andersoni* 1.5×, Kenya.

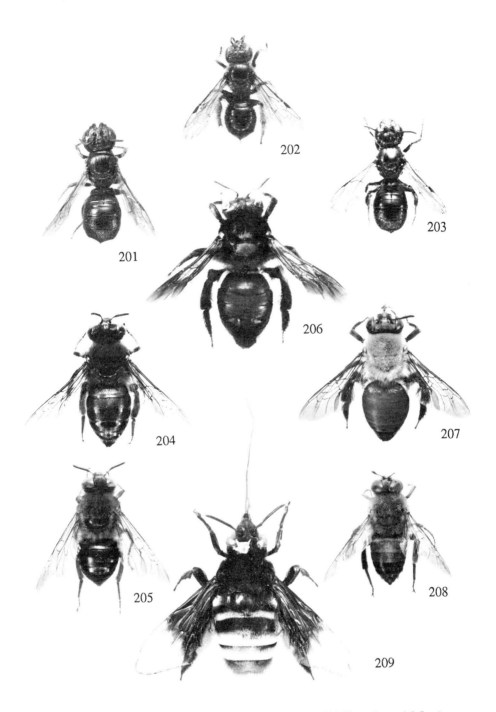

Figure B.17. ANTHOPHORIDAE: **201** *Ceratina eximia* 3×, Panama; **202** *Manuelia gayi* 3.5×, Argentina; **203** *Ceratina laeta* 3×, Panama. OXAEIDAE: **204,205** ♂ *Oxaea flavescens* 2×, Brazil; **206** *Mesoxaea* 1.5×, Mexico; **207,208** ♂ *Protoxaea ferruginea* 2×, Paraguay. APIDAE: **209** *Eulaema meriana* shown carrying two species of orchid pollinaria, 1.5×, Panama.

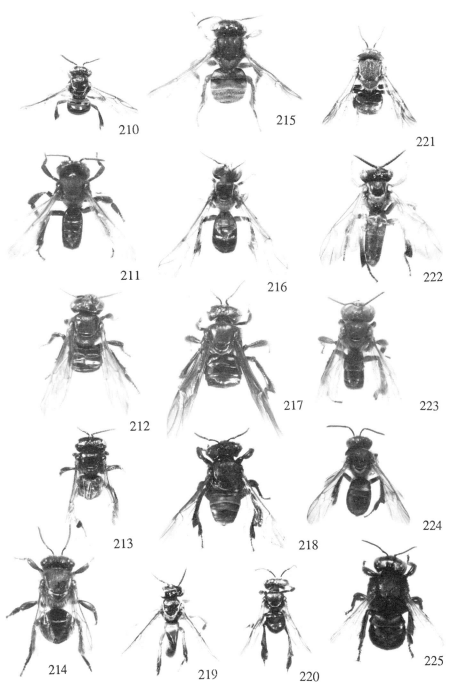

Figure B.18. APIDAE: 210 *Plebeia jatiformis* 4×, Panama; 211 *Frieseomelitta nigra* 4×, Panama; 212 *Partamona* 4×, Panama; 213 *Oxytrigona daemoniaca* 4×, Panama; 214 *Melipona marginata* 3×, Panama; 215 *Scaptotrigona pectoralis* 4×, Panama; 216 *Ptilotrigona lurida* 3×, French Guiana; 217 *Trigona amalthea silvestriana* 3×, Panama; 218 *Cephalotrigona capitata zexmeniae* 3×, Panama; 219 *Tetragonisca angustula* (=*jaty*) 4×, Panama; 220 *Scaura latitarsis* 4×, Panama; 221 *Nannotri*-

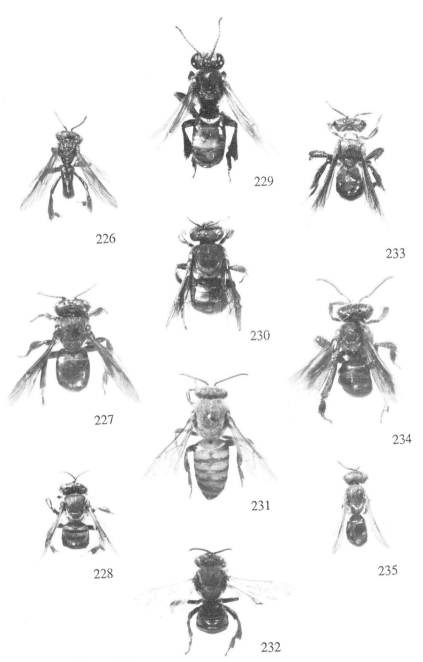

Caption to Figure B.18 (*cont.*)
gona testaceicornis 4×, Panama; **222** *Tetragona dorsalis* 4×, Panama; **223** *Trigona dallatorreana* 3×, French Guiana; **224** *Schwarziana quadripunctata* 3×, Brazil; **225** *Melipona fuliginosa* 2×, Panama.

Figure B.19. APIDAE: **226** *Dactylurina* 4×, Nigeria; **227** *Axestotrigona* 3×, Ghana; **228** *Axestotrigona erythra* 3×, Uganda; **229** ♂ *Homotrigona fimbriata* 3×, Thailand; **230,231** ♂ *Meliponula bocandei* 3×, Uganda; **232** *Meliplebeia beccarri* 3×, S. Africa; **233** *Tetrigona apicalis* 3×, Thailand; **234** *Tetragonula atripes* 3×, Thailand; **235** *Plebeilla lendliana* 4×, Nigeria.

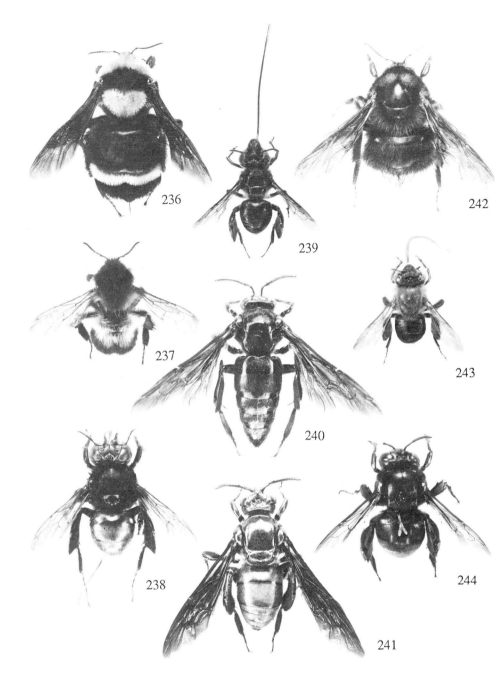

Figure B.20. APIDAE: **236** *Bombus (Fervidobombus) transversalis* 1.5×, French Guiana; **237** *Bombus (Pyrobombus) ephippiatus* 2×, Panama; **238** ♂ *Euglossa (Glossura) intersecta* 2×, French Guiana; **239** ♂ *Euglossa (Glossurella) asarophora* 2×, Panama; **240** *Aglae caerulea* 1.5×, French Guiana; **241** ♂ *Exaerete frontalis* 1.5×, French Guiana; **242** *Bombus (Robustobombus) volucelloides* 1.5×, Panama; **243** ♂ *Euglossa decorata* 2×, French Guiana; **244** ♂ *Eufriesea chrysopyga* shown carrying orchid pollinia, 2×, Panama.

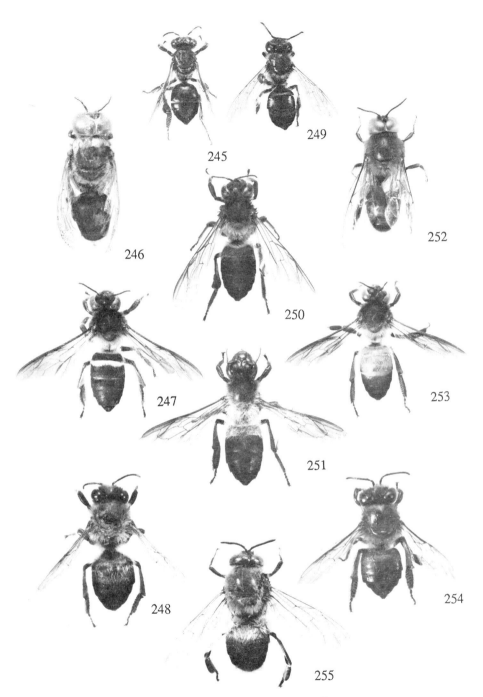

Figure B.21. APIDAE: 245 *Apis andreniformis* 3×, Thailand; 246 ♂ *Apis florea* 3×, Sri Lanka; 247 *Apis dorsata* 2×, Philippines; 248 *Apis mellifera* 3×, Panama; 249 *Apis florea* 3×, India; 250 *Apis dorsata* 2×, Sulawesi; 251 *Apis laboriosa* 2×, Nepal; 252 ♂ *Apis dorsata* 2×, Sri Lanka; 253 *Apis dorsata* 2×, India; 254 *Apis cerana* 2×, Sulawesi; 255 ♂ *Apis cerana* 2×, Sri Lanka.

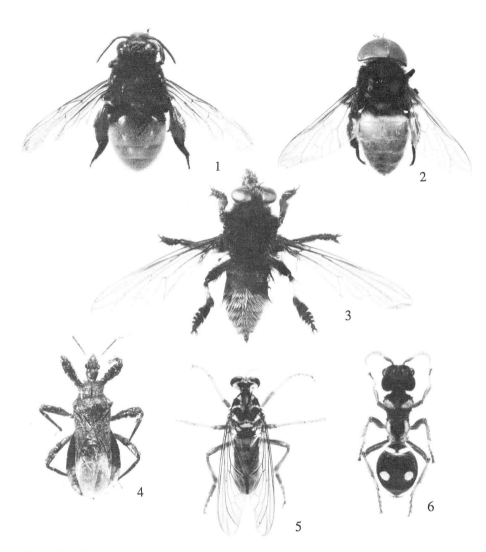

Figure B.22. Mimics, parasites, and predators of bees: 1 *Eulaema speciosa* (Apidae) and mimic, 2 Syrphidae (Diptera), 2×, Panama; 3 Asilidae (Diptera) note false pollen loads, 2×, Panama; 4 Reduviidae (Hemiptera), 3×, Thailand; 5 Asilidae 2×, Panama; 6 Mutillidae (Hymenoptera), 2×, Panama.

References

Abrams, J., and G. C. Eickwort. 1980. Biology of the communal sweat bee *Agapostemon virescens* (Hymenoptera: Halictidae) in New York State. *Search*: Agriculture 1.

Absy, M. L. 1985. Palynology of Amazonia: the history of the forests as revealed by the palynological record. Pp. 72–82 in G. T. Prance and T. E. Lovejoy, eds. *Amazonia*. Pergamon Press, Oxford.

Absy, M. L., E. B. Bezerra, and W. E. Kerr. 1980. Plantas nectaríferas utilizadas por duas espécies de *Melipona* da Amazônia. *Acta Amazonica* 10:271–281.

Absy, M. L., J. M. F. Camargo, W. E. Kerr, and I. P. A. Miranda. 1984. Espécies de plantas visitadas por Meliponinae (Hymenoptera; Apoidea) para coleta de pólen na região do médio Amazonas. *Rev. Bras. Biol.* 44:227–237.

Absy, M. L., and W. E. Kerr. 1977. Algumas plantas visitadas para obtenção de pólen por operárias de *Melipona seminigra merrillae* em Manaus. *Acta Amazonica* 7:309–315.

Ackerman, J. D. 1983a. Diversity and seasonality of male euglossine bees (Hymenoptera: Apidae) in central Panama. *Ecology* 64:274–283.

——— 1983b. Specificity and mutual dependency of the orchid–euglossine bee interaction. *Biol. J. Linn. Soc.* 20:301–314.

——— 1983c. Euglossine bee pollination of the orchid *Cochleanthes lipscombiae*: a food source mimic. *Am. J. Bot.* 70:830–834.

——— 1985. Euglossine bees and their nectar hosts. Pp. 225–233 in W. G. D'Arcy and M. D. Correa A., eds. *The botany and natural history of Panamá*. Missouri Botanical Garden, St. Louis, Mo.

——— 1986. Mechanisms and evolution of food-deceptive pollination systems in orchids. *Lindleyana* 1:108–113.

Ackerman, J. D., M. R. Mesler, K. L. Lu, and A. M. Montalvo. 1982. Food-foraging behavior of male Euglossini (Hymenoptera: Apidae): Vagabonds or trapliners? *Biotropica* 14:241–248.

Ackerman, J. D., and A. M. Montalvo. 1985. Longevity of euglossine bees. *Biotropica* 17:79–81.

Adams, J., E. D. Rothmann, W. E. Kerr, and Z. L. Paulino. 1977. Estimation of the number of sex alleles and queen-matings from diploid male frequencies in a population of *Apis mellifera*. *Genetics* 86:583–596.

Adlahka, R. L., and O. P. Sharma. 1976. *Stylops* (Strepsiptera) parasites of honey bees in India. *Am. Bee J.* 116:66.

Alcock, J. 1978. Notes on male mate-locating behavior in some bees and wasps of Arizona. *Pan-Pac. Entomol.* 54:215–225.

Alcock, J. 1979. The evolution of intraspecific diversity in male reproductive strategies in some bees and wasps. Pp. 381–402 in M. S. Blum and N. A. Blum, eds. *Sexual selection and reproductive competition in insects*. Academic Press, New York.

——— 1980. Natural selection and mating systems of solitary bees. *Am. Sci.* 68:146–153.

——— 1984. Long-term maintenance of size variation in populations of *Centris pallida* (Hymenoptera: Anthophoridae). *Evolution* 38:220–223.

Alcock, J., E. M. Barrows, G. Gordh, L. J. Hubbard, L. Kirkendall, D. W. Pyle, T. L. Ponder, and F. G. Zalom. 1978. The ecology and evolution of male reproductive behaviour in the bees and wasps. *Zool. J. Linn. Soc. London* 64:293–326.

Alcock, J., and S. L. Buchmann. 1984. The significance of post-insemination display by male *Centris pallida* (Hymenoptera: Anthophoridae). *Z. Tierphysiol.* 68:231–243.

Alcock, J., G. C. Eickwort, and K. R. Eickwort. 1977. The reproductive behavior of *Anthidium maculosum* (Hymenoptera: Megachilidae) and the evolutionary significance of multiple copulations by females. *Behav. Ecol. Sociobiol.* 2:285–296.

Alcock, J., C. E. Jones, and S. L. Buchmann. 1976. The nesting behavior of three species of *Centris* bees (Hymenoptera: Anthophoridae). *J. Kans. Entomol. Soc.* 49:469–474.

Alexander, B., and J. G. Rozen Jr. 1987. Ovaries, ovarioles, and oocytes in parasitic bees. *Pan-Pac. Entomol.* 63:155–164.

Alford, D. V. 1975. *Bumblebees*. Davis–Poynter Ltd., London.

Anasiewicz, A. 1975. The bees (Apoidea, Hymenoptera) on alfalfa (*Medicago media* Pers.) plantations. I. The species composition and variation of flights. *Ekol. Pol.* 23:147–162.

Andersen, J. F., S. L. Buchmann, D. Weisleder, R. D. Plattner, and R. L. Minckley. 1988. Identification of thoracic gland constituents from male *Xylocopa* spp. Latreille (Hymenoptera: Anthophoridae) from Arizona. *J. Chem. Ecol.* 14:1153–1162.

Anderson, G. J., and D. E. Symon. 1985. Extrafloral nectaries in *Solanum*. *Biotropica* 17:40–45.

Anderson, R. H., B. Buys, and M. F. Johannsmeier. 1983. *Beekeeping in South Africa*. Dept. Agric. Bull. No. 394, Pretoria, South Africa.

Anderson, W. R. 1979. Floral conservatism in neotropical Malpighiaceae. *Biotropica* 11:219–223.

Andrewartha, H. G., and L. C. Birch. 1984. *The ecological web*. Univ. Chicago Press, Chicago.

Anonymous. [Board on science and technology for international development] 1983a. *Calliandra: a versatile small tree for the humid tropics*. National Academy Press, Washington, D.C.

Anonymous. [Board on science and technology for international development] 1983b. *Firewood crops. Shrub and tree species for energy production*, Vol. 2. National Academy Press, Washington, D.C.

Anonymous. [Board on science and technology for international development] 1984. *Leucaena: promising forage and tree crop for the tropics* (2nd ed.). National Academy Press, Washington, D.C.

Anzenberger, G. 1977. Ethological study of African carpenter bees of the genus *Xylocopa* (Hymenoptera, Anthophoridae). *Z. Tierpsychol.* 44:337–374.

——— 1986. How do carpenter bees recognize the entrance of their nests? (An experimental investigation in a natural habitat). *Ethology* 71:54–62.

Appanah, S. 1981. Pollination in Malaysian primary forests. *Malays. For.* 44:37–42.

——— 1982. Pollination of androdioecious *Xerospermum intermedium* Radlk. (Sapindaceae) in a rain forest. *Biol. J. Linn. Soc.* 18:11–34.

——— 1985. General flowering in the climax rain forests of South-east Asia. *J. Trop. Ecol.* 1:225–240.

Appanah, S., S. Willemstein, and A. Marshall. 1986. Pollen foraging by two *Trigona* colonies in a Malaysian rain forest. *Malay. Nat. J.* 39(3):177–191.

Armbruster, W. S. 1984. The role of resin in angiosperm pollination: Ecological and chemical considerations. *Am. J. Bot.* 71:1149–1160.

——— 1986. Reproductive interactions between sympatric *Dalechampia* species: are natural assemblages random or organized? *Ecology* 67:522–533.

Armbruster, W. S., and A. L. Herzig. 1984. Partitioning and sharing of pollinators by four sympatric species of *Dalechampia* (Euphorbiaceae) in Panama. *Ann. Mo. Bot. Gard.* 71:1–16.

Armbruster, W. S., and W. R. Mziray. 1987. Pollination and herbivore ecology of an African *Dalechampia* (Euphorbiaceae): comparisons with New World species. *Biotropica* 19:64–73.

Armbruster, W. S., and G. L. Webster. 1979. Pollination of two species of *Dalechampia* (Euphorbiaceae) in Mexico by euglossine bees. *Biotropica* 11:278–283.

Arroyo, M. T. K. 1976. Geitonogamy in animal pollinated tropical angiosperms: a stimulus for the evolution of self-incompatibility. *Taxon* 25:543–548.

Arroyo, M. T. K. 1981. Breeding systems and pollination biology in Leguminosae. Pp.723–769 in R. M. Polhill and P. H. Raven, eds. *Advances in legume systematics*, Part 2. Royal Botanic Gardens, Kew, England.

Ashton, P. S. 1969. Speciation among tropical forest trees: some deductions in the light of recent evidence. *Biol. J. Linn. Soc.* 1:155–196.

Audley-Charles, M. G. 1981. Geological history of the region of Wallace's line. Pp. 24–35 in T. C. Whitmore ed. *Wallace's line and plate tectonics*. Clarendon Press, Oxford.

Audley-Charles, M. G., A. M. Hurley, and A. G. Smith. 1981. Continental movements in the Mesozoic and Cenozoic. Pp. 9–24 in T. C. Whitmore, ed. *Wallace's line and plate tectonics*. Clarendon Press, Oxford.

Augspurger, C. K. 1980. Mass-flowering of a tropical shrub (*Hybanthus prunifolius*): influence on pollinator attraction and movement. *Evolution* 34:475–488.

———. 1983a. Phenology, flowering synchrony, and fruit set of six Neotropical shrubs. *Biotropica* 15:257–267.

———. 1983b. Seed dispersal of the tropical tree, *Platypodium elegans*, and the escape of its seedlings from fungal pathogens. *J. Ecol.* 71:759–771.

Avise, J. C., J. Arnold, R. M. Ball, E. Bermingham, T. Lamb, J. E. Nigel, C. A. Reeb, and N. C. Saunders. 1987. Intraspecific phylogeography: the mitochondrial DNA bridge between population genetics and systematics. *Annu. Rev. Ecol. Syst.* 18:489–522.

Bailey, L.1981. *Honey bee pathology*. Academic Press, London.

Baker, E. W., M. Delfinado-Baker, and F. Reyes. 1983. Some laelapid mites (Laelapidae: Mesostigmata) found in nests of wasps and stingless bee. *Int. J. Acarol.* 9:3–10.

Baker, E. W., D. W. Roubik, and M. Delfinado-Baker. 1987. The developmental stages and dimorphic males of *Chaetodactylus panamensis*, n. sp. (Acari: Chaetodactylidae) associated with solitary bee (Apoidea: Anthophoridae). *Int. J. Acarol.* 13:65–73.

Baker, H. G.1976. Mistake pollination as a reproductive system, with special reference to the Caricaceae. Pp. 161–170 in J. Burley and B. T. Stiles, eds. *Tropical trees: variation, breeding and conservation*. Academic Press, London.

———. 1978. Chemical aspects of the pollination of woody plants in the tropics. Pp. 57–82 in P. B. Tomlinson and M. H. Zimmerman, eds. *Tropical trees as living systems*. Cambridge Univ. Press, London.

———. 1983. An outline of the history of anthecology, or pollination biology. Pp. 7–28 in L. Real, ed. *Pollination biology*. Academic Press, New York.

Baker, H. G., and I. Baker. 1979. Starch in angiosperm pollen grains and its evolutionary significance. *Am. J. Bot.* 66:591–600.

———. 1982a. Starchy and starchless pollen in the Onagraceae. *Ann. Mo. Bot. Gard.* 69:748–754.

———. 1982b. Chemical constituents of nectar in relation to pollination mechanisms and phylogeny. Pp. 131–172 in M. H. Nitecki, ed. *Biochemical aspects of evolutionary biology*. Univ. Chicago Press, Chicago.

———. 1983a. A brief historical review of the chemistry of floral nectar. Pp. 126–152 in B. Bentley and T. Elias, eds. *The biology of nectaries*. Columbia Univ. Press, New York.

———. 1983b. Floral nectar sugar constituents in relation to pollinator type. Pp. 117–141 in C. E. Jones and R. J. Little, eds. *Handbook of experimental pollination ecology*. Van Nostrand Reinhold, New York.

Baker, H. G., I. Baker, and P. A. Opler. 1973. Stigmatic exudates and pollination. Pp. 47–80 in N. B. M. Brantjes, ed. *Pollination and dispersal.* Dept. of Botany, Univ. of Nijmegen, the Netherlands.

Baker, H. G., K. S. Bawa, G. W. Frankie, and P. A. Opler. 1983. Reproductive biology of plants in tropical forests. Pp. 183–215 in F. B. Golley, ed. *Tropical rain forest ecosystems.* Elsevier Scientific, Amsterdam.

Baker, H. G., R. W. Cruden, and I. Baker. 1971. Minor parasitism in pollination biology and its community function. The case of *Ceiba acuminata. Bioscience* 21:1127–1129.

Baker, H. G., and P. D. Hurd. 1968. Intrafloral ecology. *Annu. Rev. Entomol.* 13:385–414.

Baker, H. G., P. A. Opler, and I. Baker. 1978. A comparison of the amino acid complements of floral and extrafloral nectars. *Bot. Gaz.* 139:322–332.

Baker, H. G., and G. L. Stebbins, eds. 1965. *The genetics of colonizing species.* Academic Press, New York.

Balduf, W. V. 1962. Life of the carpenter bee *Xylocopa virginica* (Linnaeus) *Ann. Entomol. Soc. Am.* 55:263–271.

Baltazar, C. R. 1966. A catalogue of Philippine Hymenoptera. *Pac. Insects Monogr.* 8:1–488.

Banaszak, J. 1980. Studies on methods of censusing the numbers of bees. *Pol. Ecol. Stud.* 6:355–366.

Banziger, H. 1981. Bloodsucking moths of Malaya. *Fauna* 1:4–16.

Barker, R. J. 1977. Some carbohydrates found in pollen and pollen substitutes are toxic to honey bees. *J. Nutrition* 107:1859–1862.

———. 1978. Poisoning by plants. Pp. 275–296 in R. A. Morse, ed. *Honey bee pests, predators, and diseases.* Cornell Univ. Press, Ithaca, N.Y.

Barker, R. J., and Y. Lehner. 1977. Galactose, a sugar toxic to honey bees, found in exudate. *Apidologie* 7:109–111.

Barrows, E. M. 1974. Aggregation behavior and response to sodium chloride in females of a solitary bee, *Augochlora pura* (Hymenoptera: Halictidae). *Fla. Entomol.* 57:189–193.

———. 1975. Mating behavior in halictine bees. III. Copulatory behavior and olfactory communication. *Insectes Soc..* 22:307–322.

———. 1976. Nectar robbing and pollination of *Lantana camara* (Verbenaceae). *Biotropica* 8:132–135.

———. 1980. Robbing of exotic plants by introduced carpenter and honey bees in Hawaii, with comparative notes. *Biotropica* 12:23–29.

———. 1983. Male territoriality in the carpenter bee *Xylocopa virginica. Anim. Behav.* 31:806–813.

Barrows, E. M, M. R. Chabot, C. D. Michener, and T. P. Snyder. 1976. Foraging and mating behavior of *Perdita texana* (Hymenoptera, Andrenidae). *J. Kans. Entomol. Soc.* 49:275–279.

Barth, F. G. 1985. *Insects and flowers.* Allen and Unwin, London.

Batra, L. R., and S. W. T. Batra. 1985. Floral mimicry induced by mummy-berry fungus exploits host's pollinators as vectors. *Science* 228:1011–1013.

Batra, L. R., S. W. T. Batra, and G. E. Bohart. 1973. The mycoflora of domesticated and wild bees. *Mycopathol. Mycol. Appl.* 49:13–44.

Batra, S. W. T. 1964. Behavior of the social bee *Lasioglossum zephyrum* within the nest. *Insectes Soc.* 11:159–186.

———. 1965. Organisms associated with *Lasioglossum zephyrum. J. Kans. Entomol. Soc.* 38:367–389.

———. 1966. Social behavior and nests of some nomiine bees in India. *Insectes Soc.* 13:145–154.

———. 1968. Behavior of some social and solitary halictine bees within their nests; a comparative study. *J. Kans. Entomol. Soc.* 41:120–133.

———. 1976. Nests of *Ceratina, Pithitis* and *Braunsapis* from India (Hymenoptera: Anthophoridae). *Orient. Insects* [India] 10:1–9.

1977. Bees of India, their behaviour, management and a key to the genera. *Orient. Insects* [India]. 11:289–324.

1980a. Ecology, behavior, pheromones, parasites and management of the sympatric vernal bees, *Colletes inaequalis, C. thoracicus* and *C. validus. J. Kans. Entomol. Soc.* 53:461–469.

1980b. Nests of the solitary bee, *Anthophora antiope*, in Punjab, India. *J. Kans. Entomol. Soc.* 53:112–114.

1984. Solitary bees. *Sci. Am.* 259:120–127.

1987. Ethology of the vernal eusocial bee, *Dialictus laevissimus* (Hymenoptera: Halictidae). *J. Kans. Entomol. Soc.* 60:100–108.

Batra, S. W. T., and A. Hefetz. 1979. Chemistry of the cephalic and Dufour's gland secretions of *Melissodes* bees. *Ann. Entomol. Soc. Am.* 72:514–515.

Batra, S. W. T., and J. C. Schuster. 1977. Nests of *Centris, Melissodes*, and *Colletes* in Guatemala. *Biotropica* 9:135–138.

Baumgartner, D., and D. W. Roubik. In press. Ecology of necrophilous and filth-gathering stingless bees (Apidae: Meliponinae) of Perú. *J. Kans. Entomol. Soc.*

Bawa, K. S. 1980. Evolution of dioecy in flowering plants. *Annu. Rev. Ecol. Syst.* 11:15–39.

1983. Patterns of flowering in tropical plants. Pp. 394–410 in C. E. Jones and R. J. Little, eds. *Handbook of experimental pollination biology*. Van Nostrand Reinhold, New York.

Bawa, K. S., and P. A. Opler. 1975. Dioecism in tropical forest trees. *Evolution* 29:167–179.

Beattie, A. J. 1985. *The evolutionary ecology of ant–plant mutualisms*. Cambridge Univ. Press, Cambridge, England.

Beech, J. H. 1984. The reproductive biology of the peach of "pejibaye" palm (*Bactris gasipaes*) and a wild congener (*B. porschiana*) in the lowlands of Costa Rica. *Principes* 28:107–119.

Beeson, C. F. C. 1938. Carpenter bees. *Indian For.* 64:135–137.

Beetsma, J. 1985. Feeding behavior of nurse bees, larval food composition and caste differentiation in the honey bee (*Apis mellifera*). Pp. 407–410 in B. Hölldobler and M. Lindauer, eds. *Experimental behavioral ecology*. G. Fischer Verlag, New York.

Bego, L. R. 1982. On social regulation in *Scaptotrigona postica* Latreille, with special reference to male production cycles. *Bol. Zool. Univ. São Paulo* 7:181–196.

Beiberdorf, F. W., A. L. Gross, and R. Weichlen. 1961. Free amino acid content of pollen. *Ann. Allergy* 19:867–876.

Beig, D. 1972. The production of males in queenright colonies of *Trigona* (*Scaptotrigona*) *postica. J. Apic. Res.* 11:33–39.

Bell, W. J., and R. Cardé, eds. 1984. *Chemical ecology of insects*. Sinauer Assoc. Inc., Sunderland, Mass.

Ben Mordecai, Y., R. Cohen, D. Gerling, and E. Moscovitz. 1978. The biology of *Xylocopa pubescens* Spinola (Hymenoptera: Anthophoridae) in Israel. *Isr. J. Entomol.* 12:107–121.

Bennett, C. F. 1964. Stingless beekeeping in western Mexico. *Geogr. Rev.* 54:85–92.

Bennett, F. D. 1965. Notes on a nest of *Eulaema terminata* Smith with a suggestion of the occurrence of a primitive social system. *Insectes Soc.*. 12:81–91.

1966. Notes on the biology of *Stelis (Odontostelis) bilineolata* (Spinola), a parasite of *Euglossa cordata* (Linnaeus). *J. N.Y. Entomol. Soc.* 74:72–79.

1972. Observations on *Exaerete* spp. and their hosts *Eulaema terminata* and *Euplusia surinamensis* in Trinidad. *J. N.Y. Entomol. Soc.* 80:118–124.

Bentley, B. L. 1983. Nectaries in agriculture, with an emphasis on the tropics. Pp. 204–222 in B. L. Bentley and T. S. Elias, eds. *Biology of nectaries*. Columbia Univ. Press, New York.

Bergström, G., and J. Tengö. 1973. Geranial and neral as main components in cephalic secretions of four species of *Prosopis* (Hym., Apidae [sic]). *Zool. Suppl.* 1:55–59.

1974. Studies on natural odoriferous compounds. IX. Farnesyl- and geranyl-esters as main volatile constituents of the secretion from Dufour's gland in 6 species of *Andrena*. *Chem. Scr.* 5:28–38.

1978. Linalool in mandibular gland secretion of *Colletes* bees (Hymenoptera, Apoidea). *J. Chem. Ecol.* 4:437–450.

1982. Multicomponent mandibular gland secretions in three species of *Andrena* bees. *Z. Naturforsch.* 37c:1124–1129.

Bertin, R. I. 1985. Nonrandom fruit production in *Campsis radicans:* between-year consistency and effects of prior pollination. *Am. Nat.* 126:750–759.

Bertsch, A. 1984. Foraging in male bumblebees (*Bombus lucorum* L.): maximizing energy or minimizing water load? *Oecologia* 62:325–336.

Bian, Z., H. M. Fales, M. S. Blum, T. H. Jones, T. E. Rinderer, and D. F. Howard. 1984. Chemistry of cephalic secretion of fire bee *Trigona* (*Oxytrigona*) *tataira*. *J. Chem. Ecol.* 10:451–461.

Bienvenu, R. J., F. W. Atchison, and E. A. Cross. 1968. Microbial inhibition by prepupae of the alkali bee, *Nomia melanderi*. *J. Invert. Pathol.* 1:278–282.

Bingham, C. T. 1897. Hymenoptera, Wasps and Bees. Vol. 1 in W. T. Blandford, ed. *The Fauna of British India*. Taylor and Francis, London.

Bino, R. J., and A. D. J. Meeuse. 1981. Entomophily in dioecious species of *Ephedra*: a preliminary report. *Acta Bot. Neerl.* 30:151–153.

Bischoff, F. 1923. Zur kenntnis afrikanischer schmarotzxerbienen *D. Entomol. Z.* 1923:585–603.

Blackmore, S., and I. K. Ferguson, eds. 1986. *Pollen and spores. Form and Function*. Academic Press, Orlando, Fla.

Blomquist, G. J., A. J. Chu, and S. Remaley. 1980. Biosynthesis of wax in the honey bee, *Apis mellifera* L. *Insect Biochem.* 10:313–321.

Blomquist, G. J., D. W. Roubik, and S. L. Buchmann. 1985. Wax chemistry of two stingless bees of the *Trigonisca* group (Apidae: Meliponinae). *Comp. Biochem. Physiol.* 82B:137–142.

Blum, M. S. 1966. Chemical releasers of social behavior. VIII. Citral in the mandibular gland secretion of *Lestrimelitta limao* (Hymenoptera: Apidae: Meliponinae). *Ann. Entomol. Soc. Am.* 59:962–964.

1976. Pheromonal communication in social and semisocial insects. Pp. 49–60 in *Proc. Symp. on Insect Pheromones and Their Application*. Natl. Inst. Agric. Sci., Tokyo.

1981. *Chemical defenses of arthropods*. Academic Press, New York.

1985. *Fundamentals of insect physiology*. Wiley, New York.

Blum, M. S., and H. M. Fales. 1988. Chemical releasers of alarm behavior in the honey bee: informational "plethora" of the sting apparatus signal. Pp. 141–148 in G. Needham, M. Delfinado-Baker, R. Page, and C. Bowman, eds. *Proc. Int. Conf. on Africanized Honey Bees and Bee Mites*. E. Horwood Ltd., Chichester, England.

Blum, M. S., and G. E. Bohart. 1972. Neral and geranial: identification in a colletid bee. *Ann. Entomol. Soc. Am.* 65:274–275.

Blum, M. S., R. M. Crewe, W. E. Kerr, L. H. Keith, A. W. Garrison, and M. M. Walker. 1970. Citral in stingless bees: isolation and functions in trail-laying and robbing. *J. Insect Physiol.* 16:1637–1648.

Boch, R. 1956. Die Tänze der Bienen bei nahen und fernen Trachtquellen. *Z. Verg. Physiol.* 38:136–167.

1957. Rassenmässige Unterschiede in den Tänzen der Honigbiene (*Apis mellifica* L.). *Z. Vergleich. Physiol.* 39:289–320.

Boch, R., and D. A. Shearer. 1966. Iso-pentyl acetate in stings of honeybees of different ages. *J. Apic. Res.* 5:65–70.

Boch, R., D. A. Shearer, and A. Petrasovits. 1970. Efficacies of two alarm substances of the honeybee. *J. Insect Physiol.* 16:17–24.

Boch, R., D. A. Shearer, and R. W. Shuel. 1979. Octanoic acid and other volatile acids in the mandibular glands of the honeybee and in royal jelly. *J. Apic. Res.* 28:250–253.

Bohart, G. E. 1970. The evolution of parasitism among bees. Faculty Honor Lecture, Utah State University. *Logan* 41:1–30.

Bohart, G. E., and E. A. Cross. 1955. Time relations in the nest construction and life cycle of the alkali bee. *Ann. Entomol. Soc. Am.* 48:403–406.

Bohart, G. E., and N. N. Youssef. 1976. Biology and behavior of *Evylaeus galipinsiae* Cockerell (Hymenoptera, Halictidae). *Wasmann J. Biol.* 34:185–234.

Bohart, G. E., P. F. Torchio, Y. Maeta, and R. W. Rust. 1972. Notes on the biology of *Emphoropsis pallida* Timberlake. *J. Kans. Entomol. Soc.* 45:381–392.

Bohart, R. M., and A. S. Menke. 1976. *Sphecid wasps of the world: a generic revision.* Univ. California Press, Berkeley.

Bolten, A. P., and P. Feinsinger. 1978. Why do hummingbird flowers secrete dilute nectar? *Biotropica* 10:307–309.

Bonelli, B. 1974. Osservazioni eto-ecologiche sugli Imenoteerri aculeati dell'Etiopia, VI. *Boll. Inst. Entomol. Univ. Bologna* 32:105–132.

——— 1976. Osservazioni eto-ecologiche sugli Imenotteri aculeati dell'Ethiopia, VII: *Xylocopa* (*Mesotrichia*) *combusta* (Hymenoptera-Anthophoridae). *Boll. Instit. Entomol. Univ. Bologna* 33:1–31.

Boreham, M. M., and D. W. Roubik. 1987. Population change and control of Africanized honey bees in the Panama canal area. *Bull. Entomol. Soc. Am.* 33:34–39.

Borg-Karlson, A. K., and J. Tengö. 1986. Odor mimetism? key substances in *Ophrys lutea* pollination relationship (Orchidaceae: Andrenidae). *J. Chem. Ecol.* 12:1927–1941.

Borgmeier, T. 1971. Further studies on phorid flies, mainly of the neotropical region (Diptera, Phoridae). *Stud. Entomol.* 14:1–172.

Botha, J. J. C. 1970. About enemies of bees in South Africa. *Gleanings in bee culture* [USA] 98:100–103.

Boyden, T. C. 1980. Floral mimicry by *Epidendrum ibaguense* (Orchidaceae) in Panama. *Evolution* 34:135–136.

Boyle-Makowski, R. M. D., and B. J. R. Philogène. 1985. Pollinator activity and abiotic factors in an apple orchard. *Can. Entomol.* 117:1509–1521.

Brand, D. D. In press. The honey bee in New Spain–Mexico. *Univ. Ariz. State Mus. J.*

Breed, M. D. 1975. Life cycle and behavior of a primitively social bee, *Lasioglossum rohweri* (Hymenoptera: Halictidae). *J. Kans. Entomol. Soc.* 48:64–80.

——— 1983. Nestmate recognition in honey bees. *Anim. Behav.* 31:86–91.

Breed, M. D., and A. J. Moore. 1988. The guard bee as a component of the defensive response. Pp. 105–109 in G. Needham, M. Delfinado-Baker, R. Page, and C. Bowman. eds. *Proc. Int. Conf. on Africanized Honey Bees and Bee Mites.* E. Horwood Ltd., Chichester, England.

Breed, M. D., and G. J. Gamboa. 1977. Control of worker activities by queen behavior in a primitively eusocial bee. *Science* 195:694–696.

Brian, A. D. 1952. Division of labor and foraging in *Bombus agrorum* Fabricius. *J. Anim. Ecol.* 21:233–240.

——— 1957. Differences in the flowers visited by four species of bumble bees and their causes. *J. Anim. Ecol.* 26:71–98.

Bromenshenk, J. J., S. R. Carlson, J. C. Simpson, and J. M. Thomas. 1985. Pollution monitoring of Puget Sound with honey bees. *Science* 227:632–634.

Brooks, D. R. 1985. Historical ecology: a new approach to studying the evolution of ecological associations. *Ann. Mo. Bot. Gard.* 72:660–680.

Brooks, R. W. 1983. Systematics and bionomics of *Anthophora*: the Bomboides group and species groups of the new world. *Univ. Calif. Publ. Entomol.* 98:1–84.

Brooks, R. W., and J. H. Cane. 1984. Origin and chemistry of the secreted nest entrance lining of *Halictus hesperus*. *J. Kans. Entomol. Soc.* 57:161–165.

Brooks, R. W., and D. W. Roubik. 1983. A halictine bee with distinct castes: *Halictus hesperus* (Hymenoptera: Halictidae) and its bionomics in central Panamá. *Sociobiology* 7:263–282.

Brothers, D. J. 1972. Biology and immature stages of *Pseudomethoca f. frigida*, with notes on other species (Hymenoptera: Mutillidae). *Univ. Kans. Sci. Bull.* 50:1–381.

——— 1975. Phylogeny and classification of the aculeate Hymenoptera, with special reference to Mutillidae. *Univ. Kans. Sci. Bull.* 50:483–648.

Brower, L. P., J. V. Z. Brower, and P. W. Westcott. 1960. Experimental studies of mimicry. 5. The reactions of toads (*Bufo terrestris*) to bumblebees (*Bombus americanorum*) and their robberfly mimics (*Mallophora bomboides*), with a discussion of aggressive mimicry. *Am. Nat.* 94:343–345.

Brower, L. P., F. H. Pough, and H. R. Meck. 1970. Theoretical investigations of automimicry, I. Single trial learning. *Proc. Natl. Acad. Sci. USA* 66:1059–1066.

Brown, C. A., J. F. Watkins, and D. W. Eldredge. 1979. Repression of bacteria and fungi by the army ant secretion: skatole. *J. Kans. Entomol. Soc.* 52:119–122.

Buchmann, S. L. 1983. Buzz pollination in angiosperms. Pp. 73–114 in C. E. Jones and R. J. Little, eds. *Handbook of experimental pollination biology*. Van Nostrand Reinhold, New York.

——— 1985. Bees use vibration to aid pollen collection from non-poricidal flowers. *J. Kans. Entomol. Soc.* 58:517–525.

——— 1986. Vibratile pollination in *Solanum* and *Lycopersicon*: A look at pollen chemistry. Pp. 218–252 in W. G. D'Arcy, ed. *Solanaceae: biology and systematics*. Columbia Univ. Press, New York.

——— 1987. The ecology of oil flowers and their bees. *Annu. Rev. Ecol. Syst.* 18:343–369.

Buchmann, S. L., and M. D. Buchmann. 1981. Anthecology of *Mouriri myrtilloidea* (Melastomataceae: Memecyleae), an oil flower in Panama. *Biotropica (Reproductive Botany Suppl.)* 13:7–24.

Buchmann, S. L., and C. E. Jones. 1980. Observations on the nesting biology of *Melissodes persimilis* Ckll. (Hymenoptera: Anthophoridae) *Pan-Pac. Entomol.* 56:200–206.

Buckle, G. R. 1984. A second look at queen–forager interactions in the primitively eusocial halictid, *Lasioglossum zephyrum*. *J. Kans. Entomol. Soc.* 57:1–6.

Buckle, G. R., and L. Greenberg. 1981. Nestmate recognition in sweat bees (*Lasioglossum zephyrum*): does an individual recognize its own odor or only odors of its nestmates? *Anim. Behav.* 29:802–809.

Bullock, S. H. 1981. Notes on the phenology of inflorescences and pollination of some rain forest palms in Costa Rica. *Principes* 25:101–105.

——— 1985. Breeding systems in the flora of a tropical deciduous forest in Mexico. *Biotropica* 17:287–301.

Bullock, S. H., and K. S. Bawa. 1981. Sexual dimorphism and the annual flowering pattern in *Jacaratia dolichaula* (D. Smith) Woodson (Cariaceae) in a Costa Rican rain forest. *Ecology* 62:1494–1504.

Bullock, S. H., J. H. Beach, and K. S. Bawa. 1983. Episodic flowering and sexual dimorphism in *Guarea rhopalcarpa* in a Costa Rican rain forest. *Ecology* 64:851–861.

Burgett, D. M. 1974. Glucose oxidase: a food protective mechanism in social Hymenoptera. *Ann. Entomol. Soc. Am.* 67:545–546.

Butler, C. G. 1974. *The world of the honeybee*. Collins, London.
Byrne, D. N., S. L. Buchmann, and H. G. Spangler. 1988. Relationship between wing loading, wingbeat frequency and body mass in homopterous insects. *J. Exp. Biol.* 135:9–24.
Callan, E. M. 1977. Observations on *Centris ruffosuffosa* Cockerell (Hymenoptera; Anthophoridae), and its parasites. *J. Nat. Hist.* 11:127–135.
Camargo, C. A. 1979. Sex determination in bees. XI. Production of diploid drones and sex determination in *Melipona quadrifasciata. J. Apic. Res.* 18:77–84.
——— 1982. Longevity of diploid males, haploid males and workers of the social bee *Melipona quadrifasciata* Lep. *J. Kans. Entomol. Soc.* 55:8–12.
Camargo, C. A., M. G. de Almeida, G. Nates, and W. E. Kerr. 1976. Genetics of sex determination in bees. IX. Frequencies of queens and workers from larvae under controlled conditions. *J. Kans.. Entomol. Soc.* 49:110–120.
Camargo, J. M. F. 1970. Ninhos e biologia de algumas espécies de Meliponideos da região de Pôrto Velho, Território de Rondônia, Brasil. *Rev. Biol. Trop.* 16:207–239.
——— 1974. Notas sobre a morfologia e biologia de *Plebeia (Schwarziana) quadripunctata quadripunctata* (Hym., Apidae). *Stud. Entomol.* 17:433–470.
——— 1980. O grupo *Partamona (Partamona) testacea* (Klug): suas espécies, distribuição e diferenciação geográfica (Meliponinae, Apidae, Hymenoptera). *Acta Amazonica* Suppl. 10(4).
——— 1984a. Notas sobre o gênero *Oxytrigona* (Meliponinae, Apidae, Hymenoptera). *Bol. Mus. Para. Emilio Goeldi Nova Ser. Zool.* 1:115–124.
——— 1984b. Notas sobre hábitos de nidificação de *Scaura (Scaura) latitatrsis* (Friese) (Hymenoptera, Apidae, Meliponinae). *Bol. Mus. Para. Emilio Goeldi Nova Ser. Zool.* 1:89–95.
———, ed. 1972. *Manual de apicultura*. Editorial Ceres. São Paulo, Brazil.
Camargo, J. M. F., G. Gottsberger, and I. Silberbauer-Gottsberger. 1984. On the phenology and flower visiting behavior of *Oxaea flavescens* (Klug) (Oxaeinae, Andrenidae, Hymenoptera) in São Paulo, Brazil. *Beitr. Biol. Pflanz.* 59:159–179.
Camargo, J. M. F., and J. S. Moure. 1983. *Trichotrigona*, um novo gênero de Meliponinae (Hymenoptera, Apidae) do Rio Negro, Amazonas, Brasil. *Acta Amazonica* 13:421–429.
Camargo, J. M. F., and J. S. Moure. In press. Duas novas espécies de *Lestrimelitta* Friese (Meliponinae, Apidae, Hymenoptera) da região Amazônica. *Bol. Mus. Para. Emilio Goeldi Nova Ser. Zool.*
Camargo, J. M. F., J. S. Moure, and D. W. Roubik. 1988. *Melipona yucatanica* n. sp. (Apidae, Meliponinae), stingless bee dispersal across the Caribbean arc and post Eocene vicariance. *Pan-Pac. Entomol.* 64:147–157.
Camargo, J. M. F., and M. Muzucato. 1986. Inventário da apifauna e flora apícola de Ribeirão Preto, São Paulo, Brasil. *Dusenia* 14:55–87.
Camargo, J. M. F., and H. H. W. Velthuis. 1979. Sobre o comportamento de *Xylocopa (Megaxylocopa) frontalis* (Olivier) (Hymenoptera-Anthophoridae). *Dusenia* 11:35–39.
Camargo, J. M. F., R. Zucchi, and S. F. Sakagami. 1975. Observations on the bionomics of *Epicharis (Epicharana) rustica flava* (Olivier) including notes on its parasite, *Rathymus* sp. (Hymenoptera, Apoidea: Anthophoridae). *Stud. Entomol.* 18:313–340.
Cameron, S. A. 1985. Brood care by male bumble bees. *Proc. Natl. Acad. Sci., USA.* 82:6371–6373.
Camillo, E., and C. A. Garófalo. 1982. On the bionomics of *Xylocopa frontalis* (Olivier) and *X. grisescens* (Lepeletier) in southern Brazil. I. Nest construction and biological cycle. *Rev. Bras. Biol.* 42:571–582.
Camillo, E., C. A. Garófalo, L. A. O. Campos, and J. C. Serrano. 1983. Preliminary notes on the biology of *Lithurgus huberi* (Hymenoptera, Megachilidae). *Rev. Bras. Biol.* 43:151–156.

Campos, L. A. O., F. M.Velthuis-Kluppell, and H. H. W. Velthuis. 1975. Sex determination in bees. VII. Juvenile hormone and caste determination in a stingless bee. *Naturwissenschaften* 62:98–99.

Campos, L. A. O., W. E. Kerr, and D. L. N. da Silva. 1979. Sex determination in bees. VIII. Relative action of genes X/a and X/b on sex determination in *Melipona* bees. *Rev. Bras. Genet.* 2:267–280.

Cane, J. H. 1981. Dufour's gland secretion in the cell linings of bees. *J. Chem. Ecol.* 7:403–410.

——— 1983a. Preliminary chemosystematics of the Andrenidae and exocrine lipid evolution of the short-tongued bees. *Syst. Zool.* 32:417–430.

——— 1983b. Chemical evolution and chemosystematics of the Dufour's gland secretions of the lactone-producing bees (Hymenoptera: Colletidae, Halictidae, and Oxaeidae). *Evolution* 37:657–674.

——— 1983c. Olfactory evaluation of *Andrena* host nest suitability by kleptoparasitic *Nomada* bees. *Anim. Behav.* 31:138–144.

——— 1986. Predator deterrence by mandibular gland secretions of bees. *J. Chem. Ecol.* 12:1295-1309.

Cane, J. H., and R. W. Brooks. 1983. Dufour's gland lipid chemistry of three species of *Centris* bees (Hymenoptera: Apoidea, Anthophoridae). *Comp. Biochem. Physiol.* 76:895–897.

Cane, J. H., and R. G. Carlson. 1984. Dufour's gland triglycerides from *Anthophora, Emphoropsis* (Anthophoridae), and *Megachile* (Megachilidae) bees (Hymenoptera: Apoidea). *Comp. Biochem. Physiol.* 78:769–772.

Cane, J. H., and G. C. Eickwort. In press. Oligolecty and solitary bees: ecology and evolution of coadaptive mutualism. In S. L. Buchmann, ed. *Experimental studies in pollination and foraging efficiency.* Univ. Ariz. Press, Tucson.

Cane, J. H., G. C. Eickwort, F. R. Wesley, and J. Spielholz. 1983. Foraging, grooming and mate-seeking behaviors of *Macropis nuda* (Hymenoptera, Melittidae) and use of *Lysimachia ciliata* (Primulaceae) oils in larval provisions and cell linings. *Am. Midl. Nat.* 110:257–264.

——— 1985. Pollination ecology of *Vaccinium stamineum* (Ericaceae: Vaccinioideae). *Am. J. Bot.* 72:135–142.

Cane, J. H., S. Gerdin, and G. Wife. 1983. Mandibular gland secretions of solitary bees: potential for nest cell disinfection. *J. Kans. Entomol. Soc.* 56:199–204.

Cane, J. H., and C. D. Michener. 1983. Chemistry and function of mandibular gland products of bees of the genus *Exoneura* (Hymenoptera, Anthophoridae). *J. Chem. Ecol.* 9:1525–1531.

Cane, J. H., and J. O. Tengö. 1981. Pheromonal cues direct mate-seeking behavior of male *Colletes cunicularis* (Hymenoptera: Colletidae). *J. Chem. Ecol.* 7:701–708.

Cardé, R. T. 1984. Chemo-orientation in flying insects. Pp. 11–124. in W. J. Bell and R. T. Cardé, eds. *Chemical ecology of insects.* Sinauer Assoc. Inc., Sunderland, Mass.

Carlson, D. A. 1988. Africanized and European honey bees, drones and comb waxes. Pp. 264–274 in G. Needham, M. Delfinado-Baker, R. Page, and C. Bowman. eds. *Proc. Int. Conf. on Africanized Honey Bees and Bee Mites.* E. Horwood Ltd., Chichester, England.

Carlson, D. A., D. W. Roubik, and S. K. Milstrey. In press. Distinctive hydrocarbons among Asian giant honey bees, the *Apis dorsata* group. *J. Kans. Entomol. Soc.*

Caron, D. M. 1978a. Marsupials and mammals. Pp. 227–256 in R. A. Morse, ed. *Honey bee pests, predators, and diseases.* Cornell Univ. Press, Ithaca, N.Y.

——— 1978b. Other insects. Pp. 158–185 in R. A. Morse, ed. *Honey bee pests, predators, and diseases.* Cornell Univ. Press, Ithaca, N.Y.

Case, T. J., and M. L. Cody. 1987. Testing theories of island biogeography. *Am. Sci.* 75:402–411.

Casey, T. M., M. L. May, and K. R. Morgan. 1985. Flight energetics of euglossine bees in relation to morphology and wing stroke frequency. *J. Exp. Biol.* 116:271–289.
Castro, P. R. C. 1975. Mutualismo entre *Trigona spinipes* (Fabricius, 1793) e *Aethalion reticulatum* (L., 1767) em *Cajanus indicus* Spreng. na presença de *Camponotus* spp. *Ciencia e Cultura* 27:537–539.
Caufield, K. 1985. *In the rainforest.* Knopf, New York.
Caughley, G. 1977. *Analysis of vertebrate populations.* John Wiley and Sons, New York.
Cazier, M. A., and E. G. Linsley. 1963. Territorial behavior among males of *Protoxaea gloriosa* (Fox) (Hymenoptera: Andrenidae). *Can. Entomol.* 95:547–556.
Cederberg, B., Bo. G. Svensson, G. Bergström, M. Applegren, and I. Groth. 1983. Male marking pheromones in north European cuckoo bumble bees, *Psithyrus* (Hymenoptera, Apidae). *Nova Acta Regiae Soc. Sci. Ursaliensis* 3:161–166.
Chance, M. M. 1983. Honeybees observed feeding on the blood of a bear. *Bee World* 64:177.
Chandler, M. T. 1975. Apiculture in Madagascar. *Bee World* 56:149–153.
Cheverton, J., A. Kacelnic, and J. R. Krebs. 1985. Optimal foraging: constraints and currencies. Pp. 109–126 in B. Hölldobler and M. Lindauer, eds. *Experimental behavioral ecology.* G. Fischer Verlag, New York.
Chino, M., T. Fukumorita, S. Kawabe, and Y. Ando. 1982. Chemical composition of rice phloem sap collected by insect technique. Pp. 105–110. *Plant Nutr., Proc. 9th Int. Plant Nutr. Colloq.* Univ. Tokyo.
Clemson, A. 1985. *Honey and pollen flora.* Department of Agriculture, New South Wales. Inkata Press, Melbourne, Australia.
Clutton-Brock, T. H., and P. H. Harvey. 1984. Comparative approaches to investigating adaptation. Pp. 7–29 in J. R. Krebs and N. B. Davies, eds. *Behavioural ecology: an evolutionary approach.* Sinauer Assoc, Inc., Sunderland, Mass.
Cockerell, T. D. A. 1926. The black bees of Peru. *Entomologist* 59:28–29.
Cole, L. K., and M. S. Blum. 1975. Antifungal properties of the insect alarm pheromones, citral, 2-heptanone and 4-methyl-3-heptanone. *Mycologia* 67:701–708.
Colin, L. J., and C. E. Jones. 1980. Pollen energetics and pollination modes. *Am. J. Bot.* 67:210–215.
Collins, A. M. 1981. Effects of temperature and humidity on honeybee response to alarm pheromones. *J. Apic. Res.* 20:13–18.
—— 1987. Overview of honey bee colony defenses. Pp.10–17 in *Proc. of the Africanized Honeybee Symp.* American Farm Bureau Research Foundation, Park Ridge, Ill.
Collins, A. M., and M. S. Blum. 1982. Bioassay of compounds derived from the honeybee sting. *J. Chem. Ecol.* 8:463–470.
—— 1983. Alarm responses caused by newly identified compounds derived from the honeybee sting. *J. Chem. Ecol.* 9:57–65.
Collins, A. M., T. E. Rinderer, J. R. Harbo, and A. B. Bolten. 1982. Colony defense by Africanized and European honey bees. *Science* 218:72–74.
Connor, E. F., and D. Simberloff. 1986. Competition, scientific method, and null models in ecology. *Am. Sci.* 74:155–162.
Cooper, P. D., W. M. Schaffer, and S. L. Buchmann. 1985. Temperature regulation of highly social bees (*Apis mellifera*) foraging in the Sonoran desert. *J. Exp. Biol.* 114:1–15.
Corbet, S. A., C. J. C Kerslake, D. Brown, and N. E. Morland. 1984. Can bees select nectar-rich flowers in a patch? *J. Apic. Res.* 23:234–247.
Corbet, S. A., D. M. Unwin and O. E. Prys-Jones. 1979. Humidity, nectar and insect visits to flowers, with special reference to *Crategus, Tilia* and *Echinum. Ecol. Entomol.* 4:9–22.
Corbet, S. A., and P. G. Willmer. 1981. The nectar of *Justicia* and *Columnea*: composition and concentration in a humid tropical climate. *Oecologia* 51:412–418.

Cornaby, B. W. 1974. Carrion reduction by animals in contrasting tropical habitats. *Biotropica* 6:51–63.

Cortopassi-Laurino, M. 1982. Divisão de recursos tróficos entre abelhas sociais, principalmente em *Apis mellifera* Linné e *Trigona (Trigona) spinipes* Fabricius (Apidae, Hymenoptera). Dissertation, Instituto de Biociências. Universidade de São Paulo.

Costa Leonardo, A. M. 1980. Estudos morfológicos do ciclo secretor das glândulas mandibulares de *Apis mellifera* L. (Hymenoptera, Apidae). *Rev. Bras. Entomol.* 24:143–151.

Coville, R. E., G. W. Frankie, and S. B. Vinson. 1983. Nests of *Centris segregata* (Hymenoptera: Anthophoridae) with a review of the nesting habits of the genus. *J. Kans. Entomol. Soc.* 56:109–122.

Coville, R. E., G. W. Frankie, S. L. Buchmann, S. B. Vinson, and H. J. Williams. 1986. Nesting and male behavior of *Centris heithausi* in Costa Rica with chemical analysis of the hindleg glands of males. *J. Kans. Entomol. Soc.* 59:325–336.

Craig, R. 1982. Evolution of eusociality by kin selection: the effect of inbreeding between siblings. *J. Theor. Biol.* 94:119–128.

Crane, E. 1980. The scope of tropical apiculture. *Bee World* 61:19–28.

——— ed. 1975. *Honey: a comprehensive survey.* Heinemann, London.

Crane, E., and P. Walker. 1984. Important honeydew sources and their honeys. *Bee World* 65:105–112.

——— 1985. *The impact of pest management on bees and pollination.* Int. Bee Res. Assn., London.

Crane, E., P. Walker, and R. Day. 1984. *Directory of important world honey sources.* Int. Bee Res. Assn., London.

Crepet, W. L. 1983. The role of insect pollination in the evolution of the angiosperms. Pp. 31–50 in L. Real, ed. *Pollination biology.* Academic Press, New York.

Crepet, W. L., and D. W. Taylor. 1985. The diversification of the Leguminosae: first fossil evidence of the Mimosoideae and Papilionoideae. *Science* 228:1087–1089.

Crewe, R. M. 1985. Bees observed foraging on an impala carcass. *Bee World* 66:8.

Crewe, R. M., and D. J. C. Fletcher. 1976. Volatile secretions of two Old World stingless bees. *S. Afr. J. Sci.* 72:119–120.

Crewe, R. M., and H. H. W. Velthuis. 1980. False queens: a consequence of mandibular gland signals in worker honey bees. *Naturwissenschaften* 67:467–469.

Croat, T. B. 1978. *Flora of Barro Colorado Island.* Stanford Univ. Press, Palo Alto, Calif.

——— 1980. Flowering behavior of the neotropical genus *Anthurium* (Araceae). *Am. J. Bot.* 67:888–904.

Cross, E. A., and G. E. Bohart. 1969. Phoretic behavior of four species of alkali bee mites as influenced by season and host sex. *J. Kans. Entomol. Soc.* 42:195–219.

Crowson, R. A. 1981. *The biology of the Coleoptera.* Academic Press, London.

Crozier, R. H. 1977. Evolutionary genetics of the Hymenoptera. *Annu. Rev. Entomol.* 22:263–288.

Crozier, R. H., B. H. Smith, and Y. C. Crozier. 1987. Relatedness and population structure of the primitively eusocial bee *Lasioglossum zephyrum* (Hymenoptera: Halictidae) in Kansas. *Evolution* 41:902–910.

Cruden, R. W., and S. M. Hermann. 1983. Studying nectar? Some observations on the art. Pp. 223–242 in B. Bentley and T. S. Elias, eds. *The biology of nectaries.* Columbia Univ. Press, New York.

Cruden, R. W., and D. L. Lyon. 1985. Patterns of biomass allocation to male and female functions in plants with different mating systems. *Oecologia* 66:299–306.

Cruz-Landim, C. 1963. Evolution of the wax and scent glands of the Apinae. *J. N.Y. Entomol. Soc.* 71:2–13.

1967. Estudo comparativo de algumas glândulas das abelhas e respectivas implicações evolutivas. *Arq. Zool. São Paulo* 15:177–290.

Cruz-Ladim, C., A. C. Stort, M. A. Costa Cruz, and E. W. Kitajima. 1965. Orgão tibial dos machos de Euglossini. Estudo ao microscopio optico e electronico. *Rev. Bras. Biol.* 25:323–42.

Cruz-Landim, C., and A. Ferreira. 1968. Mandibular gland development and communication in field bees of *Trigona (Scaptotrigona) postica* (Hymenoptera, Apidae). *J. Kans. Entomol. Soc.* 41:474–481.

Cruz-Landim, C., A. Höfling, and V. L. Imperatriz-Fonseca. 1980. Tergal and mandibular glands in queens of *Paratrigona subnuda* (Moure) (Hymenoptera: Apidae). Morphology and associated behavior. *Naturalia* 58:121–133.

Cruz-Landim, C., and L. Rodrigues. 1967. Comparative anatomy and histology of the alimentary canal of adult Apinae. *J. Apic. Res.* 6:17–28.

Cruz-Landim, C., S. M. F. dos Santos, and M. C. Aparecida Höfling. 1980. Sex determination in bees. XV. Identification of queens of *Melipona quadrifasciata anthdioides* with the worker phenotype by a study of the tergal glands. *Rev. Bras.Genet.* 3:295–302.

Culliney, T. W. 1983. Origin and evolutionary history of the honeybees *Apis*. *Bee World* 64:29–38.

Dafni, A. 1984. Mimicry and deception in pollination. *Annu. Rev. Ecol. Syst.* 15:259–278.

Daly, H. V. 1966. Biological studies on *Ceratina dallatorreana*, an alien bee in California which reproduces by parthenogenesis. *Ann. Entomol. Soc. Am.* 59:1138–1154.

——— 1983. Taxonomy and ecoiogy of Ceratinini of North Africa and the Iberian Peninsula (Hymenoptera; Apidae). *Syst. Entomol.* 8:29–62.

——— 1988. Bees of the new genus *Ctenoceratina* in Africa south of the Sahara (Hymenoptera: Apoidea). *Univ. Calif. Publ. Entomol.* 108.

Daly, H. V., and S. S. Balling. 1978. Identification of Africanized honeybees in the Western Hemisphere by discriminant analysis. *J. Kans. Entomol. Soc.* 51:1057–1069.

Daly, H. V., and R. Coville. 1982. *Hylaeus pubescens* and associated arthropods at Kilauea, Hawaii Volcanoes National Park (Hymenoptera: Apoidea and Chalcidoidea; Mesostigmata: Ameroseiidae) *Proc. Hawaii. Entomol. Soc.* 24:75–81.

Daly, H. V., G. I. Stage, and T. B. Brown. 1967. Natural enemies of bees of the genus *Ceratina*. *Ann. Entomol. Soc. Am.* 60:1273–1282.

Danka, R. G., R. L. Hellmich II, T. E. Rinderer, and A. M. Collins. 1987. Diet-selection ecology of tropically and temperately adapted honey bees. *Anim. Behav.* 35:1858–1863.

Danks, H. V. 1971. Nest mortality factors in stem-nesting aculeate Hymenoptera. *J. Anim. Ecol.* 40:79–82.

——— 1983. Differences between generation in the sex ratio of aculeate Hymenoptera. *Evolution* 37:414–416.

——— 1987. *Insect dormancy: an ecological perspective*. Biol. Survey Canada Monogr. Series 1. Biol. Survey Canada (Terrestrial Arthropods), Ottawa.

Darchen, R. 1962. Observation directe du dèveloppement d'un rayon de cire. Le rôle des chaines d'abeilles. *Insectes Soc.* 9:103–120.

——— 1966. Sur l'éthologie de *Trigona (Dactylurina) staudingeri* Gribodoi [sic]. *Biol. Gabonica* 2:37–45.

——— 1972a. Ethologie comparative de l'économie des matériaux de construction chez divers apides sociaux, *Apis, Trigona* et *Melipona*. *Rev. Comp. Anim.* 6:201–215.

——— 1972b. Ecologie de quelques trigones (*Trigona* sp.) de la savane de Lamto (Côte D'Ivoire). *Apidologie* 3:341–367.

——— 1977a. L'essaimage chez les hypotrigones au Gabon: dynamique de quelques populations. *Apidologie* 8:33–59.

1977b. L'acclimatation des trigones Africaines (Apidae, Trigonini) en France. *Apidologie* 8:147–154.
Darchen, R., and B. Delage-Darchen. 1971. Le déterminisme des castes chez les Trigones (Hyménoptères, Apidés). *Insectes Soc.* 10:337–357.
——— 1977. Sur le déterminisme des castes chez les mélipones (Hyménoptères Apidés). *Bull. Biol. France* 61:91–109.
Darwin, C. 1876. *Cross and self-fertilisation in the vegetable kingdom.* J. Murray, London.
Daumer, K. 1958. Blumenfarben, wie sie die Bienen sehen. *Zeitschr. Vgl. Physiol.* 41:49–110.
DeGrandi-Hoffman, G., R. A. Hoopingarner, and K. K. Baker. 1986. Influence of honey bee in-hive pollen transfer on cross-pollination and fruit set in apple. *Environ. Entomol.* 15:723–725.
DeJong, D. 1978. Insects: Hymenoptera (ants, wasps, and bees). Pp. 138–157 in R. A. Morse, ed. *Honey bee pests, predators, and diseases.* Cornell Univ. Press, Ithaca, N.Y.
——— 1984. Africanized bees now preferred by Brazilian beekeepers. *Am. Bee J.* 124:116–118.
DeJong, D., R. A. Morse, and G. C. Eickwort. 1982. Mite pests of honeybees. *Annu. Rev. Entomol.* 27:229–252.
Delage-Darchen, B., and R. Darchen. 1982. Les enzymes digestives des glandes salivaires et de l'intestin moyen d'une abeille sociale du Mexique, *Melipona beecheii* (B.). *Ann. Sci. Nat. Zool., Paris.* 4:91–96.
Delage-Darchen, B., S. Talec, and R. Darchen. 1979. Sérétion enzymatique des glandes salivaires et de l'intestin moyen d'une abeille sans dard *Apotrigona nebulata* (Smith) (Hyménoptères, Apidés). *Ann. Sci. Nat. Zool., Paris.* 1:261–267.
Delfinado-Baker, M., E. W. Baker, and D. W. Roubik. 1983. A new genus and species of Hypoaspidinae (Acari: Laelaplidae) from nests of stingless bees. *Internat. J. Acarol.* 9:195–203.
Delfinado-Baker, M., and W. E. Styer. 1983. Mites of honeybees as seen by scanning electron microscope (SEM). *Am. Bee J.* 123:813–819.
Delgado, M., and S. del Amo. 1984. Dinámica de poblaciones de *Apis mellifera* L. en una zona tropical húmeda. *Biotica* 9:351–365.
Denlinger, D. L. 1986. Dormancy in tropical insects. *Annu. Rev. Entomol.* 31:239–264.
Deshmukh, I. 1986. *Ecology and tropical biology.* Blackwell Scientific Publications, Oxford.
Deuth, D. 1980. The function of extra-floral nectaries in *Aphelandra deppeana* Schl. & Cham. (Acanthaceae). *Brenesia* 10/11:135–145.
Dhaliwal, H. S., and P. L. Sharma. 1974. Foraging range of the Indian honeybee. *J. Apic. Res.* 13: 137–141.
Diamond, J. M. 1978. Niche shifts and the rediscovery of interspecific competition. *Am. Sci.* 66:322–331.
Diamond, J. M., and T. J. Case, eds. 1986. *Community ecology.* Harper and Row, New York.
Dias, D. 1958. Contribuição para o conhecimento da bionomia de *Bombus incarum* Franklin da Amazônia. *Rev. Bras. Entomol.* 8:1–20.
Dicklow, M. B., R. D. Firman, D. B. Rupert, K. L. Smith, and T. E. Ferrari. 1986. Controlled endopollination of honeybees (*Apis mellifera*): bee-to-bee and bee-to-tree pollen transfer. Pp. 449–454 in D. L. Mulcahy, G. B. Mulcahy, and E. Ottaviano, eds. *Biotechnology and ecology of pollen.* Springer–Verlag, New York.
Dietz, A. 1982. Honey bees. Pp. 323–360 in H. R. Hermann, ed. *Social insects*, Vol. 3. Academic Press, New York.
Dietz, A., R. Krell, and F. A. Eischen. 1985. Preliminary investigation on the distribution of Africanized honey bees in Argentina. *Apidologie* 16:99–108.

Dietz, A., R. Krell, and J. Pettis. 1987. The potential limit of survival for Africanized bees in the United States. Pp. 87–100 in *Proc. of the Africanized Honey Bee Symp.* Am. Farm Bureau, Park Ridge, Ill.

Dobat, K., and T. Peikert-Holle, eds. 1985. *Blüten und Fledermäuse.* Kramer, Frankfurt.

Dobson, H. E. M. 1984. Pollen lipids in bee-visited flowers. Pp. 61–64 in J. N. Tasei, ed., Ve *Symposium internationale sur la pollinisation,* Versailles. Les colloques INRA No. 21, Paris.

Dodson, C. H. 1962. The importance of pollination in the evolution of the orchids of tropical America. *Am. Orchid Soc. Bull.* 31:525–534, 641–649, 731–735.

——— 1966. Ethology of some bees of the tribe Euglossini. *J. Kans. Entomol. Soc.* 39:607–629.

——— 1967. Relationships between pollinators and orchid flowers. *Atlas Simpós. Biota Amazônica (Zool.)* 5:1–72.

——— 1975. Coevolution of orchids and bees. Pp. 91–99 in L. Gilbert and P. H. Raven, eds. *Coevolution of animals and plants.* Univ. Texas Press, Austin.

Dodson, C. H., R. L. Dressler, H. G. Hills, R. M. Adams, and N. H. Williams. 1969. Biologically active compounds in orchid fragrances. *Science* 164:1243–1249.

Dodson, C. H., and G. P. Frymire. 1961. Natural pollination of orchids. *Mo. Bot. Gard. Bull.* 49:133–152.

Dotimas, E. M., and R. C. Hider. 1987. Honeybee venom. *Bee World* 68:51–70.

Dover, C. 1924. Some observations on the bionomics of *Xylocopa aestuans* Linn., with a note on beetle larvae by K. G. Blair. *Trans Entomol. Soc. London.* 144–150.

Dressler, R. L. 1979. *Eulaema bombiformis, E. meriana* and Müllerian mimicry in related species (Hymenoptera: Apidae). *Biotropica* 11:144–151.

——— 1981. *The orchids: natural history and classification.* Harvard Univ. Press, Cambridge, Mass.

——— 1982. Biology of the orchid bees (Euglossini). *Annu. Rev. Ecol. Syst.* 13:373–394.

Ducke, A. 1901. Beobachtungen über blutenbesuch, erscheiungszeit, etc. der bei Pará vorkommenden bienen. *Vgl. Z.. Syst. Hymenoptera, Diptera.* 1(8):49–67.

——— 1902. Beobachtungen über blutenbesuch, erscheiungszeit, etc. det bei Pará vorkommenden. *Allg. Z. Entomol.* 7:321–325, 360–368, 400–404, 417–421.

——— 1906. Neue beobachtungen über die bienen der Amazonslander. *Allg. Z. Entomol.* 2:51–60.

——— 1925. Die stachellosen bienen Brasiliens. *Zool. Jahrb. Abt. Syst. Oekol. Geogr. Tiere* 49:355–448.

Duffield, R. M., A. Fernandes, C. Lamb, J. W. Wheeler, and G. C. Eickwort. 1981. Macrocyclic lactones and isopentenyl esters in the Dufour's gland secretion of halictine bees. *J. Chem. Ecol.* 7:319–331.

Duffield, R. M., A. Fernandes, S. McKay, J. W. Wheeler, and R. R. Snelling. 1980. Chemistry of the exocrine secretions of *Hylaeus modestus* (Hymenoptera: Colletidae). *Comp. Biochem. Physiol.* 67B:159–162.

Duffield, R. M., W. E. LaBerge, J. H. Cane, and J. W. Wheeler. 1982. Exocrine secretions of bees. IV. Macrocyclic lactones and isopentenyl esters in Dufour's gland secretions of *Nomia* bees (Hymenoptera: Halictidae). *J. Chem. Ecol.* 8:535–543.

Duffield, R. M., J. W. Wheeler, and G. C. Eickwort. 1984. Sociochemicals of bees. Pp. 387–428 in W. J. Bell and R. T. Cardé, eds. *Chemical ecology of insects.* Chapman and Hall, Ltd., London.

DuPraw, E. J. 1967. The honeybee embryo. Pp. 183–217 in F. Wilt and N. Wessels, eds. *Methods of developmental biology.* Crowell Co., New York.

Durkee, L. T. 1983. The ultrastructure of floral and extrafloral nectaries. Pp. 1–29 in B. Bentley and T. S. Elias, eds. *The biology of nectaries.* Columbia Univ. Press, New York.

Dutta, T. R., R. Ahmed, and S. R. Abbas. 1983. The discovery of a plant in the Andaman Islands that tranquilizes *Apis dorsata. Bee World* 64:158–163.

Dutton, R. W., F. Ruttner, A. Berkeley, and M. J. D. Manley. 1981. Observations on the morphology, relationships and ecology of *Apis mellifera* of Oman. *J. Apic. Res.* 20:201–213.

Dyer, F. C. 1985. Nocturnal orientation by the Asian honey bee, *Apis dorsata*. *Anim. Behav.* 33:769–774.

—— 1987. New perspectives on the dance orientation of the Asian honey bees. Pp. 234–246 in R. Menzel and A. Mercer, eds. *The neurobiology and behavior of honey bees*. Springer-Verlag, Berlin.

Dyer, F. C., and J. L. Gould. 1983. Honey bee navigation. *Am. Sci.* 71:587–597.

Dyer, F. C., and T. D. Seeley. 1987. Interspecific comparisons of endothermy in honey bees (*Apis*): deviations from the expected size-related patterns. *J. Exp. Biol.* 127:1–26.

Eberhard, W. G. 1985. *Sexual selection and animal genitalia*. Harvard Univ. Press, Cambridge, Mass.

Eckert, J. E. 1933. The flight range of the honeybee. *J. Agric. Res.* 47:257–285.

Edwards, L. H., J. H. Chen, T.-L. Ku, and G. J. Wasserburg. 1987. Precise timing of the last interglacial period from mass spectrometric determination of Thorium-230 in corals. *Science* 236:1547–1553.

Eguiarte, L., C. M. del Rio, and H. Arita. 1987. El néctar y polen como recursos: el papel ecológico de los visitantes a las flores de *Pseudobombax ellipticum* (H. B. K.) Dugand. *Biotropica* 19:74–82.

Eickwort, G. C. 1975. Gregarious nesting of the mason bee *Hoplitis anthocopoides* and the evolution of parasitism and sociality among megachilid bees. *Evolution* 29:142–150.

—— 1979. Mites associated with sweat bees (Halictidae). Pp. 575–581 in J. Rodriguez, ed. *Recent Adv. Acar.* 1.

—— 1981. Presocial insects. Pp. 199–280 in H. Hermann, ed. *Social insects*, Vol. 2. Academic Press, New York.

—— 1986. The first steps into eusociality: the sweat bee *Dialictus lineatulus*. *Fla. Entomol.* 69:742–754.

—— 1988a. Distribution patterns and biology of West Indian sweat bees (Hymenoptera: Halictidae). Pp. 231–254 in J. K. Leibherr, ed. *Zoogeography of Caribbean insects*. Cornell Univ. Press, Ithaca, N.Y.

—— 1988b. The origins of mites associated with honey bees. Pp. 327–338 in G. Needham, M. Delfinado-Baker, R. Page, and C. Bowman. eds. *Proc. Int. Conf. on Africanized Honey Bees and Bee Mites*. E. Horwood Ltd., Chichester, England.

Eickwort, G. C., and K. R. Eickwort. 1972. Aspects of the biology of Costa Rican halictine bees. IV. *Augochlora* (*Oxystoglosella*) (Hymenoptera: Halicitadae). *J. Kans. Entomol. Soc.* 45:18–45.

—— 1973. Aspects of the biology of Costa Rican bees. II. *Sphecodes kathleenae*, a social cleptoparasite of *Dialictus umbrepennae*. *J. Kans. Entomol. Soc.* 45:529–541.

Eickwort, G. C., K. R. Eickwort, and E. G. Linsley. 1977. Observations on nest aggregations of the bees *Diadasia olivacea* and *D. diminuta* (Hymenoptera: Anthophoridae). *J. Kans. Entomol. Soc.* 50:1–17.

Eickwort, G. C., and H. S. Ginsberg. 1980. Foraging and mating behavior in Apoidea. *Annu. Rev. Entomol.* 25:421–446.

Eickwort, G. C., and E. G. Linsley. 1978. The species of the parasitic bee genus *Protepeolus* (Hymenoptera: Anthophoridae). *J. Kans. Entomol. Soc.* 51:14–21.

Eickwort, G. C., R. W. Matthews, and J. Carpenter. 1981. Observations on the nesting behavior of *Megachile rubi* and *M. texana* with a discussion of the significance of soil nesting in the evolution of megachilid bees. *J. Kans. Entomol. Soc.* 54: 557–570.

Eisner, T., M. Deyrup, R. Jacobs, and J. Meinwald. 1986. Necrodols: anti-insectan terpenes from

defensive secretion of carrion beetle (*Necrodes surinamensis*). *J. Chem. Ecol.* 12:1407–1415.
Elias, T. S. 1983. Extrafloral nectaries: Their structure and distribution. Pp. 174–203 in B. L. Bentley and T. S. Elias, eds. *The biology of nectaries.* Columbia Univ. Press, New York.
Elton, C. 1958. *The ecology of invasions by animals and plants.* Methuen, London.
Engel, M. S., and F. Dingemans-Bakels. 1980. Nectar and pollen resources for stingless bees in Surinam. *Apidologie* 11:341–350.
Engels, E., and W. Engels. 1984. Drohnen-Absammlungen bei nestern der stachellosen biene *Scaptotrigona postica. Apidologie* 15:315–328.
Engels, E., W. Engels, W. Schroder, and W. Francke. 1987. Intranidal worker reactions to volatile compounds identified from cephalic secretions in the stingless bee, *Scaptotrigona postica* (Hymenoptera, Meliponinae). *J. Chem. Ecol.* 13:371–386.
Erickson, E. H. 1975. Surface electric potentials on worker honey bees leaving and entering the hive. *J. Apic. Res.* 14:141–147.
Erickson, E. H., W. R. Enns, and F. G. Werner. 1976. Bionomics of the bee-associated Meloidae (Coleoptera): bee and plant hosts of some nearctic meloid beetles – a synopsis. *Ann. Entomol. Soc. Am.* 69:959–970.
Erwin, T. L. 1982. Tropical forests: their richness in Coleoptera and other arthropod species. *Coleopt. Bull.* 36:74–75.
Esch, H. 1967. Die bedeutung der lauterzeugung für die verständigung der stachellosen bienen. *Z. vergl. Physiol.* 56:199–220.
——— 1976. Body temperature aand flight performance of honey bees in a servo-mechanically controlled wind tunnel. *J. Comp. Physiol.* 109:265–277.
Esch, H, I. Esch, and W. E. Kerr. 1965. An element common to communication of stingless bees and to dances of honey bees. *Science* 149:320–321.
Evans, H. E. 1955. *Philanthus sanbornii* Cresson as a predator on honeybees. *Bull. Brooklyn Entomol. Soc.* 50:47.
Evans, H. E., and E. G. Linsley. 1960. Notes on a sleeping aggregation of solitary bees and wasps. *Bull. Calif. Acad. Sci.* 59:30–37.
Fabre, J. H. 1921. The adventures of a grub. *Fabre's book of insects.* Dodd, Mead and Co., New York.
Faegri, K., and L. van der Pijl. 1979. *The principles of pollination ecology,* 3rd ed. Pergamon Press, New York.
Fahn, A. 1979. *Secretory tissues in plants.* Academic Press, New York.
Farenhorst, H. 1977. Nachweis übereinstimmender Chromosomenzahlen ($n = 16$) bei allen 4 *Apis*-Arten. *Apidologie* 8:89–100.
Farish, D. J. 1972. The evolutionary implications of qualitative variation in the grooming behavior of the Hymenoptera. *Anim. Behav.* 20:662–676.
Feinsinger, P. 1983. Coevolution and pollination. Pp. 282–310 in D. J. Futuyma and M. Slatkin, eds. *Coevolution.* Sinauer Assoc. Inc., Sunderland, Mass.
Feinsinger, P., K. G. Murray, S. Kinsman, and W. H. Busby. 1986. Floral neighborhood and pollination success in four hummingbird-pollinated cloud forest plant species. *Ecology* 67:449–464.
Fent, K., and R. Wehner. 1985. Ocelli: a celestial compass in the desert ant *Cataglyphis. Science* 228:192–194.
Fisher, E. M. 1983. *Pilica formidolosa* (Mosca asesina, robber fly). Pp. 755–758 in D. H. Janzen, ed. *Costa Rican Natural History.* University of Chicago Press.
Fisher, R. A. 1958. *The genetical theory of natural selection.* Dover, New York.
Fisher, R. M. 1987. Queen–worker conflict and social parasitism in bumble bees (Hymenoptera: Apidae). *Anim. Behav.* 35:1026–1036.

Flechtmann, C. H. W., and C. A. Camargo. 1974. Acari associated with stingless bee (Meliponidae [sic], Hymenoptera) from Brazil. Pp. 315–319 in *Proc. 4th Int. Congr. Acarol.* Saalfelder, Austria.

Flessa, K. W., K. V. Powers, and J. L. Cisne. 1975. Specialization and evolutionary longevity in the Arthropoda. *Paleobiol.* 1:71–81.

Fletcher, D. J. C. 1978. The African bee: *Apis mellifera adansonii,* in Africa. *Annu. Rev. Entomol.* 23:151–171.

In press. Behavior and ecology of *Apis mellifera scutellata* in Africa. In D. J. F. Fletcher and M. D. Breed, eds. *The 'African' honey bee.* Westview Press, Boulder, Colo.

Fletcher, D. J. C., and M. D. Breed, eds. In press. *The 'African' honey bee.* Westview Press, Boulder, Colo.

Fletcher, D. J. C., and C. D. Michener, eds. 1987. *Kin recognition in animals.* Wiley and Sons, Chichester, England.

Fletcher, D. J. C., and K. G. Ross. 1985. Regulation of reproduction in eusocial Hymenoptera. *Annu. Rev. Entomol.* 30:319–343.

Ford, D. M., H. R. Hepburn, F. B. Moseley, and R. J. Rigby. 1981. Displacement sensors in the honey bee pollen basket. *J. Insect. Physiol.* 27:339–346.

Foster, R. A. 1977. *Tachigalia versicolor* is a suicidal neotropical tree. *Nature* 268:624–626.

1980. Heterogeneity and disturbance in tropical vegetation. Pp. 75–92 in M. E. Soulé and B. A. Wilcox, eds. *Conservation biology.* Sinauer Assoc. Inc., Sunderland, Mass.

1982. Famine on Barro Colorado Island. Pp. 201–212 in E. G. Leigh, A. S. Rand and D. M. Windsor, eds. *The ecology of a tropical forest: seasonal and longer-term rhythms.* Smithsonian Institution Press, Washington, D.C.

Fowler, H. G. 1979. Responses by a stingless bee to a subtropical environment. *Rev. Biol. Trop.* 27:111–118.

Fox, M. D., and B. J. Fox. 1986. The susceptibility of natural communities to invasion. Pp. 57–66 in R. H. Groves and J. J. Burdon, eds. *Ecology of biological invasions.* Cambridge Univ. Press, Cambridge, England.

Francke, W., W. Reith, G. Berström, and J. Tengö. 1980. Spiroketals in the mandibular glands of *Andrena* bees. *Naturwiss.* 67:149–150.

Francke, W., W. Schröder, G. Bergström, and J. Tengö. 1984. Esters in the volatile secretion of bees. Pp. 127–136 in B. Kullenberg, G. Bergström, B. G. Svensson, J. Tengö, and L. Ågren, eds. *The ecological station of Uppsala University on Öland 1963–1983.* Almqvist and Wiksell, Int., Stockholm.

Francke, W., W. Schröder, E. Engels, and W. Engels. 1983. Variation in cephalic volatile substances in relation to worker age and behavior in the stingless bees, *Scaptotrigona postica. Z. Naturforsch.* 38c:1066–1068.

Frankie, G. W. 1973. A simple field technique for marking bees with fluorescent powders. *Ann. Entomol. Soc. Am.* 57:296–301.

Frankie, G. W., and H. G. Baker. 1974. The importance of pollinator behavior in the reproductive biology of tropical trees. *An. Inst. Biol. Univ. Nat. Autón. Méx. Ser. Bot.* 45(1):1–10.

Frankie, G. W., H. G. Baker, and P. A. Opler. 1974. Comparative phenological studies of trees in tropical wet and dry forests in the lowlands of Costa Rica. *J. Ecol.* 62:881–919.

Frankie, G. W., and R. Coville. 1979. An experimental study on the foraging behavior of selected solitary bee species in the Costa Rican dry forest. *J. Kans. Entomol. Soc.* 52:591–602.

Frankie, G. W., and W. A. Haber. 1983. Why bees move among mass-flowering neotropical trees. Pp. 360–372. in C. E. Jones and R. J. Little, eds. *Handbook of experimental pollination biology.* Van Nostrand Reinhold, New York.

Frankie, G. W., W. A. Haber, P. A. Opler, and K. S. Bawa. 1983. Characteristics and organization of the large bee pollination system in the Costa Rican dry forest. Pp. 411–448 in C. E. Jones and R. J. Little, eds. *Handbook of experimental pollination biology*. Van Nostrand Reinhold, New York.

Frankie, G. W., P. A. Opler, and K. S. Bawa. 1976. Foraging behavior of solitary bees: Implications for outcrossing of a neotropical forest tree species. *J. Ecol.* 64:1049–1057.

Frankie, G. W., and S. B. Vinson. 1977. Scent marking of passion flowers in Texas by females of *Xylocopa virginica texana* (Hymenoptera: Anthophoridae). *J. Kans. Entomol. Soc.* 50:613–625.

———. 1984. Morphology, chemical contents and possible function of the tibial gland of males of the Costa Rican solitary bees *Centris nitida* and *Centris trigonoides subtarsata* (Hymenoptera: Anthophoridae). *J. Kans. Entomol. Soc.* 57:50–54.

Frankie, G. W., S. B. Vinson, and R. E. Coville. 1980. Territorial behavior of *Centris adani* and its reproductive function in the Costa Rican dry forest (Hymenoptera: Apidae). *J. Kans. Entomol. Soc.* 53:837–857.

Frankie, G. W., S. B. Vinson, and A. Lewis. 1979. Territorial behavior in male *Xylocopa micans* (Hymenoptera: Anthophoridae). *J. Kans. Entomol. Soc.* 52:313–323.

Franks, N. R., and W. H. Bossert. 1983. The influence of swarm raiding army ants on the patchiness and diversity of a tropical leaf litter ant community. Pp. 151–163 in S. L. Sutton, T. C. Whitmore, and A. C. Chadwick, eds. *Tropical rain forest: ecology and management*. Blackwell, Oxford.

Free, J. B. 1956. A study of the stimuli which release the food begging and offering responses of worker honey bees. *Br. J. Anim. Behav.* 4:94–101.

———. 1961. Hypopharyngeal gland development and division of labour in honeybee (*Apis mellifera* L.) colonies. *Proc. Roy. Entomol. Soc. London* (A)36:5–8.

———. 1967. Factors determining the collection of pollen by honeybee foragers. *Anim. Behav.* 15:134–144.

———. 1987. *Pheromones of social bees*. Comstock Pub. Assoc., Ithaca, N.Y.

Free, J. B., and C. G. Butler. 1959. *Bumblebees*. Collins, London.

Free, J. B., J. A. Pickett, A. W. Ferguson, J. R. Simpkins, and C. Williams. 1984. Honeybee Nasonov pheromone lure. *Bee World* 65:175–181.

Freire, J. A. H., and R. I. Gara. 1970. Algumas observações sobre o comportamento de algumas especies do gênero *Trigona* (Apidae-Meliponini). *Turrialba* 20:351–356.

Fretwell, S. D., and H. L. Lucas. 1970. On territorial behaviour and other factors influencing habitat distribution in birds. *Acta Biotheor.* 19:16–36.

Friedmann, H. 1974. The Asian honeyguides. *J. Bombay Nat. Hist. Soc.* 71:426–437.

Frisch, K. von. 1967. *The dance language and orientation of bees*. Belknap Press of Harvard Univ. Press, Cambridge, Mass.

Frolich, D. R. 1983. On the nesting biology of *Osmia* (*Chenosmia*) *bruneri* (Hymenoptera: Megachilidae). *J. Kans. Entomol. Soc.* 56:123–130.

Frolich, D. R., and V. J. Tepedino. 1986. Sex ratio, parental investment, and interparent variability in nesting success in a solitary bee. *Evolution* 40:142–151.

Frumhoff, P. C., and S. Schneider. 1987. The social consequences of honey bee polyandry: the effects of kinship on worker interactions within colonies. *Anim. Behav.* 35:255–262.

Fuchs, S., and N. Koeniger. 1974. Schallerzegung im Dienst der Verteidigung des Bienenvolkes (*Apis cerana* Fabr.). *Apidologie* 5:271–287.

Futuyma, D. J. 1983. Evolutionary interactions among herbivorous insects and plants. Pp. 207–231. In D. J. Futuyma and M. Slatkin, eds. *Coevolution*. Sinauer Assoc. Inc., Sunderland, Mass.

Gadagkar, R. 1985. Kin recognition in social insects and other animals – a review of recent findings and a consideration of their relevance for the theory of kin selection. *Proc. Indian Acad. Sci.* 94:587–621.

Garófalo, C. A. 1973. Occurrence of diploid drones in a neotropical bumble bee. *Experientia* 29:726.

―― 1985. Social structure of *Euglossa cordata* nests (Hymenoptera: Apidae: Euglossini). *Entomol. Gener.* 11:77–83.

Garófalo, C. A., E. Camillo, M. J. O. Campos, R. Zucchi and J. C. Serrano. 1981. Bionomical aspects of *Lithurgus corumbae* (Hymenoptera, Megachilidae), including evolutionary considerations on the nesting behavior of the genus. *Rev. Bras. Genet.* 4(2):165–182.

Garófalo, C. A., R. Zucchi, and G. Muccillo. 1986. Reproductive studies of a Neotropical bumblebee, *Bombus atratus* (Hymenoptera, Apidae). *Rev. Bras. Genet.* 9(2):231–243.

Gary, N. E. 1960. A trap to quantitatively recover dead and abnormal honeybees from the hive. *J. Econ. Entomol.* 53:782–785.

―― 1962. Chemical mating attractants in the queen honeybee. *Science* 136:773.

―― 1971. Magnetic retrieval of ferrous labels in a capture–recapture system for honey bees and other insects. *J. Econ. Entomol.* 64:961–965.

―― 1974. Pheromones that affect the behavior and physiology of honey bees. Pp. 200–221 in M. C Birch, ed. *Pheromones*. North-Holland Publ. Co., Amsterdam.

Gary, N. E., P. C. Witherell, K. Lorenzen, and J. M. Marston. 1977. Area fidelity and intra-field distribution of honey bees during the pollination of onions. *Environ. Entomol.* 6:303–310.

Gary, N. E., H. V. Daly, S. Locke, and M. Race.1985. The Africanized honey bee: ahead of schedule. *Calif. Agric.* 39:4–7.

Gates, D. M. 1980. *Biophysical ecology*. Springer–Verlag, New York.

Gentry, A. H. 1974a. Flowering phenology and diversity in tropical Bignoniaceae. *Biotropica* 6:64–68.

―― 1974b. Coevolutionary patterns in Central American Bignoniaceae. *Ann. Mo. Bot. Gard.* 61:728–759.

―― 1978. Anti-pollinators for mass flowering plants? *Biotropica* 10:68–69.

―― 1983. Neotropical floristic diversity: phytogeographic connections between Central and South America, Pleistocene climatic fluctuations, or an accident of the Andean orogeny? *Ann. Mo. Bot. Gard.* 69:557–593.

―― 1985. Contrasting phytogeographic patterns of upland and lowland Panamanian plants. Pp. 147–160, in W. G. D'Arcy and M. D. Correa A., eds. *The botany and natural history of Panamá*. Missouri Botanical Garden, St. Louis, Mo.

Gerber, H. S., and E. C. Klostermeyer. 1970. Sex control by bees: a voluntary act of egg fertilization during oviposition. *Science* 167:82–84.

Gerling, D., and H. R. Hermann. 1978. Biology and mating behaviour of *Xylocopa virginica* L. *Behav. Ecol. Sociobiol.* 3:99–111.

Gerling, D., P. D. Hurd, and A. Hefetz. 1981. In-nest behavior of the carpenter bee *Xylocopa pubescens* Spinola (Hymenoptera: Anthophoridae). *J. Kans. Entomol. Soc.* 54:209–218.

―― 1983. Comparative behavioral biology of two middle east species of carpenter bees (*Xylocopa* Latreille). Smithsonian Institution Press, Washington, D.C.

Gerling, D., T. Orion, and M. Ovadia. 1979. Morphology, histochemistry and ultrastructure of the yellow glands of *Xylocopa pubescens* Spinola (Hymenoptera: Anthophoridae). *Int. J. Insect Morphol. Embryol.* 8:123–134.

Getz, W. M., D. Brückner, and T. R. Parisian. 1982. Kin structure and the swarming behavior of the honey bee, *Apis mellifera*. *Behav. Ecol. Sociobiol.* 10:265–270.

Getz, W. M , and K. B. Smith. 1983. Genetic kin recognition: honey bees discriminate between full and half sisters. *Nature* 302:147–148.

Ghatge, A. 1951. The bees of India. *Indian bee J.* 13:88.
Ghisalberti, E. L. 1979. Propolis: A review. *Bee World* 60:59–84.
Giblin, R. M., H. K. Kaya, and R. W. Brooks. 1983. Occurrence of *Huntaphelenchoides* sp. (Aphelenchoididae) and *Acrostichus* sp. (Diplogasteridae) in the reproductive tracts of soil-nesting bees. *Nematologica* 27:20–27.
Gilbert, W. M.1973. Foraging behavior of *Trigona fulviventris* in Costa Rica. *Pan-Pac. Entomol.* 49:21–25.
Gill, F. B., A. L. Mack, and R. T. Ray. 1982. Competition between hermit hummingbirds and insects for nectar in a Costa Rican rain forest. *Ibis* 124:44–49.
Gilliam, M. 1978. Fungi. Pp. 78–101 in R. E. Morse, ed. *Honeybee pests, predators and diseases.* Cornell Univ. Press, Ithaca, N.Y.
———1979. Microbiology of pollen and bee bread: the genus *Bacillus. Apidologie* 10:269–274.
Gilliam, M., S. L. Buchmann, and B. J. Lorenz. 1984. Microbial flora of the larval provisions of the solitary bees, *Centris pallida* and *Anthophora* sp. *Apidologie* 15:1–10.
Gilliam, M., S. L. Buchmann, B. J. Lorenz, and D. W. Roubik. 1985. Microbiology of the larval provisions of the stingless bee, *Trigona hypogea*, an obligate necrophage. *Biotropica* 17:28–31.
Gilliam, M., and D. B. Prest. 1987. Microbiology of feces of the larval honey bee, *Apis mellifera. J. Invert. Path.* 49:70–75.
Giorgini, J. F., and A. B. Gusman. 1972. A importancia das abelhas na polinização. Pp. 155–180 in J. M. F. Camargo, ed. *Manual de apicultura.* Editôra Agronômica Ceres Ltda., São Paulo.
Glasser, J. W. 1984. Is conventional foraging theory optimal? *Am. Nat.* 124:900–905.
Gonnet, M., P. Lavie, and P. Nogueira-Neto. 1964. Étude de quelques caractéristiques des miels récoltés par certains méliponines brésiliens. *C. R. Séances Acad. Sci. Sér. D Sci. Nat.* 258:3107–3109.
González, J. A. 1983. Acerca de la regionalización de la nomenclatura maya de las abejas sin aguijón en Yucatán. *Rev. Geogr. Agríc.* 5/6:190–193.
Gottsberger, G. 1985. Floral ecology, report on the years 1981 (79) to 1985. *Progress in Botany.* 47:384–417. Springer-Verlag, Heidelberg.
———1986. Some pollination strategies in neotropical savannas and forests. *Plant Syst. Evol.* 152:29–45.
Gottsberger, G., J. Schrauwen, and H. F. Linskens. 1984. Amino acids and sugars in nectar, and their putative evolutionary significance. *Plant Syst. Evol.* 145:55–77.
Gould, J. L. 1982. Why do honeybees have dialects? *Behav. Ecol. Sociobiol.* 10:53–56.
———1986. The locale map of honey bees: do insects have cognitive maps? *Science* 232:861–863.
———1987. Landmark learning by honey bees. *Anim. Behav.* 35:26–34.
Gould, J. L., F. C. Dyer, and W. F. Towne. 1985. Recent progress in the study of the dance language. Pp. 141–162 in B. Hölldobler and M. Lindauer, ed. *Experimental behavioral ecology.* G. Fischer Verlag, New York.
Gould, J. L., and W. F. Towne. 1987. Evolution of the dance language. *Am. Nat.* 130:317–338.
Gould, S. J. 1985. Evolution and the triumph of homology, or why history matters. *Am. Sci.* 74:60–69.
Gould, S. J., and R. C. Lewontin. 1979. The spandrels of San Marco and the Panglossian paradigm: a critique of the adaptationist program. *Proc. Roy. Soc. London* 205:581–598.
Graham, R. W. 1986. Response of mammalian communities to environmental changes during the late Quaternary. Pp. 300–313 in J. Diamond and T. J. Case, eds. *Community Ecology.* Harper and Row, New York.
Grandi, G. 1961. Studi di un entomologo sugli imenotteri superiori. Vol. XXV, *Bol. Inst. Entomol. Univ. Bologna.*

Graur, D. 1985. Gene diversity in Hymenoptera. *Evolution* 39:190–199.
Greenberg, L. 1979. Genetic component of bee odor in kind recognition. *Science* 206:1075–1079.
Grimaldi, D., and B. A. Underwood. 1986. *Megabraula*, a new genus for two new species of Braulidae (Diptera) and a discussion of braulid evolution. *Syst. Entomol.* 11:427–438.
Griswold, T. 1983. Revision of *Proteriades* subgenus *Acrosmia* Michener (Hymenoptera; Megachilidae). *Ann. Entomol. Soc. Am.* 76:707–714.
Griswold, T., and C. D. Michener. 1988. Taxonomic observations on Anthidiini of the Western Hemisphere (Hymenoptera: Megachilidae). *J. Kans. Entomol. Soc.* 61:22–45.
Guirguis, G. N., and W. A. Brindley. 1974. Insecticide susceptibility and response to selected pollens of larval leafcutting bees, *Megachile pacifica* (Panzer) (Hymenoptera: Megachilidae). *Environ. Entomol.* 3:691–694.
Haas, A. 1960. Vergleichende Verhaltensstudien zum Paarungsschwarz solitärer Apiden. *Zeitschr. Tierpsychol.* 17:402–416.
Haber, W. A., and G. W. Frankie. 1982. Pollination of *Luehea* (Tiliaceae) in Costa Rican deciduous forest. *Ecology* 63:1740–1750.
Haffer, J. 1969. Speciation in Amazonian forest birds. *Science* 165:131–137.
Hagen, K. S. 1986. Ecosystem analysis: plant cultivars (HPR), entomophagous species and food supplements. Pp. 151–187 in D. J. Boethel and R. D. Eikenbary, eds. *Interactions of plant resistance and parasitoids and predators of insects*. E. Horwood Ltd., Chichester, England.
Hamburg, S. P., and R. L. Sanford 1986. Disturbance, *Homo sapiens*, and ecology. *Bull. Ecol. Soc. Am.* 67: 169–171.
Hamilton, W. D. 1972. Altruism and related phenomena, mainly in social insects. *Annu. Rev. Ecol. Syst.* 3:193–232.
Hamrick, J. L., and M. D. Loveless. 1986. Isozyme variation in tropical trees: procedures and preliminary results. *Biotropica* 18:201–207.
Handel, S. N. 1983. Pollination ecology, plant population structure and gene flow. Pp 163–211. in L. Real, ed. *Pollination biology*. Academic Press, New York.
Harbo, J. R., A. B. Bolten, T. E. Rinderer, and A. M. Collins. 1981. Development periods for eggs of Africanized and European honeybees. *J. Apic. Res.* 20:156–159.
Harder, L. D. 1982. Measurement and estimation of functional proboscis length in bumble bees (Hymenoptera: Apidae). *Can. J. Zool.* 60:1073–1079.
——— 1983. Functional differences of the proboscides of short- and long-tongued bees (Hymenoptera: Apoidea). *Can. J. Zool.* 61:1580–1586
Harper, J. L. 1977. *Population biology of plants*. Academic Press, New York.
Hartfelder, K. 1986. Trophogene Basis und endokrine Reaktion in der Kastenentwicklung bei Stachellosen Bienen. Dissertation, Universität Tübingen.
——— 1987. Rates of juvenile hormone synthesis control caste differentiation in the stingless bee *Scaptotrigona postica depilis*. *Dev. Biol.* 196:522–526.
Hasselrot, T. B. 1960. Studies on Swedish bumblebees (Genus *Bombus* Latr.): their domestication and biology. *Opuscula Entomol. Suppl.* 17:1–203.
Heaney, L. R. 1986. Biogeography of mammals in SE Asia: estimates of rates of colonization, extinction and speciation. *Biol. J. Linn. Soc.* 28:127–165.
Hedrick, P. W. 1986. Genetic polymorphisms in heterogeneous environments: A decade later. *Annu. Rev. Ecol. Syst.* 17:535–566.
Hedrick, P. W., M. E. Ginevan, and E. P. Ewing. 1976. Genetic polymorphism in heterogeneous environments. *Annu. Rev. Ecol. Syst.* 7:1–32.
Hefetz, A. 1983. Function of secretion of mandibular gland of males in territorial behavior of *Xylocopa sulcatipes* (Hymenoptera: Anthophoridae). *J. Chem. Ecol.* 9: 923–931.
——— 1987. The role of Dufour's gland secretions in bees. *Physiol. Entomol.* 12:243–253.

Hefetz, A., S. W. T. Batra, and M. S. Blum. 1979a. Linalool, neral and geranial in the mandibular glands of *Colletes* bees – an aggregation pheromone. *Experientia* 35:319–320.

1979b. Chemistry of the mandibular gland secretion of the Indian bee *Pithitis smaragdula*. *J. Chem. Ecol.* 5:753–758.

Hefetz, A., G. Bergström, and J. Tengö. 1986. Species, individual and kin specific blends in Dufour's gland secretions of halictine bees. *J. Chem. Ecol.* 12:197–208.

Hefetz, A., M. S. Blum, G., C. Eickwort, and J. W. Wheeler. 1978. Chemistry of the Dufour's gland secretion of halictine bees. *Comp. Biochem Physiol.* 61B:129–132.

Hefetz, A., G. C. Eickwort, J. H. Cane, M. Blum, and A. Shinn. 1982. A comparative study of the exocrine products of clepto-parasitic bees (*Holcopasites*) and their hosts (*Calliopsis*) (Hymenoptera: Anthophoridae, Andrenidae). *J. Chem. Ecol.* 8:1389–1397.

Hefetz, A., H. M. Fales, and S. W. T. Batra. 1979. Natural polyesters: Dufour's gland macrocyclic lactones form brood cell laminesters in *Colletes* bees. *Science* 204:415–417.

Heinrich, B. 1972. Energetics of temperature regulation and foraging in a bumble bee, *Bombus terricola* Kirby. *J. Comp. Physiol.* 77:49–64.

1975. Energetics of pollination. *Annu. Rev. Ecol. Syst.* 6:139–170.

1976a. Foraging specializations of individual bumble bees. *Ecol. Monogr.* 46:105–128.

1976b. Resource partitioning among some eusocial insects: Bumble bees. *Ecology* 57:874–889.

1979a. *Bumble bee economics.* Harvard Univ. Press, Cambridge, Mass.

1979b. Keeping a cool head: honey bee thermoregulation. *Science* 205:1269–1271.

1979c. Thermoregulation of African and European honey bees during foraging, attack and hive exits and returns. *J. Exp. Biol.* 80:217–229.

1980. Mechanisms of body-temperature regulation in honeybees, *Apis mellifera*. I. Regulation of head temperature. *J. Exp. Biol.* 85:61–72.

Heinrich, B., and S. L. Buchmann. 1986. Thermoregulatory physiology of the carpenter bee, *Xylocopa varipuncta*. *J. Comp. Physiol. B Metab. Transp. Funct.* 156:557–562.

Heinrich, B., and P. H. Raven. 1972. Energetics and pollination ecology. *Science* 176:597–602.

Heithaus, E. R. 1974. The role of plant–pollinator interactions in determining community structure. *Ann. Mo. Bot. Gard.* 61:675–691.

1979a. Community structure of neotropic flower visiting bees and wasps: diversity and phenology. *Ecology* 60:190–202.

1979b. Flower-feeding specialization in wild bee and wasp communities in seasonal neotropical habitats. *Oecologia* 42:179–194.

1979c. Flower visitation records and resource overlap of bees and wasps in northwest Costa Rica. *Brenesia* 16:9–52.

Henderson, A. 1986. A review of pollination studies in the Palmae. *Bot. Rev.* 52:221–259.

Hepburn, H. R. 1986. *Honeybees and wax.* Springer–Verlag, Heidelberg.

Heyneman, A. J. 1983. Optimal sugar concentrations of floral nectars – dependence on sugar intake efficiency and foraging costs. *Oecologia* 60:198–213.

Hockings, H. J. 1884. Notes on two Australian species of *Trigona*. *Trans. Entomol. Soc. London*, 32:149–157.

Hodges, C. M., and R. B. Miller. 1981. Pollinator flight directionality and the assessment of pollen returns. *Oecologia* 50:376–379.

Hodges, C. M., and L. L. Wolf. 1981. Optimal foraging bumblebees: why is nectar left behind in flowers? *Behav. Ecol. Sociobiol.* 9:41–44.

Hölldobler, B., and C. D. Michener. 1980. Mechanisms of identification and discrimination in social Hymenoptera. Pp. 35–38 in H. Markl, ed. *Evolution of social behavior: hypotheses and empirical tests.* Dahlem Konferenzen, Verlag Chemie, Weinheim.

Horvitz, C. C., and D. W. Schemske. 1988. A test of the pollinator limitation hypothesis for a neotropical herb. *Ecology* 69:200–206.

Hoshiba, H., I. Okada, and A. Kusanagi. 1981. The diploid drone of *Apis cerana japonica* and its chromosomes. *J. Apic. Res.* 20:143–147.

Houston, A., P. Schmid-Hempel, and A. Kacelnik. 1988. Foraging strategy, worker mortality, and the growth of the colony in social insects. *Am. Nat.* 131:107–114.

Houston, T. F. 1970. Discovery of an apparent male soldier caste in a nest of a halictine bee, with notes on the nest. *Aust. J. Zool.* 18:245–351.

———. 1981. Alimentary transport of pollen in a paracolletine bee (Hymenoptera: Colletidae). *Aust. Entomol. Mag.* 8:57–59.

———. 1983. An extraordinary new bee and adaptation of palpi for nectar-feeding in some Australian Colletidae and Pergidae (Hymenoptera). *J. Aust. Entomol. Soc.* 22:263–270.

———. 1984. Biological observations of bees in the genus *Ctenocolletes* (Hymenoptera: Stenotritidae). *Rec. West. Aust. Mus.* 11:153–172.

———. 1987. The symbiosis of acarid mites, genus *Ctenocolletacarus* (Acarina: Acaraiformes), and stenotritid bees, genus *Ctenocolletes* (Insecta: Hymenoptera). *Aust. J. Zool.* 35:459–468.

Houston, T. F., and R. W. Thorp. 1984. Bionomics of the bee *Stenotritus greavesi* and ethological characteristics of Stenotritidae. *Rec. West. Aust. Mus.* 11:375–385.

Howard, J. J. 1985. Observations on resin collecting by six interacting species of stingless bees (Apidae: Meliponinae). *J. Kans. Entomol. Soc.* 58:337–345.

Howe, H. F. 1984. Constraints on the evolution of mutualisms. *Am. Nat.* 123:764–777.

Hubbell, S. P. 1979. Tree dispersion, abundance and diversity in a tropical dry forest. *Science* 203:1299–1309.

Hubbell, S. P., and R. B. Foster. 1986. Biology, chance, and history and the structure of tropical rain forest tree communities. Pp. 314–329 in J. Diamond and T. J. Case, eds. *Community ecology*. Harper and Row, New York.

Hubbell, S. P., and L. K. Johnson. 1977. Competition and nest spacing in a tropical stingless bee community. *Ecology* 58:949–963.

———. 1978. Comparative foraging behavior of six stingless bee species exploiting a standardized resource. *Ecology* 59:1123–1136.

Huheey, J. E. 1984. Warning coloration and mimicry. Pp. 257–300 in W. J. Bell and R. T. Cardé, eds. *Chemical ecology of insects*. Chapman and Hall Ltd., London.

Hung, A. C. F., and S. B. Vinson. 1977. Electrophoretic techniques and genetic variability in Hymenoptera. *Heredity* 38:409–411.

Hurd, P. D. 1958. Observations on the nesting habits of some new world carpenter bees with remarks on their importance in the problem of species formation. *Ann. Entomol. Soc. Am.* 51:365–375.

———. 1978. *An annotated catalog of the carpenter bees (genus Xylocopa Latreille) of the Western Hemisphere*. Smithsonian Institution Press, Washington, D.C.

———. 1979. Superfamily Apoidea. Pp. 1741–2209 in K. V. Krombein, P. D. Hurd, D. R. Smith, and B. D. Burks, eds. *Catalog of Hymenoptera in America north of Mexico*, Vol. 2. Smithsonian Institution Press, Washington, D.C.

Hurd, P. D., and E. G. Linsley. 1959. Observations on the nest-site behavior of *Melissodes composita* Tucker and its parasites, with notes on the communal use of nest entrances (Hymenoptera: Apoidea). *Entomol. News* 70:141–146.

———. 1963. Pollination of the unicorn plant (Martyniaceae) by an oligolectic, corolla-cutting bee. *J. Kans. Entomol. Soc.* 36:248–252.

———. 1975. The principal *Larrea* bees of the southwestern United States. *Smithsonian Contrib. Zool.* 293:1–74.

1976. The bee family Oxaeidae with a revision of the North American species (Hymenoptera: Apoidea). *Smithsonian Contrib. Zool.* No. 220.
Hurd, P. D., and J. S. Moure. 1960. A New World subgenus of bamboo-nesting carpenter bees belonging to the genus *Xylocopa* Latreille. *Ann. Entomol. Soc. Am.* 53:809–821.
   1961. Some notes on sapygid parasitism in the nests of carpenter bees belonging to the genus *Xylocopa* Latreille. *J. Kans. Entomol. Soc.* 34:19–22.
   1963. A classification of the large carpenter bees (Xylocopini). *Univ. Calif. Publ. Entomol.* 29:1–365.
Hüttinger, E. 1974. Dickkopffliegen (Conopidae, Diptera) wenig beachtete Bienenparasiten. *Bienenvater* 95:102–103.
Hydak, M. H. 1970. Honey bee nutrition. *Annu. Rev. Entomol.* 15:143–156.
Ihering, H. von. 1903. Biologia das abelhas melliferas do Brasil [translation from German] 1930. *Bol. Agric.* [São Paulo], 31 (5,6) 435–507; (7,8) 649–714.
Imperatriz-Fonseca, V. L. 1977. Studies on *Paratrigona subnuda* (Moure) Hymenoptera, Apidae, Meliponinae. II. Behaviour of the virgin queen. *Bol. Zool. Univ. São Paulo* 2:169–182.
   1978. Studies on *Paratrigona subnuda* (Moure) Hymenoptera, Apidae, Meliponinae. III. Queen supersedure. *Bol. Zool. Univ. São Paulo* 3:153–162.
Imperatriz-Fonseca, V. L., S. C. F de Souza, and P. Nogueira-Neto. 1972. Subterranean nest structure of a stingless bee (*Paratrigona subnuda* Moure) (Meliponinae, Apidae, Hymenoptera). *Ciência e Cultura* 24:662–666.
Inoue, T., S. F. Sakagami, S. Salmah, and N. Nukmal. 1984. Discovery of successful absconding in the stingless bee *Trigona* (*Tetragonula*) *laeviceps*. *J. Apic. Res.* 23:136–142.
Inoue, T., S. F. Sakagami, S. Salmah, and S. Yamane. 1984. The process of colony multiplication in the Sumatran stingless bee *Trigona* (*Tetragonula*) *laeviceps*. *Biotropica* 16:100–111.
Inoue, T., S. Salmah, I. Abbas, and E. Yusuf. 1985. Foraging behavior of individual workers and foraging dynamics of colonies of three Sumatran stingless bees. *Res. Popul. Ecol.* 27:373–392.
Inouye, D. W. 1975. Flight temperatures of male euglossine bees (Hymenoptera: Apidae: Euglossini). *J. Kans. Entomol. Soc.* 48:366–373.
   1978. Resource partitioning in bumble bees: Experimental studies of foraging behavior. *Ecology* 59:672–678.
   1980. The terminology of floral larceny. *Ecology* 61:1251–1253.
   1983. Ecology of nectar robbing. Pp. 153–173 in B. L. Bentley and T. S. Elias, eds. *The biology of nectaries.* Columbia Univ. Press, New York.
Inouye, D. W., N. A. Favre, J. A. Lanum, D. M. Levine, J. B. Meyers, M. S. Roberts, F. C. Tsao, and Y. Y. Wang. 1980. The effects of nonsugar nectar constituents on estimates of nectar energy content. *Ecology* 61:992–996.
Inouye, D. W., and R. S. Inouye. 1980. The amino acids of extrafloral nectar from *Helianthella quinquenervis* (Asteraceae). *Am. J. Bot.* 67:1394–1396.
Inouye, D. W., and G. D. Waller. 1982. Responses of honey bees (*Apis mellifera*) to amino acid solutions mimicking floral nectars. *Ecology* 65:618–625.
Ishay, J., H. Virinsky-Salz, and A. Shulov. 1967. Contributions to the bionomics of the oriental hornet (*Vespa orientalis* Fab.) *Isr. J. Entomol.* 2:45–106.
Itô, M. 1985. Supraspecific classification of bumblebees based on the characters of male genitalia. *Contrib. Inst. Low. Temp. Sci.* No. 20. Hokkaido Univ., Sapporo.
Itô, M., and S. F. Sakagami. 1985. Possible synapomorphies of the parasitic bumblebees (*Psithyrus*) with some nonparasitic bumblebees (*Bombus*). *Sociobiology* 10:105–120.
Itô, M., T. Matsumura, and S. F. Sakagami. 1984. A nest of the Himalayan bumblebee *Bombus* (*Festivobombus*) *festivus*. *Kontyû* 52:537–539.

Itô, Y. 1980. *Comparative ecology.* [translated from Japanese, 1978] Cambridge Univ. Press, London.
Iwama, S., and T. S. Melhem. 1979. The pollen spectrum of the honey of *Tetragonisca angustula angustula* Latreille (Apidae, Meliponinae). *Apidologie* 10:275–295.
Iwata, K. 1938. Habits of some bees in Formosa (II) and (IV). *Trans. Nat. Hist. Soc. Formosa* 28:205–215, 373–379.
——— 1976. *Evolution of instinct* [translated from Japanese, 1971]. Amerind Publishing Co. Pvt. Ltd., New Delhi, India.
Iwata, K., and S. F. Sakagami. 1966. Gigantism and dwarfism in bee eggs in relation to the modes of life, with notes on the number of ovarioles. *Jpn. J. Ecol.* 16:4–16.
Jackson, J. F. 1973. Mimicry of *Trigona* bees by a reduviid (Hemiptera) from British Honduras. *Fla. Entomol.* 56:200–202.
Jander, R. 1975. Ecological aspects of spatial orientation. *Annu. Rev. Ecol. Syst.* 6:171–188.
——— 1976. Grooming and pollen manipulation in bees (Apoidea): The nature and evolution of movements involving the foreleg. *Physiol. Entomol.* 1:179–184.
Jander, R., and U. Jander. 1978. Wing grooming in bees (Apoidea) and the evolution of wing grooming in insects. *J. Kans. Entomol. Soc.* 51:653–664.
Janzen, D. H. 1964. Notes on the behavior of four subspecies of the carpenter bee *Xylocopa* (*Notoxylocopa*) *tabaniformis*, in Mexico. *Ann. Entomol. Soc. Am.* 57:296–301.
——— 1966. Notes on the behavior of the carpenter bee *Xylocopa fimbriata* in Mexico. *J. Kans. Entomol. Soc.* 39:633–641.
——— 1967a. Syncronization of sexual reproduction of trees within the dry season in Central America. *Evolution* 21:620–637.
——— 1967b. Notes on nesting and foraging behavior of *Megalopta* (Hymenoptera: Halictidae) in Costa Rica. *J. Kans. Entomol. Soc.* 41:342–350.
——— 1967c. Why mountain passes are higher in the tropics. *Am. Nat.* 101:233–249.
——— 1968. Reproductive behavior in the Passifloraceae and some of its pollinators in Central America. *Behavior* 32:33–48.
——— 1971. Euglossine bees as long-distance pollinators of tropical plants. *Science* 171:203–205.
——— 1975. *Ecology of plants in the tropics.* E. Arnold, London.
——— 1977. A note on optimal mate selection by plants. *Am. Nat.* 11:365–371.
——— 1980. When is it coevolution? *Evolution* 34:611–612.
——— 1981a. Bee arrival at two Costa Rican female *Catasetum* orchid inflorescences, and a hypothesis on euglossine population structure. *Oikos* 36:177–183.
——— 1981b. Reduction in euglossine bee species richness on Isla del Caño, a Costa Rican offshore island. *Biotropica* 13:238–240.
——— 1983. Seed and pollen dispersal by animals: convergence in the ecology of contamination and sloppy harvest. *Biol. J. Linn. Soc.* 20:103–113.
——— 1984. Two ways to be a tropical big moth: Santa Rosa saturniids and sphingids. *Oxford Surv. Evol. Biol.* 1:85–140.
——— 1985. What parasites of animals and plants do not have in common. Pp. 83–99 in K. C. Kim, ed. *Coevolution of parasitic arthropods and mammals.* J. Wiley and Sons, New York.
——— 1986. The future of tropical ecology. *Ann. Rev. Ecol. Syst.* 17:305–324.
——— ed. 1983. *Costa Rican Natural History.* Univ. Chicago Press, Chicago.
Janzen, D. H. , P. J. DeVries, M. L. Higgins, and L. S. Kimsey. 1982. Seasonal and site variation in Costa Rican euglossine bees at chemical baits in lowland deciduous and evergreen forests. *Ecology* 63:66–74.
Janzen, D. H., and R. Liesner. 1980. Annotated check-list of plants of lowland Guanacaste Province, Costa Rica. exclusive of grasses and non-vascular cryptogams. *Brenesia* 18:15–90.

Jay, S. C. 1959. Factors affecting the laboratory rearing of honeybee larvae (*Apis mellifera* L.). Dissertation, University of Toronto.

Jayasingh, D. B., and B. E. Freeman. 1980. The comparative population dynamics of eight solitary bees and wasps trap-nested in Jamaica. *Biotropica* 12:214–219.

Jaycox, E. R. 1967. Territorial behavior among males of *Anthidium banningense* (Hymenoptera: Megachilidae). *J. Kans. Entomol. Soc.* 40:565–570.

Jeanne, R. L., and D. W. Davidson. 1984. Population regulation in social insects. Pp. 560–585 in C. B. Huffaker and R. L. Rabb, eds. *Ecological entomology.* J. Wiley and Sons, New York.

Johnson, L. K. 1974. The role of agonistic behavior in the foraging strategies of *Trigona* bees. Dissertation, Univ. of California, Berkeley.

―― 1980. Alarm response of foraging *Trigona fulviventris* (Hymenoptera: Apidae) to mandibular gland components of competing bee species. *J. Kans. Entomol. Soc.* 53:357–362.

―― 1981. Effect of flower clumping on defense of artificial flowers by aggressive stingless bees. *Biotropica* 13:151–157.

―― 1983a. Foraging strategies and the structure of stingless bee communities in Costa Rica. Pp. 31–58 in P. Jaisson, editor. *Social Insects in the Tropics*, Vol. 2. Univ. Paris–Nord.

―― 1983b. *Apiomerus pictipes* (Reduviidae, Chinche asesina, assassin bug). Pp. 684–687 in D. H. Janzen ed. *Costa Rican Natural History.* Univ. Chicago Press, Chicago.

Johnson, L. K., L. W. Haynes, M. A. Carlson, H. A. Fortnum, and D. L. Gorgas. 1985. Alarm substances of the stingless bee, *Trigona silvestriana. J. Chem. Ecol.* 11:409–416.

Johnson, L. K., and S. P. Hubbell. 1974. Aggression and competition among stingless bees: field studies. *Ecology* 55:120–127.

―― 1975. Contrasting foraging strategies and coexistence of two bee species on a single resource. *Ecology* 56:1398–1406.

―― 1984. Nest tree selectivity and density of stingless bee colonies in a Panamanian forest. Pp.147–154 in A. C. Chadwick and S. L. Sutton, eds. *Tropical rain-forest: the Leeds symposium.* Leeds Phil., Lit. Soc., Leeds, England.

Johnson, L. K., S. P. Hubbell, and D. H. Feener, Jr. 1987. Defense of food supply by eusocial colonies. *Am. Zool.* 27:347–358.

Johnson, L. K., and D. F. Wiemer. 1982. Nerol: An alarm substance of the stingless bee *Trigona fulviventris* (Hymenoptera: Apidae). *J. Chem. Ecol.* 9:1167–1181.

Johnson, M. D. 1984. The pollen preferences of *Andrena (Melandrena) dunningi* Cockerell (Hymenoptera: Andrenidae). *J. Kans. Entomol. Soc.* 57:34–43.

Jolly, A., P. Oberlé, and R. Albignac, eds. 1984. *Madagascar.* Pergamon Press, Oxford.

Jones, C. E. 1978. Pollinator constancy as a pre-pollination isolating mechanism between sympatric species of *Cercidium. Evolution* 32:189–198.

Jones, C. E., and S. L. Buchmann. 1974. Ultraviolet floral patterns as functional orientation cues in Hymenopterous pollination systems. *Anim. Behav.* 22:481–485.

Jones, C. E., and R. J. Little, eds. 1983. *Handbook of experimental pollination biology.* Van Nostrand Reinhold, New York.

Jones, C. G. 1985. Microorganisms as mediators of plant resource exploitation by insect herbivores. Pp. 54–99 in P. W. Price, C. N. Slobodchikoff, and W. S. Gaud, eds. *A new ecology: novel approaches to interactive systems.* Wiley–Interscience, New York.

Jones, D., M. L. Williams, and G. Jones. 1980. The biology of *Stylops* spp. in Alabama, with emphasis on *S. bipunctatae. Ann. Entomol. Soc. Am.* 73:448–451.

Joshi, N. V., and R. Gadagkar. 1985. Evolution of sex ratios in social Hymenoptera: kin selection, local mate competition, polyandry and kin recognition. *J. Genet.* [India] 64:41–58.

Kacelnik, A., A. I. Houston, and P. Schmid-Hempel. 1986. Central-place foraging in honey bees: the effect of travel time and nectar flow on crop filling. *Behav. Ecol. Sociobiol.* 19:19–24.

Kalmus, H. 1954. The clustering of honeybees at a food source. *Br. J. Anim. Behav.* 2:63–71.
Kamm, D. 1974. Effects of temperature, day length, and number of adults on the sizes of cells and offspring in a primitively social bee (Hymenoptera: Halictidae). *J. Kans. Entomol. Soc.* 47:8–18.
Kammer, A. E., and B. Heinrich. 1978. Insect flight metabolism. *Adv. Insect Physiol.* 13:133–228.
Kapil, R. P., and J. S. Daliwahl. 1968a. Defense of nest by the female of *Xylocopa fenestrata* Fab. *Insectes Soc.* 15:419–422.
———1968b. Biology of *Xylocopa* species. I. Seasonal activity, nesting behaviour and life cycle. *J. Res.* (Ludhiana Univ.). 5:406–419.
———1969. Biology of *Xylocopa* species. II. Field activities, flight range and trials on transportation of nests. *J. Res.* (Ludhiana Univ.). 6:262–271.
Kapil, R. P., and K. L. Jain. 1980. *Biology and utilization of insect pollinators for crop production.* Haryana Agric. Univ., Hissar, India.
Kapil, R. P., and S. Kumar. 1969. Biology of *Ceratina binghami* Cockerell. *J. Res.* (Ludhiana Univ.). 6:359–371.
Kauffeld, N. M. 1980. Chemical analysis of Louisiana pollen and colony conditions during a year. *Apidologie* 11:47–55.
Keeler, K. H. 1978. Insects feeding at extrafloral nectaries of *Ipomoea carnea* (Convolvulaceae) *Entomol. News* 89:163–168.
Keeping, M. G., R. M. Crewe, and B. I. Field. 1982. Mandibular gland secretions of the Old World stingless bee, *Trigona gribodoi* Magretti: Isolation, identification, and compositional changes with age. *J. Apic. Res.* 21:65–73.
Kempff Mercado, N. 1962. Mutualism between *Trigona compressa* Latr., and *Crematogaster stolli* Forel. *J. N.Y. Entomol. Soc.* 70:215–217.
Kennedy, J. S., and I. H. M. Fosbrooke. 1972. The plant in the life of an aphid. Pp. 129–140 in H. F. van Emden, ed. *Insect–plant relationships.* Blackwell, Oxford.
Kerfoot, W. B. 1967a. Correlation between ocellar size and the foraging activities of bees. *Am. Nat.* 101:65–70.
———1967b. The lunar periodicity of *Specodogastra texana*, a nocturnal bee. *Anim. Behav.* 15:479–486.
Kerr, W. E. 1951. Bases para o estudo da genética dos Hymenoptera em general e dos Apinae sociais em particular. *An. Esc. Super. Agric. Luiz de Queiroz Univ. São Paulo* 8:220–354.
———1957. Multiple alleles and genetic load in bees. *J. Apic. Res.* 6:61–64.
———1959. Bionomy of meliponids. VI. Aspects of food gathering and processing in some stingless bees. Mimeo pp. 24–31 in Food gathering in Hymenoptera, *Symp. Entomol. Soc. Am.*
———1967. The history of the introduction of African bees to Brazil. *South African Bee J.* 39:3–5.
———1969. Some aspects of the evolution of social bees. *Evol. Biol.* 3:119–175. Appleton–Century–Crofts, New York.
———1972. Effect of low temperature on male meiosis in *Melipona marginata. J. Apic. Res.* 11:95–99.
———1974. Advances in cytology and genetics of bees. *Annu. Rev. Entomol.* 19:253–268.
———1975. Evolution of the population structure in bees. *Genetics* 79:73–84.
———1976. Population genetic studies in bees. II. Sex–limited genes. *Evolution* 30:94–99.
———1978. Papel das abelhas sociais na Amazônia. *Apimondia Int. Symp. on Apiculture in Hot Climates.* Pp. 119–129 (Florianópolis, Brazil).
———1984a. Virgilio de Portugal Brito Araújo (1919–1983). *Acta Amazonica* 13:327–328.
———1984b. História parcial da ciência apícola no Brasil. Pp. 47–60 in L. S. Gonçalves, A. E. E. Soares, D. DeJong, J. Steiner, M. R. Martinho, and N. Message, eds. *V Cong. Bras. Apic.*,

and III Cong. Latino-Ibero-Americ. de Apic. Imprensa Universitária, Viçosa, Minas Gerais, Brasil.
1987. Determinação do sexo nas abelhas. XVI. Informações adicionais sobre os genes XO, XA, e XB. *Rev. Bras. Biol.* 47:111–113.
Kerr, W. E., Y. Akahira, and C. A. Camargo. 1975. Sex determination in bees. IV. Genetic control of juvenile hormone production in *Melipona quadrifasciata*. *Genetics* 81:749–756.
Kerr, W. E., and M. G. de Almeida. 1981. Estudos em genética de populações em abelhas (Apidae, Hymenoptera) 3. Carga genética em *Melipona scutellaris* Latreille, 1811. *Rev. Bras. Biol.* 41: 137–139.
Kerr, W. E., A. Ferreira, and N. S. de Mattos. 1963. Communication among stingless bees – additional data. *J. N.Y. Entomol. Soc.* 71:80–90.
Kerr, W. E., and H. H. Laidlaw. 1956. General genetics of bees. *Advances in genetics* 8:109–153. Academic Press, New York.
Kerr, W. E., and E. Lello. 1962. Sting glands in stingless bees – a vestigal character (Hymenoptera: Apidae). *J. N.Y. Entomol. Soc.* 70:190–214.
Kerr, W. E., M. R. Martinho, and L. S. Gonçalves. 1980. Kinship selection in bees. *Rev. Bras. Genet.* 3:339–344.
Kerr, W. E., and V. Maule. 1964. Geographic distribution of stingless bees and its implications. *J. N.Y. Entomol. Soc.* 72:2–18.
Kerr, W. E., and D. Posey. In press. Um cipó que mata abelhas. *Rev. Bras. Zool.*
Kerr, W. E., S. del Rio, and M. D. Barrionuevo. 1982. The southern limits of the distribution of the Africanized honey bee in South America. *Am. Bee J.* 123:193–194.
Kerr, W. E., S. F. Sakagami, R. Zucchi, V. de Portugal-Araújo, and J. M. F. Camargo. 1967. Observações sôbre arquitetura dos ninhos e comprtamento de algumas espécies de abelhas sem ferrão dos vizenhanças de Manaus, Amazonas. *Atlas Simp. Biot. Amaz.* 3:255–309.
Kerr, W. E., and Z. V. da Silveira. 1972. Karyotype evolution in bees and corresponding taxonomic implications. *Evolution* 26:197–202.
Kerr, W. E., A. C. Stort, and M. J. Montenegro. 1966. Importância de alguns facôres ambientais na determinação das castas do gênero *Melipona*. *An. Acad. Bras. Cien.* 38:149–168.
Kerr, W. E., R. Zucchi, J. T. Nakadaira, and J. E. Butolo. 1962. Reproduction in the social bees. *J. N.Y. Entomol. Soc.* 70:265–276.
Kevan, P. G. 1975. Sun tracking solar furnaces in high arctic flowers: significance for pollination and insects. *Science* 189:723–726.
1978. Floral coloration, its colorimetric analysis and significance in anthecology. Pp. 51–79 in A. J. Richards, ed. *The pollination of flowers by insects.* Academic Press, New York.
1983. Floral colors through the insect eye: what they are and what they mean. Pp. 3–30 in C. E. Jones and R. J. Little, eds. *Handbook of experimental pollination ecology.* Van Nostrand Reinhold, New York.
Kevan, P. G., and H. G. Baker. 1983. Insects as flower visitors and pollinators. *Annu. Rev. Entomol.* 28:407–453.
Kevan, P. G., S. St. Helena, and I. Baker. 1983. Honey bees feeding from honeydew exudate of diseased Gambel's oak in Colorado. *J. Apic. Res.* 22:53–56.
Khoo, S.-G., and H.-S. Yong. 1987. Nest structure and colony defence in the stingless bee *Trigona terminata* Smith. *Nature Malaysiana* 12:4–15.
Kigatiira, I. K. 1984. Aspects of the ecology of the African honeybee. Dissertation, Cambridge University, Cambridge, England.
1988. Amalgamation in tropical honey bees. Pp. 62–71 in G. Needham, M. Delfinado-Baker, R. Page, and C. Bowman. eds. *Proc. Int. Conf. on Africanized Honey Bees and Bee Mites.* E. Horwood Ltd., Chichester, England.

Kigatiira, I. K., J. W. L. Beament, J. B. Free, and J. A. Pickett. 1986. Using synthetic pheromone lures to attract honeybee colonies in Kenya. 1986. *J. Apic. Res.* 25:85–86.

Kimsey, L. S. 1979. An illustrated key to the genus *Exaerete* with descriptions of male genitalia and biology (Hymenoptera: Euglossini, Apidae). *J. Kans. Entomol. Soc.* 52:735–746.

— 1980. The behavior of male orchid bees (Apidae, Hymenoptera, Insecta) and the question of leks. *Anim. Behav.* 28:996–1004.

— 1982a. Systematics of bees of the genus *Eufriesea*. *Univ. Calif. Publ. Entomol.* 95: 1–125.

— 1982b. Abstract. International Union for the Study of Social Insects Meetings at Boulder, Colo.

— 1984a. Re-evaluation of the phylogenetic relationships in the Apidae. *Syst. Entomol.* 9:435–441.

— 1984b. The behavioural and structural aspects of grooming and related activities in euglossine bees. *J. Zool., London* 204:541–550.

— 1987. Generic relationships within the Euglossini (Hymenoptera: Apidae). *Syst. Entomol.* 12:63–72.

Kimsey, L. S., and R. L. Dressler. 1986. Synonymic species list of Euglossini. *Pan-Pac. Entomol.* 62:229–336.

Kislow, C. J. 1976. The comparative biology of two species of small carpenter bees, *Ceratina strenua* F. Smith and *C. calcarata* Robertson. Dissertation, Univ. of Georgia, Athens.

Kistner, D. H. 1982. The social insects' bestiary. Pp. 2–244 in H. R. Hermann, ed. *Social insects*, Vol. 3. Academic Press, New York.

Kleinert-Giovannini, A. 1984. Aspectos do nicho trófico de *Melipona marginata marginata* Lepeletier (Apidae, Meliponinae). Dissertation, Inst. Biociências. Univ. São Paulo.

Kleinert-Giovannini, A., L. S. Guibu, and V. L. Imperatriz-Fonseca. 1983. Foraging activity of *Melipona marginata marginata* Lepeletier and *Melipona quadrifasciata quadrifasciata* Lepeletier. *Apimondia,* p. 130. Bucharest.

Klungness, L. M., and Y.-S. Peng. 1984. A histochemical study of pollen digestion in the alimentary canal of honey bees (*Apis mellifera* L.). *J. Insect Physiol.* 30:511–521.

Knerer, G. 1980. Evolution of halicine castes. *Naturwissenschaften* 67:133–135.

Knerer, G., and M. Schwarz. 1976. Halictine social evolution: the Australian enigma. *Science* 194:445–448.

Knuth, P. 1906–1909. *Handbook of flower pollination.* Clarendon Press, Oxford.

Knutson, L. V. 1978. Insects: Diptera (flies). Pp.128–137 in R. A. Morse, ed. *Honey bee pests, predators, and diseases.* Cornell Univ. Press, Ithaca, N.Y.

Kodric-Brown, A., J. H. Borwn, G. S. Byers, and D. F. Gori. 1984. Organization of a tropical island community of hummingbirds and flowers. *Ecology* 65:1358–1368.

Koeniger, G. 1983. Die Entfernung des Begattungszeichens bei der Mehrfachpaarung der Bienenkönigin. *Allg. Dtsch. Bienenztg.* 17:244–245.

— 1986. Reproductive and mating behavior. Pp. 255–282 in T. E. Rinderer, ed. *Bee genetics and breeding.* Academic Press, Orlando, Fla.

Koeniger, N. 1970. Factors determining the laying of drone and worker eggs by the queen honeybee. *Bee World* 51:166–169.

Koeniger, N. , G. Koeniger, and M. Delfinado-Baker. 1983. Observations on mites of the Asian honey bee species *Apis cerana, Apis dorsata* and *Apis florea. Apidologie* 14:197–204.

Koeniger, N., G. Koeniger, S. Tingek, M. Mardan, and T. E. Rinderer. 1988. Reproductive isolation by different time of drone flight between *Apis cerana* Fabricius, 1973 and *Apis vechti* (Maa, 1953). *Apidologie* 19:103–106.

Koeniger, N., G. Koeniger, and N. Wijayagunasekera. 1981. Observations on the adaptation of *Varroa jacobsoni* to its natural host *Apis cerana* in Sri Lanka. *Apidologie* 12:37–40.

Koeniger, N., and H. N. P. Wijayagunasekera. 1976. Time of drone flight in three Asiatic honeybee species (*A. cerana, A. florea, A. dorsata*) of Sri Lanka. *J. Apic. Res.* 15:67–71.
Kornberg, H., and M. H. Williamson, eds. 1987. *Quantitative aspects of the ecology of biological invasions.* Royal Society, London.
Korst, P. J. A. M., and H. H. W. Velthuis. 1982. The nature of trophallaxis in honeybees. *Insectes Soc.* 29:209–221.
Krebs, J. R., and R. H. McCleery. 1984. Optimization in behavioral ecology. Pp. 91–121 in J. R. Krebs and N. B. Davies, eds. *Behavioural ecology, an evolutionary approach.* Sinauer Assoc. Inc., Sunderland, Mass.
Krell, R., A. Dietz, and F. A. Eischen. 1985. A preliminary study on winter survival of Africanized and European honey bees in Cordoba, Argentina. *Apidologie* 16:109–118.
Krombein, K. V. 1967. *Trap-nesting wasps and bees: life histories, nests and associates.* Smithsonian Press, Washington, D.C.
———. 1969. *Life history notes on some Egyptian solitary bees and their associates.* Smithsonian Institution Press, Washington, D.C.
Kronenberg, S., and A. Hefetz. 1984. Comparative analysis of Dufour's gland secretions of two carpenter bees (Xylocopinae: Anthophoridae) with different nesting habits. *Comp. Biochem. Physiol.* 79B:421–425.
Kugler, H. 1942. Raphidenpollen bei Bromeliaceen. *Dtsch. Bot. Ges.* 30: 388–393.
Kukuk, P. F. 1985. Evidence for an antiaphrodisiac in the sweat bee *Lasioglossum (Dialictus) zephyrum. Science* 227:656–657.
Kukuk, P. F., M. D. Breed, A. Sobti and W. J. Bell. 1977. The contributions of kinship and conditioning to nest recognition and colony member recognition in a primitively eusocial bee, *Lasioglossum zephyrum* . *Behav. Ecol. Sociobiol.* 2:319–327.
Kukuk, P. F., and M. Schwarz. 1987. Intranest behavior of the communal sweat bee *Lasioglossum (Chialictus) erythrurum* (Hymenoptera: Halictidae). *J. Kans. Entomol. Soc.* 60:58–64.
Kullenberg, B., A.-K. Borg-Karlson, and A.-L. Kullenberg. 1984. Field studies on the behaviour of the *Eucera nigrilabris* male in the odor flow from flower labellum extract of *Ophrys tenthredinifera.* Pp. 79–117 in B. Kullenberg, G. Bergström, B. G. Svensson, J. Tengö, and L. Ågren, eds. *The ecological station of Uppsala University on Öland 1963–1983.* Almqvist and Wiksell, Int., Stockholm.
Kumbkarni, C. G. 1964. Cytological studies in Hymenoptera Part I: Cytology of parthenogenesis in the honeybees – *Apis dorsata. Indian J. Exp. Biol.* 2:65–68.
Kulinčević, J. M. 1986. Breeding accomplishments with honey bees. Pp. 391–414 in T. E. Rinderer, ed. *Bee genetics and breeding.* Academic Press, Orlando, Fla.
Kunkel, H., and W. Kloft. 1977. Fortschritte auf dem Gebiet der Honigtauforschung. *Apidologie* 8:369–391.
Kurihara, M., Y. Maeta, K. Chiba, and S. F. Sakagami. 1981. The relation between ovarian conditions and life cycle in two small carpenter bees, *Ceratina flavipes* and *C. japonica. J. Fac. Agric. Iwate Univ.* 15: 131–153.
Lacher, V. 1967. Verhaltensreaktionen der Bienenarbeiterin beei Dressur auf Kohlendioxid. *Z. Vrgl. Physiol.* 54:75–84.
Laidlaw, H. H., Jr. 1974. Relationships of bees within a colony. *Apiacta* 9:49–52.
Laidlaw, H. H., Jr., and R. E. Page Jr. 1986. Mating designs. Pp. 323–344 in T. E. Rinderer, ed. *Bee genetics and breeding.* Academic Press, Orlando, Fla.
Langenheim, J. H. 1969. Amber: A botanical inquiry. *Science* 163:1157–1169.
———. 1973. Leguminous resin-producing trees in Africa and South america. Pp. 89–104 in B. J.

Meggers, E. S. Ayensu, and W. D. Duckworth, eds. *Tropical forest ecosystems in Africa and South America*. Smithsonian Institution Press, Washington, D.C.

Langenheim, J. H., D. E. Lincoln, and C. E. Foster. 1978. Implications of variation in resin composition among organs, tissues and populations in the tropical legume *Hymenaea*. *Biochem. Syst. Ecol.* 6:299–313.

Langenheim, J. H., D. E. Lincoln, W. H. Stubblebine, and A. C. Gabrielli. 1982. Evolutionary implications of leaf resin pocket patterns in the tropical tree *Hymenaea* (Caesalpinioideae, Leguminosae). *Am. J. Bot.* 69:595–607.

Langenheim, J. H., and W. H. Stubblebine. 1983. Variation in leaf resin composition between parent tree and progeny in *Hymenea* – implications for herbivory in the humid tropics. *Biochem. Syst. Ecol.* 11:97–106.

Laroca, S. 1970. Contribuição para o cohecimento das relações entre abelhas e flôres: coleta de pólen das anteras tubulares de certas Melastomataceae. *Floresta* 2:69–74.

———. 1983. Biocoenotics of wild bees (Hymenoptera, Apoidea) at three nearctic sites, with comparative notes on some neotropical assemblages. Dissertation, Univ. Kansas, Lawrence.

Laroca, S., and M. C. Almeida. 1985. Adaptação dos palpos labiais de *Niltonia virgilii* (Hymenoptera, Apoidea, Colletidae) para coleta de néctar em *Jacaranda puberula* (Bignoniaceae), com descrição do macho. *Rev. Bras. Entomol.* 29:289–297.

Laroca, S., and S. Lauer. 1973. Adaptação comportamental de *Scaura latitarsis* para coleta de pólen (Hymenoptera, Apoidea). *Acta Biol. Para., Curitiba* 2:147–152.

Laroca, S., and A. I. Orth. 1984. Pilagem de um ninho de *Plebeia catamarcensis meridonalis* por *Lestrimelitta limao* (Apidae, Meliponinae) em Itapiranga, Santa Catarina, sul do Brasil. *Dusenia* 14:123–127.

Laroca, S., and A. M. Sakakibara. 1976. Mutualismo entre *Trigona hyalinata branneri* (Apidae) e *Aconophora flavipes* (Membracidae). *Rev. Bras. Entomol.* 20:71–72.

Laurie, A., and J. Sidensticker. 1979. Behavioural ecology of the sloth bear (*Melursus ursinus*). *J. Zool. London* 182:187–204.

Laverty, T. M. 1980. The flower-visiting behavior of bumblebees: floral complexity and learning. *Can. J. Zool.* 58:1324–1335.

Lee, G. L. 1974. The effect of gene dosage on variability in the honeybee. 2. Wing hook number. *J. Apic. Res.* 13:257–263.

Lehnert, T. 1978. Nematodes. Pp. 102–104 in R. A. Morse, ed. *Honey bee pests, predators, and diseases*. Cornell Univ. Press, Ithaca, N.Y.

Lello, E. 1976. Adnexal glands of the sting apparatus in bees: anatomy and histology, V. (Hymenoptera: Apidae). *J. Kans. Entomol. Soc.* 49:85–99.

Lenko, A. 1964. *Hoplomutilla triumphans* Mickel, 1939 (Hymenoptera, Mutillidae) como parasito de abelhas do gênero *Euplusia* (Hymenoptera, Apoidea). *Pap. Avulsos Depto. Zool. São Paulo.* 16:199–205.

Lensky, Y., and Y. Slabezki. 1981. The inhibitory effect of the queen bee (*Apis mellifera* L.) foot-print pheromone on the construction of swarming queen cups. *J. Insect Physiol.* 27:313–323.

Lester, L. J., and R. K. Selander. 1979. Population genetics of haplo-diploid insects. *Genetics* 92:1329–1345.

Levin, D. A. 1979. Pollinator foraging behavior: genetic implications for plants. Pp. 131–153 in O. T. Solbrig, S. Jain, G. B. Johnson, and P. H. Raven, eds. *Topics in plant population biology*. Columbia Univ. Press, New York.

Levin, M. D., and M. H. Hydak. 1957. Comparative value of different pollens in the nutrition of *Osmia lignaria*. *Bee World* 38:221–226.

Lewis, A. C. 1986. Memory constraints and flower choice in *Pieris rapae*. *Science* 232: 863–865.

Lewontin, R. C. 1974. *The genetic basis of evolutionary change*. Columbia Univ. Press, New York.
— 1985. Population genetics. Pp. 3–18 in P. J. Greenwood, P. H. Harvey, and M. Slatkin, eds. *Evolution: essays in honour of John Maynard Smith*. Cambridge Univ. Press, Cambridge, England.
Leyden, B. W. 1984. Guatemalan forest synthesis after Pleistocene aridity. *Proc. Natl. Acad. Sci. USA* 81:4856–4859
Lieberman, D., M. Lieberman, R. Peralta, and G. S. Hartshorn. 1985. Mortality patterns and stand turnover rates in a wet tropical forest in Costa Rica. *J. Ecol.* 73:915–924.
Lieftinck, M. A. 1955. The carpenter-bees (*Xylocopa* Latr.) of the Lesser Sunda Islands and Tanimbar. *Mus. Volkerkunde und des Naturhistorischen Museums*, Basel.
— 1972. Further studies on Old World melectine bees, with stray notes on their distribution and host relationships (Hymenoptera, Anthophoridae). *Tijdsch. Entomol.* 115:253–325.
— 1977. Notes on the melectine genus *Paracrocisa* Alfken, with a new record of *P. sinaitica* Alfken (Hymenoptera, Anthophoridae). *Entomol. Ber. Berl.* 37:125–127.
— 1983. Notes on the nomenclature and synonymy of the Old World melectine and anthophorine bees (Hymenoptera, Anthophoridae). *Tijdsch. Entomol.* 126:269–284.
Liévano, A., and R. Ospina. 1984. Contribución al conocimiento de los abejorros sociales de Cundinamarca (*Bombus*, Latreille) (Hymenoptera: Apidae). Dissertation, Univ. Nac. Colombia, Bogotá.
Lin, N., and C. D. Michener. 1972. Evolution and selection in social insects. *Q. Rev. Biol.* 47:131–159.
Lindauer, M. 1954. Temperaturregulierung und Wasserhaushalt im Bienenstaat. *Z. Vrgl. Physiol.* 36:391–432.
— 1955. Schwarmbienen auf Wohnungssuche. *Z. Vrgl. Physiol.* 37:263–324.
— 1961. *Communication among social bees*. Harvard Univ. Press, Cambridge, Mass.
— 1967. Recent advances in bee communication and orientation. *Annu. Rev. Entomol.* 12:439–470.
Linsley, E. G. 1958. The ecology of solitary bees. *Hilgardia* 27:543–599.
— 1960. Ethology of some bee and wasp-killing robber flies of southeastern Arizona and western New Mexico (Diptera: Asilidae). *Univ. Calif. Publ. Entomol.* 16:357–392.
— 1962a. The colletids *Ptiloglossa arizonensis* Timberlake, a matinal pollinator of *Solanum*. *Pan-Pac. Entomol.* 38:75–82.
— 1962b. Sleeping aggregations of aculeate Hymenoptera – II. *Ann. Entomol. Soc. Am.* 55:148–164.
— 1966. Pollinating insects of the Galápagos Islands. Pp. 225–232 in R. I. Bowman, ed. *The Galápagos Proceedings of the Symposia of the Galápagos International Scientific Project*. Univ. California Press, Berkeley.
— 1978. Temporal patterns of flower visitation by solitary bees, with particular reference to the southwestern United States. *J. Kans. Entomol. Soc.* 51:531–546.
Linsley, E. G., and M. A. Cazier. 1963. Further observations on bees which take pollen from plants of the genus *Solanum* (Hymenoptera: Apoidea). *Pan-Pac. Entomol.* 39:1–18.
Linsley, E. G., and J. W. MacSwain. 1958. The significance of floral constancy among bees of the genus *Diadasia* (Hymenoptera: Anthophoridae). *Evolution* 12:219–223.
Linsley, E. G., J. W. MacSwain, and C. D. Michener. 1980. Nesting biology and associates of *Melitoma* (Hymenoptera, Anthophoridae). *Univ. Calif. Publ. Entomol.* 90:1–46.
Linsley, E. G., J. W. MacSwain, and P. H. Raven. 1963. Comparative behavior of bees and Onagraceae. I. *Oenothera* bees of the Colorado desert; II. *Oenothera* bees of the Great Basin. *Univ. Calif. Publ. Entomol.* 33: 1–58.

Little, R. J. 1983. A review of floral food deception mimicries with comments on floral mutualism. Pp. 294–309 in C. E. Jones and R. J. Little, eds. *Handbook of experimental pollination biology.* Van Nostrand Reinhold, New York.

Lloyd, 1984. Gourmet insect behavioral ecology: stalking the wild speculation. *Fla. Entomol.* 67:1–5.

Lobreau-Callen, R. Darchen, and A. L. Thomas. 1986. Apport de la palynologie a la connaissance des relations abeilles/plantes en savanes arborées du Togo et du Bénin. *Apidologie* 17:279–306.

Louveaux, J., A Maurizio, and G. Vorwohl. 1978. Methods of melissopalynology. *Bee World* 59:139–157.

Louw, G. N., and S. W. Nicholson. 1983. Thermal, energetic and nutritional considerations in the foraging and reproduction of the carpenter bee *Xylocopa capitata. J. Entomol. Soc. S. Afr.* 46:227–240.

Luby, J. M., F. E. Regnier, E. T. Clark, E. C. Weaver, and N. Weaver. 1973. Volatile cephalic substances of the stingless bees, *Trigona mexicana* and *Trigona pectoralis. J. Insect Physiol.* 19:1111–1127.

Lutz, F. E. 1931. Light as a factor in controlling the start of daily activity of a wren and stingless bees. *Am. Mus. Novit.* No. 468.

Maa, T. C. 1953. An inquiry into the systematics of the tribus Apidini or honeybees. *Trebuia* 21:525–640.

MacArthur, R. H. 1972. *Geographical ecology.* Harper and Row, New York.

MacArthur, R. H., and E. O. Wilson. 1967. *The theory of island biogeography.* Princeton Univ. Press, Princeton, N.J.

McDade, L. A., and P. Davidar. 1984. Determinants of fruit and seed set in *Pavonia dasypetala* (Malvaceae). *Oecologia* 64:61–67.

McDade, L. A., and S. Kinsman. 1980. The impact of floral parasitism in two neotropical hummingbird-pollinated plant species. *Evolution* 34:944–958.

Macevicz, S. 1979. Some consequences of Fisher's sex ratio principle for social Hymenoptera that reproduce by colony fission. *Am. Nat.* 113:363–371.

McEvoy, M. V., and B. A. Underwood. 1988. The drone and species status of the Himalayan honey bee, *Apis laboriosa* (Hymenoptera: Apidae). *J. Kans. Entomol. Soc.* 61:246–249.

McGinley, R. J. 1980. Glossal morphology of the Colletidae and recognition of the Stenotritidae at the family level (Hymenoptera: Apoidea). *J. Kans. Entomol. Soc.* 53:539–552.

——— 1986. Studies of Halictinae (Apoidea: Halictidae), I: Revision of New World *Lasioglossum* Curtis. *Smithsonian Contrib. Zool.* No. 429.

McGinley, R. J., and J. G. Rozen, Jr. 1987. Nesting biology, immature stages, and phylogenetic placement of the palaearctic bee *Pararhophites* (Hymenoptera: Apoidea). *Am. Mus. Novit.* No. 2903.

Machado, J. O. 1971. Simbiôse entre as abelhas sociais brasileiras (Meliponinae, Apidae) e uma espécie de bacteria. *Ciência e Cultura* 23:625–633.

McIntosh, R. P. 1987. Pluralism in ecology. *Annu. Rev. Ecol. Syst.* 18:321–342.

Mackensen, O. 1943. The occurrence of parthenogenetic females in some strains of honey bees. *J. Econ. Entomol.* 36:465–467.

McLellan, A. R. 1978. Growth and decline of honeybee colonies and inter-relationships of adult bees, brood, honey and pollen. *J. Appl. Ecol.* 15:155–161.

Maeta, Y., N. Kubota, and S. F. Sakagami. 1987. *Nomada japonica* as a thelytokous cleptoparasitic bee, with notes on egg size and egg complement in some cleptoparasitic bees. *Kontyû* 55:21–31.

Maeta, Y., S. F. Sakagami, and C. D. Michener. 1985. Laboratory studies on the life cycle and nesting biology of *Braunsapis sauteriella,* a social xylocopine bee. *Sociobiology* 10:17–42.

Maeta, Y., M. Shiokawa, S. F. Sakagami, and C. D. Michener. 1984. Field studies in Taiwan on nesting behavior of a social xylocopine bee, *Braunsapis sauteriella. Kontyû* 52:266–277.

Malagodi, M., W. E. Kerr, and A. E. E. Soares. 1986. Introdução de abelhas na ilha de Fernando de Noronha. 2. População de *Apis mellifera ligustica. Ciência e Cultura* 38:1700–1704.

Malyshev, S. I. 1935. The nesting habits of solitary bees: a comparative study. *Eos* 11:201–309.

Manning, J. C., and D. J. Brothers. 1986. Floral relations of four species of *Rediviva* in Natal (Hymenoptera: Apoidea: Mellitidae). *J. Entomol. Soc. S. Afr.* 49:107–114.

Marden, J. H. 1984. Remote perception of floral nectar by bumble bees. *Oecologia* 64:232–240.

——— 1987. Maximum lift production during takeoff in flying animals. *J. Exp. Biol.* 130:235–258.

Marden, J. H., and K. D. Waddington. 1981. Floral choices by honey bees in relation to the relative distances to flowers. *Physiol. Entomol.* 6:431–435.

Mares, M. A. 1986. Conservation in South America: problems, consequences, and solutions. *Science* 233:734–739.

Martin, H., and M. Lindauer. 1977. The effect of the earth's magnetic field on gravity orientation in the honey bee. *J. Comp. Physiol.* 122:145–187.

Martinho, M. R. 1975. Digestion in vitro of pollen grains collected by workers of *Melipona quadrifasciata anthidioides.* Pp. 165–170. *Anais 3 Cong. Bras. Apic.*

Masson, C. 1982. Physiologie sensorielle et comportement de l'abeille. *C. R. Acad. Agric.* 1982:1350–1361.

Matsuura, M., and S. F. Sakagami. 1973. A bionomic sketch of the giant hornet, *Vespa mandarinia,* a serious pest for Japanese apiculture. *J. Fac. Sci. Hokkaido Univ. Ser. VI Zool.* 19:125–162.

Matthewson, J. A. 1968. Nest construction and life history of the eastern cucurbit bee, *Peponapis pruinosa. J. Kans. Entomol. Soc.* 41:255–261.

Maurizio, A. 1934. Über die Kalkbrut (*Pericystis*-Mykose) der Bienen. *Archiv. f. bienenkunde* 15:165–193.

——— 1953. Weitere Untersuchungen an Pollenhöschen. *Beihefte zür Schweizerischen Bienen-Zeitung.* 20: 485–556.

May, D. G. K. 1972. Water uptake during larval development of a sweat bee, *Augochlora pura. J. Kans. Entomol. Soc.* 45:439–449.

——— 1974. An investigation of the chemical nature and origin of the waxy lining of the brood cells of a sweat bee, *Augochlora pura* (Hymenoptera, Halictidae). *J. Kans. Entomol. Soc.* 47:504–516

May, M. L., and T. M. Casey. 1983. Thermoregulation and heat exchange in euglossine bees. *Physiol. Zool.* 56:541–551.

May, R. M., and J. Seger. 1986. Ideas in ecology. *Am. Sci.* 74:256–267.

Maynard Smith, J. 1982. *Evolution and the theory of games.* Cambridge Univ. Press, London.

Medler, J. T. 1980. Insects of Nigeria, check list and bibliography. *Mem. Am. Entomol. Institute* 30.

Medway, Lord. 1972. Phenology of a tropical rain forest in Malaya. *Bio. J. Linn. Soc.* 4:117–146.

Meeuse, B., and S. Morris. 1984. *The sex life of flowers.* Faber and Faber Ltd., London.

Mello, M. L. S., and B. C. Vidal. 1971. Histochemical and histophysical aspects of silk secretion in *Melipona quadrifasciata* (Hymenoptera, Apidae). *Z. Zellforsch. Mikrosk. Anat. Abt. Histochem.* 188:555–560.

Mello, M. L. S., and B. C. Vidal. 1979. A mucous secretion in the Malpighian tubes of a neotropical bumble bee, *Bombus atratus* Franklin. *Protoplasma* 99:147–158.

Messer, A. C. 1984. *Chalicodoma pluto*: The world's largest bee rediscovered living communally in termite nests (Hymenoptera: Megachilidae). *J. Kans. Entomol. Soc.* 57:165–168.

Mestriner, M. A., and E. P. B. Contel. 1972. The P-3 and EST loci in the honey bee, *Apis mellifera. Genetics* 72:733–738.

Michelsen, A., W. H. Kirchner, and M. Lindauer. 1986. Sound and vibrational signals in the dance language of the honeybee, *Apis mellifera. Behav. Ecol. Sociobiol.* 18:207–212.

Michener, C. D. 1946. Notes on some Panamanian stingless bees. *J. N.Y. Entomol. Soc.* 54:179–197.

―― 1954. Bees of Panamá. *Bull. Am. Mus. Nat. Hist.* 104:1–175.

―― 1961. Probable parasitism among Australian bees of the genus *Allodapula. Ann. Entomol. Soc. Am.* 54:532–534.

―― 1962. An interesting method of pollen collecting by bees from flowers with tubular anthers. *Rev. Biol. Trop.* 10:167–175.

―― 1964a. Reproductive efficiency in relation to colony size in hymenopterous societies. *Insectes Soc.* 11:317–342.

―― 1964b. Evolution of the nests of bees. *Am. Zool.* 4:227–239.

―― 1965a. The life cycle and social organization of bees of the genus *Exoneura* and their parasite, *Inquilina. Univ. Kans. Sci. Bull.* 46:317–358.

―― 1965b. A classification of the bees of the Australian and South Pacific regions. *Bull. Am. Mus. Nat. Hist.* 130:1–362.

―― 1968a. Nests of some African megachilid bees, with description of a new *Hoplitis. J. Entomol. Soc. S. Afr.* 31:337–359.

―― 1968b. Biological observations on primitively social bees (*Allodapula*) from Cameroon (Hymenoptera, Xylocopinae). *Insectes Soc.* 15:423–434.

―― 1969. Immature stages of a chalcidoid parasite tended by allodapine bees (Hymenoptera: Perilampidae and Anthophoridae). *J. Kans. Entomol. Soc.* 42:247–250.

―― 1970a. Social parasites among African allodapine bees. *Zool. J. Linn. Soc.* 49:199–215.

―― 1970b. Nest sites of stem and twig inhabiting African bees. *J. Entomol. Soc. S. Afr.* 33:1–22.

―― 1971. Biologies of African allodapine bees. *Bull. Am. Mus. Nat. Hist.* 145:221–301.

―― 1972a. Activities within artificial nests of an allodapine bee. *J. Kans. Entomol. Soc.* 45:263–268.

―― 1972b. Direct food transferring behavior in bees. *J. Kans. Entomol. Soc.* 45:373–376.

―― 1974a. *The social behavior of the bees: a comparative study.* Belknap Press of Harvard Univ. Press, Cambridge, Mass.

―― 1974b. Further notes on nests of *Ancyloscelis* (Hymenoptera: Anthophoridae). *J. Kans. Entomol. Soc.* 47:19–27.

―― 1975. The Brazilian bee problem. *Annu. Rev. Entomol.* 20:399–416.

―― 1977a. Allodapine bees of Madagascar. *Am. Mus. Novit.* No. 2622.

―― 1977b. Discordant evolution and the classification of allodapine bees. *Syst. Zool.* 26:32–56.

―― 1978. The parasitic groups of Halictidae (Hymenoptera, Apoidea). *Univ. Kans. Sci. Bull.* 51:291–339.

―― 1979. Biogeography of the bees. *Ann. Mo. Bot. Gard.* 66:277–347.

―― 1980. The large species of *Homalictus* and related Halictinae from the New Guinea area (Hymenoptera, Apoidea). *Am. Mus. Novit.* No. 2693.

―― 1981. Classification of the bee family Melittidae with a review of species of Meganomiinae. *Contrib. Am. Entomol. Inst. Ann Arbor* 18:1–135.

―― 1982a. A new interpretation of fossil social bees from the Dominican Republic. *Sociobiology* 7:37–46.

―― 1982b. Early stages in insect social evoltuion: individual and family odor differences and their functions. *Entomol. Soc. Am. Bull.* 28:7–11.

1983. The parasitic Australian allodapine genus *Inquilina* (Hymenoptera: Anthophoridae). *J. Kans. Entomol. Soc.* 56:555–559.

1984. A comparative study of the mentum and lorum of bees (Hymenoptera: Apoidea). *J. Kans. Entomol. Soc.* 57:705–714.

1985. From solitary to eusocial: need there be a series of intervening species? Pp. 293–305 in B. Hölldobler and M. Lindauer, eds. *Experimental behavioral ecology.* G. Fischer Verlag, New York.

1986. Family-group names among bees. *J. Kans. Entomol. Soc.* 59:219–234.

Michener, C. D., and M. Amir. 1977. The seasonal cycle and habitat of a tropical bumble bee. *Pac. Insects* 17:237–240.

Michener, C. D., and F. D. Bennett. 1977. Geographical variation in nesting biology and social organization of *Halictus ligatus*. *Univ. Kans. Sci. Bull.* 51:233–260.

Michener, C.D., M. D. Breed, and W. J. Bell. 1979. Seasonal cycles, nests and social behavior of some Colombian halictine bees. *Rev. Biol. Trop.* 27:13–34.

Michener, C. D., and R. W. Brooks. 1984. Comparative study of the glossae of bees. *Contrib. Am. Entomol. Inst. Ann Arbor* 22:1–73.

Michener, C. D., and L. Greenberg. 1980. Ctenoplectridae and the origin of long-tongued bees. *Zool. J. Linn. Soc.* 69:183–203.

Michener, C. D., and D. A. Grimaldi. 1988. A *Trigona* from Late Cretaceous amber of New Jersey (Hymenoptera: Apidae: Meliponinae). *Am. Mus. Novit.* No. 2917.

Michener, C. D., and R. B. Lange. 1958a. Observations on the ethology of neotropical anthophorine bees (Hymneoptera, Anthophoridae). *Univ. Kans. Sci. Bull.* 39:69–96.

1958b. Observations on the behavior of Brazilian halictid bees (Hymenoptera, Apoidea) IV. *Augochloropsis*, with note on extralimital forms. *Am. Mus. Novit.* No. 1924.

Michener, C. D., R. B. Lange, J. J. Bigarella, and R. Salamuni. 1958. Factors influencing the distribution of bees' nests in earth banks. *Ecology* 39:207–217.

Michener, C. D., and R. J. McGinley. In press. The genera of bees found north of Central America. *Smithsonian Contrib. Zool.*

Michener, C. D., and C. W. Rettenemeyer. 1956. The ethology of *Andrena erythronii* with comparative data on other species (Hymenoptera, Andrenidae). *Univ. Kans. Sci. Bull.* 37:645–684.

Michener, C. D., and A. Wille. 1961. The bionomics of a primitively social bee, *Lasioglossum inconspicuum*. *Univ. Kans. Sci. Bull.* 42:1123–1202.

Michener, C. D., M. L. Winston, and R. Jander. 1978. Pollen manipulation and related activities and structures in bees of the family Apidae. *Univ. Kan. Sci. Bull.* 51:575–601.

Mickel, C. E. 1928. Biological and taxonomic investigations on the mutillid wasps. *Bull. U.S. Natl. Mus.* 143:1–351.

Miller, N. C. E. 1956. *The biology of the Heteroptera.* Leonard Hill Ltd., London.

Montalvo, A. M., and J. D. Ackerman. 1986. Relative pollinator effectiveness and evolution of floral traits in *Spathiphyllum friedrichsthalii* (Araceae). *Am. J. Bot.* 73:1665–1676.

Moore, A. J., M. D. Breed, and M. J. Moor. 1987. The guard honey bee: ontogeny and behavioural variability of workers performing a specialized task. *Anim. Behav.* 35:1159–1167.

Moreno, J. E., and W. Devia. 1982. Estudio del origen botánico del polen y la miel almacenados por las abejas *Apis mellifera*, *Melipona eburnea* y *Tetragonisca angustula* en el municipio de Arbelaz, Colombia. Dissertation, Univ. Nac. Colombia, Bogotá.

Morgan, K. R., and G. A. Bartholomew. 1982. Homeothermic response to reduced ambient temperature in a scarab beetle. *Science* 216:1409–1410.

Mori, S. A., and J. J. Pipoli. 1984. Observations on the big bang flowering of *Miconia minutiflora* (Melastomataceae). *Brittonia* 36:337–341.

Moritz, R. F. A. 1983. Homogeneous mixing of honeybee semen by centrifugation. *J. Apic. Res.* 22:249–255.

———. 1985. The effects of multiple mating on the worker–queen conflict in *Apis mellifera* L. *Behav. Ecol. Sociobiol.* 16:375–377.

———. 1986. Genetics of bees other than *Apis mellifera*. Pp. 121–154 in T. E. Rinderer, ed. *Bee genetics and breeding.* Academic Press, Orlando, Fla.

Moritz, R. F. A., C. F. Hawkins, R. H. Crozier, and A. G. McKinlay. 1986. A mitochondrial DNA polymorphism in honeybees (*Apis mellifera* L.). *Experientia* 42:322–324.

Moritz, R. F. A., E. E. Southwick, and J. B. Harbo. 1987. Genetic analysis of defensive behaviour of honeybee colonies (*Apis mellifera* L.) in a field test. *Apidologie* 18:27–42.

Morse, D. H. 1977. Estimating proboscis length from wing length in bumble bees (*Bombus* spp.). *Ann. Entomol. Soc. Am.* 70:311–315.

———. 1978. Interactions among bumble bees on roses. *Insectes Soc.* 25:365–371.

———. 1979. Prey capture by the crab spider *Misumena vatia* (Clerck) (Thomisidae). *Oecologia* 39:309–319.

———. 1982. Behavior and ecology of bumble bees. Pp. 245–322 in H. R. Hermann, ed. *Social insects.* Vol. III. Academic Press, New York.

———. 1985. Costs in a milkweed–bumblebee mutualism. *Am. Nat.* 125:903–905.

Morse, R. A., ed. 1978. *Honey bee pests, predators and diseases.* Cornell Univ. Press, Ithaca, N.Y.

Morse, R. A., and F. M. Laigo. 1969. *Apis dorsata* in the Philippines. Monogr. No. 1. Philippine Assn. Entomologists, Inc.

Morse, R. A., D. A. Shearer, R. Boch, and A. W. Benton. 1967. Observations on alarm substances in the genus *Apis. J. Apic. Res.* 6:113–118.

Moure, J. S. 1950. Contribuição para o conhecimento do genero *Eulaema* Lepeletier (Hymenoptera, Apoidea). *Dusenia* 1:181–200.

———. 1961. A preliminary supraspecific classification of the Old World meliponine bees. *Stud. Entomol.* [Rio de Janeiro]. 4:181–242.

———. 1964. A key to the parasitic euglossine bees and a new species of *Exaerete* from Mexico. *Rev. Biol. Trop.* 12:15–18.

———. 1967. A checklist of the known euglossine bees. *Atlas Simpos. Biota Amazônica (Zool.)* 5:372–394.

Moure, J. S., and P. D. Hurd. 1987. *Halictidae of the Western Hemisphere.* Smithsonian Institution Press, Washington, D.C.

Moure, J. S., and W. E. Kerr. 1950. Sugestões para a modificação da sistematica do género *Melipona* (Hymenoptera, Apidae). *Dusenia* 1:105–131.

Moure, J. S., P. Nogueira-Neto, and W. E. Kerr. 1958. Evolutionary problems among Meliponinae. *Proc. X Int. Congr. Entomol.* [Montreal] 2:481–493.

Mower, R. L., and J. G. Hancock. 1975. Mechanism of honeydew formation by *Claviceps* species. *Can. J. Bot.* 53:2826–2834.

Mulcahy, D. L. 1979. The rise of the angiosperms: a genecological factor. *Science* 206:903–905.

Mulcahy, D. L., and G. B. Mulcahy. 1987. The effects of pollen competition. *Am. Sci.* 75:44–50.

Mulcahy, D. L., G. B. Mulcahy, and E. Ottaviano, eds. 1986. *Biotechnology and ecology of pollen.* Springer-Verlag, New York.

Muller, J. 1981. Fossil pollen records of extant angiosperms. *Bot. Rev.* 47:1–142.

Murillo, R. M. 1984. Uso y manejo actual de las colonias de *Melipona beecheii* Bennett (Apidae: Meliponini) en el estado de Tabasco, Mexico. *Biotica* 9:423–428.

Murrell, D. C., and W. T. Nash. 1981. Nectar secretion by toria (*Brassica campestris* L. v. *toris*) and foraging behaviour of three *Apis* species in Bangladesh. *J. Apic. Res.* 20:34–38.

Myers, J., and M. D. Loveless. 1976. Nesting aggregations of the euglossine bee *Euplusia surinamensis* (Hymenoptera: Apidae): Individual interactions and the advantages of living together. *Can. Entomol.* 108:1–6.

Myers, N. 1984. *The primary source: tropical forests and our future earth.* W. W. Norton and Co., New York.

Nates, G., and O. Cepeda. 1983. Comportamiento defensivo en algunas especies de meliponínos colombianos. *Bol. Dept. Biol. Univ. Nac. Colombia, Bogotá.* 1:65–81.

Needham, G. R., M. Delfinado-Baker, R. E. Page, Jr., and C. E. Bowman, eds. 1988. *Proc. Int. Conf. on Africanized Honey Bees and Bee Mites.* E. Horwood, Chichester, England.

Neff, J. L. 1984. Observations on the biology of *Eremapis parvula* Ogloblin, an anthophorid bee with a metasomal scopa (Hymenoptera: Anthophoridae). *Pan-Pac. Entomol.* 60:155–162.

Neff, J. L., and B. B. Simpson. 1981. Oil-collecting structures in the Anthophoridae (Hymenoptera): Morphology, function and use in systematics. *J. Kans. Entomol. Soc.* 54:95–123.

―――. 1988. Pollen-harvesting by *Megachile mendica* Cresson (Hymenoptera: Megachilidae). *J. Kans. Entomol. Soc.* 61:242–244.

Nelson, G., and N. Platnick. 1981. *Systematics and biogeography: cladistics and vicariance.* Columbia Univ. Press, New York.

Nevo, E., A. Beiles, and R. Ben-Shlomo. 1984. The evolutionary significance of genetic diversity: ecology, demography and life-history correlates. Pp. 13–213 in G. S. Mani, ed. *Evolutionary dynamics of genetic diversity.* Lecture Notes in Biomathematics, Vol. 53. Springer-Verlag, W. Berlin.

Nicholson, S. W., and G. N. Louw. 1982. Simultaneous measurement of evaporative water loss, oxygen consumption and thoracic temperature during flight in a carpenter bee. *J. Exp. Zool.* 222:287–296.

Nightingale, J. 1983. *A lifetime's recollection of Kenya tribal beekeeping.* International Bee Research Association, London.

Ninjhout, H. E., and D. E. Wheeler. 1982. Juvenile hormone and the physiological basis of insect polymorphisms. *Q. Rev. Biol.* 57:109–133.

Nogueira-Neto, P. 1948. Notas bionômicas sôbre meliponíneos. I. Sôbre a ventilação dos ninhos e as construções com ela relacionadas. *Rev. Bras. Biol.* 8:465–488.

―――. 1949. Notas bionômicas sôbre meliponíneos. II. Sôbre a pilhagen. *Pap. Avulsos Dep. Zool. São Paulo* 9:13–31.

―――. 1950. Notas bionômicas sôbre meliponíneos. IV. Colonias mistas e questões relacionadas. *Rev. Entomol. Rio de J.* 21:305–367.

―――. 1954. Notas bionômicas sôbre meliponíneos. III. Sôbre a enxameagem. *Arq. Mus. Nac. Rio de J.* 42:219–452.

―――. 1957. Colheita de pólen e seu armazenamento provisorio na região ventral do torax de muitos Trigonini. *Chacaras e Quintais.* 11:880.

―――. 1962. The scutellum nest structure of *Trigona (Trigona) spinipes* Fab. (Hymenoptera, Apidae). *J. N.Y. Entomol. Soc.* 70:239–264.

―――. 1964. The spread of a fierce African bee in Brazil. *Bee World* 45:119–121.

―――. 1970a. Behavior problems related to the pillages made by some parasitic stingless bees (Meliponinae, Apidae). Pp. 416–434 in *Development and evolution in behavior, essays in memory of T. C. Schneirla.* W. H. Freeman and Co., San Francisco.

―――. 1970b. *A criação de abelhas indígenas sem ferrão.* Chacaras e Quintais, São Paulo.

Nogueira-Neto, P., V. L. Imperatriz-Fonseca, A Kleinert-Giovannini, B. F. Viana, and M. S. de Castro. 1986. *Biologia e manejo das abelhas sem ferrão.* Edição Tecnapis, São Paulo.

Nogueira-Neto, P., and S. F. Sakagami. 1966. Nest structure of a subterranean stingless bee – *Geotrigona mombuca* Smith (Meliponinae, Apidae, Hymenoptera). *An. Acad. Bras. Cien.* [Rio de Janeiro] 38:187–194.

Norden, B., and S. W. T. Batra. 1985. Male bees sport black moustaches for picking up parsnip perfume (Hymenoptera: Anthophoridae). *Proc. Entomol. Soc. Wash.* 87:317–322.

Norden, B., S. W. T. Batra, H. F. Fales, A. Hefetz, and G. J. Shaw. 1980. *Anthophora* bees: unusual glycerides from maternal Dufour's glands serve as larval food and cell lining. *Science* 207:1095–1097.

Nuñez, J. A. 1973. Estudio cuantitativo del comportamiento de *Apis mellifera ligustica* Spinola y *Apis mellifera adansonii:* Factores energeticos e informacionales condicionates y estratégia del trabajo recolector. *Apiacta* 8:151–154.

——— 1979. Time spent on various components of foraging activity: comparison betrween European and Africanized honeybees in Brazil. *J. Apic. Res.* 18:110–115.

——— 1982. Honey bee foraging strategies at a food source in relation to its distance from the hive and the rate of sugar flower. *J. Apic. Res.* 21:139–150.

Nunnamaker, R. A., W. T. Wilson, and R. Ahmad. 1984. Malate dehydrogenase and non-specific esterase isoenzymes of *Apis florea, A. dorsata* and *A. cerana* as detected by isoelectric focusing. *J. Kans. Entomol. Soc.* 57:591–595.

O'Connor, B. M. 1982. Evolutionary ecology of astigmatid mites. *Annu. Rev. Entomol.* 27:385–409.

——— 1988. Coevolution in astigmatid mite–bee associations. Pp. 339–346 in G. R. Needham, M. Delfinado-Baker, R. E. Page, Jr., and C. E. Bowman, eds., *Proc. Int. Conf. on Africanized Honey Bees and Bee Mites.* E. Horwood, Chichester, England.

Odum, E. P. 1971. *Fundamentals of Ecology* (3rd ed.). Saunders, Philadelphia.

O'Dowd, D. J. 1979. Foliar nectar production and ant activity on a neotropical tree, *Ochroma pyramidale. Oecologia* 43:233–248.

Okumura, G. T. 1966. The dried-fruit moth (*Vitula edmandsae serratilineella* Ragonot), pest of dried fruits and honeycombs. *Bull. Calif. Dept. Agric.* 55:180–186.

Olesen, J. M. 1985. The Macronesian bird–flower element and its relation to bird and bee opportunists. *Bot. J. Linn. Soc.* 91:395–414.

O'Neill, K., and L. Bjostad. 1987. The male mating strategy of the bee *Nomia nevadensis* (Hymenoptera: Halictidae): leg structure and mate guarding. *Pan-Pac. Entomol.* 63:207–217.

Ono, M., I. Okada, and M. Sasaki. 1987. Heat production by balling in the Japanese honeybee, *Apis cerana japonica* as a defensive behavior against the hornet, *Vespa simillima xanthoptera* (Hymenoptera: Vespidae). *Experientia* 43:1031–1032.

Opler, P. A. 1983. Nectar production in a tropical ecosystem. Pp. 30–79 in B. L. Bentley and T. S. Elias, eds. *The biology of nectaries.* Columbia Univ. Press, New York.

Opler, P. A., H. G. Baker, and G. W. Frankie. 1975. Reproductive biology of some Costa Rican *Cordia* species (Boraginaceae). *Biotropica* 7:234–247.

Opler, P. A., G. W. Frankie, and H. G. Baker. 1980a. Comparative phenological studies of treelet and shrub species in tropical wet and dry forests in the lowlands of Costa Rica. *J. Ecol.* 68:167–188.

——— 1980b. Plant reproductive characteristics during secondary succession in neotropical lowland forest ecosystems. Symp. on tropical succession, *Biotropica* Suppl. 12:40–46.

Örösi-Pal, Z. 1966. Die bienenlaus-Arten. *Angew. Parasitol.* 7:138–171.

Ospina, H. M. 1969. Los antipolinizadores. *Orquideología* 4:23–27.

Oster, G., and B. Heinrich. 1976. Why do bumble bees major? A mathematical model. *Ecol. Monogr.* 46:129–133.

Oster, G., and E. O. Wilson. 1978. *Caste and ecology in the social insects.* Princeton Monogr. Popul. Biol., Princeton Univ. Press, Princeton, N.J.

Otis, G. W. 1982. Population biology of the Africanized honey bee. Pp. 209–219 in P. Jaisson ed. *Social insects in the tropics.* Vol. 1. Univ. Paris–Nord.

Otis, G. W., M. L. Winston, and O. R. Taylor. 1981. Engorgement and dispersal of Africanized honeybee swarms. *J. Apic. Res.* 20:3–12.

Otis, G. W., R. L McGinley, L. Garling, and L. Malaret. 1982. Biology and systematics of the bee genus *Crawfordapis* (Colletidae, Diphaglossinae). *Psyche* 89:279–296.

Owen, D. F. 1978. Abundance and diversity of bumblebees and cuckoo bees in a suburban garden. *Entomol. Rec. J. Var.* 90:242–244.

Owen, R. E. 1983. Chromosome numbers of 15 North American bumble bee species. *Can. J. Genet. Cytol.* 25:26–29.

Packer, L. 1985. Two social halictine bees from southern Mexico with a note on two bee hunting philanthine wasps (Hymenoptera: Halictidae and Sphecidae). *Pan-Pac. Entomol.* 61:291–298.

―――. 1986. The biology of a subtropical population of *Halictus ligatus.* IV: A cuckoo-like caste. *J. N.Y. Entomol. Soc.* 94:458–466.

Packer, L., and G. Knerer. 1987. The biology of a subtropical population of *Halictus ligatus* Say (Hymenoptera; Halictidae). III. The transition between annual and continuously brooded colony cycles. *J. Kans. Entomol. Soc.* 60:510–516.

Page, R. E., Jr. 1980. The evolution of multiple mating behavior by honey bee queens (*Apis mellifera* L.). *Genetics* 96:263–273.

―――. 1986. Sperm utilization in social insects. *Annu. Rev. Entomol.* 31:297–320.

Page, R. E., Jr., and E. H. Erickson, Jr. 1986. Kin recognition and virgin queen acceptance by worker honey bees (*Apis mellifera* L.). *Anim. Behav.* 34:1061–1069.

Page, R. E., Jr., and R. A. Metcalf. 1982. Multiple mating, sperm utilization, and social evolution. *Am. Nat.* 119:263–281.

Palacios Chavez, R. 1985. Lluvia de polen moderno en el bosque tropical caducifolio de la estación de biología de Chamela, Jalisco (México). *An. Esc. Nac. Cienc. Biol. Méx.* 29:43–55.

Pamilo, P., A. Pekkarinen, and S.-L.Varvio-Aho. 1981. Phylogenetic relationship and the origin of social parasitism in Vespidae and in *Bombus* and *Psithyrus* as revealed by enzyme genes. Pp. 37–48 in P. E. Howse and J. L. Clement, eds. *Biosystematics of social insects.* Academic Press, New York.

Pamilo, P., S.-L. Varvio-Aho, and A. Pekkarinen. 1978. Low enzyme gene variability in Hymenoptera as a consequence of haplodiploidy. *Hereditas* 88:93–99.

Parker, F. D., S. W. T. Batra, and V. J. Tepedino. 1987. New pollinators for our crops. *Agric. Zool. Rev.* 2:279–304.

Parker, F. D., and R. M. Bohart. 1966. Host–parasite associations in some twig-nesting Hymenoptera from western North America. *Pan-Pac. Entomol.* 42:91–98.

―――. 1968. Host–parasite associations in some twig-nesting Hymenoptera from western North America. *Pan-Pac. Entomol.* 44:1–6.

Parker, F. D., J. H. Cane, G. W. Frankie, and S. B. Vinson. 1987. Host records and nest entry by *Dolichostelis*, a kleptoparasitic anthidiine bee (Hymenoptera: Megachilidae). *Pan-Pac. Entomol.* 63:172–177.

Parker, G. A. 1984. Evolutionary stable strategies. Pp. 30–61 in J. R. Krebs and N. B. Davies, eds. *Behavioral ecology, an evolutionary approach.* Sinauer Assoc. Inc., Sunderland, Mass.

Parker, G. A., and D. I. Rubenstein. 1981. Role assessment reserve strategy, and acquisition of information in asymmetric animal contests. *Anim. Behav.* 29:221–240.

Parsons, P. A. 1983. *The evolutionary biology of colonizing species.* Cambridge Univ. Press, Cambridge, England.

Pasteels, J. M. 1977a. The Megachilini parasites (*Coelioxys*) of tropical Africa. Genera and subgeneric subdivisions. *Rev. Zool. Afr.* 91:161–197.

1977b. Une revue comparative de L'èthologie des Anthidiinae nidificateurs de L'ancien monde (Hymenoptera, Megachilidae). *Ann. Soc. Entomol. Fr.* 13:651–658.

Pasteels, J. M., and J. J. Pasteels. 1975. Etude au microscope electronique a balayage des scopas collectrices de pollen chez les Fideliidae (Hymenoptera, Apoidea). *Arch. Biol.* 86:453–466.

Pate, V. S. L. 1947. Neotropical Sapygidae, with a conspectus of the family (Hymenoptera: Aculeata). *Acta Zool. Lilloana* 4:393–426.

Pearson, D. L., and R. L. Dressler. 1985. Two-year study of male orchid bee (Hymenoptera: Apidae: Euglossini) attraction to chemical baits in lowland southeastern Perú. *J. Trop. Ecol.* 1:37–54.

Pearson, J. F. W. 1933. Studies on ecological relations of bees in the Chicago region. *Ecol. Monogr.* 3:373–441.

Pekkarinen, A. 1979. Morphometric, color and enzyme variation in bumble bees in Fennoscandia and Denmark. *Acta Zool. Fenn.* 158:1–60.

Pekkarinen, A., S.-L. Varvio-Aho, and S. L. Pamilo. 1978. Evolutionary relationships in North European *Bombus* and *Psithyrus* species (Hymenoptera: Apidae) studied on the basis of allozymes. *Ann. Entomol. Fenn.* 45:77–80.

Peng, Y.-S. C. 1988. The resistance mechanism of the Asian honey bee (*Apis cerana* Fab.) to *Varroa jacobsoni*. Pp. 426–429 in G. Needham, M. Delfinado-Baker, R. Page, and C. Bowman, eds. *Proc. Int. Conf. on Africanized Honey Bees and Bee Mites.* E. Horwood Ltd., Chichester, England.

Peng, Y.-S. C., and Y. Fang. 1988. Removal of mite *Varroa jacobsoni* from European honey bee *Apis mellifera* L. brood by the foster Asian honey bee *Apis cerana* Fabr. Pp. 430–433 in G. Needham, M. Delfinado-Baker, R. Page, and C. Bowman, eds. *Proc. Int. Conf. on Africanized Honey Bees and Bee Mites.* E. Horwood Ltd., Chichester, England.

Percival, M. S. 1961. Types of nectar in angiosperms. *New phytol.* 60:235–281.

——— 1965. *Floral biology.* Pergamon, Oxford.

Pereira, M. R., and G. Gottsberger. 1980. A polinização de *Aspilla floribunda* (Asteraceae) e *Cochlospermum regium* (Cochlospermaceae) e a relação das abelhas visitantes com outras plantas do cerrado de Botucatu, estado de São Paulo. *Rev. Braz. Bot.* 3:67–77.

Perkins, R. C. L., and A. Forel. 1899. *Fauna Hawaiiensis.* Hymenoptera, Aculeata, Vol. 1, Part 1. Cambridge Univ. Press, London.

Perry, D. A., and A. Starrett. 1980. The pollination ecology and blooming strategy of a neotropical emergent tree, *Dipteryx panamensis. Biotropica* 12:307–311.

Phadke, R. P. 1961. Some physio-chemical constants of Indian beeswaxes. *Bee World* 42:149–153.

Pham-Delegue, M. H., C. Masson, P. Etievant, and M. Azar. 1986. Selective olfactory choices of the honeybee among sunflower aromas: a study by combined olfactory conditioning and chemical analysis. *J. Chem. Ecol.* 12:781–793.

Pianka, E. R. 1986. *Ecology and natural history of desert lizards.* Princeton Univ. Press, Princeton, N.J.

Pickel, D. B. 1928a. Contribuição para a biologia do phorídeos. *Bol. Mus. Nac. Rio de J.* 4:67–68.

——— 1928b. Contribuição para a biologia de *Centris* (*Melanocentris*) *sponsa* e *Acanthopus excellens* (Hymenoptera). *Bol. Biol. Mus. Paulista* 6:135–143.

Pickett, C. H., and W. D. Clark. 1979. The function of extrafloral nectaries in *Opuntia acanthocarpa* (Cactaceae). *Am. J. Bot.* 66:618–625.

Pickett, J. A., I. H. Williams, A. P. Martin, and M. C. Smith. 1980. Nasanov pheromone of the honey bee, *Apis mellifera* L. (Hymenoptera: Apidae) Part I. Chemical characterization. *J. Chem. Ecol.* 6:425–434.

Pijl, L. van der. 1954. *Xylocopa* and flowers in the tropics. I–III. *K. Nederlandse ak. weterns. Proc.* 57, Ser. C., 4133–4123, 541–562.
Plateaux-Quénu, C. 1959. Un nouveau type de societe d'insectes: *Halictus marginatus* Brulle (Hymenoptera, Apoidea). *Année Biol.* 35:326–444.
——— 1972. *La biologie des abeilles primitives.* Masson, Paris.
Pleasants, J. M. 1983. Structure of plant and pollinator communities. Pp. 375–393 in C. E. Jones and R. J. Little, eds. *Handbook of experimental pollination biology.* Van Nostrand Reinhold, New York.
Plowright, R. C., and T. M. Laverty. 1984. The ecology and sociobiology of bumble bees. *Annu. Rev. Entomol.* 29:175–199.
Plowright, R. C., and R. E. Owen. 1980. The evolutionary significance of bumble bee color patterns: a mimetic interpretation. *Evolution* 34:622–637.
Plowright, R. C., and M. J. Pallett. 1979. Worker–male conflict and inbreeding in bumble bees. *Can. Entomol.* 111:289–294.
Plowright, R. C., B. A. Peridrel, and I. A. McLaren. 1978. The impact of aerial Fenitrothion spraying upon the population biology of bumble bees in Southwestern New Brunswick. *Can. Entomol.* 110:1145–1156.
Pohl, F. 1941. Uber raphidenpollen und seine blutenokologische Bedeutung. *Oesterr. Bot. Z.* 90:81–96.
Poinar, G. O. Jr. 1983. *The natural history of nematodes.* Prentice-Hall, Englewood Cliffs, N.J.
Pomeroy, N. 1979. Brood bionomics of *Bombus ruderatus* in New Zealand (Hymenoptera: Apidae). *Can. Entomol.* 111:865–874.
Portugal-Araújo, V. de. 1955. Notas sôbre colônias del meliponíneos de Angola, Africa. *Dusenia* 6:97–114
——— 1958. A contribution to the bionomics of *Lestrimelitta cubiceps. J. Kans. Entomol. Soc.* 31:203–211.
——— 1963. Subterranean nests of two African stingless bees. *J. N.Y. Entomol. Soc.* 71:130–141.
——— 1971. The central African bee in South America. *Bee World* 7:91–102.
——— 1977. The biology of *Pseudohypocera kerteszi* (Enderlein, 1912). Its mating behavior and capture. *Acta Amazonica* 7:153–166.
Posey, D. A. 1983. Folk apiculture of the Kayapó Indians of Brazil. *Biotropica* 15:154–158.
Posey, D. A., and J. M. F. Camargo. 1985. Additional notes on the classification and knowledge of stingless bees by the Kayapó indiana of Gorotire, Pará, Brazil. *Ann. Carnegie Mus.* 54:247–274.
Post, D. C., R. E. Page, and E. H. Erickson. 1987. Honeybee (*Apis mellifera* L.) queen feces: source of a pheromone that repels worker bees. *J. Chem. Ecol.* 13:583–591.
Poulton, E. B. 1924. The relation between the larvae of the asilid genus *Hyperechia* and those of xylocopid bees. *Trans. Entomol. Soc. London* 121–133.
Pouvreau, A. 1974. Le comportement alimentaire des bourdons (*Bombus* Latr.): la consommation de solutions sucrées. *Apidologie* 5:247–270.
Powell, A., and G. Powell. 1987. Population dynamics of male euglossine bees in Amazonian forest fragments. *Biotropica* 19:176–179.
Powell, J. A., and G. I. Stage. 1962. Prey selection by robberflies of the genus *Stenopogon*, with particular observations on *S. engelhardti* Bromley (Diptera: Asilidae). *Wasmann J. Biol.* 20: 139–157.
Prance, G. T. 1985. The pollination of Amazonian plants. Pp. 166–191 in G. T. Prance and T. E. Lovejoy, eds. *Key environments: Amazonia.* Pergamon Press, Oxford.
Price, P. W. 1980. *Evolutionary biology of parasites.* Princeton Univ. Press, Princeton, N.J.

1984. Communities of specialists: vacant niches in ecological and evolutionary time. Pp. 510–524 in D. R. Strong, D. Simberloff, L. G. Arele, and A. B. Thistle, eds. *Ecological communities: conceptual issues and the evidence.* Princeton Univ. Press, Princeton, N.J.
Price, P. W., C. N. Slobodchikoff, and W. S. Gaud. eds. 1984. *A new ecology: novel approaches to interactive systems.* Wiley–Interscience, New York.
Primack, R. B. 1985. Longevity of individual flowers. *Annu. Rev. Ecol. Syst.* 16:15–38.
Proctor, M., and P. Yeo. 1973. *The pollination of flowers.* Collins, London.
Pyke, G. H. 1978a. Optimal foraging: Movement patterns of bumble bees between inflorescences. *Theor. Popul. Biol.* 13:72–98.
1978b. Optimal body size in bumble bees. *Oecologia* 34:255–266.
1982. Local geographic distributions of bumble bees near Crested Butte Colorado: competition and community structure. *Ecology* 63:555–573.
1984. Optimal foraging theory: A critical review. *Annu. Rev. Ecol. Syst.* 15:523–575.
Pyke, G. H., and L. Blazer. 1982. The effects of the introduced honey bee *Apis mellifera* on Australian native bees. Report to the New South Wales National Parks and Wildlife Service.
Qayyum, H. A., and N. Ahmad. 1967. Biology of *Apis dorsata. Pak. J. Sci.* 19:109–113.
Ramalho, M., V. I. Imperatriz-Fonseca, A. Kleinert-Giovannini, and M. Cortopassi-Laurino. 1985. Exploitation of floral resources by *Plebeia remota* Holmberg (Apidae, Meliponinae). *Apidologie* 16:307–330.
Ramberg, F. B., P. Kukuk and W. L. Brown. 1984. Karyotypes of three species of Halictidae (Hymenoptera: Apoidea). *J. Kans. Entomol. Soc.* 57:159–161.
Ramírez, W. 1982. *Bombus mexicanus* Cresson, un hospedante de *Melaloncha*, moscas parásitas de la abeja de miel en el nuevo mundo. *Rev. Biol. Trop.* 30:177.
Ranta, E. 1982. Species structure of North European bumblebee communities. *Oikos* 38:202–209.
Rathcke, B. 1983. Competition and facilitation among plants for pollination. Pp. 305–330 in L. Real, ed. *Pollination biology.* Academic Press, New York.
Rau, P. 1926. The ecology of a sheltered clay bank: a study in insect sociology. *Trans. Acad. Sci. St. Louis* 25:159–276.
1929. Experimental studies on the homing of carpenter and mining bees. *J. Comp. Psychol.* 9:35–70.
1933. *The jungle bees and wasps of Barro Colorado Island (with notes on other insects).* Kirkwood, Mo.
Raven, P. H., and D. I. Axelrod. 1974. Angiosperm biogeography and past continental movements. *Ann. Mo. Bot. Gard.* 61:539–673.
Raw, A. 1975. Territorality and scent marking by *Centris* males (Hymenoptera, Anthophoridae) in Jamaica. *Behaviour* 54:311–321.
1976. Seasonal changes in the numbers and foraging activities of two Jamaican *Exomalopsis* species (Hymenoptera, Anthophoridae). *Biotropica* 8:270–277.
1977. The biology of two *Exomalopsis* species (Hymenoptera: Anthophoridae) with remarks on sociality in bees. *Rev. Biol. Trop.* 25:1–11.
1984. The nesting biology of nine species of Jamaican bees. *Rev. Bras. Entomol.* 28:497–506.
1985. The ecology of Jamaican bees (Hymenoptera). *Rev. Bras. Entomol.* 29:1–16.
Raw, A., and C. O'Toole. 1979. Errors in the sex of eggs laid by the solitary bee *Osmia rufa* (Megachilidae). *Behaviour* 70:168–171.
Real, L. A. 1981. Nectar availability and bee-foraging on *Ipomoea* (Convolvulaceae). *Biotropica (Reproductive Botany Suppl.)*, pp. 64–69.
ed. 1983. *Pollination biology.* Academic Press, New York.
Reddy, C. C. 1980. Observations on the annual cycle of foraging and brood rearing by *Apis cerana indica* colonies. *J. Apic. Res.* 19:17–20.

Reich, P. B., and R. Borchert. 1984. Water stress and tree phenology in a tropical dry forest in the lowlands of Costa Rica. *J. Ecol.* 72:61–74.
Renner, S. 1983. The widespread occurrence of anther destruction by *Trigona* bees in Melastomataceae. *Biotropica* 15:251–256.
———. 1984. *Phänologie, Blütenbiologie und rekimbinationssysteme einiger zentralamazonischer Merlastomataceen.* Dissertation, Univ. Hamburg.
———. In press. A survey of reproductive biology in neotropical Melastomataceae and Memecylaceae. *Ann. Mo. Bot. Gard.*
Rettenmeyer, C. W. 1970. Insect mimicry. *Annu. Rev. Entomol.* 15:43–74.
Reyes, F. 1983. A new record of *Pseudohypocera kerteszi*, a pest of honey bees in Mexico. *Am. Bee J.* 123:119–120.
Ribbands, C. R. 1953. *The behaviour and social life of honey bees.* Bee Res. Assn. Ltd., London.
Richards, P. W. 1980. *The tropical rain forest: an ecological study.* Cambridge Univ. Press, Cambridge, England.
Richardson, P. T., D. A. Baker, and L. C. Ho. 1982. The chemical composition of cucurbit vascular exudates. *J. Exp. Bot.* 33:1239–1247.
Ricklefs, R. E., R. M. Adams, and R. L. Dressler. 1969. Species diversity of *Euglossa* in Panama. *Ecology* 50:713–716.
Rinderer, T. E. 1982. Regulated nectar harvesting by the honeybee. *J. Apic. Res.* 21:74–87.
———. 1986. Selection. Pp. 305–321 in T. E. Rinderer, ed. *Bee breeding and genetics.* Academic Press, New York.
———. 1988. Evolutionary aspects of the Africanization of honey-bee populations in the Americas. Pp. 13–28 in G. Needham, M. Delfinado-Baker, R. Page, and C. Bowman, eds. *Proc. Int. Conf. on Africanized Honey Bees and Bee Mites.* E. Horwood Ltd., Chichester, England.
———, ed. 1986. *Bee breeding and genetics.* Academic Press, Orlando, Fla.
Rinderer, T. E., A. B. Bolten, A. M. Collins, and J. R. Harbo. 1984. Nectar-foraging characteristics of Africanized and European honey bees in the neotropics. *J. Apic. Res.* 23:70–79.
Rinderer, T. E., A. M. Collins, A. B. Bolten, and J. R. Harbo. 1981. Size of nest cavities selected by swarms of Africanized honeybees in Venezuela. *J. Apic. Res.* 20:160–164.
Rinderer, T. E., A. M. Collins, and K. W. Tucker. 1985. Honey production and underlying nectar harvesting activities of Africanized and European honeybees. *J. Apic. Res.* 23:161–167.
Rinderer, T. E., R. L. Hellmich, R. G. Danka, and A. M. Collins. 1985. Male reproductive parasitism: a factor in the Africanization of European honey-bee populations. *Science* 228:1119–1121.
Roberts, D. R., W. D. Alecrim, J. M. Heller, S. R. Ehrhardt, and J. B. Lima. 1982. Male *Eufriesea purpurata*, a DDT-collecting bee in Brazil. *Nature* 297:62–63.
Roberts, R. B. 1971. Biology of the crepuscular bee *Ptiloglossa guinnae* n. sp. with notes on associated bees, mites and yeasts. *J. Kans. Entomol. Soc.* 44:283–294.
Roberts, R. B., and R. Brooks. 1987. Agapostemonine bees of Mesoamerica (Hymenoptera: Halictidae). *Univ. Kans. Sci. Bull.* 53(7):357–392.
Roberts, R. B., and C. H. Dodson. 1967. Nesting biology of two communal bees, *Euglossa imperialis* and *Euglossa ignita* including description of larvae. *Ann. Entomol. Soc. Am.* 60:1007–1014.
Roberts, R. B., and S. R. Vallespir. 1978. Specialization of hairs bearing pollen and oil on the legs of bees (Hymenoptera: Apoidea). *Ann. Entomol. Soc. Am.* 71:619–627.
Röhl, E. 1942. *Fauna descriptiva de Venezuela (vertebrados).* Tipografía Americana, Caracas.
Romero, G. A., and C. E. Nelson. 1986. Sexual dimorphism in *Catasetum* orchids: forcible pollen emplacement and male flower competition. *Science* 232:1538–1540.

Röseler, P. F. 1970. Unterschiede in der Kastendetermination zwichen den Hummelarten *Bombus hynorum* und *Bombus terrestris*. *Z. Naturforsch*. 25:543–548.

Rosenthal, G. A., and D. H. Janzen, eds. 1979. *Herbivores: their interaction with secondary plant metabolites*. Academic Press, New York.

Rossel, S., and R. Wehner. 1986. Polarization vision in bees. *Nature* 323:128–131.

Rothenbuhler, W. C., J. M. Kulinčević, and W. E. Kerr. 1968. Bee genetics. *Annu. Rev. Genet.* 2:413–438.

Rothney, G. A. J. 1903. The aculeate Hymenoptera of Barrackpore, Bengal. *Trans. Entomol. Soc. London* 1903:93–116.

Rothschild, G. H. L. 1963. A checklist of insects in the Sarawak museum collections (mimeo). Smithsonian Institution Libraries.

Roubik, D. W. 1978. Competitive interactions between neotropical pollinators and Africanized honeybees. *Science* 201:1030–1032.

———. 1979a. Africanized honey bees, stingless bees, and the structure of tropical plant–pollinator communities. Pp. 403–417 in D. Caron, ed. *Proc. IV Int. Symp. Poll. Maryland Agric. Exp. Sta. Misc. Publ. 1.*

———. 1979b. Nest and colony characteristics of stingless bees from French Guiana (Hymenoptera: Apidae). *J. Kans. Entomol. Soc.* 52:443–470.

———. 1980a. Foraging behavior of competing Africanized honey bees and stingless bees. *Ecology* 61:836–845.

———. 1980b. New species of *Trigona* and cleptobiotic *Lestrimelitta* from French Guiana (Hymenoptera: Apidae). *Rev. Biol. Trop.* 28:263–269.

———. 1981a. A natural mixed colony of *Melipona* (Hymenoptera: Apidae). *J. Kans. Entomol. Soc.* 54:263–268.

———. 1981b. Comparative foraging behavior of *Apis mellifera* and *Trigona corvina* (Hymenoptera: Apidae) on *Baltimora recta* (Compositae). *Rev. Biol. Trop.* 29:177–184.

———. 1982a. Obligate necrophagy in a social bee. *Science* 217:1059–1060.

———. 1982b. The ecological impact of nectar-robbing bees and pollinating hummingbirds on a tropical shrub. *Ecology* 63:354–360.

———. 1982c. Ecological impact of Africanized honeybees on native neotropical pollinators. Pp. 233–247 in P. Jaisson ed. *Social insects in the tropics*, Vol. 1. Univ. Paris–Nord.

———. 1982d. Seasonality in colony food storage, brood production and adult survivorship: studies of *Melipona* in tropical forest (Hymenoptera: Apidae). *J. Kans. Entomol. Soc.* 55:789–800.

———. 1983a. Nest and colony characteristics of stingless bees from Panamá. *J. Kans. Entomol. Soc.* 56:327–355.

———. 1983b. Experimental community studies: time-series tests of competition between African and neotropical bees. *Ecology* 64:971–978.

———. 1987a. Long-term consequences of the African honey bee invasion: implications for the United States. Pp. 46–54. in *Proc. of the Africanized Honey Bee Symp.* American Farm Bureau, Park Ridge, Ill.

———. 1987b. Notes on the biology of *Tetrapedia* (Hymenoptera: Anthophoridae) and *Chaetodactylus panamensis* Baker, Roubik and Delfinado-Baker (Acari: Chaetodactylidae). *Int. J. Acarol.* 65:75.

———. 1988. An overview of Africanized honey bee populations: reproduction, diet and competition. Pp. 45–54 in G. Needham, M. Delfinado-Baker, R. Page, and C. Bowman, eds. *Proc. Int. Conf. on Africanized Honey Bees and Bee Mites.* E. Horwood Ltd., Chichester, England.

———. In press a. Aspects of Africanized honey bee ecology in tropical America. In D. J. C. Fletcher and M. D. Breed, eds. *The 'African' honey bee.* Westview Press, Boulder, Colo.

———. In press b. Destructive flower visitors and pollination efficiency: a tropical perspective. In S.

L. Buchmann, ed. *Experimental studies in pollination and foraging efficiency.* Univ. Ariz. Press, Tucson.
———. In press c. Learning to live with Africanized honeybees: a tropical scenario. In S. W. T. Batra, ed. *Proc. AAAS Symp. on the Africanized Honeybee: Projected Impact in the United States.* Westview Press, Boulder, Colo.
Roubik, D. W., and J. D. Ackerman. 1987. Long-term ecology of euglossine orchid-bees in Panamá. *Oecologia* 73:321–333.
Roubik, D. W., J. D. Ackerman, C. Copenhaver, and B. H. Smith. 1982. Stratum, tree and flower selection by tropical bees: implications for the reproductive biology of outcrossing *Cochlospermum vitifolium* in Panamá. *Ecology* 63:712–720.
Roubik, D. W., and M. Aluja. 1983. Flight ranges of *Melipona* and *Trigona* in tropical forest. *J. Kans. Entomol. Soc.* 56:217–222.
Roubik, D. W., and S. L. Buchmann. 1984. Nectar selection by *Melipona* and *Apis mellifera* (Hymenoptera: Apidae) and the ecology of nectar intake by bee colonies in a tropical forest. *Oecologia* 61:1–10.
Roubik, D. W., N. M. Holbrook, and G. Parra V. 1985. Roles of nectar robbers in the reproduction of a tropical treelet, *Quassia amara* (Simaroubaceae). *Oecologia* 66:161–167.
Roubik, D. W., and L. K. Johnson. 1982. The aggressive foraging syndrome in highly-social bees. Abstr. in M. D. Breed, C. D. Michener, and H. E. Evans, eds. *The biology of social insects.* Westview Press, Boulder, Colo.
Roubik, D. W., and C. D. Michener. 1980. The seasonal cycle and nests of *Epicharis zonata*, a bee whose cells are below the wet-season water table (Hymenoptera, Anthophoridae). *Biotropica* 12:56–60.
———. 1984. Nesting biology of *Crawfordapis* in Panamá (Hymenoptera: Colletidae). *J. Kans. Entomol. Soc.* 57:662–671.
Roubik, D. W., J. E. Moreno, C. Vergara, and D. Wittmann. 1986. Sporadic food competition with the African honey bee: projected impact on neotropical social bees. *J. Trop. Ecol.* 2:97–111.
Roubik, D. W., and F. J. A. Peralta. 1983. Thermodynamics in nests of two *Melipona* species in Brasil. *Acta Amazonica* 13:453–466.
Roubik, D. W., and F. Reyes. 1984. African honey bees have not brought acarine mite infestations to Panama. *Am. Bee J.* 124:665–667.
Roubik, D. W., S. F. Sakagami, and I. Kudo. 1985. A note on the distribution and nesting of the Himalayan honey bee, *Apis laboriosa* Smith. *J. Kans. Entomol. Soc.* 58:746–749.
Roubik, D. W., R. J. Schamlzel, and E. Moreno. 1984. *Estudio apibotanico de Panamá: cosecha y fuentes de polen y nectar usados por Apis mellifera y sus patrones estacionales y anuales.* Tec. Bull. 24. Organismo internacional regional de sanidad agropecularia, Mexico–Centroamerica–Panamá.
Roubik, D. W., B. H. Smith, and R. L. Carlson. 1987. Formic acid in caustic cephalic secretions of stingless bee *Oxytrigona* (Hymenoptera: Apidae). *J. Chem. Ecol.* 13:1079–1086.
Roubik, D. W., and Q. D. Wheeler. 1982. Flightless beetles and stingless bees: phoresy of scotocryptine beetles (Leiodidae) on their meliponine hosts (Apidae). *J. Kans. Entomol. Soc.* 55:125–135.
Rowley, F. A. 1976. The sugars of some common Philippine nectars. *J. Apic. Res.* 15:19–22.
Rozen, J. G., Jr. 1958. Monographic study of the genus *Nomadopsis* Ashmead (Hymenoptera: Andrenidae). *Univ. Calif. Publ. Entomol.* 15:1–202.
———. 1965. The biology and immature stages of *Melitturga clavicornis* (Latreille) and of *Sphecodes albilabris* (Kirby) and the recognition of the Oxaeidae at the family level. *Am. Mus. Novit.* No. 2224.

1966. The larvae of the Anthophoridae (Hymenoptera, Apoidea) Part 2. The Nomadinae. *Am. Mus. Novit.* No. 2244.

1967. Review of the biology of panurgine bees, with observations on North American forms. *Am. Mus. Novit.* No. 2297.

1968. Review of the South African cuckoo-bee genus *Pseudodichroa*. *Am. Mus. Novit.* No. 2347.

1969a. The biology and description of a new species of African *Thyreus*, with life history notes on two species of *Anthophora* (Hymenoptera, Anthophoridae). *J. N.Y. Entomol. Soc.* 78:51–60.

1969b. The larvae of the Anthophoridae (Hymenoptera, Apoidea). Part 3. The Melectini, Ericrocini, and Rhathymini. *Am. Mus. Novit.* No. 2382.

1969c. Biological notes on the bee *Tetralonia minuta* and its cleptoparasite, *Morgania histrio transvaalensis*. *Proc. Entomol. Soc. Wash.* 71:102–107.

1977a. The ethology and systematic relationships of fideliine bees, including a description of the mature larva of *Parafidelia*. *Am. Mus. Novit.* No. 2637.

1977b. Biology and immature stages of the bee genus *Meganomia* (Hymenoptera, Melittidae). *Am. Mus. Novit.* No. 2630.

1977c. Immature stages of and ethological observations on the cleptoparasitic bee tribe Nomadini (Apoidea, Anthophoridae). *Am. Mus. Novit.* No. 2638.

1978. The bionomics and immature stages of the cleptoparasitic bee genus *Protepeolus* (Anthoiphoridae, Nomadinae). *Am. Mus. Novit.* No. 2640.

1984a. Comparative nesting biology of the bee tribe Exomalopsini (Apoidea, Anthophoridae). *Am. Mus. Novit.* No. 2798.

1984b. Nesting biology of diphaglossine bees (Hymenoptera, Colletidae). *Am. Mus. Novit.* No. 2786.

1986. The natural history of the Old World Nomadine parasitic bee *Pasites maculatus* (Anthophoridae: Nomadinae) and its host *Pseudapis diversipes* (Halictidae: Nomiinae). *Am. Mus. Novit.* No. 2861.

Rozen, J. G., Jr., K. R. Eickwort, and G. C. Eickwort. 1978. The bionomics and immature stages of the cleptoparasitic bee genus *Protepeolus* (Anthophoridae, Nomadinae). *Am. Mus. Novit.* No. 2640.

Rozen, J. G., Jr., and N. R. Jacobson. 1980. Biology and immature stages of *Macropis nuda*, including comparisons to related bees (Apoidea, Melittidae). *Am. Mus. Novit.* No. 2702.

Rozen, J. G., Jr., and G. D. MacNeill. 1957. Biological observations on *Exomalopsis* (*Anthophorula*) *chionura* Cockerell, including a comparison of the biology of *Exomalopsis* with that of other anthophorid groups. *Ann. Entomol. Soc. Am.* 50:522–529.

Rozen, J. G., Jr., and C. D. Michener. 1968. The biology of *Scrapter* and its cuckoo bee, *Pseudodichroa* (Hymenoptera: Colletidae and Anthophoridae). *Am. Mus. Novit.* No. 2335.

Rust, R., A. Menke, and D. Miller. 1985. A biogeographic comparison of the bees, sphecid wasps, and mealybugs of the California Channel Islands (Hymenoptera, Homoptera). Pp. 29–59 in A. S. Menke and D. R. Miller, eds. *Entomology of the California Channel Islands.* Santa Barbara Mus. Nat. Hist., Santa Barbara, Calif.

Ruttner, F. 1975. African races of honey bees. Pp. 325–344. *Proc. 25th Intl. Beekeeping Congr.,* Grenoble. Apimondia, Bucharest, Romania.

1980. Breeding honeybees in the German Federal Republic. *Imkerfreund* 35:247–254.

1985. Reproductive behavior in honeybees. Pp. 225–336 in B. Hölldobler and M. Lindauer, eds. *Experimental behavioral ecology.* G. Fischer Verlag, New York.

1988. *Biogeography and taxonomy of honeybees.* Springer–Verlag, Berlin.

Ruttner, F., and B. Hesse. 1981. Rassenspezifische unterschiede in ovarentwicklung und eiablage von wiesellosen arbeiterinnen der honigbiene *Apis mellifera* L. *Apidologie* 12:159–183.

Ruttner, F., and V. Maul. 1983. Experimental analysis of reproductive interspecies isolation of *Apis mellifera* L., and *Apis cerana* Fabr. *Apidologie* 14:309–328.

Ruttner, F., D. Pourasghar, and D. Kauhausen. 1985. Die honigbienen des Iran 1. *Apis florea* Fabricius. *Apidologie* 16:119–138.

Ruttner, F., and H. Ruttner. 1966. Untersuchungen über die Flugaktivität und das Paarungsverhalten der Drohnen. 3. Flugweite und Flugrichtung det Drohnen. *Z. Bienenforsch.* 8:1–9

——— 1972. Untersuchungen über die Flugaktivität und das Paarungsverhalten der Drohnen. V. Drohnensammelplätze und Paarungsdistanz. *Apidologie* 3:203–232.

Ruttner, F., J. Woyke, and N. Koeniger. 1973. Reproduction in *Apis cerana* II: Reproductive organs and natural insemination. *J. Apis. Res.* 12:21–34.

Sage, R. D. 1968. Observations on feeding, nesting and territorial behavior of carpenter bees genus *Xylocopa* in Costa Rica. *Ann. Entomol. Soc. Am.* 61:884–889.

Sakagami, S. F. 1976. Specific differences in the bionomic characters of bumblebees. A comparative review. *J. Fac. Sci. Hokkaido Univ. Zool.* 20:390–447.

——— 1978. *Tetragonula* stingless bees of the continental Asia and Sri Lanka (Hymenoptera, Apidae). *J. Fac. Sci. Hokkaido Univ. Ser. VI Zool.* 21:165–247.

——— 1982. Stingless bees. Pp. 361–423 in H. R. Hermann, ed. *Social insects*, Vol. 3. Academic Press, New York.

Sakagami, S. F., and Y. Akahira. 1960. Studies on the Japanese honeybee, *Apis cerana cerana* Fabricius. VIII. Two opposing adaptations in the post-stinging behavior of honeybees. *Evolution* 14:29–40.

Sakagami, S. F., Y. Akahira, and R. Zucchi. 1967. Nest architecture and brood development in a neotropical bumblebee, *Bombus atratus*. *Insectes Soc.* 14:389–414.

Sakagami, S. F., D. Beig, R. Zucchi, and Y. Akahira. 1963. Occurrence of ovary-developed workers in queenright colonies of stingless bees. *Rev. Bras. Biol.* 23:115–129.

Sakagami, S. F., and J. M. F. Camargo. 1964. Cerumen collection accompanied by thieving and attacking in a stingless bee, *Scaptotrigona postica* (Latreille), with a consideration of territoriality in social insects. *Rev. Biol. Trop.* 12:197–207.

Sakagami, S. F., and K. Fukushima. 1961. Female dimorphism in a social halictine bee, *Halictus (Seladonia) aerarius* (Smith). *Jpn. J. Ecol.* 11:118–124.

Sakagami, S. F., K. Hoshikawa, and H. Fukuda. 1984. Overwintering ecology of two social halictine bees, *Lasioglossum duplex* and *L. problematicum*. *Res. Popul. Ecol.* 26:363–378.

Sakagami, S. F., T. Inoue, and S. Salmah. 1985. Key to the stingless bee species found or expected from Sumatra. Pp. 37–43 in *Evolutionary ecology of insects in humid tropics, especially in central Sumatra*. Sumatra Nature Study (Entomology), Kanazawa Univ.

Sakagami, S. F., T. Inoue, S. Yamane, and S. Salmah. 1983a. Nest architecture and colony composition of the Sumatran stingless bee *Trigona (Tetragonula) laeviceps*. *Kontyû* 51:100–111.

——— 1983b. Nesting habits of Sumatran stingless bees. *Bull. Fac. Educ. Ibaraki Univ.* 32:1–21.

——— In press. Nests of the myrmecophilous stingless bee *Trigona moorei*: How do bees initiate their nest within an arboreal ant nest? *Biotropica*.

Sakagami, S. F., and M. Itô. 1981. Specific and subgeneric variation in tibial corbiculation of male bumble bees (Hymenoptera: Apidae): An apparently functionless character. *Entomologica Scandinavica* Suppl. 15:365–376

Sakagami, S. F., and S.-G. Khoo. 1987. Taxonomic status of the Malesian stingless bee *Trigona reepeni*, with discovery of *Trigona pagdeni* from northern Malaya. *Kontyû* 55:207–214.

Sakagami, S. F., and S. Laroca. 1963. Additional observations on the habits of the cleptobiotic stingless bees, the genus *Lestrimelitta* Friese. *J. Fac. Sci. Hokkaido Univ. Ser. VI Zool.* 15:319–339.

1971. Observations on the bionomics of some neotropical xylocopine bees, with comparative and biofaunistic notes. *J. Fac. Sci. Hokkaido Univ. Ser. VI Zool.*18:57–127.

Sakagami, S. F., S. Laroca, and J. S. Moure. 1967. Wild bee biocoenotics in São José dos Pinhais. *J. Fac. Sci. Hokkaido Univ. Ser. VI Zool.* 16:253–291.

Sakagami, S. F., and Y. Maeta. 1977. Some presumably presocial habits of Japanese *Ceratina* bees, with notes on various social types in Hymenoptera. *Insectes Soc.* 24:319–343.

1984. Multifemale nests and rudimentary castes in the normally solitary bee *Ceratina japonica* (Hymenoptera: Xylocopinae). *J. Kans. Entomol. Soc.* 57:639–656.

1986. Socialities induced and/or naturally found in *Ceratina*, the basically solitary bees (abstract). *Symp. on biological aspects of optimal strategy and social structure.* Kyoto, Japan.

1987. Sociality, induced and/or natural, in the basically solitary small carpenter bees (*Ceratina*) in Y. Itô, J. L. Brown, and J. Kikkawa, eds. *Animal societies: theories and facts.* Japan Sci. Soc. Press, Tokyo.

Sakagami, S. F., T. Matsumura, and K. Itô. 1980. *Apis laboriosa* in Himalaya, the little known world's largest honey bee. *Insecta Matsumurana* 19:47–77.

Sakagami, S. F., T. Matsumura, and Y. Maeta. 1985. Bionomics of halictine bees in northern Japan III. *Lasioglossum* (*Evylaeus*) *allodalum*, with remarks on the serially arranged cells in the halictine nests. *Kontyû* 53:409–419.

Sakagami, S. F., and C. D. Michener. 1962. *The nest architecture of the sweat bees.* Univ. Kans. Press, Lawrence, Kansas.

1987. Tribes of Xylocopinae and origin of the Apidae (Hymenoptera: Apoidea). *Ann. Entomol. Soc. Am.* 80:439–450.

Sakagami, S. F., and J. S. Moure. 1965. Cephalic polymorphism in some neotropical halictine bees. *An. Acad. Bras. Cien.* 37:303–313.

1967. Additional observations on the nesting habits of some Brazilian halictine bees (Hymenoptera, Apoidea) *Mushi* 40:119–138.

Sakagami, S. F., and T. Okazawa. 1985. A populous nest of the halictine bee *Halictus* (*Seladonia*) *lutescens* from Guatemala (Hymenoptera; Halictidae). *Kontyû* 53:645–651.

Sakagami, S. F., D. W. Roubik, and R. Zucchi. In press. Foraging behavior of the robber stingless bee, *Lestrimelitta limao* (Hymenoptera, Apidae). *J. Linn. Soc. London.*

Sakagami, S. F., and H. Strum.1965. *Euplusia longipennis* (Friese) und ihre merkwürdigen brutzellen aus kolumbien (Hymenoptera: Apidae). *Insecta Matsumurana* 28:83–92.

Sakagami, S. F., K. Tanno, H. Tsutsui, and K. Honma. 1985. The role of cocoons in overwintering of the soybean pod borer *Leguminivora glycinivorella* (Lepidoptera: Tortricidae). *J. Kans. Entomol. Soc.* 58: 240–247.

Sakagami, S. F., and M. Ushi. 1976. Occurrence of communal nests in *Eucera sociabilis*. *Kontyû* 44:354–357.

Sakagami, S. F., and S. Yamane. 1984. Notes on taxonomy and nest architecture of the Taiwanese stingless bee *Trigona* (*Lepidotrigona*) *ventralis hoozana*. *Bull. Fac. Educ. Ibaraki Univ.* 33:37–48.

1987. Oviposition behavior and related notes of the Taiwanese stingless bee *Trigona* (*Lepidotrigona*) *ventralis hoozana*. *J. Ethol.* 5:17–27.

Sakagami, S. F., S. Yamane, and G. G. Hambali. 1983. Nests of some Southeast Asian stingless bees. *Bull. Fac. Educ. Ibaraki Univ.* (*Nat. Sci.*) 32:1–21.

Sakagami, S. F., S. Yamane, and T. Inoue. 1983. Oviposition behavior of two Southeast Asian stingless bees *Trigona* (*Tetragonula*) *laeviceps* and *T.* (*T.*) *pagdeni*. *Kontyû* 51:441–457.

Sakagami, S. F., and K. Yoshikawa. 1961. Bees of Xylocopinae and Apinae collected by the Osaka City University biological expedition to southeast Asia 1957–58, with some biological notes. *Nat. Life Southeast Asia* 1:409–444.

1973. Additional observation of the nest of the dwarf honeybee, *Apis florea. Kontyû* 41:217–219.
Sakagami, S. F., and R. Zucchi. 1963. Oviposition process in a stingless bee, *Trigona (Scaptotrigona) postica* Latreille. *Stud. Entomol.* [Rio de Janeiro] 6:497–510.
  1965. Winterhaverhalten einer neotropischen Hummel, *Bombus atratus*, innerhalb des Beobachtungskastens. Ein Beitrag zur Biologie der Hummeln. *J. Fac. Sci. Hokkaido Univ. Ser. VI Zool.* 15:712–762.
  1974. Oviposition behavior of two dwarf stingless bees, *Hypotrigona (Leurotrigona) muelleri* and *H. (Trigonisca) duckei*, with notes on the temporal articulation of the oviposition process in stingless bees. *J. Fac. Sci. Hokkaido Univ.Ser. VI Zool.* 19:361–421
  1978. Nests of *Hylaeus (Hylaeopsis) tricolor*: the first record of non-solitary life in colletid bees, with notes on communal and quasisocial colonies. *J. Kans. Entomol. Soc.* 51:597–614.
Salden, F. W. L. 1912. *The humble-bee*. Macmillan, London.
Salt, G. 1929. A contribution to the ethology of the Meliponinae. *Trans. Entomol. Soc. London* 77:431–470.
Sanford, R. L., J. Saldarriaga, K. E. Clark, C. Uhl, and R. Herrera. 1985. Amazon rain-forest fires. *Science* 227:53–55.
Saunders, D. S. 1982. *Insect clocks* (2nd ed.). Pergamon Press, Oxford.
Schaffer, W. M., D. B. Jensen, D. E. Hobbs, J. Gurevitch, J. R. Todd, and M. V. Schaffer. 1979. Competition, foraging energetics and the cost of sociality in three species of bees. *Ecology* 60:976–987.
Schemske, D. W. 1980. Evolution of floral display in the orchid *Brassavola nodosa. Evolution* 34:489–493.
  1981. Floral convergence and pollinator sharing in two bee-pollinated tropical herbs. *Ecology* 62:946–954.
  1983. Limits to specialization and coevolution in plant–animal mutualisms. Pp. 67–109 in M. H. Nitecki, ed. *Coevolution*. Univ. of Chicago Press, Chicago.
Schemske, D. W., and C. C. Horvitz. 1984. Variation among floral visitors in pollination ability: a precondition for mutualism specialization. *Science* 225:519–521.
Schemske, D. W., and R. Lande. 1984. Fragrance collection and territorial display in male orchid bees. *Anim. Behav.* 32:935–937.
  1985. The evolution of self-fertilization and inbreeding depression in plants. II. empirical observations. *Evolution* 39:41–52.
Schemske, D. W., and L. P. Pautler. 1984. The effects of pollen composition on fitness components of a neotropical herb. *Oecologia* 62:31–36.
Schemske, D. W., M. F. Willson, M. N. Melampy, L. J. Miller, L. Verner, K. M. Schemske, and L. B. Best.1978. Flowering phenology of some spring woodland herbs. *Ecology* 59:351–366.
Schlising, R. A.1970. Sequence and timing of bee foraging in flowers of *Ipomoea* and *Aniseia* (Convolvulaceae). *Ecology* 51:1061–1067.
Schmid-Hempel, P. 1984. The importance of handling time for the flight directionality in bees. *Behav. Ecol. Sociobiol.* 15:303–309.
Schmid-Hempel, P., A. Kacelnik and A. I. Houston. 1985. Honeybees maximize efficiency by not filling their crop. *Behav. Ecol. Sociobiol.* 17:61–66.
Schmidt, J. O. 1982. Biochemistry of insect venoms. *Annu. Rev. Entomol.* 27:339–368.
  1984. Feeding preferences of *Apis mellifera* L.: individual vs. mixed pollen species. *J. Kans.. Entomol. Soc.* 57:323–327.
Schmidt, J. O., M. S. Blum, and W. L. Overal. 1980. Comparative lethality of venoms from stinging Hymenoptera. *Toxicon* 18:469–474.

Schmidt, J. O., and S. L. Buchmann. 1985. Pollen digestion and nitrogen utilization by *Apis mellifera* L. *Comp. Biochem. Physiol.* 82:499–503.
—— 1986. Floral biology of the saguaro (*Cereus giganteus*). I. Pollen harvest by *Apis mellifera*. *Oecologia* 69:491–498.
Schmidt, J. O., S. L. Buchmann, and M. Glaiim. In press. Is *Typha* pollen a good pollen source for honey bees? *Apidologie*.
Schneider, S., J. Stamps, and N. Gary. 1986. The vibration dance of the honey bee. I. Communication regulating foraging on two time scales. *Anim. Behav.* 34:377–385.
Schoener, T. W. 1979. Generality of the size–distance relation in models of optimal feeding. *Am. Nat.* 114:902–914.
—— 1983. Field experiments on interspecific competition. *Am. Nat.* 122:240–285.
Scholz, E., and D. Wittmann. In press. Bionomy of five xylocopine bee species in Rio Grande do Sul, Southern Brazil. *Proc. IUSSI München (1986)*.
Schremmer, F. 1941. Eine Bauchsammelbiene (*Megachile circumcincta* K.) als Zerstörer der blüten von *Salvia glutinosa*. *Zool. Anz.* 133:230–232.
—— 1955. Über anormalen Blütenbesuch und das Lernvermögen blütenbesuchender Insekten. *Oesterr. Bot. Z.* 102:551–571.
—— 1972. Der Stechsaugrüssel, der nektarraub, das pollensammeln und der blütenbesuch der holzbienen (*Xylocopa*). *Z. Morphol. Tiere* 72:263–294.
Schwarz, H. F. 1932. The genus *Melipona*. *Bull. Am. Mus. Nat. Hist.* 63:231–480.
—— 1937. Results of the Oxford Sarawak Expedition: Bornean stingless bees of the genus *Trigona*. *Bull. Am. Mus. Nat. Hist.* 73:281–328.
—— 1939. The Indo-Malayan species of *Trigona*. *Bull. Am. Mus. Nat. Hist.* 76:83–141.
—— 1943. New *Trigona* bees from Peru. *Am. Mus. Novit.* No. 1243.
—— 1945. The wax of stingless bees (Meliponidae) and the uses to which it has been put. *J. N.Y. Entomol. Soc.* 53:137–144.
—— 1948. Stingless bees (Meliponidae) of the Western Hemisphere. *Bull. Am. Mus. Nat. Hist.* 90:1–546.
Schwarz, M. P., O. Scholz, and G. Jensen. 1987. Ovarian inhibition among nestmates of *Exoneura bicolor* Smith (Hymenoptera: Xylocopinae). *J. Aust. Entomol. Soc.* 26: 355–359.
Seabra, C. A., and J. S. Moure. 1961. Sobre duas espécies mimeticas de *Centris*. *Rev. Bras. Entomol.* 10:25–29.
Seeley, T. D. 1974. Atmospheric carbon dioxide regulation in honey-bee (*Apis mellifera*) colonies. *J. Insect Physiol.* 20:2301–2305.
—— 1977. Measurement of nest cavity volume by the honey bee (*Apis mellifera*). *Behav. Ecol. Sociobiol.* 2:201–227.
—— 1982. Adaptive significance of the age polyethism schedule in honeybee colonies. *Behav. Ecol. Sociobiol.* 11:287–293.
—— 1985. *Honey bee ecology*. Princeton Univ. Press, Princeton, N.J.
—— 1986. Social foraging by honeybees: how colonies allocate foragers among patches of flowers. *Behav. Ecol. Sociobiol.* 19:343–354.
—— 1987. The effectiveness of information collection about food sources by honey bee colonies. *Anim. Behav.* 35:1572–1575.
Seeley, T. D., and B. Heinrich. 1981. Regulation of temperature in the nests of social insects. Pp. 159–234 in B. Heinrich, ed. *Insect thermoregulation*. Wiley, New York.
Seeley, T. D., and R. A. Morse. 1976. The nest of the honey bee (*Apis mellifera*). *Insectes Soc.* 23:495–512.
—— 1978. Nest site selection by the honey bee. *Insectes Soc.* 25:323–337.

Seeley, T. D., J. W. Nowicke, M. Meselson, J. Guillemin, and P. Akratanakul. 1985. Yellow rain. *Sci. Am.* 253: 128–137.
Seeley, T. D., R. H. Seeley, and P. Akratanakul. 1982. Colony defense strategies of the honeybees in Thailand. *Ecol. Monogr.* 52:43–63.
Seeley, T. D., and P. K. Visscher. 1985. Survival of honeybees in cold climates: the critical timing of colony growth and reproduction. *Ecol. Entomol.* 10:81–88.
——— 1988. Assessing the benefits of cooperation in honeybee foraging: search costs, forage quality, and competitive ability. *Behav. Ecol. Sociobiol.* 22:229–238.
Sekiguchi, K., and S. F. Sakagami. 1966. Structure of foraging population and related problems in the honeybee, with considerations on the division of labour in bee colonies. *Hokkaido Natl. Agric. Exp. Stn. Rep.* No. 69.
Selander, R. B. 1965. The systematic position of *Meloetyphlus*, a genus of blind blister beetles (Coleoptera: Meloidae). *J. Kans. Entomol. Soc.* 38:45–55.
——— 1983. An annotated catalog of blister beetles of the tribe Tetraonylini (Coleoptera, Mloidae). *Trans. Am. Entomol. Soc.* 109:277–293.
——— 1985. A new genus of blister beetles linking *Meloetyphlus* with *Tetraonyx* (Coleoptera: Meloidae). *J. Kans. Entomol. Soc.* 58:611–619.
——— 1987. Biological observations in *Cyaneolytta* and a description of the triungulin larva of *C. fryi* (Coleoptera: Meloidae). *J. Kans. Entomol. Soc.* 60:288–304.
Severinghaus, L. L., B. M. Kurtak, and G. C. Eickwort. 1981. The reproductive behavior of *Anthidium manicatum* (Hymenoptera: Megachilidae) and the significance of size for territorial males. *Behav. Ecol. Sociobiol.* 9:51–58.
Severson, D. W., and J. E. Parry. 1981. A chronology of pollen collection by honeybees. *J. Apic. Res.* 20:97–103.
Shanks, S. S. 1986. A revision of the neotropical bee genus *Osiris* (Hymenoptera: Anthophoridae). *Wasmann J. Biol.* 44:1–56.
Shaw, D. E., and D. F. Robertson. 1980.Collection of *Neurospora* by honeybees. *Trans. Br. Mycol. Soc.* 74:459–464.
Shelly, T. E. 1984. Comparative foraging behavior of Neotropical robber flies (Diptera: Asilidae). *Oecologia* 62:188–195.
——— 1985. Ecological comparisons of robber fly species (Diptera: Asilidae) coexisting in a neotropical forest. *Oecologia* 67:57–70.
Sheppard, W. S. 1985. Electrophoretic variation in the honey bees, *Apis. Isozyme Bull.* 18:69.
Sheppard, W. S., and S. H. Berlocher. 1984. Enzyme polymorphisms in *Apis mellifera mellifera* from Norway. *J. Apic. Res.* 23:64–69.
Shimron, O., A. Hefetz, and J. Tengö. 1985. Structural and communicative functions of Dufour's gland secretion in *Eucera palestinae* (Hymenoptera: Anthophoridae). *Insect Biochem.* 15:635–638.
Shinn, A. F. 1967. A revision of the bee genus *Calliopsis* and the biology and ecology of *C. andreniformis* (Hymenoptera, Andrenidae). *Univ. Kans. Sci. Bull.* 46:753–936.
Shorrocks, B., ed. 1985. *Evolutionary ecology.* Blackwell Scientific, Palo Alto, Calif.
Silberglied, R. E. 1979. Communication in the ultraviolet. *Annu. Rev. Ecol. Syst.* 10:373–398.
——— 1984. Visual communication and sexual selection among butterflies. Pp. 207–223 in R. I. Vane-Wright and P. R. Ackery, eds. *The biology of butterflies.* Academic Press, London.
Silberrad, R. E. M. 1976. *Bee-keeping in Zambia.* Apimondia, Bucharest.
Silva, A. G. A., C. R. Gonçalves, D. M. Galvão, A. J. L. Gonçalves, J. Gomes, et al. 1968. *Quarto catálogo dos insetos que vivem nas plantas do Brasil, seus parasitos e predadores.* Parte II. Ministério das Agricultura, Rio de Janeiro.

Simberloff, D. 1981. Community effects of exotic species. Pp. 58–83 in M. H. Nitecki, ed. *Biotic crises in ecological and evolutionary time*. Academic Press, New York.

Simmonds, M. S. J., W. M. Blaney, F. D. Monache, M. M. Mac-Quhae, and G. B. Marini Bettolo. 1985. Insect antifeedant properties of anthranoids from the genus *Vismia*. *J. Chem. Ecol*. 11:1593–1599.

Simões, D., L. R. Bego, R. Zucchi, and S. F. Sakagami. 1980. *Melaloncha sinistra* Borgmeier, an endoparasitic phorid fly attacking *Scaptotrigona postica* Latrieille (Hymenoptera, Apidae). *Rev. Bras. Entomol*. 24:137–142.

Simpson, B. B., and J. L. Neff. 1981. Floral rewards: alternatives to pollen and nectar. *Ann. Mo. Bot. Gard*. 68:301–322.

―――. 1983. Evolution and diversity of floral rewards. Pp. 142–159 in C. E. Jones and R. J. Little, eds. *Handbook of experimental pollination biology*. Van Nostrand Reinhold, New York.

―――. 1985. Plants, their pollinating bees, and the great American interchange. Pp. 427–452 in F. G. Stehli and S. David Webb, eds. *The great American biotic interchange*. Plenum, New York.

Simpson, G. G. 1953. *The major features of evolution*. Columbia Univ. Press, New York.

―――. 1980. *Splendid isolation: the curious history of South American mammals*. Yale Univ. Press, New Haven, Conn.

Singh, S. 1962. *Beekeeping in India*. Indian council of agricultural research, New Delhi.

Skaife, S. H. 1952. The yellow-banded carpenter bee, *Mesotrichia caffra* Linn., and its symbiotic mite, *Dinogamasus braunzi* Vitzhum. *J. Entomol. Soc. S. Afr*. 15: 63–76.

―――. 1954. *African Insect Life*. Longmans Green and Co., London.

Skou, L. J., and K. A. Holm. 1980. Occurrence of melanosis and other diseases in the queen honeybee, and risk of their transmission during instrumental insemination. *J. Apic. Res*. 19:133–143.

Smith, B. H. 1983. Recognition of female kin by male bees through olfactory signals. *Proc. Natl. Acad. Sci. USA*. 80:4551–4553.

―――. 1986. Effects of genealogical relationship and colony age on the dominance hierarchy in the primitively eusocial bee *Lasioglossum zephyrum*. *Anim. Behav*. 35:211–217.

―――. 1988. Ethological aspects of olfactory learning and memory in the honey bee. Pp. 97–104 in G. Needham, M. Delfinado-Baker, R. Page, and C. Bowman, eds. *Proc. Int. Conf. on Africanized Honey Bees and Bee Mites*. E. Horwood Ltd., Chichester, England.

Smith, B. H., R. G. Carlson, and J. Frazier. 1985. Identification and bioassay of macrocyclic lactone sex pheromone of the halictine bee *Lasioglossum zephyrum*. *J. Chem. Ecol*. 11:1447–1456.

Smith, B. H., and D. W. Roubik. 1983. Mandibular glands of stingless bees (Hymenoptera: Apidae): Chemical analysis of their contents and biological function in two species of *Melipona*. *J. Chem. Ecol*. 9:1465–1472.

Smith, B. H., and C. Weller. In press. What information do social halictine bees encode in the amount of the macrocyclic lactone pheromones? *Proc. IUSSI München (1986)*.

Smith, F. G. 1960. *Beekeeping in the tropics*. Longmans, London.

Smith, K. G. V. 1966. The larva of *Thecophora occidensis*, with comments upon the biology of Conopidae (Diptera). *J. Zool. Soc. London* 149:263–276.

Smith, N. G. 1980. Some evolutionary, ecological, and behavioral correlates of communal nesting by birds with wasps or bees. Pp. 1199–1205 in *Proc. 13th Int. Ornithol. Congr*. Deutsche Ornithologen-Gesellschaft, Berlin.

Snelling, R. R. 1984. Studies on the taxonomy and distribution of American centridine bees (Hymenoptera: Anthophoridae). *Contrib. Sci*. No. 327. Nat. His. Mus. of Los Angeles County.

Snelling, R. R., and R. W. Brooks. 1986. A review of the genera of cleptoparasitic bees of the

tribe Ericrocini (Hymenoptera: Anthophoridae). *Contrib. Sci.* No. 369. Nat. Hist. Mus. of Los Angeles County.
Snow, A. A. 1982. Pollination intensity and potential seed set in *Passiflora vitifolia*. *Oecologia* 55:231–237.
Snow, A. A., and D. W. Roubik. 1987. Pollen deposition and removal by bees visiting two tree species in Panamá. *Biotropica* 19:57–63.
Snow, D. W. 1965. A possible selective factor in the evolution of fruiting seasons in tropical forest. *Oikos* 15:274–281.
Snyder, T. P. 1974. Lack of allozymic variability in three bee species. *Evolution* 28:687–689.
——— 1977. A new electrophoretic approach to biochemical systematics of bees. *Biochem. Syst. Ecol.* 5:133–150.
Sobrevila, C., and M. T. K. Arroyo. 1982. Breeding systems in a montane tropical cloud forest in Venezuela. *Plant Syst. Evol.* 140:19–37.
Soderstrom, T. R., and C. E. Calderón. 1971. Insect pollination of tropical rain forest grasses. *Biotropica* 3:1–16.
Sohal, R. S. 1986. The rate of living theory: a contemporary interpretation. Pp. 22–44 in K. G. Collatz and R. S. Sohal, eds. *Insect aging*. Springer-Verlag, Berlin.
Sommeijer, M. J. 1984. Distribution of labour among workers of *Melipona favosa* F.: age-polyethism and worker oviposition. *Insectes Soc.* 31:171–184.
——— 1985. The social behavior of *Melipona favosa:* some aspects of the activity of the queen in the nest. *J. Kans. Entomol. Soc.* 58:386–396.
Sommeijer, M. J., F. T. Beuvens, and H. J. Verbeek. 1982. Distribution of labour among workers of *Melipona favosa* F.: construction and provisioning of brood cells. *Insectes Soc.* 29:222–237.
Sommeijer, M. J., L. L. M. de Bruijn, and C. van de Guchte. 1985. The social food-flow within the colony of a stingless bee, *Melipona favosa* (F.). *Behaviour* 92:39–58.
Sommeijer, M. J., G. A. DeRooy, W. Punt, and L. L. M. De Bruijn. 1983. A comparative study of foraging behavior and pollen resources of various stingless bees and honeybees in Trinidad, West Indies. *Apidologie* 14:205–224.
Sommeijer, M. J., J. L. Houtekamer, and W. Bos. 1984. Cell construction and egg-laying in *Trigona nigra paupera* with a note on the adaptive significance of oviposition behaviour of stingless bees. *Insectes Soc.* 31:199–217.
Sommeijer, M. J., and H. H. W. Velthuis. 1977. Worker oviposition in orphan colonies of *Melipona favosa* (F.). *Proc. 8th Int. Congr. IUSSI*. Wageningen, The Netherlands, 315–316.
Sommeijer, M. J., M. van Zeijl, and M. R. Dohmen. 1984. Morphological differences between worker-laid eggs from a queenright colony and a queenless colony of *Melipona rufiventris paraensis* (Hymenoptera: Apidae). *Entomol. Ber.* 44:91–95.
Southwick, A. K., and E. E. Southwick. 1983. Aging effect on nectar production in two clones of *Asclepias syriaca*. *Oecologia* 56:121–125.
Southwick, E. E. 1984. Photosynthate allocation to floral nectar: a neglected energy investment. *Ecology* 65:1775–1779.
——— 1985a. Bee hair structure and the effect of hair on metabolism at low temperature. *J. Apic. Res.* 24:144–149.
——— 1985b. Allometric relations, metabolism and heat conductance in clusters of honey bees at cool temperatures. *J. Comp. Physiol. B Metab. Transp. Funct.* 156:143–149.
——— 1985c. Thermal conductivity of wax comb and its effect on heat balance in colonial honey bees (*Apis mellifera* L.). *Experientia* 41:1486–1487.
——— 1987. Cooperative metabolism in honey bees: an alternative to antifreeze and hibernation. *J. Therm. Biol.* 12:155–158.

Southwick, E. E., G. M. Loper, and S. E. Sadwick. 1981. Nectar production, composition, energetics and pollinator attractiveness in spring flowers of western New York. *Am. J. Bot.* 68:994–1002.

Southwick, E. E., and R. F. A. Moritz. 1987a. Social synchronization of circadian rhythms of metabolism in honeybees (*Apis mellifera*). *Physiol. Entomol.* 12:209–212.

———. 1987b. Effects of meteorological factors on defensive behaviour of honey bees. *Int. J. Biometeor.* 31:259–265.

Southwick, E. E., and D. Pimentel. 1981. Energy efficiency of honey production by bees. *Bioscience* 31:730–732.

Southwood, T. R. E. 1981. *Ecological methods.* 3rd ed. Methuen, London.

Spangler, H. G. 1984. Attraction of female lesser wax moths (Lepidoptera: Pyralidae) to male-produced and artificial sounds. *J. Econ. Entomol.* 77:346–349.

Spiess, E. B. 1977. *Genes in populations.* J. Wiley and Sons, New York.

Sprengel, C. K. 1793. *Das entdeckte Geheimnis der Natur in Dau und in der Befruchtung der Blumen.* Berlin.

Stanley, R. G., and H. F. Linskens. 1974. *Pollen: Biology, biochemistry, management.* Springer–Verlag, New York.

Stanton, M. L., A. A. Snow, and S. N. Handel. 1986. Floral evolution: attractiveness to pollinators increases male fitness in a hermaphroditic angiosperm. *Science* 232:1625–1627.

Starr, C. K. 1979. Origin and evolution of insect sociality, a review of modern theory. Pp. 35–79 in H. R. Hermann, ed. *Social insects*, Vol. 1. Academic Press, New York.

———. 1984. Sperm competition, kinship and sociality in the aculeate Hymenoptera. Pp. 427–464 in R. L. Smith, ed. *Sperm competition and the evolution of animal mating systems.* Academic Press, New York.

Starr, C. K., and S. F. Sakagami. 1987. An extraordinary concentration of stingless bee colonies in the Philippines, with notes on nest structure (Hymenoptera: Apidae: *Trigona* spp.). *Insectes Soc.* 34:96–107.

Starr, C. K., P. J. Schmidt, and J. O. Schmidt. 1987. Nest-site preferences of the giant honey bee, *Apis dorsata* in Borneo. *Pan-Pac. Entomol.* 63:37–42.

Stebbins, G. L. 1974. *Flowering plants.* Belknap Press of Harvard Univ. Press, Cambridge, Mass.

Steiner, K. E. 1984. Nectarivory and potential pollination by a neotropical marsupial. *Ann. Mo. Bot. Gard.* 68:505–513.

———. 1985. The role of nectar and oil in the pollination of *Drymonia serrulata* (Gesneriaceae) by *Epicharis* bees (Anthophoridae) in Panama. *Biotropica* 17:217–229.

Steinmann, E. 1976. Über die Nahorientierung solitärer Hymenopteren: individuelle Markierung der Nesteingänge. *Mitt. Schweiz. Entomol. Ges.* 49:253–258.

Stemmer, W. P. C., D. B. Archer, M. J. Daniels, A. M. C. Davies, and S. J. Eden-Green. 1982. Effects of lethal yellowing on the composition of the phloem sap from coconut palms *Cocos nucifera* in Jamaica, West Indies. *Phytopathology* 72:672–675.

Stephen, W. P., G. E. Bohart, and P. F. Torchio. 1969. *The biology and external morphology of bees.* Oregon State Univ. Press, Corvallis.

Stephens, D. W., and J. R. Krebs. 1986. *Foraging theory.* Princeton Univ. Press, Princeton, N.J.

Stephens, D. W., J. F. Lynch, A. E. Sorenson, and C. Gordon. 1986. Preference and profitability: theory and experiment. *Am. Nat.* 127:533–553.

Stephenson, A. G. 1979. An evolutionary examination of the floral display of *Catalpa speciosa* (Bignoniaceae). *Evolution* 33:1200–1209.

———. 1981. Toxic nectar deters nectar thieves of *Catalpa speciosa. Am. Midl. Nat.* 105:381–383

Stephenson, A. G., J. A. Winsor, and L. E. Davis. 1986. Effects of pollen load size on fruit maturation and sporophyte quality in zucchini. Pp. 429–435 in D. L. Mulcahy, G. B.

Mulcahy, and E. Ottaviano, eds. *Biotechnology and ecology of pollen.* Springer-Verlag, New York.

Stiles, E. W. 1976. Comparison of male bumblebee flight paths: temperate and tropical. *J. Kans. Entomol. Soc.* 49:266–274.

——— 1979. Evolution of color pattern and pubescence characteristics in male bumblebees: automimicry vs. thermoregulation. *Evolution* 33:940–957.

Stonehouse, B., and D. Gilmore, eds. 1977. *The biology of marsupials.* University Park Press, Baltimore, Md.

Stork, N. E. 1987. Arthropod faunal similarity of Bornean rain forest trees. *Ecol. Entomol.* 12:219–226.

Stort, A. C., and O. C. Bueno. 1985. Are *Apis mellifera* bees morphologically Africanized in Brazil? *Rev. Bras. Biol.* 45:393–397.

Stort, A. C., and C. Cruz-Landim. 1965. Glandulas dos apendices locomotores do genero *Centris* (Hymenoptera, Anthophoridae). *Bol. Inst. Angola* 21:5–14.

Strickler, K. 1979. Specialization and foraging efficiency of solitary bees. *Ecology* 60:998–1009.

——— 1982. Parental investment per offspring by a specialist bee: does it change seasonally? *Evolution* 36:1098–1100.

Strong, D. R.1985. Density-vague ecology and liberal population regulation in insects. Pp. 313–327 in P. W. Price, C. N. Slobodchikoff, and W. S. Gaud, eds. 1984. *A new ecology: novel approaches to interactive systems.* Wiley–Interscience, New York.

Strong, D. R., J. A. Lawton, and T. R. E. Southwood. 1984. *Insects on plants.* Harvard Univ. Press, Cambridge, Mass.

Strong, D. R., D. Simberloff, L. G. Avele, and A. B. Thistle, eds. 1984. *Ecological communities.* Princeton Univ. Press, Princeton, N.J.

Stubblefield, J. W., and E. L. Charnov. 1986. Some conceptual issues in the origin of eusociality. *Heredity* 57:181–187.

Sutherland, S. 1986. Patterns of fruit-set: what controls fruit–flower ratios in plants? *Evolution* 40:117–128.

Sylvester, H. A. 1986. Biochemical genetics. Pp. 177–204 in T. E. Rinderer, ed. *Bee genetics and breeding.* Academic Press, Orlando, Fla.

Taber, S. 1955. Sperm distribution in the spermathecae of multiply-mated queen honeybees. *J. Econ. Entomol.* 48:522–525.

Tangkanasing, P., S. Vongsamanode, and S. Wongsiri. 1988. Integrated control of *Varroa jacobsoni* and *Tropilaelaps clareae* in Thailand. Pp. 408–412 in G. Needham, M. Delfinado-Baker, R. Page, and C. Bowman. eds. *Proc. Int. Conf. on Africanized Honey Bees and Bee Mites.* E. Horwood Ltd., Chichester, England.

Tasei, J. N., and M. M. Masure. 1978. Sur quelques facteurs influensant le developpement de *Megachile pacifica* Panz. (Hymenoptera, Megachilidae). *Apidologie* 9:273–290.

Tauber, M. J., C. A. Tauber, and S. Masaki. 1986. *Seasonal adaptations of insects.* Oxford Univ. Press, New York.

Taylor, D. W., and W. L. Crepet. 1987. Fossil flora evidence of Malpighiaceae and an early plant–pollinator relationship. *Am. J. Bot.* 74:274–286.

Taylor, O. R. 1977. Past and possible future spread of Africanized honeybees in the Americas. *Bee World* 58:19–30.

——— 1985. African bees: potential impact in the United States. *Bull. Entomol. Soc. Am.* 31:15–24.

——— 1988. Ecology and economic impact of African and Africanized honey bees. Pp. 29–44 in M. Delfinado-Baker, G. Needham, M. Delfinado-Baker, R. Page, and C. Bowman, eds. *Proc. Int. Conf. on Africanized Honey Bees and Bee Mites.* E. Horwood Ltd., Chichester, England.

Taylor, O. R., and M. Spivak. 1984. Climatic limits of tropical African honeybees in the Americas. *Bee World* 65:38–47.

Tengö, J., and G. Bergström. 1975. All-trans farnesyl hexanoate and geranyl octanoate in the Dufour's gland secretion of *Andrena* (Hymenoptera: Apidae). *J. Chem. Ecol.* 1:253–268.

——— 1976. Odor correspondence between *Melitta* females and males of their nest parasite *Nomada flavopicta* K. (Hymenoptera: Apoidea). *J. Chem. Ecol.* 2:57–65.

——— 1977. Cleptoparasitism and odor mimetism in bees: Do *Nomada* males initate the odor of *Andrena* females? *Science* 196:1117–1119.

——— 1978. Identical isoprenoid esters in the Dufour's gland secretions of North American and European *Andrena* bees (Hymenoptera: Andrenidae). *J. Kans. Entomol. Soc.* 51:521–526.

Tengö, J., G. Bergström, A.-K. Borg-Karlson, I. Groth, and W. Francke. 1982. Volatile compounds from cephalic secretions of females in two cleptoparasite bee genera, *Epeolus* and *Coelioxys*. *Z. Naturforsch.* 37:376–380.

Tepedino, V. J. 1981. Notes on the reproductive biology of *Zigadenus paniculatus*, a toxic range plant. *Great Basin Nat.* 41:427–430.

Tepedino, V. J., and D. R. Frolich. 1984. Fratricide in *Megachile rotundata*, a non-social megachilid bee: impartial treatment of sibs and non-sibs. *Behav. Ecol. Sociobiol.* 15:19–23.

Tepedino, V. J., L. L. McDonald, and R. Rothwell. 1979. Defense against parasitization in mud-nesting Hymenoptera: can empty cells increase net reproductive output? *Behav. Ecol. Sociobiol.* 6:99–104.

Tepedino, V. J., and F. D. Parker. 1984. Nest selection, mortality and sex ratio in *Hoplitis fulgida* (Cresson) (Hymenoptera: Megachilidae). *J. Kans. Entomol. Soc.* 57:181–189.

——— 1986. Effect of rearing temperature on mortality, second-generation emergence, and size of adult in *Megachile rotundata* (Hymenoptera: Megachilidae). *J. Econ. Entomol.* 79:974–977.

Terada, Y. 1972. Enxameagem em *Frieselmelitta varia* Lepeletier (Hymenoptera, Apidae). Pp. 293–299 in S. P. Hebling, E. de Lello, and C. S. Takahashi, eds. *Homenagem à Warwick E. Kerr*. Rio Claro, São Paulo.

Terada, Y. C., Garófalo, C. A., and S. F. Sakagami. 1975. Age-survival curves for workers of two eusocial bees (*Apis mellifera* and *T. (Plebeia) droryana*). *J. Apic. Res.* 14:161–170.

Thomson, J. D. 1986. Pollen transport and deposition by bumble bees in *Erythronium*: influences of floral nectar and bee grooming. *J. Ecol.* 74:329–341.

Thomson, J. D., R. C. Plowright, and G. R. Thaler. 1985. Maticil insecticide spraying, pollinator mortality and plant fecundity in New Brunswick forests. *Can. J. Bot.* 63:2056–2062.

Thornhill, R. 1979. Male and female sexual selection and the evolution of mating strategies in insects. Pp. 81–122 in M. S. Blum and N. A. Blum, eds. *Sexual selection and reproductive competition in insects*. Academic Press, New York.

——— 1984. Scientific methodology in entomology. *Fla. Entomol.* 67:74–96.

Thornhill, R., and J. Alcock. 1983. *The evolution of insect mating systems*. Harvard Univ. Press, Cambridge, Mass.

Thorp, R. W. 1979. Structural, behavioral and physiological adaptations of bees for collecting pollen. *Ann. Mo. Bot. Gard.* 66:788–812.

Thorp, R. W., and J. R. Estes. 1975. Intrafloral behavior of bees on flowers of *Cassia fasciculata*. *J. Kans. Entomol.Soc.* 48:175–184.

Tingek, S., M. Mardan, T. E. Rinderer, N. Koeniger, and G. Koeniger. 1988. Rediscovery of *Apis vechti* (Maa, 1953): the Saban honey bee. *Apidologie* 19:97–102.

Tong, S. C., R. A. Morse, C. A. Bache, and D. J. Lisk. 1975. Elemental analysis of honey as an indicator of pollution. *Arch. Environ. Health* 30:332.

Torchio, P. F. 1984. The nesting biology of *Hylaeus bisinuatus* Forster and development of its immature forms (Hymenoptera: Colletidae). *J. Kans. Entomol. Soc.* 57:276–297.

Torchio, P. F., and V. J. Tepedino. 1980. Sex ratio, body size and seasonality in a solitary bee, *Osmia lignaria propinque* Cresson (Hymenoptera, Megachilidae). *Evolution* 34:993–1003.

Torchio, P. F., and G. E. Trostle. 1986. Biological notes on *Anthophora urbana urbana* and its parasite, *Xeromelecta californica* (Hymenoptera: Anthophoridae), including descriptions of late embryogenesis and hatching. *Ann. Entomol. Soc. Am.* 79:434–447.

Towne, W. F. 1985. Acoustic and visual cues in the dances of four honey bee species. *Behav. Ecol. Sociobiol.* 16:185–187.

Townes, H. K., S. Momoi, and M. Townes. 1965. A catalogue and reclassification of the eastern Palearctic Ichneumonidae. *Mem. Am. Entomol. Inst. Ann Arbor* 5.

Townes, H. K., and M. Townes. 1966. A catalogue and reclassification of the neotropic Ichneumonidae. *Mem. Am. Entomol. Inst. Ann Arbor* 8.

Townes, H. K., M. Townes, and V. K. Gupta. 1961. Ichneumon-flies of America north of Mexico. 4. Subfamily Gelinae, tribe Hemigasterini. *Mem. Am. Entomol. Inst. Ann Arbor* 2.

Trivers, R. L., and H. Hare. 1976. Haplodiploidy and the evolution of social insects. *Science* 191:249–263.

Tulloch, A. P. 1980. Beeswax – composition and analysis. *Bee World* 61:47–62.

Turrell, M. J. 1976. Observation of the mating behavior of *Anthidiellum notatum* and *Anthidiellum perplexum*. *Fla. Entomol.* 59:686–688.

Undurraga, J., and W. P. Stephen. 1980. Effect of temperature on development and survival in post-diapausing alfalfa leafcutting bee prepupae and pupae (*Megachile rotundata* [F.] Hymenoptera: Megachilidae). II. Low temperature. *J. Kans. Entomol. Soc.* 53:677–682.

Unwin, D. M. 1980. *Microclimate measurement for ecologists*. Academic Press, London.

Unwin, D. M., and S. A. Corbett. 1984. Wingbeat frequency, temperature and body size in bees and flies. *Physiol. Entomol.* 9:115–121.

Usinger, R. L. 1958. Harwanzen or resin bugs in Thailand. *Pan-Pac. Entomol.* 34:52–53.

Vecchi, M. A. 1959. La microflora dell'ape mellifica. *Ann. Microbiol. Enzimol.* 9:73–86.

Velthuis, H. H. W. 1972. Observations on the transmission of queen substances in the honey bee colony by the attendants of the queen. *Behaviour* 41:105–129.

— 1985. The honeybee queen and social organization of her colony. Pp. 343–358 in B. Hölldobler and M. Lindauer, eds. *Experimental behavioral ecology and sociobiology*. G. Fischer Verlag, New York.

— 1987. The evolution of sociality: ultimate and proximate factors leading to primitive social behavior in carpenter bees. Behavior in social insects. *Experientia Supplementum* 54:405–430.

Velthuis, H. H. W., and J. M. F. Camargo. 1975a. Observations on male territories in a carpenter bee, *Xylocopa* (*Neoxylocopa*) *hirsutissima* Maidl. (Hymenoptera, Anthophoridae). *Z. Tierphysiol.* 38:409–418.

— 1975b. Further observations on the function of male territories in the carpenter bee *Xylocopa* (*Neoxylocopa*) *hirsutissima* Maidl. *Neth. J. Zool.* 25:516–528.

Velthuis, H. H. W., and D. Gerling. 1980. Observations on territoriality and mating behaviour of the carpenter bee *Xylocopa sulcatipes*. *Entomol. Exp. Appl.* 28:82–91.

Vergara, C. 1983. Rango de vuelo y cuantificación de los recursos colectados por abejas Africanizadas en un bosque tropical de Panama. Unpublished data report, *Smithsonian Tropical Research Institute*, Balboa, Panama.

Vergara, C., and O. Pinto. 1981. Primer registro para Colombia de abehas sin aguijón encontradas a más de dos mil metros de altura (Hymenoptera, Apidae). *Lozania* No. 35.

Vesey-Fitzgerald, D. 1939. Observations on bees in Trinidad, B.W.I. *Proc. R. Entomol. Soc. London. Ser. A Gen. Entomol.* 14:107–110.

Villa, J. D. 1987. Africanized and European colony conditions at different elevations in Colombia. *Am. Bee J.* 127:53–57.

Villanueva, R. 1984. Plantas de importancia apícola en el ejido de Plan de Río, Veracruz, México. *Biotica* 9:279–340.

Vinson, S. B., G. W. Frankie, M. S. Blum, and J. W. Wheeler. 1978. Isolation, identification and function of the Dufour's gland secretion of *Xylocopa virginica texana* (Hymenoptera: Anthophoridae). *J. Chem. Ecol.* 4:315–323.

Vinson, S. B., G. W. Frankie, and R. E. Coville. 1987. Nesting habits of *Centris flavofasciata* Friese (Hymenoptera: Apoidea: Anthophoridae) in Costa Rica. *J. Kans. Entomol. Soc.* 60:249–263.

Vinson, S. B., G. W. Frankie, and H. J. Williams. 1986. Description of a new dorsal mesosomal gland in two *Xylocopa* species (Hymenoptera: Anthophoridae) from Costa Rica. *J. Kans. Entomol. Soc.* 59:185–189.

Vinson, S. B., H. J. Williams, G. W. Frankie, J. W. Wheeler, M. S. Blum, and R. E. Coville. 1982. Mandibular glands of male *Centris adani* (Hymenoptera: Anthophoridae): their morphology, chemical constituents, and function in scent marking and territorial behavior. *J. Chem. Ecol.* 4:319–327.

Visscher, P. K., and T. D. Seeley. 1982. Foraging strategy of honey bee colonies in a temperate deciduous forest. *Ecology* 63:1790–1801.

Viswanathan, H. 1950. Temperature readings on *Apis dorsata* combs. *Indian Bee J.* 12:72.

Vitousek, P. M., and R. L. Sanford, Jr. 1986. Nutrient cycling in moist tropical forest. *Annu. Rev. Syst. Ecol.* 17:137–167.

Vogel, S. P. 1963. Duftdrusen im Dienste der Bestaubung: Über Bau und Funktion der Osmophoren. *Akad. Wiss. Lit. Abh. Math. Naturwiss. Kl. [Mainz]* 1962:599–763.

—— 1969. Flowers offering fatty oil instead of nectar. Abstr. *XI Int. Bot. Congr.*, p 229.

—— 1974. Ölblumen und ölsammelnde Bienen. *Akad. Wiss. Lit. Abh. Math. Naturwiss. Kl. [Mainz] Trop. Subtrop. Pflanz.* 7:1–267.

—— 1981. Die klebstoffeharre an den antheren von *Cyclanthera pedata* (Cucurbitaceae). *Plant Syst. Evol.* 137:291–316.

—— 1984. Abdominal oil-mopping – a new type of foraging in bees. *Naturwissenschaften* 68:627–628.

Vogel, S., and C. D. Michener. 1985. Long bee legs and oil-producing floral spurs, and a new *Rediviva* (Hymenoptera, Melittidae; Scrophulariaceae). *J. Kans. Entomol. Soc.* 58:359–364.

Waddington, K. D. 1976. Foraging patterns of halictid bees at flowers of *Convolvulus arvensis*. *Psyche* 83:112–119.

—— 1979. Divergence of inflorescence height: An evolutionary response to pollinator fidelity. *Oecologia* 40:43–50.

—— 1981. Factors influencing pollen flow in bumble bee-pollinated *Delphinium virescens*. *Oikos* 37:153–159.

—— 1983. Foraging behavior of pollinators. Pp. 213–241 in L. Real, ed. *Pollination biology*. Academic Press, New York

—— 1985. Cost-intake information used in foraging. *J. Insect Physiol.* 31:891–897.

Waddington, K. D., L. Herbst, and D. W. Roubik. 1986. Relationship between recruitment systems of stingless bees and within-nest worker size variation. *J. Kans. Entomol. Soc.* 59:95–102.

Waddington, K. D., and L. R. Holden. 1979. Optimal foraging: On flower selection by bees. *Am. Nat.* 114:179–196.

Wagner, A. E., and D. A. Briscoe. 1983. An absence of enzyme variability within two species of *Trigona* (Hymenoptera). *Heredity* 50:97–103.

Wagner, D. L., and S. A. Cameron. 1985. *Bombus bifarius* foraging at aphid honeydew. *Pan-Pac. Entomol.* 61:266.
Waldbauer, G. P. 1968. The consumption and utilization of food by insects. *Adv. Insect Physiol.* 12:711–730.
Waldbauer, G. P., and W. E. LaBerge. 1985. Phenological relationships of wasps, bumblebees, their mimics and insectivorous birds in northern Michigan. *Ecol. Entomol.* 10:99–110.
Waldbauer, G. P., and J. K. Sheldon. 1971. Phenological relationships of some aculate Hymenoptera, their dipteran mimics, and insectivorous birds. *Evolution* 25:371–382.
Waldbauer, G. P., J. G. Sternburg, and C. T. Maier. 1977. Phenological relationships of wasps, bumblebees, their mimics, and insectivorous birds in an Illinois sand area. *Ecology* 58:583–591.
Walker, E. P., F. Warnick, S. E. Hamlet, K. I. Lange, M. A. Davis, H. E. Uible, and P. F. Wright. 1975. *Mammals of the world.* Johns Hopkins Press, Baltimore.
Walker, F. 1871. *A list of Hymenoptera in Egypt, in the neighborhood of the Red Sea, and in Arabia.* Smithsonian Institution Libraries.
Waller, G. D., D. M. Caron, and G. M. Loper. 1981. Pollen-patties maintained brood rearing when pollen was trapped from honeybee colonies. *Am. Bee J.* 121:101–105.
Waser, N. M. 1979. Pollinator availability as a determinant of flowering time in Ocotillo (*Fouquieria splendens*). *Oecologia* 39:107–121.
———. 1983. The adaptive nature of floral traits: ideas and evidence. Pp. 242–286 in L. K. Real, ed. *Pollination biology.* Academic Press, Orlando, Fla.
———. 1986. Flower constancy: definition, cause and measurement. *Am. Nat.* 127:593–603.
Waser, N. M., and M. V. Price. 1983. Optimal and actual outcrossing in plants, the nature of plant–pollinator interaction. 1983. Pp. 341–359 in C. E. Jones and R. John Little, eds. *Handbook of experimental pollination biology.* Van Nostrand Reinhold, New York.
Watanabe, H., and Y. Tanada. 1972. Infection of nuclear-polyhedrosis virus in armyworm, *Pseudaletia unipunctata* Haworth (Lepidoptera: Noctuidae), reared at a high temperature. *Appl. Entomol. Zool.* 7:43–51.
Watmough, R. H. 1974. Biology and behaviour of carpenter bees in southern Africa. *J. Entomol. Soc. S. Afr.* 37:261–281.
Way, M. J. 1963. Mutualism between ants and honeydew-producing Homoptera. *Annu. Rev. Entomol.* 8:307–344.
Wcislo, W. T. 1986. Host nest discrimination by a cleptoparasitic fly, *Metopia campestris* (Fallen) (Diptera: Sarcophagidae: Miltogramminae). *J. Kans. Entomol. Soc.* 59:82–88.
———. 1987a. The role of learning in the mating biology of a sweat bee *Lasioglossum zephyrum* (Hymenoptera: Halictidae). *Behav. Ecol. Sociobiol.* 20:179–185.
———. 1987b. The roles of seasonality, host synchrony, and behaviour in the evolutions and distributions of nest parasites in Hymenoptera, with special reference to bees. *Biol. Rev.* 62:515–543.
Wehner, R. 1982. The bee's celestial map – A simplified model of the outside world. Pp. 375–379 in M. D. Breed, C. D. Michener, and M. H. Evans, eds. *The biology of social insects.* Westview Press, Boulder, Colo.
———. 1984. Astronavigation in insects. *Annu. Rev. Entomol.* 29:277–298.
Wehner, R., and S. Rossel. 1985. The bee's celestial compass: a case study in behavioural neurobiology. Pp. 11–54 in B. Hölldobler and M. Lindauer, eds. *Experimental behavioral ecology and sociobiology.* G. Fisher Verlag, New York.
Wells, H., and P. H. Wells. 1983. Honey bee foraging ecology: optimal diet, minimal uncertainty or individual constancy? *J. Anim. Ecol.* 52:829–836.
Wells, P. H., and J. Giacchino. 1968. Relationship between the volume and the sugar concentration of loads carried by honey bees. *J. Apic. Res.* 20:77–82.

West-Eberhard, M. J. 1981. Intragroup selection and the evolution of insect societies. Pp. 3–17 in R. D. Alexander and D. W. Tinkle, eds. *Natural selection and social behavior.* Chiron Press, New York.

———. 1983. Sexual selection, sexual competition, and speciation. *Q. Rev. Biol.* 58:155–183.

Wetherwax, P. B. 1986. Why do honeybees reject certain flowers? *Oecologia* 69:567–570.

Weygoldt, P. 1969. *The biology of pseudoscorpions.* Harvard Univ. Press, Cambridge, Mass.

Wheeler, J. W., J. Avery, F. Birmingham, and R. M. Duffield. 1984. Exocrine secretions of bees. VIII. 8-acetoxy-2,6-dimethyl-2-octenal: a novel mandibular gland component from *Panurginus* bees (Hymenoptera: Andrenidae). *Insect Biochem.* 14:391–392.

Wheeler, J. W., M. S. Blum, H. V. Daly, C. J. Kislow, and J. M. Brand. 1977. Chemistry of mandibular gland secretions of small carpenter bees (*Ceratina* spp.). *Ann. Entomol. Soc. Am.* 70:635–636.

Wheeler, J. W., S. L. Evans, M. S. Blum, H. H. W. Velthuis, and J. M. F. Camargo. 1976. Cis-2-methyl-5-hydroxyhexanoic acid lactone in the mandibular gland of a carpenter bee. *Tetrahedron Lett.* 45:4029–4032.

Wheeler, J. W., M. T. Shamim, O. Ekpa, G. C. Eickwort, and R. M. Duffield. 1985. Exocrine secretions of bees. VI. Unsaturated ketones and aliphatic esters in the Dufour's gland secretion of *Dufourea novaeangliae*. *J. Chem. Ecol.* 11:353–361.

Wheeler, Q. D., and M. Blackwell, eds. 1984. *Fungus–insect relationships: perspectives in ecology and evolution.* Columbia Univ. Press, New York.

Whitehead, A. T. 1978. Electrophysiological response of honey bee labial palp contact chemoreceptors to sugars and electrolytes. *Physiol. Entomol.* 3:241–248.

Whitehead, D. R. 1983. Wind pollination: Some ecological and evolutionary perspectives. Pp. 97–109 in L. Real, ed. *Pollination biology.* Academic Press, New York.

Whitmore, T. C. 1984. *Tropical rain forests of the Far East.* Clarendon Press, Oxford.

Whitten, A. J., S. J. Damanik, J. Anwzr, and N. Hisyam. 1987. *The Ecology of Sumatra.* Gadjah Mada Univ. Press.

Whitten, W. M., N. H. Williams, W. S. Armbruster, M. A. Battiste, L. Stekowski, and N. Lindquist. 1986. Carvone oxide: an example of convergent evolution in euglossine pollinated plants. *Syst. Bot.* 11:222–228.

Wickerman, L. T., and K. A. Burton. 1952. Occurrence of yeast mating types in nature. *J. Bacteriol.* 63:449–451.

Wickler, W. 1968. *Mimicry in plants and animals.* McGraw-Hill, New York.

Wiens, D. 1978. Mimicry in plants. Pp. 365–403 in M. K. Hecht, W. C. Steere, and B. Wallace, eds. *Evolutionary biology,* Vol. 11. Plenum, New York.

Wiens, J. A. 1977. On competition and variable environments. *Am. Sci.* 65:590–597.

———. 1978. Population responses to patchy environments. *Annu. Rev. Ecol. Syst.* 7:81–120.

Wightman, J. A., and V. M. Rogers. 1978. Growth, energy and nitrogen budgets and efficiencies of the growing larvae of *Megachile pacifica* (Panzer) (Hymenoptera, Megachilidae). *Oecologia* 36:245–257.

Wiley, E. O. 1981. *Phylogenetics: the theory and practice of phylogenetic systematics.* Wiley-Interscience, New York.

Wille, A. 1958. Comparative studies of the dorsal vessels of bees (Hymenoptera: Apioidea). *Ann. Entomol. Soc. Am.* 51:538–546.

———. 1963. Behavioral adaptations of bees for pollen collecting from *Cassia* flowers. *Rev. Biol. Trop.* 11:205–210.

———. 1966. Notes on two species of ground-nesting stingless bees (*Trigona mirandula* and *T. buchwaldi*) from the Pacific rain forest of Costa Rica. *Rev. Biol. Trop.* 14:251–277.

1969. A new species of stingless bee *Trigona (Plebeia)* from Costa Rica, with descriptions of its general behavior and cluster-type nest. *Rev. Biol. Trop.* 15:299–313.

1976. Las abejas jicótes del género *Melipona* (Apidae: Meliponini) de Costa Rica. *Rev. Biol. Trop.* 24:123–147.

1977. A general review of the fossil stingless bees. *Rev. Biol. Trop.* 25:43–46.

1979a. A comparative study of the pollen press and nearby structures in bees of the family Apidae. *Rev. Biol. Trop.* 27:217–221.

1979b. Phylogeny and relationships among the genera and subgenera of the stingless bees (Meliponinae) of the world. *Rev. Biol. Trop.* 27:241–277.

1983a. Biology of the stingless bees. *Annu. Rev. Entomol.* 28:41–64.

1983b. *Corcovado. Meditaciones de un biólogo.* Universidad estatal a distancia, San José, Costa Rica.

Wille, A., and L. C. Chandler. 1964. A new stingless bee from the tertiary amber of the Dominican Republic. *Rev. Biol. Trop.* 12:187–196.

Wille, A., and G. Fuentes. 1975. Efecto de la ceniza del volcán Irazú (Costa Rica) en algunos insectos. *Rev. Biol. Trop.* 23:165–175.

Wille, A., and C. D. Michener. 1971. Observations on the nests of Costa Rican *Halictus* with taxonomic notes on neotropical speicies (Hymenoptera: Halictidae). *Rev. Biol. Trop.* 18:17–31.

1973. The nest architecture of stingless bees with special reference to those of Costa Rica. *Rev. Biol. Trop.* 21:1–278.

Wille, A., and E. Orozco. 1970. The life cycle and behavior of the social bee *Lasioglossum (Dialictus) umbripennae. Rev. Biol. Trop.* 17:199–245.

1975. Observation on the founding of a new colony by *Trigona cupira* (Hymenoptera: Apidae) in Costa Rica. *Rev. Biol. Trop.* 22:253–287.

Williams, G. C. 1966. *Adaptation and natural selection.* Princeton Univ. Press, Princeton, N.J.

Williams, H. J., G. W. Elzen, M. R. Strand, and S. B. Vinson. 1983. Chemistry of Dufour's gland secretions of *Xylocopa virginica texana* and *Xylocopa micans* (Hymenoptera: Anthophoridae) – a comparison and reevaluation of previous work. *Comp. Biochem. Physiol.* 74B:759–761.

Williams, H. J, M. R. Strand, G. W. Elzen, S. B. Vinson, and S. J. Merritt. 1986. Nesting behavior, nest architecture, and use of Dufour's gland lipids in nest provisioning by *Megachile integra* and *M. mendica mendica* (Hymenoptera: Megachilidae). *J. Kans. Entomol. Soc.* 59:588–597.

Williams, H. J., S. B. Vinson, and G. W. Frankie. 1987. Chemical content of the dorsal mesosomal gland of two *Xylocopa* species (Hymenoptera: Anthophoridae) from Costa Rica. *Comp. Biochem. Physiol.* 86B:311–312.

Williams, H. J., S. B. Vinson, G. W. Frankie, R. E. Coville, and G. W. Ivie. 1984. Morphology, chemical contents and possible function of the tibial gland of males of the Costa Rican solitary bees *Centris nitida* and *Centris trigonoides subtarsata. J. Kans. Entomol. Soc.* 57:50–54.

Williams, J. L. 1978. Insects: Lepidoptera (moths). Pp. 105–127 in R. A. Morse, ed. *Honey bee pests, predators, and diseases.* Cornell Univ. Press, Ithaca, N.Y.

Williams, N. H. 1982. The biology of orchids and euglossine bees. Pp. 119–171 in J. Arditti, ed. *Orchid Biology: Reviews and perspectives,* II. Cornell Univ. Press, Ithaca, N.Y.

Williams, N. H., and R. L. Dressler. 1976. Euglossine pollination of *Spathiphyllum* (Araceae). *Selbyana* 1:349–356.

Williams, N. H., and W. M. Whitten. 1983. Orchid floral fragrances and male euglossine bees: methods and advances in the last sesquidecade. *Biol. Bull.* 164:355–395.

Williams, P. H. 1985. A preliminary cladistic investigation of relationships among the bumble bees. *Syst. Entomol.* 10:239–255.
Williamson, M. H. 1984. The measurement of population variability. *Ecol. Entomol.* 9:239–241.
Willmer, P. G., and S. A. Corbet. 1981. Temporal and microclimatic partitioning of the floral resources of *Justicia aurea* amongst a concourse of pollen vectors and nectar robbers. *Oecologia* 51:67–78.
Willson, M. F. 1983. *Plant reproductive ecology.* Wiley–Interscience, New York.
Willson, M. F., and N. Burley. 1983. *Mate choice in plants.* Princeton Univ. Press, Princeton, N.J.
Wilson, E. O. 1971. *The insect societies.* Belknap Press of Harvard Univ. Press, Cambridge, Mass.
   1975. *Sociobiology: the new synthesis.* Belknap Press of Harvard Univ. Press, Cambridge, Mass.
   1985a. Invasion and extinction in the West Indian ant fauna: evidence from the Dominican amber. *Science* 229:265–267.
   1985b. The sociogenesis of insect colonies. *Science* 228:1489–1495.
Winston, M. L. 1979a. The proboscis of the long-tongued bees: A comparative study. *Univ. Kans. Sci. Bull.* 51:631–667.
   1979b. Intra-colony demography and reproductive rate of the Africanized honeybee in South America. *Behav. Ecol. Sociobiol.* 4:279–292.
   1981. Seasonal patterns of brood rearing and worker longevity in colonies of Africanized honey bees in South America. *J. Kans. Entomol. Soc.* 53:157–165.
Winston, M. L., and S. J. Katz. 1982. Foraging differences between cross-fostered honeybee workers (*Apis mellifera*) of European and Africanized races. *Behav. Ecol. Sociobiol.* 10:125–129.
Winston, M. L., and C. D. Michener. 1977. Dual origin of highly social behavior among bees. *Proc. Natl. Acad. Sci. USA* 74:1136–1137.
Winston, M. L., and G. O. Otis. 1978. Ages of bees in swarms and afterswarms of the Africanized honeybee. *J. Apic. Res.* 17:123–129.
Winston, M. L., G. W. Otis, and O. R. Taylor. 1979. Absconding behaviour of the Africanized honeybee in South America. *J. Apic. Res.* 18:85–94.
Winston, M. L., O. R. Taylor, and G. W. Otis. 1983. Some differences between temperate European and tropical African and South American honeybees. *Bee World* 64:12–21.
Wirtz, P., M. Szabados, H. Pethig, and J. Plant. 1988. An extreme case of interspecific territoriality: male *Anthidium manicatum* (Hymenoptera: Megachilidae) wound and kill intruders. *J. Ethol.* 4:159–167.
Witherell, P. C. 1975. Other products of the hive. Pp. 531–558 in *The hive and the honey bee.* Danant and Sons, Hamilton, Ill.
Wittmann, D. 1985. Aerial defense of the nest by workers of the stingless bee *Trigona (Tetragonisca) angustula* (Lat.) (Hymenoptera: Apidae). *Behav. Ecol. Sociobiol.* 16:111–114.
Wolda, H. 1978. Fluctuations in abundance of tropical insects. *Am. Nat.* 112:1017–1045.
   1983. Long-term stability of tropical insect populations. *Res. Popul. Ecol. Suppl.* 3:112–126.
Wolda, H., and D. W. Roubik. 1986. Nocturnal bee abundance and seasonal bee activity in a Panamanian forest. *Ecology* 67:426–433.
Wong, M. 1984. Behavioural indication of an African origin for the Malaysian honeyguide *Indicator archipelagicus*. *Bull. Brit. Ornithol. Club* 104:57–60.
Woyke, J. 1963. Drone larvae from fertilized eggs of the honeybee. *J. Apic. Res.* 2:19–24.

1975. Natural and instrumental insemination of *Apis cerana indica* in India. *J. Apic. Res.* 14:153–159.

1979. Sex determination in *Apis cerana indica*. *J. Apic. Res.* 18:122–127.

1980. Evidence and action of cannibalism substance in *Apis cerana indica*. *J. Apic. Res.* 19:6–16.

1986. Sex determination. Pp. 91–119 in T. E. Rinderer, ed. *Bee genetics and breeding.* Academic Press, Orlando, Fla.

Wu, Y.-R., and Kuang, B.-Y. 1986. A study of the genus *Micrapis* (Apidae). *Zool. Res.* 7:99–102.

Wyatt, R. 1983. Pollinator–plant interactions and the evolution of breeding systems. Pp. 51–96 in L. Real, ed. *Pollination biology.* Academic Press, New York.

Yamane, S. 1983. The Aculeate fauna of the Krakatau islands (Insecta, Hymenoptera). *Rep. Fac. Sci. Kagoshima Univ. (Earth Sci. Biol.)* 16:75–107.

Yong, H.-S. 1976. Flower mantis. *Nat. Malay.* 1:32–35.

1986. Allozyme variation in the stingless bee *Trigona fuscobalteata* (Hymenoptera, Apidae) from peninsular Malaysia. *Comp. Biochem. Physiol.* 83B:627–628.

Yoshikawa, K., R. Ohgushi, and S. F. Sakagami. 1969. Preliminary report on entomology of the Osaka city university 5th scientific expedition to southeast Asia. Pp. 153–199 in T. Kira and K. Iwata, eds. *Nature and life in southeast Asia*, Vol. VI. Japan.

Young, A. M. 1985. Notes on the nest structure and emergence of *Euglossa turbinifex* Dressler (Hymenoptera: Apidae) in Costa Rica. *J. Kans. Entomol. Soc.* 58:538–542.

Zeuner, F. E., and F. J. Manning. 1976. A monograph on fossil bees. *Bull. Br. Mus. Nat. Hist.* 27:1–268.

Zimmerman, J. K., D. W. Roubik, and J. D. Ackerman. In press. Asynchronous phenologies of a neotropical orchid and its euglossine pollinator. *Ecology.*

Zimmerman, M. 1979. Optimal foraging: A case for random movement. *Oecologia* 43:261–267.

Zimmerman, M. H., and H. Ziegler. 1975. List of sugars and sugar alcohols in sieve-tube exudates. Pp. 480–503 in M. H. Zimmerman and J. A. Milburn, eds. *Encyclopedia of plant physiology.* Springer, New York.

Zucchi, R. 1973. Aspectos bionômicos de *Exomalopsis aureopilosa* e *Bombus atratus* incluindo considerações sobre a evolução do comportamento social. Dissertation, Univ. São Paulo, Ribeirão Preto.

Zucchi, R., and S. F. Sakagami. 1972. Capicidade termo-reguladora em *Trigona spinipes* e em algumas outras espécies de abelhas sem ferrão. Pp. 301–309 in S. P. Hebling, E. de Lello, and C. S. Takahashi, eds. *Homenagem à Warwick E. Kerr.* Rio Claro, São Paulo.

Zucchi, R., S. F. Sakagami, and J. M. F. Camargo. 1969. Biological observations on a neotropical parasocial bee *Eulaema nigrita,* with a review of the biology of Euglossinae. A comparative study. *J. Fac. Sci. Hokkaido Univ. Ser. VI Zool.* 17:271–380.

Zucoloto, F. S. 1979. Utilization of carbohydrates by *Scaptotrigona postica*. *J. Apic. Res.* 18:36–39.

Zucoloto, F. S., and M. C. T. Penedo. 1977. Physiological effects of mannose in *Scaptotrigona postica*. *Bol. Zool.* 2:129–134.

Subject index

(Page numbers in italics indicate material in tables or figures.)

abdomen, heat regulation and, 93–4
abundance
  peaks, 338, 345
  stability index as log-transformed variance of, 340
Ackerman, J. D., 115, 335, 340
activity patterns of social and solitary bees, 131, *132, 133*, 134–8
adaptationist interpretations, 2–3
adaptive radiation of ancient bee fauna, 8
adults, emergence of solitary bee, 286
Africanized honeybees
  advantage over native bees in tropics, 366
  biology of, 361–2
  change in traits of, 359
  changing relative bee abundance following invasion by, *362*
  compared with European honeybees, 376
  fighting between workers of, 194
  flight and foraging range of, 87, 90
  foraging in rain by, 99
  foraging success of, 104
  high metabolic activity and early morning foraging by, 99
  limits to spread of, 359
  low survival rate of European honeybees in nests of, 223
  nest of, *371*
  nest site competition and, 364
  pollen utilization by, *323, 326*
  pollination by, 366
  possible waning of, 362
  predators of, 223–4
  preference for nocturnally dehiscent flowers, 379
  reproduction and seasonal activity along Panama Canal, 333–4, *335*
  reproductive isolation and, 361
  stability of population in Panama, 340
  stinging and defense behavior of, 218, 224, 354
  "swamping" of other varieties of *Apis mellifera* by, 361
  swarm movements of feral, *289*
  temperature and time of day and foraging by, 135
  tranquilizing plants and, 369
  in tropical America, 357–66
  winter temperatures and, 360
  see also honeybees
aggression, 106–7
  during foraging by highly social bees, 104–5
  between host and parasite, 262
  mating and, 270–1, 275
  by meliponines against *Lestrimelitta*, 214
  parasitism and subdued, 261
  at potential nest sites, 193–4
  ritualized threat display of stingless bees and, 105
  by robbers toward pollinators, 159
  by stingless bees, 112
  by *Xylocopa* females, 173
aggressive foraging syndrome, 112–13
alarm attractants, 218
alarm behavior, 76, 217
alimentary structures, 57
Allen's rule, 94
allodapine bees
  nests of, *170*
  removal of feces from nest by, 173
allomones, 214, 218, 266
altitudinal limits, 341
Aluja, M., 85
amber, bees in, 10, 11
amino acids
  in honey, 36
  in nectar, 33, 34, 36, 103
  in pollen, 36, 40
  resorption by rectal pads, 57
annual variance (AV), 336, 338, 340
anteaters, 225

488 Subject index

anthophorid bees, 180
  foraging by, 115
  hindlegs of, *46*
  odors at nest entrances and, 76
ants
  as bee enemies, 233
  as nest attackers, 222
  nests of stingless bees in nests of, 196, 218, 220
  as predators of small bees, 123, 130
  secretions repellent to, 213
  stingless bees' response to raids by, 388
apid bees
  cleptobiotic (robbing), 45
  defenses of, 206–8, 210, 215–18, 222–4
  destructive to flowers, *156*
  evaporative nest cooling by highly eusocial, 95
  evolution of corbicula in, 63
  feces collected by, 43
  grooming movements of, 61
  hindleg specializations used for pollen manipulation by, *62*
  nest sites of, 185
  sugar solution imbibing rates of, 68
Appanah, S., 104–5, *316*
apples, European honeybees and outcrossing varieties of, 149
armadillo, giant, 225
army ants, 220
aroids
  calcium oxalate from, 40
  secretions from stigmas of, 36
  volatile oils of, 32
asilid flies as mimics and predators of bees, *126–7, 420*
automimicry, 127, 346
avoidance behavior, 82

baboons, 225
bacteria
  as bee pathogens, 227
  mutualism with bees, 227–8
  see also microbes; Taxonomic index
baits, see chemical baiting studies
Baker, H. G., 33
Baker, I., 33
bamboo, *Xylocopa* nests in, 192
barometric pressure, sensitivity to, 98
bats, 364
  nectar taken by, 34
batumen, 199
Bawa, K. S., 115, 148
bears, bees and, 184, 207, 222, 224
bee anatomy, *45, 201*

bee assemblages
  composition of, 347–51
  in neotropics, 348–9
  in paleotropics, 349–50
bee families, 6
bee–flower coadaptation, 55
bee–flower community, 314
bee genera, 6; *see also* Appendix A
beekeeping
  impact of Africanized honeybees on, 358
  with introduced honeybees, 369
  role of bees in communities and tropical, 366–9, *370*, 371–3, *374, 375*–80
bee populations
  intraspecific competition and size of, 355
  stability of, 336, *337, 338,* 339, 340–1
bees
  identification of, 6–7
  origin of, 325
  as parasites, 57, 82, 254 (*see also* cleptoparasites)
bee seasonality, 315, 327–36
  body size and euglossine, 345–6
  community studies at flowering plants and, 317–27
  flower preference and, 316, 317–27
  Müllerian mimicry in euglossines and, 346–7
  techniques in studying, 328
bee species at flower patches, changing composition during the day, 113–14
beetles
  as bee commensals, 228
  as bee mutualists, 229
  as parasites, 238–40
bee traits, quick changes in, 15
behavior
  guarding, 384
  tropical and temperate differences in same species, 375–6
behavioral dominance, 22
Bergmann's rule, 97–8
Bertsch, A., 91
biogeography, ages of bee groups and, 9–15
bird nests, bee nests associated with, 196, 352
birds, 123, 225, 364
body color
  dark, 95–8
  high elevations and, 97
  light, 96–8
  seasonal dimorphism of, 98
  sex differences in *Xylocopa,* 96
  thermoregulation and, 95–6
  time of day of activity and, 95–7, 137
bombine bees, wax of, 41

division of labor, 383, 387
  in *Ceratina,* 211
  colony success and, 24
DNA, variation in honeybee mitochondrial, 305
dormancy
  adaptive responses involved in, 176
  dry weather and euglossine pupal, 346
  see also diapause
dragonflies, 123
drones
  population of stingless-bee (compared to workers), 295
  regulation of *Apis* production of, 296–7
  weight of *Apis mellifera,* 283
Dufour's gland, 164, 175, 200
  antimicrobial secretions of, 232, 282
  attraction of parasites of *Andrena* by host's, 192
  brood cell lining and, 162, 177
  defensive secretion of, 23, 213
  kin group recognition and products of, 75
  nest aggregation pheromones and, 191–2
  nest ownership recognition and products of, 75
Dyer, F. C., 94

ecological diversity within a bee family, 9
ecological factors, influence on foraging, 137
edendates, 225
egg consumption, 21, 22, *295*
egg laying, 173
eggs
  adhesives on, 178
  diploid from unfertilized female, 307
  hatching of, 280
  of honeybees, 204, 297
  of parasitic bees, 249
  queens and development of, 294–5
  size of, 280
  unfertilized worker, 307
  see also brood cells; development
Eickwort, G. C., 271, 274–5, 277
elaiophores, 29–31
electrophoresis, genetic studies with, 302–5
embryo development, 280; *see also* development
enemies of bees, 123
  apoid parasites, 249, *250–2,* 253–67
  insect, 123–30, 233–44
  microbial, 229–33
  mite, 244–5, *246–7,* 248
  nematode, 248–9
  see also parasites; parasitoids; predators
environment, phenotypic variation and, 302

equilibrium theory, 309–10
euglossine bees
  abundance peaks of, *338*
  altitudinal limits of, 341
  ambient temperature and size of, 345
  assemblages of species of, 341
  asynchronous grooming pattern in, 63
  foraging range of, 90
  foraging structures of male, 57
  generation length of, 287–8
  heat loss in, 94
  in-depth studies of, 341–7
  life span of, 287
  long-term population studies of, 335–6, *337*
  Müllerian mimicry in, 346
  nectar sources of, 342
  in nest with mother, 18
  orchids and, 341, 342–5
  parasitism of, 346
  plants visited, 342
  pollen sources of, 342
  pupal dormancy and, 346
  resin sources of, 342
  size in tropical and subtropical American forests, 341
  thoracic temperature during hovering, 93
  tongues of, 65
eusocial bees, 6
  adaptive significance of guarding behavior at advanced age in, 212–13
  ecological limitations on, 381–3
  evolution of sterility in workers of, 384
  foraging styles of, 109
  requirements for development of highly, 380–1
eusocial colony(-nies)
  evolution of, 23
  formation of, 21
  initiation in temperate climates, 383
  temporarily, 384
eusociality
  delayed, 21
  evolution of, 16–19
  loss of, 382
  permanent, 381
evolution of bees, 4–6, 8, 9
excavating soil from burrows, 188
exocrine gland system, *201*
eyes, 74
  mating and, 268–9, *270*

feces
  collected by tropical apids and meliponines, 43
  ecological significance of, 354

feces (*cont.*)
  removal from nest, 173
feeding canal of adult bees, 65
feeding rate and efficiency, 68–71
female bees
  cohabiting, 18
  facultative association of, 383
  loss of foraging ability in reproductive, 19
  mating of, 279–80
feral populations
  absence of *Apis mellifera* (in Asia and tropical America), 207, 358
  of Africanized honeybees, 357–9
  of European honeybees, 360
  reproductive advantage of drones of, 361
fieldwork
  data from, 2
  experiments and, 3
flabellum(-la), 65–6, *67*
flavonoid pigments, 27
flexibility in foraging behavior, 123
flies (Diptera), 240–2
  as bee enemies, 123, 126
  as bee mimics, 123, 126–7
flight
  meteorological factors and, 91
  physiology of heat distribution and, 91, *92*, 93–5
flight direction
  foraging ranges of *Apis cerana*, 86
  influences on, 90
flight muscles
  energy metabolized by, 91
  preflight warming of, 92
  weight of (as proportion of body weight), 118
flight range, 82–90, 114, *116*, 121
floral constancy, 144–6
floral lipids, 29–31
  collection of, 53–6
  in desert regions, 327
floral morphology, feeding efficiency and, 70
floral syndromes, 110
flowering
  "big bang" and "steady state," 112–13
  synchronous, 148
flowering episodes ("mast years") in Southeast Asian dipterocarp forests, 142
flowering plants
  bees as commensals of, 139
  bee seasonality and, 317–27
  coevolution between bees and, 149–53
  dependence of bees on a species of, 142
  ecological relationships with bees, 68–73
  preferred by tropical bees, 321, 324–5

flowers
  bee-mimicking odors of, 32
  bee mouthparts and structure of, 6
  bee species and total bees visiting, 320
  bee species turnover at, 113–14
  contrasts in breeding systems and pollination by small and large tropical bees, 148, *319*
  damage by bees to, 152, 154, *155*, 156–60
  food from, 25
  fossil, 10
  at ground level, 111
  mimics of, 129
  percent robbed (Panama), 160
  scent-marking of, 122
  size of floral display and number of visits to, 121
  very small, 111
  see also plant products; plants; pollination
food
  abundance and quality of, 308–9
  assimilation of nutrients of, 151
  energy per foraging trip from, 119
  flow through honeybee and stingless bee colonies, 297–8
  flow through *Melipona favosa* colony, 293, *294*
  mouth-to-mouth transfer of, 102 (*see also* trophallaxis)
  storage of, 385–7
foragers
  aggressive social, 104–5
  independence of solitary bee, 121
  predators of, 123–30, 243
  seemingly anomalous behavior of, 121–2
  spiders and, 129
  unaggressive social, 105
foraging
  adverse conditions for, 98
  choices in, 73
  cold weather and, 134
  costs and benefits and, 71–2
  determinants of specialization in, 111
  diurnal trends in, 136, 137
  duration of, 131
  ecological factors and, 137
  equilibrium with resources and competitors and dynamics of, 114
  extending area of, 118
  fragrances and, 31
  heat production and regulation during, 91–5
  hypothesis on efficiency of, 119–20
  mimicry and, 129
  nest sites and, 188
  orientation during, 73

Subject index 493

patch selection, 120–1
relationship between nectar quality and effort in, *117–18*
seasonal differences in diurnal, 98
solitary, 109
weather conditions and, 91, 134
foraging ability, loss of (in reproductive females), 19
foraging behavior, 2, 71–3, 87–90, *116*
  light and, 94
  optimality and patterns of, 113–23
  over a range of conditions, 131
  relative transit and feeding costs and, *117*
  samples brought to nest by returning bees, 102
  shifts from trees to shrubs, 324
  of solitary vs. highly eusocial bees, 108
  styles and syndromes of, 107–13
  thermal environment and, 91–9
  unpredictability of patterns of, 114
  worker recruitment to resources and, 103
foraging choices, 120–1, 324
foraging and flight range, 82–90, *116*
  bee size and, 121
  gauging, 114
foraging intensity, model of, 87–90
foraging load, weight of, 118
foraging profitability
  bee assessment of, 119, 120
  colony assessment of, 378
foraging region, 72
foraging structures present in males but absent in females, 57
formic acid, 80, 207–8
fossil bees, 4–5, 9–10, 12
fragrances gathered by bees, 31–2
Frankie, G. W., 113, 115, 148
free zones on bodies of bees, 51
Frisch, K. von, 2, 87, 99, 102
frogs, 123
fruit initiation and seed maturation, bees and, 140
fungi, 173–4, 176, 230; *see also* Taxonomic index

Gary, N. E., 114
gathering and unloading behavior, evolution of, 61–5
generalized flower foraging syndrome, 110
gene(s)
  deleterious recessive, 195, 305
  electrophoretic detection of products of amino acid substitutions in, 302
  sex-limited, 305
genetic deaths, 306

genetic load, 305–6
genetic relatedness of worker to sister queen, 299
genetic studies of bees, biochemical, 302–5
genetic variability, 302–6
geographic distribution of bees, dispersal and, 10
geographic races, 5
giant honeybees, 12–14
  aggregations of nests of, 195
  reaction to predators, 222
Ginsberg, H. S., 271, 277
glacial periods, bee distribution during and following, 7–8
glossae, 65
  feeding and, 68–70
  functions of, 67–8
  in male collectid bee, *61*
  specialization of, 65, *67*
  see also labium(-bia); mouthparts; proboscis; tongue
Gondwana, 11
gorillas, 225
Gould, J. L., 73, 194–5
Greenberg, L., 57
grooming by bees, 51–2, 63
ground-nesting bees, *see* hypogeous (ground-nesting) bees
guard bees, 208–13, 384, 388
gums, 29
  transporting, 63
gymnosperms, nectarlike substances of, 36

hairs, 45, 46
  electrostatic charge of, 47
  on labial palpus used to remove pollen from anthers, *59*
  melittid and andrenid scopal, 63
  thermoregulation and, 94, 96–8
  used in nest construction, 202, *203*
halictid bees
  brood chambers of, 169
  flight ranges of, 86
  larval care by, 173
  mating of, 210
  nest burrows of, *190*
  nesting patterns of, 191
  nest recognition by, 75
hamulus(-li), 83
haploid male bee, retention of apparently useless traits in, 56
heat distribution, flight and physiology, 91, *92*, 93–5
Heinrich, B., 93–4, 119
helper castes, 19

Hemiptera (bugs) as predators of foraging bees, 123, 124–5, 243; *see also* reduviids (Hemiptera)
hibernation (diapause) and sociality, 20
hoarding, 82
Homoptera, honeydew from, 26–7
honey, 36
   characteristics of *Melipona* and Trigonini, 372–3
   of giant honeybees, 373
   nocturnal harvest of, *370*
   robbing by *Apis* species, 375
   species excelling in storage of, 373, 375
   storage of, 205, 375
   transfer from mother nest of meliponines, 205
honey badger, 223
honeybee *e*-vector compass, 74
honeybees
   alarm pheromones of, 217
   body temperature regulation in, 93
   cold tolerance of, 184
   colonies composed only of sisters, 17
   communication of, 2, 99, 100–2
   control in colonies of, 292
   diseases of, 230–1
   dispersal distance of swarm of, 195
   electrostatic charge and pollen–holding by, 47
   European and African compared, 376
   evolution and spread of, 12, 14–15
   external anatomy of, *45*
   failure in tropics of European, 369
   flight ranges of, 86–7
   foraging success of, 104
   highly eusocial life cycles and, 16
   honey of, 36
   humidity and, 94
   internal anatomy of, *201*
   management strategies for, 378–80
   number of colonies supported by cactus, 327
   odor attachment and stinging by, 216
   pollen diets of, *323*, *326*
   predators of, 266 (*see also* parasites)
   reaction to predator attacks on nest, 222–3
   reproductive isolation and, 12
   responses to odors left on flowers, 116
   return to flowers by, 114
   smoke to subdue, 367
   spatial separation of nests of, 194
   species of, 5–6
   species in danger of extinction, 367
   stinging by, 216
   tropical *Apis mellifera*, 377
   unusual materials collected by, 39–40
   as visitors of obligately outcrossing dioecious plants, 149
   wax of, 41
   see also Africanized honeybees; giant honeybees; little honeybees
honey buzzards, 222
honeydew, 25, 26–7
   toxic, 32–3
honey-making bees
   exploitation of native, 372
   percent cultivated, 367
   problems with imported, 371–2
   utilization in tropics, 367
hooks on wings, 83–4, 297; *see also* hamulus
Hopkins host-selection principle, 153
host-seeking behavior, 153
hovering, 96
Hubbell, S. P., 104–5, 108–9, 192–3
humans, bees and, 224
humidity
   detection by bees, 98
   honeybee sensitivity to, 94
hummingbirds, 36, 145
   competition with robbing stingless bees, 154
hydrogen peroxide in honey, 36
Hymenoptera, 233, *234–6*, 237–8
   low genetic variability in, 304–5
   see also ants; bees; wasps
hyperparasitism, 237
hypogeous (ground-nesting) bees
   cocoons of, 180
   long prepupal stage of, 180
   nests of, *168*, *169*–73
hypopharyngeal gland products, 284

identification of bees, 6–7
immature bees, mortality of, 307; *see also* parasites; parasitoids
inbreeding, fitness costs of, 305, 308
insect enemies of bees, 242–3
   Coleoptera (beetles), 238–40
   Diptera (flies), 240–2
   Hymenoptera (wasps and ants), 233, *234–6*, 237–8
insects
   as bee enemies, 233, *234–6*, 237–43
   parasitic, 226, 237–42
   see also parasites; predators
island species, 356

Johnson, L. K., 104–5, 108–9, 193–4
juvenile hormone (JH), queen development and, 284

kairomones, 214
Kapil, R. P., 84
Kerr, W. E., 303, 308
Kimsey, L. S., 61, 279
kin recognition, guard bees and, 208–13
Kirchner, W. H., 101

labium(-bia), *59, 64*, 65, *66*, 69
landmarks, 73, 74
larvae, 203
  care of, 173–4
  development and nutrition of, 151, 280–4
  diapause of mature, 176
  digestion of, 282
  eaten by burrowing beetle larvae, 239–40
  eaten by flies, 241–2
  feces and defecation of, 173, 179, 281
  food of, 151, 177–8
  food abundance and quality and, 308–9
  host bee feeding of parasitic, 239
  killed by termites, 242
  molting of, 281
  of parasitic bees, 239, 257
  pollen diet of, 151
  proportion of diseased, 231
  provisions for, 281
  queen, 283
leaf-cutting bees, 202
learning, 74, 144; *see also* memory
legs, special structures
  for antennal cleaning (strigil), *45*
  for floral lipid handling, 53–4
  for pollen handling, 43, *44, 45, 46*, 47–53, 62
legumes
  capture of pollen from, 51
  in Mexican tropical caducifolious forest, 327
  resin from, 28
Lepidoptera, 243
life cycles
  evolution of highly eusocial, 16
  social bees, 15–24
  solitary bees, 15–24
life span
  difference between bees in mother and daughter colonies, 300
  increased activity and shorter, 119
light intensity, influence on stingless bees, 134; *see also* darkness; sunlight
light trap studies, 328, 329–33
Lindauer, M., 2, 101
lipids, *see* floral lipids
little honeybees, 12
lizards, 123, 128

lodger bees, 161
long-tongued bees, 65, *66*
  nectar collecting by, 70

magnetic recapture experiments, 85–6
Malay honey bears, 222
male bees
  balanced mortality hypothesis and, 311
  buzzing flight of, 267
  development of stingless bee, 294, 295
  display and precopulatory behavior in euglossine, 278–9
  eyes of, 268, 270
  genitalia of honeybee, 280
  mating aggression by, 270–1, 275
  preference for unfamiliar female pheromones, 268
  produced by unmated females, 172
  ratio to queens (*Apis*), 310
  size of, 274
  structural modifications in, 268–70
  temporal partitioning of foraging and mating by, 277
  work of meliponine, 296
mammals as predators of social bees, 224
mandibles
  modifications and diversity of, *60*, 175
  resin mining by stingless bees using, 63
  secretions of, 76, 128, 207
mandibular gland odors of stingless bees, 76, 128
mandibular teeth, 59, 154
mandibular wear, nest excavation and, 17
mantids, 123
mantisipids (Neuroptera), 243
Marden's rule, 118
marginal value theorem, 117
mason bees, 161, 185
mating
  chemical features of, 268, *269*
  display and precopulatory behavior in male euglossines, 278–9
  encounter sites for, 271–2, 275
  interspecific, 273
  searching for mate and, 272–3
  selecting a mate, 267–75, 277–80
  sites for, 274–5, *276*
  of solitary bees, 171, 270–1
  of stingless bees, 171–2
  structural modifications and, 268
  success of, *274*
  visual signals and, 280
mating aggregations, 272
  of stingless bees, 76
matrifilial associations, 15, 16

megachilid bees
 damage to plants by, 154
 foraging by *Chelostomoides*, 115
 nest sites of, 185
meliponine bees
 damage to plants by, 154
 feces collected by, 43
 flight range of, 86
 fossil, 9, 10
 honey transfer and storage by, 205
 lack of rastellum in some, 62
 mandibular gland odor trails of, 2, 76, 80–1, 105, 109
 most primitive, 12, 387
 opportunistic use of related species' foraging trails by, 82
 species visited by, 152–3
 wax of, 41
 see also stingless bees
memory, 144, 147
 of European honeybees, 131
 see also learning
metabolic efficiency, 282
metasocial bees, 17
Michelsen, A., 101
Michener, C. D., 5, 11, 57, 61, 161, 262, 388
microbes
 bee defense against, 231–3
 bees and nonpathogenic, 227, 228–9, 388
 pathogenic to bees, 227, 230–3
 resins and, 27
 see also Taxonomic index
migration, resource shortages inducing, 123
mimicry
 bee, 124, 126, 128, 129, 241
 euglossines and, 346–7
 between flowers, 145
 foraging and, 129
 systems of, 146
mimics of bees, 125–6, 128, *420*
 Batesian, 124, 127
 Müllerian, 124, 128
 of stingless bees, *126*, 128
mites, 51
 as bee commensals, 228
 as bee mutualists, 229, 244–5
 damaging to bees, 14, 226, 245, *246–7*, 248
 giant honeybee parasitic, 14
 phoretic, 248
 tracheal, 226
molds, *see* Taxonomic index
monitoring studies of bees, 327–8
monkeys, 222, 225
monomorphism at a genetic locus, 304
moonlight, 73–4

mouth parts, *66*
 robbing and, 154
 used to gather pollen, 48–50
 see also glossae; labium; proboscis
mud in nest construction, 202
multifemale associations, 21
multifoundress associations, 15
Murrell, D. C., 114
mustelids, 225
mutualism, 218, 227

Nasanov gland, 217
Nash, W. T., 114
natural history
 evolutionary biology and tropical, 2
 importance of, 313
 tropical studies and, 3
navigation, 73–5
 chemical ecology and, 75–6, *77–81*, 82
necrophagic bees, 27, 41, 42
nectar, 32–7
 amino acids in, 33–4, 36, 103
 changes in, 34, 136, 147
 composition of extrafloral, 35–6
 correlations between bee family and composition of, 33, 35
 fed to returning pollen forager, 103
 foraging effort and quality of, 117–18
 honey and, 36
 ingredients of, 32–3, 35–6, 103, 112, 118, 277
 loads of, 118–19
 marginal value theorem and feeding on, *117*
 mixed with pollen, 51
 preferences for different sugar concentrations in, 118
 rate and amounts secreted, 34–5
 regurgitation by nonparasitic apids, 230
 regurgitation by overheated foragers, 95
 robbers of, 154, 160
 sap and composition of, 35
 substances similar to, 36–7
 sugar concentrations of, 32–3, 35–6, 103, 112, 118, 277
 sugars in, 32, 33, 34, 35
 sweetness as quality indicator of, 103
 time of day collected, 136
 times of secretion of, 34, 95
 toxicity of, 32–3
 variable availability of, 146
nectar harvesting, 65, 67–8, 70
nectaries, extrafloral, 35
nectarless tropical flora, 145
nematodes, 245, 248–9
neotropics, bee family representation in, *348*

Subject index   497

nesting ability, loss of (in reproductive
    females), 19
nesting associations of nonsolitary bees, 21
nesting behavior of *Xylocopa*, 173
nesting cavities, bee measuring of potential,
    202
nesting habits of tropical and temperate bees,
    8–9
nesting materials, 25, 28, 162
    cerumen, 200
    collection of, 134
    Dufour's gland, 162, 177
    gums, 29
    hairs, 202
    mandibular gland secretions, 199
    mud, 202
    propolis, 27
    regurgitated material, 199
    resins, 27, 29, 63, 162, 199, 202, 204–5
    wax, 200
nest mates
    guard bee recognition of, 208–13
    influence of foraging experience on, 109–10
nests, *163, 219*
    of allodapines, *170*
    animals attacking, 222
    atmospheric gases in, 177
    behavior in building, 174–5
    construction of, 199–200, 201–6
    construction patterns of, 166
    defense of, 76, 161, 206–8, *209*, 210–18,
        220–5
    entrances to, 75–6, 161, 164, 165, *169*,
        173, 202
    evaporative cooling of, 95, 185
    food and occupants of old and new, 300
    humidity in, 176, 180
    hypogeous, *168*, 169–73
    initiation of new, 300
    largest, 162
    laterals of, 161
    neighboring, 194
    patching of, 199
    pheromones and dispersion of, 193
    pollen storage in, 102
    protection against water, 178–9
    receiver bees in, 103
    recognition of, 75, 161
    repellent secretions at entrances to, 173
    resin in construction of, 27, 29, 63, 162,
        200, 202
    reuse by female progeny of parental, 51
    of solitary bees, 171
    of stingless bees, 165, 180, 182
    temperature in, 95, 177, 180–5, 233

    of *Tetragona clavipes, 374*
    time required to build, 204, 205
    usurpation of, 84, 173
    vestibule of, 161
    xylophilous, *166*
nest sites, *163,* 185, *186,* 187
    availability of, 356
    competition for, 364
    dispersion of, *276*
    dispersion of eusocial colonies', 192–7
    dispersion of solitary bees', 189–92
    foraging and, 188
    nesting microhabitat, 187–9
    slope of excavated, 188–9
    of stingless bees, 197–9
neuropterans, 243
nighttime, social bees in field at, 82, 137
niche preemption, 357
nitrogen assimilation, 151
nocturnal flight, 73–4, 95, 96
nutrients, assimilation of, 151; *see also* diets;
    food

ocelli, 74
odors, 75
    defensive, 214
    to lead nest mates to new nest site, 194
    mimicked by flowers, 76
    nest sharing by different species and, 218,
        220
    perception of, 75
    recognized by parasites, 76
    responses of foragers to bee, 122–3
    used to mark flowers, 116, 122
    see also allomones; kairomones;
        pheromones
oil-collecting species, 53
    in desert regions, 327
oils, *see* floral lipids
ommatidia, 74; *see also* eyes
onion flowers, return of honeybees and
    bumbleees to, 114
Opler, P. A., 115
opposums, 225, 364
orchids, 139, 149, 150, 365
    euglossines and, 341, 342–5
    fragrances of, 32, 335
    sites of pollinaria placement and, 145
orientation, 74–5
    of euglossine bees, 85
orientation flights, 131
origin of bees, 9
ovary development
    in halictid workers after removal of
        reproductive female, 23

ovary development (*cont.*)
  inhibition in *Apis dorsata*, 296
  inhibition in offspring by maternal care, 18
ovary(ies) and ovarioles
  in *Apis* workers in absence of queen, 20
  in parasitic bees, 249, 253
  of semisocial or primatively eusocial females, 19
  in workers in colonies of stingless bees, 20, 295
overheating, avoidance of, 94; *see also* thermoregulation
oxygen consumption, 91

palms, 150
  abundance of bee species at dehiscence of, *108*
  calcium oxalate from, 40
  secretions from stigmas of, 36
  succession of highly eusocial bees at, 109
  visited during wet season, 324
parakeets, 364
parasites, 226
  beetles as, 238–40
  defined, 225
  of euglossines, 346
  flies as, 240–2
  of giant honeybees, 14
  host bee feeding of larvae of, 239
  host nest odors and, 76
  megachilid bee, 238
  mites as, 14, 226, 245, *246–7*, 248
  nematodes as, 248–9
  phroetic, 51
  Scrapter, 167
  of stingless bees, 199
  wasps as, 233, 237
  see also cleptoparasites; parasitoids
parasitic bees, 57
  adult biology of, 261–3
  chemical cues used by, 263
  colors and markings of, 264–5
  eggs of, 261
  eusocial, 254
  frequency of occurence of, 259
  host selection by, 263–7
  larval biology of, 260–1
  loss or modification of structures in, 253, 260
  resting aggregations with hosts, 82
  specialization among, 265
  taxonomic relationships and, 255–6, 257–9
  types of, 253–9
  see also cleptobiosis; cleptoparasites; robbers

parasitic life style
  glossal characteristics and, 67
  predicting, 57
parasitism, sociality and, 17–18
parasitoids, 226
  deceptive nest architecture as defense against, 221
  defined, 225
  wasps as, *234–6*
parasocial bees, 17
parental care of offspring, 20
parthenogenesis, 307–8
patch selection, 120–1
pesticides, 278
pheromones, *77–81*
  alarm, 130, 213, 214, 215–16, 266
  "alert and assembly," 217
  Apis sex, 12
  defined, 214
  foraging, 213
  of honeybee, 12, 217
  marking of mating and feeding territories with, 76
  masking, 266
  mate attracting, 273
  multiple functions of, 75
  of Nasanov gland, 217
  nest aggregation, 191
  nest dispersion and, 193
  perception of, 75
  release of lactone, 23
plant families used by bees in Amazon basin, 327
plant products used by bees, 25
plants
  Africanized honeybee impact on, 365
  bee influences on reproductive fitness of, 138–9
  bees as enemies of, 152, 154, *155*, 156–60
  bee-tranquilizing and repellent, 368–9
  genetic variability among tropical, 149
  groups most important to tropical bees, 321–2
  nectarless, 38
  visited for nectar and honey, 153
  see also coevolution; pollination; resources gathered by bees; Taxonomic index
polarized light, 73–4
pollen
  bacteria in, 227–8
  bee species that never use, 57–8
  bumblebee gathering of, 122
  buzz collecting of, 48
  capture and transport of, 43, *44*, 45, *46*, 47–51

cleaning, manipulation, and storage of, 51–3
composition of, 37–40
corn, 37
deposited on bee notum or sternum, 145
deposited in storage areas of nest, 102
eaten by colletid females while collecting, 47
electrostatic charge of, 47
evolutionary appearance of angiosperm, 325
foraging behavior and methods of collecting, 64–5
glandular development dependent on, 213
intranidal transfer of, 149
legume, 327
lipid covering grains of, 30
of nectarless plants, 38
of nonporicidal plants, 39
poricidal dehiscence of, 38
release of, 37
size of, 38, 39
starchy, 39
on stigmas, 47
time of day of collecting, 136
toxic, 39, 151
viability of, 141
pollen basket (corbicula), 49
pollen collecting structures, 49, 62
  loss of, 58
  retention of no-longer employed, 59
pollenkit, 30, 38
pollen manipulation, hindleg specializations of apid bees and, 62
pollen rain, 325, *326*, 327
pollen robbing, 50
pollen storage pots, 205
pollination ecology, 138–43, 144–54, *155*, 156–60
pollination specialization and coevolution, 149–53
pollinators
  effectiveness of, 141, 143–9
  mistakes of, 145
  plant adjustment to probability of visits by, 148
  robbers and, 160
pollinaria, *38*
pollinia, 37, *38*
polymorphism, genetic, 302–3
population data, 313, 328
predators, 23, 387
  of bumblebees, 124, 127
  chemical adaptations against, 207
  of foraging bees, 123–30, 243
  nest architecture as defense against, 221
preflight warming-up, 93, 95, 99

prereproductive offspring, toleration by "queen" Xylocopinae, 20
primates, 222, 225
proboscis(-scides), 65, 70
propolis, 25, 27
pseudoscorpions, 229

queens
  colony reproduction and, 288–302
  control of development and reproduction by, 291
  corpora allata and development of, 284
  determination of development of, 308–9
  distinguishing from workers in semisocial and subsocial colonies, 21–2
  final generation prior to demise of eusocial halictid colonies, 19
  influence on worker behavior, 291
  killing of virgin, 309
  larval, 283
  mating flights of, 288
  obligate mutualism with workers, 20
  ratio to males (*Apis*), 310
  regulation of *Apis* production of, 296–7
  reproduction by colonies and, 288–302
  selection by *Melipona* workers, 309
  stings of honeybee, 301
  suppression of ovarian development among offspring of, 20
  trigger for development of, 284

radish, wild, 143–4
rain
  Africanized *Apis mellifera* and, 99
  foraging and, 98–9
raphides, 40
rare species, population variance of, 340
rastellum, 61–2
rats, 225
receiver bees in nests, 103
  ranking of resources by, 120
recessive alleles, mortality from, 305–6
reconnaissance activity, 104
recruitment of foragers, 103–4, 105, 106, 113
rectal pads, resorption of amino acids, ions, and water by, 57
reduviids (Hemiptera), 124–6
reforestation, 371
reproduction
  frequency in highly eusocial colonies, 298
  patterns among different groups of, bees, 301
  by queens and colonies, 288–302
  of solitary bees, 284–8

reproduction (cont.)
  see also brood production; genetic variability; mating
reproductive females, loss of foraging and nesting abilities in, 19
reproductive success
  ecological constraints on female, 22
  selection for female, 305
resin, 27–8, *29, 30*
  apid corbicula for transporting, 63
  enlarged labrum in megachilids for carrying, *64*
  foraging behavior and methods of collecting, 64–5
  mandibles of stingless bees to mine, 63
  mixed with wax, 25
  for nest construction, 27, *29*, 63, 162, 200, 202
  as protection against ants, 222
  removal from legs, 63
  social evolution in tropics and, 388–9
  stealing of, 259
resources gathered and used by bees, 25
  animal, 43
  carrion, 41, 42,
  floral lipids, 29–31, 53–6, 327
  fragrances, 31–2
  honeybee communication and exploitation of, 2
  honeydew, 26
  mechanisms and methods used in collecting, 43, *44, 45, 46,* 47–61, 65, *66,* 67–8, 71–6, *77–81*, 82
  nectar, 32–7, 51, 103, 112, 117–19, 136, 147, 277
  pollen, 37–41
  ranking by receiving bees in nest, 120
  resin, 25, 27–8, *29, 30*, 63, 162, 200, 202, 222, 259, 388, 389
  sap, 26–7
  shortages of, 123
  spatial partioning of, 363
  waxes, 25, 41, 174, 200
Rinderer, T. E., 104
robber flies (asilids) as predators of bees, 127, *420*
robbers, 106, 159, 267
  of building materials, 259
  defense against, 214
  of honey, 375
  of honey-making bees, 47, 197
  of nectar, 154, 160
  of pollen, 50
  secondary, 154
  stingless bee, 197

  see also cleptobiosis; Taxonomic index (*Cleptotrigona; Lestrimelitta*)
Roubik, D. W., 85, 93, 104–5, 109, 112, 117–18, 266, 335, 340
Rozen, J. G., Jr., 260, 262–3

Sakagami, S. F., 4, 13, 16, 18, 21, 266–7, 292–5
salivary glands of larvae, 282
salts, 41–2
sap, 26–7
  extraction by stingless bees, 37
  nectar and, 35
scavengers, bees as, 41, 58, 139
Schmid-Hempel, P., 119
Schwarz, H. F., 84
scopae, 43, 50, 53
scout bees, independence of colony's, 121
scouting, 378
secretions, volatile, *77–81*
Seeley, T. D., 87, 94, 103–4, 120, 222, 377
Seeley's rule, 299
self-pollination, 148–9
setae, 50, 52–3, 56
sex alleles, 307
sex determination, 307, 308
sex differences in flight range, 82–3
sex ratio, 306–7, 310–12
sexual dimorphism, 127–8
Sharma, P. L., 90
short-tongued bees, 6, 65, *66*
silk production, 282
size of bees
  Bergmann's rule and, 97–8
  body mass to wing area ratio, 94
  daily succession at flowers and, 137
  plant breeding systems and, 148
  pollination and, 145
  significance of differences in, 102
  thermoregulation and, 95, 97–8
skatole, 220, 279
skunks, 225
smoke to control bees, 367, 369
snakes, 123
Snyder, T. P., 303
social bees
  aggression during foraging by highly, 104–5
  communication of, 2, 74, 99, 102
  life cycle of, 15–24
  tongue length of, 70
  tropical–temperate distribution of, 380
  in the tropics, 8
social behavior compatible with solitary life cycle, 21
social evolution in bees, 18

social foraging, 99–107
sociality, 16
  comparative stages of, 18
  field data to describe, 17
  hibernation and, 20
social role, physiological changes and reversals in, 19
social species
  bee families, tribes, and genera with, 22
  flexibility of, 123
soil-nesting bees, 165, 175; see also hypogeous bees
soil removal from burrow, 175
solitary bees, 6
  body color and diurnal activity of, 95–6
  diurnal staggering of foraging of, 137
  flower damage by, *157*
  generations per tropical year, 286
  life cycle of, 15–24
  life spans of, 286
  mating of, 171, 270–1
  mimetic, 129
  nest-construction behavior, 175
  nests of, 161, 171
  possible group foraging by, 107
  range of conditions under which activity occurs, 94
  rarity of aggressive foraging behavior in, 108
  reproduction of, 284–8
  reuse of central burrow of nest by, 191
  use of nests of other species, 190
sounds of bees, 101–2, 175, 207, 264
  mating and, 267–8
  returning foragers and, 106
specialized flower foraging syndrome, 110–11
sperm, 279–80
  uneven mixing of queen's stored, 301
spermatids, pupal temperature and development of polypoid, 182
spiders as bee predators, 123, 129, *130*
stability index (SI), 336, 338, 340
sterility, evolution in workers of eusocial bees, 384
stimuli
  behavior regulating, 175
  combined visual and olfactory, 75
  see also chemical signals; pheromones; tactile signals
stingless bees, 367
  aggressive species of, 104–5, 109, 113, 158–9
  begging sounds of, 102
  brood cell construction and provisioning of, 292–3
  colonies composed only of sisters, 17
  colony survival without food, 353
  crop damage by, 37
  defense of nests by, 206–7, 210, 216–17
  egg-eating and, 294
  evolution and dispersion of, 10, 11, 14, 15
  flight ranges of, 85–6
  food storage containers of, 165, *169, 374*
  foraging behavior of, 104, 109, 112, 137–8, 158–9
  foraging trails shared with related species of, 82
  fossil, 11
  highly eusocial life cycles among, 16
  honey of, 36
  hummingbirds and, 154
  larval food of, 178
  magnetic recapture experiments with, 85–6
  males from worker eggs in queenless colonies of, 294
  management of, 380
  mandibular gland odors of, 76, *80–1*, 125
  mandibular teeth reduced or absent in males of, 59
  mating aggregations of, 76
  mimicry by reduviids, 124
  nests of, *163*, 165, 169, 192–3, 195, 196, 197–9, 218, *221, 374*
  odor production and attachment by workers of, 216
  parasites of, 199
  predation by reduviids, 125
  queen–worker interaction and, 291, 293–4
  removal of floral tissue by, 58–9, 154
  resin mining by, 63
  resin sources of, 28
  response to raids by ants, 388
  robbing (cleptobiosis), 253–5, 265–7
  robbing (destructive flower visitation), 154
  role of queen among, 292
  termites and, 193, 196, 198, 218
  turnover of workers among, 354
  unusual materials collected by, 39
  as visitors of obligately outcrossing dioecious plants, 149
  wax of, 41
  workers and colony reproduction among, 292
  see also meliponine bees
stings
  of honeybee queens, 301
  of honeybees, 216
  of parasitic bees, 262
stratum fidelity foraging syndrome, 111–12
Strickler, K., 283

subcolonies formed after departure of mother queen and swarm, 301
subsocial allodapine bees, 18
sucrose solutions, imbibing of, 68–9
sugars
  concentration in hummingbird- and bee-pollinated flowers, 159
  in nectar, 159
  in pollen, 40
  sensitivity to different, 65
sun, time of day and, 74
sun compass, 73, 74
sunlight
  foraging and, 94
  nectar secretion and, 95
  visitation to flowers and temperature and, 96
  see also light intensity
superparasitism, 237, 261
swarming, 19
  of feral Africanized honeybees, 289
  food storage as prerequisite to reproductive, 386
  food stored before meliponine, 300
  frequency per reproductive cycle, 301
  subcolonies formed after mother queen departs during, 301
  survival of dispersing honeybees during, 300
sweat bees, 42
Sylvester, H. A., 302
symbionts, 192, 227, 244
syndromes
  aggressive foraging, 112–13
  foraging vs. floral, 110
  generalized flower foraging, 110–11
  specialized flower foraging, 110–11
  trap-line foraging, 112–13, 146
  see also foraging behavior

tactile signals, 174
tamanduas, 225
tectonic plates, positions 180 million years ago, *10*
temperature
  as defense against microbes, 233
  as defense against wasps, 207
  influence on development, 173, 176
  mortality from high nest, 180
  in nests, 95, 177, 180–5, 233
  tolerance to low, 183–4
temperature regulation, *see* thermoregulation
Tepedino, V. J., 237
termites
  bees' use of arboreal nests of, 190, 364
  stingless bee nests in nests of, 193, 196, 218, 220, *221*

stingless bees as mutualists with, 218
thermoregulation
  of body, 93
  body color and, 95, 96
  forced air convection and, 95
  hair and, 94, 95
  metabolism and response to weather and, 91, 93–9
  of nest, 95, 177, 180, 181–5
threat display by aggressive stingless bees, 105
time of day, resources gathered at different, 131
toads, 123
tolerance
  of prereproductive offspring by Xylocopine "queen," 20
  between resident bees and new arrivals, 17
tongues
  adaptive significance of lengths of, 69–70
  Euglossine long, 65
  measurement of, 70
  see also long-tongued bees; mouth parts; short-tongued bees
toxic substances(s)
  from plants, 39, 40
  pollen as a, 39, 151
training, 74
traits, retention or loss of, 56–60
trap-lining foraging syndrome, 112–13, 146
trap nest studies, 328, 333–5
trash from colonies, 353, 354–5
tree shrews, 222
trophallaxis, 102, 211
tropical bees
  biogeography and ages of bee groups and, 9–15
  differences from bees of other areas, 8–9
  diversity of, 4–25
  ecological and evolutionary studies of, 3
  life cycles of social and solitary, 15–24
  natural groups and distribution of, 4–9
  number of species of, 4–6
  valid names of, 4
tropical biology, limitations in study of, 1–2
tropical forest bees
  assemblages of, 9
  stratum fidelity of, 111
tropical habitats, information from remaining, 1
tumulus, 165

ultraviolet light, 40

venoms, comparative toxicity of, 207
vibrations, communicative, 101–2; *see also* buzz collecting of pollen; sounds of bees

Villa, J. D., 360
viruses, 230
Visscher, P. K., 87, 103

wasps
  as bee enemies, 233, *234–6*, 237–8
  honey of, 36
  predation of bees by, 123, 129, 207, 222
water
  bees living under, 284
  for cooling, 95, 185
  to dilute honey and larval provisions, 185
  jettisoned in flight, 91
  metabolic, 41
  physiological budget of, 91
  protection against, 178–9
wax, 41
  combs and, 200
  evolution of production of, 174
  mixed with resin, 25
wax glands, 200, 385
wax moth, 243
wind conditions, 98
wing-grooming behavior, 52
wings, coupling of fore- and hind-, 83–4
wood-nesting bees
  brood number per year, 280
  dispersal of, 15
  life span of adult, 286
  nesting behavior of *Xylocopa*, 173

  see also xylophilous bees
woodpeckers, 225
workers
  age and tasks of, 212
  development of ovaries, 20
  distinguishing from queens in semisocial
    and subsocial colonies, 21–2
  genetic relatedness to reproductive females
    and behavior of, 299
  longevity of, 288
  nest founding, 386
  obligate mutualism with queen, 20
  queen harassment of, 23
  queen influence on egg development of, 295
  regulation of *Apis* production of, 296–7
  seasonal survivorship of, *290*

xylocopine bees
  damage to plants by, 154
  larval care of, 173
  in same nest as mother, 18
xylophilous (twig- and stem-nesting) bees,
    nests and brood cells of, *166*

yeasts, 230
yellow flowers, results of insects' preference
    for (over white flowers), 143–4

Zimmerman, J. K., 345

Taxonomic index

Abromelissa, 250
Acacia, 190, 323
Acanthaceae, 36, 96, 342
Acanthopus, 250, 264
Acarapis, 247
  woodi, 226
Acari, 227, 244, 245
Acaridae, 245, 246
Achroia grisella, 243
Acromyrmex, 193
Acrosmia, 269
Acrostichus, 249
Africanized honeybees, 87, 88, 90, 104, 135, 153, 184, 218, 266, 280, 288, 289, 298, 315, 323, 333–5, 339, 340, 357–66, 371, 372, 376, 378, 379
Agalomelissa, 250
Agamermes, 249
Agapostemon, 168, 251, 284
  virescens, 69
Agavaceae, 39
Aglae, 250, 258, 264
Allodape, 250, 255
  mucronata, 260
Allodapini (allodapine bees), 22, 173, 174, 188, 210, 213, 250, 256, 257, 261, 281, 311, 315, 384,
Allodapula, 166, 250, 255
  melanopus, 170
Aloe, 175
Alpigenobombus, 156
Amaranthaceae, 318, 322, 325
Amegilla, 49, 239, 252
Amitermes hastatus, 242
Amonum aculeatum, 369
Amphylaeus, 247
Anacardiaceae, 28, 322, 323, 342
Anacardium excelsum, 34
Anagasta, 243
Andira inermis, 320
Andrena, 22, 76, 78, 149, 192, 232, 238, 241, 251, 269
  alleghaniensis, 263
  carlini, 69
  erythronii, 276
  geranii, 69
Andrenidae (andrenid bees), 6, 7, 9, 21, 22, 43, 49, 53, 65, 67, 69, 77, 154, 186, 191, 231, 232, 239, 245, 249, 263, 270, 271, 275, 284, 320, 336, 348, 350
Angiospermae (angiosperm plants), 4, 27, 37, 40
Annonaceae, 39, 146, 342, 369,
Anoetidae, 245, 246
Anoetus, 246
Antherophagus, 228
Anthidiellum, 276
Anthidiinae, 64, 260, 319
Anthidiini, 150
Anthidium, 234, 236, 239, 241, 246, 250, 273, 276
  manicatum, 270
Anthomyiidae (anthomyiid flies), 241
Anthophora, 49, 54, 82, 149, 178, 213, 227, 230, 234–6, 239, 241, 242, 249, 250, 252,
  abrupta, 277
Anthophoridae (anthophorid bees), 6–9,18, 21, 22, 26, 35, 38, 42, 46, 48, 50, 65, 67, 69, 162, 166, 167, 170, 173, 175, 178, 180, 186, 190, 191, 227, 232, 237, 240, 245, 249, 256, 261–3, 271, 275, 280, 284, 297, 306, 310, 320, 321, 336, 347, 348, 350
Anthophorinae, 128, 179, 256
Anthrax, 240, 333
Aparatrigona, 158, 182
  isopterophila, 220
Aphelandra, 36, 156
Aphomia, 243
Apiaceae (Umbelliferae), 28
Apidae (apid bees), 6–9, 21, 22, 36, 43, 44, 50, 53, 61, 63, 65, 69, 79, 168, 175, 185, 186, 200, 249, 259, 306, 320, 337, 348, 350, 366
Apinae (honeybees), 8, 12, 14, 16, 20, 25, 33, 36, 41, 42, 47, 63, 65, 91, 114, 116, 119, 149, 174, 178, 200, 225, 242, 265, 277, 288, 289, 322, 326, 347, 350, 353, 357, 382, 387, 389

504

Taxonomic index

Apiomerus, 125
  pictipes, 124
Apis, 5, 12, 14, 19, 20, 22, 26, 31, 34, 37, 41, 43, 51, 80, 84, 91, 93, 94, 95, 99, 100, 101, 106, 107, 109, 116, 121, 128, 135, 137, 153, 154, 174, 175, 180, 182, 185, 195, 202, 207, 210, 213, 215, 222, 225–30, 233, 235, 241, 243, 244, 248, 254, 255, 272, 279, 280, 284, 296, 298, 300, 306, 308, 309, 319, 366, 367
  andreniformis, 12, 98, 223, 247
  cerana, 86, 90, 98, 101, 114, 129, 130, 132, 183, 184, 207, 212, 216, 222, 223, 226, 233, 246–8, 280, 299, 304, 307, 367, 373, 375, 380, 381
  cerana indica, 86
  dorsata, 12, 14, 73, 97, 101, 105, 137, 183, 202, 204, 207, 222, 246, 247, 283, 296, 304, 355, 369, 370, 373, 387
  dorsata binghami, 13
  florea, 12, 13, 98, 100, 114, 185, 207, 217, 222, 223, 246–8, 304, 367–9, 373, 387
  laboriosa, 13, 14, 97, 183, 242, 246, 373, 381
  mellifera, 12, 15, 20, 23, 26, 27, 36, 40, 45, 60, 66, 68, 73, 75, 84, 87, 88, 94, 97, 99, 101, 103, 106, 108, 118, 119, 120, 126, 129, 135, 137, 145, 152, 175, 177, 183, 184, 194, 199, 201, 202, 207, 209, 212, 217, 222, 223, 231, 233, 241, 245–9, 254, 259, 266, 273, 275, 280, 282, 283, 291, 297, 299–305, 307, 310, 312, 315, 319, 323, 336, 339, 354, 358–61, 365, 369, 372, 373, 375, 377, 378, 380, 381, 385 (*see also* European *Apis mellifera*)
  mellifera capensis, 20, 273, 307
  mellifera carnica, 87, 194
  mellifera intermissa, 301
  mellifera lamarcki, 301
  mellifera ligustica, 134, 194, 288
  mellifera litorea, 15
  mellifera monticola, 97
  mellifera scutellata, 134, 333, 357
  mellifera sicula, 301
  mellifera syriaca, 207, 301
  mellifera unicolor, 15
  vechti, 183, 184, 367, 373, 380
Apocynaceae, 342
Apotrigona nebulata, 282
Aprostocetus, 234
Araceae, 39, 342
Araucariaceae, 28
Arctictis, 222
Arctogalidia, 222

Arecaceae (Palmae, palms), 28, 36, 40, 146, 322, 323, 325, 327
Argyamoeba distigma, 241
Araceae, 32, 36, 40
Aritranis, 234
Arrabidaea, 157
Asaropoda, 252
Asclepiadaceae, 37, 38
Ascosphaera, 230
Asilidae (asilid flies), 126, 241
Aspergillis, 230
Asteraceae (Compositae), 28, 39, 321
Astigmata, 244
Asystasia, 156
Augochlora, 125, 169, 252, 303
Augochlorella, 22, 235, 251
  edentata, 276
  striata, 69
Augochloropsis, 22, 252, 319
Austroplebeia, 303
Austrosphecodes, 250
Axestotrigona, 12, 182, 387
  oyani, 196
Axima, 234
Azteca, 210, 220, 233

Bacillus, 227, 230, 282
  larvae, 227
Bactris, 108, 323
Baltimora, 323
Bambusa stenotachya, 192
Beauveria, 230
Bellucia, 156
Betulaceae, 325
Bidens, 323
Bignoniaceae, 157, 318, 321, 342
Bisternalis, 246
Bixaceae, 342
Bombacaceae, 36, 163, 323–5
Bombinae, 239, 297, 306, 336,
Bombini (bumblebees), 4, 36, 41, 71, 91, 94, 98, 114, 118, 120, 129, 185, 187, 200, 265, 301, 306, 317, 385, 387, 389
Bombus, 19, 22, 26, 31, 48, 49, 51, 70, 71, 76, 80, 84, 91, 92, 93, 95, 98, 99, 106, 114, 116, 122, 127, 144, 156, 168, 174, 175, 206, 207, 210, 212, 226, 228, 229, 234, 235, 239, 241, 246, 254–6, 258, 259, 262, 265, 269, 272, 280, 284, 299, 303, 308, 309, 312, 319, 336, 366, 383
  americanorum, 60
  agrorum, 339
  atratus, 195, 308
  dahlbomi, 98
  hortorum, 339

Bombus (cont.)
  hypnorum, 299
  lapidarus, 339
  lucorum, 339
  mexicanus, 242
  pratorum, 339
  pullatus, 277
  ternaris, 69
  tuerrestris, 339
  transversalis, 129
  volucelloides, 128, 129
Bombyliidae (bombyliid flies), 240, 241
Bombylius, 240
Boraginaceae, 318, 321, 323, 325
Braconidae, 234
Braula, 241, 242
Braunsapis, 24, 211, 235, 238, 250, 255
Bromeliaceae, 40, 323
Bufo, 225
Burseraceae, 28
Bursera, 323
  simaruba, 323

Caesalpinia, 323
Caesalpinioideae, 28, 35, 321, 325, 327, 342
Caiusa indica, 241
Caiusa testacea, 241
Calathea, 140
  latifolia, 115
Callanthidium, 235, 269
Calliandra, 371
Callicarpa, 316
Calliopsis, 78, 177, 239, 251, 269
  andreniformis, 276
Callophoridae, 240
Calosta, 234
Camponotus, 196, 233
Candida, 229
Carabidae (carabid beetles), 213
Cardiospermum halicacabum, 323
Cassia (see also *Senna*), 37, 156, 342
Catasetum, 345
Caupolicana, 232, 250
  yarrowi, 276
Cavanillesia platanifolia, 86, 323
Cecropia, 323
Ceiba asculifolia, 323
Ceiba pentandra, 163
Cedrela, 323
Centridini (centridine bees), 8, 53, 56, 89, 128
Centris, 43, 48, 49, 53, 56, 76, 78, 115, 171, 182, 190, 227, 232, 234, 239, 240, 242, 243, 250, 269, 271–6, 284, 319, 332, 333
  adani, 83, 107

  aethyctera, 107
  analis, 53, 204, 237, 285, 333, 334, 339, 340
  flavofasciata, 264, 277
  heithausi, 152
  inermis, 333
  longimana, 42
  obsoleta, 251
  pallida, 274
Centrosmia, 42
Cephalosporium, 230
Cephalotrigona, 76, 193, 199, 294
  capitata, 60, 69, 85, 132
Ceratina, 18, 21, 22, 24, 47, 71, 79, 154, 166, 174, 188, 190, 211, 232, 234–6, 240, 243, 246, 252, 281, 283, 286, 287, 308, 319, 383, 384, 388
  dallatorreana, 307, 308
  duplex, 69
  flavipes, 18
  japonica, 18, 21
  laeta, 211
Ceratinini (little carpenter bees), 315
Chaemedorea, 323
Chaetodactylidae, 245, 246
Chaetodactylus, 246, 248
Chalcidoidea (chalcidoid wasps), 233
Chalepogenus, 53
Chalicodoma, 22, 182, 234–6, 238–40, 311
  pluto, 64, 190
Chelifer, 229
Chelostoma, 246
Chelostomoides, 115
Chelynia, 250
Cheyletidae, 245
Chilalictus erythrurum, 211
Chilicola, 317
Chloranthaceae, 325
Chrysididae, 234
Chrysis, 234
Chrysura, 234
Cicadellidae, 26
Cissites maculata, 238–40
Citrus, 156
Cladosporium, 230–1
Cledostethus, 229
Cleptotrigona, 47, 57, 214, 250, 254, 255, 265, 266
Clibadium, 323
Clidemia, 156
Clusia, 7, 342
Clusiaceae, 28, 342
Coccidae, 26
Coccinellidae, 229
Cochlospermaceae, 342

Cochlospermum, 144, 145, 156
Coelioxys, 43, 78, 167, 250, 257, 333
Coelopencyrtus, 234
Coleoptera, 228, 233, 238
Collembola, 228
Colletes, 61, 77, 82, 168, 175, 178, 230, 232, 239, 240, 241, 246, 249, 250
  cunicularis, 276
Colletidae (colletid bees), 6, 7, 9, 10, 21, 22, 44, 49, 53, 61, 77, 134, 168, 175, 179, 191, 192, 232, 239, 245, 249, 257, 262, 263, 271, 275, 281, 320, 336, 348, 350
Comolia, 156
Compositae (Asteraceae), 318, 321, 323–5
Connarus, 316
Conopidae, 241
Convolvulaceae, 28, 322, 323, 342
Convolvulus, 121
Cordia, 323
  inermis, 320
Corynebacterium, 230
Corynura, 235, 251
Costus, 145
Coumarouna, 198
Coussapoa, 323
Crawfordapis luctuosa, 70, 77, 97, 134, 135, 152, 239, 276
Crematogaster, 196, 197, 220, 233
Crithidia, 230
Croton niveaus, 323
Cryptocellus gamboa, 229
Cryptophagidae (cryptophagid beetles), 228
Crytophleba, 241
Ctenioschelini, 256
Ctenioschelus, 250, 264
Ctenoceratina, 234, 235
Ctenocolletes, 234, 239
Ctenocolletacarus, 245
Ctenoplectra, 56, 5, 250
Ctenoplectridae (ctenoplectrid bees), 6–9, 44, 48, 65, 186, 249, 350
Ctenoplectrina, 5, 250
Cucurbitaceae, 39, 56, 146
Curculionidae (curculionid beetles), 322
Cyanoderes, 245
Cyclopes, 225
Cymatodera, 239

Dactylurina, 200
  staudingerii, 182
Dalechampia, 28, 30, 145, 150, 342
Dasychernes, 229
Dasymutilla, 235
Dasypoda, 251
Dasypus, 225

Dermaptera, 228
Dermestidae (dermestid beetles), 240
Desmoscelis, 156
Diacamma, 222
Diadasia, 144, 235, 238, 239, 241, 262
Dianthidium, 234–6
Diascia, 150
Dibrachys, 235
Dictyophora, 39
Didelphis, 225
Dilleniaceae, 342
Dinogasmus, 245
Dioclea, 156
Diomorus, 235
Diptera, 233, 240
Dipterocarpaceae, 28
Dipteryx, 198
  panamensis, 111
Doeringiella, 250
Dolichoderus, 222
Dorylus, 233
Drosophila, 303, 306
Drymonia serrulata, 35
Duckeola, 198, 295
Dufourea, 77
Dynastinae (dynastine beetles), 322

Echium vulgare, 283
Echthralictus, 250
Echthrodape, 235
Eciton, 220, 233
Ectinoderus, 243
Effractapis, 250, 255
Eira barbara (formerly *Tayra barbara*), 224
Elaeis, 323
Elaphropoda, 252
Ellingsenius, 229
Emphoropsis pallida, 26
Encyrtidae, 234
Encyrtus, 234, 235
Enterolobium cyclocarpum, 193
Epeolus, 9, 82, 250, 280
Ephedra, 36
Ephutomorpha, 235
Epicharis, 35, 49, 53, 63, 98, 167, 190, 239, 251, 284, 319, 332, 333
  rustica, 177, 262
  zonata, 64, 189
Eremapis, 50
Ericaceae, 33, 38, 39
Ericrocini, 256
Ericrocis, 250
Erythrina, 156, 317, 323
Euaspis, 250, 280
Eucalyptus, 324

Eucera, 22, 76, 78, 241, 251
Eucondylops, 250, 255
Eufriesea, 43, 65, 126, 172, 182, 235, 239, 250, 258, 262, 265, 287, 336, 340, 343, 344, 346, 347, 359
　anisochlora, 337
　ornata, 346
　purpurata, 278
　surinamensis, 84, 287
Eugenia, 110
Euglossa, 22, 65, 80, 91, 115, 127, 235, 252, 258–60, 262, 269, 278, 287, 319, 336, 344
　championi, 337, 338
　cordata, 204, 288
　deceptrix, 339
　decorata, 339
　dissimula, 337
　dodsoni, 172
　gorgonensis, 339
　ignita, 204
　imperialis, 17, 70, 114, 128, 145, 172, 180, 204, 258, 279, 339, 342, 342
　sapphirina, 337, 339, 341
　townsendii, 204
　tridentata, 337, 339
　turbinifex, 172, 204
Euglossini (euglossine bees), 8, 18, 22, 36, 51, 65, 84, 90, 91, 93, 123, 126, 137, 145, 149, 179, 182, 187, 190, 202, 265, 272, 277, 301, 306, 328, 335, 337, 339, 340–7, 350, 389
Euherbstia, 5
Eulaema, 22, 43, 65, 80, 91, 114, 126, 220, 221, 235, 239, 250, 258, 262, 265, 269, 278, 279, 287, 319, 336, 346, 359
　cingulata, 38, 287, 338, 339
　leucopyga, 129, 279
　meriana, 30, 337, 339, 341
　nigrita, 121, 164, 339
Eulophidae, 234
Eupelmidae, 234
Eupetersia, 250
Euphorbiaceae, 28, 39, 145, 146, 150, 322–4, 327, 342, 369
European *Apis mellifera*, 34, 36, 98, 135, 149, 366
Euryglossinae, 44, 47
Eurytomidae, 234
Eusapyga, 235
Euvarroa, 247, 248
Evylaeus, 75, 77
Exaerete, 250, 258, 262, 264, 287, 336, 344, 346
　frontalis, 115, 337
Exomalopsini (exomalopsine bees), 8, 53, 56, 58, 128, 149, 258
Exomalopsis, 22, 49, 54, 118, 167, 169, 232, 257, 258, 322
　aureopilosa, 162
　solani, 258
Exoneura, 79, 130, 211, 251, 254

Fabaceae (*see also* Leguminosae), 28, 323, 342
Fagaceae, 325
Faramea, 156
Fervidobombus, 127
Ficus 196
　elastica, 29
Fidelidae (fideliid bees), 6, 7, 9, 10, 44, 48, 50, 52, 62, 65, 167, 173, 187, 249
Flacourtiaceae, 342
Florilegus, 319
Formicinae (formicine ants), 24
Frieseomelitta, 169, 193, 198, 295, 255, 266, 292, 293, 295
　savannensis, 165
Fuchsia, 156
Fulgoridae, 26
Fusarium, 230

Gaesischia, 49, 115
　exul, 107
Galbulidae, 124
Galleria, 243
Gasteruptiidae, 234
Gasteruption, 234
Genipa, 323
Gentianaceae, 39, 342
Geotrigona, 81, 178, 293
Gesneriaceae, 31, 35, 342
Gnathoxylocopa, 175
Gramineae (Poaceae, grasses), 149, 323
Grotea, 234
Guttiferae, 28

Habitrys, 235
Habralictus, 251
Habropoda, 48, 49, 251, 252
Halictidae (halictid bees, sweat bees), 6–9, 19–22, 42, 44, 49, 67, 69, 77, 94, 128, 134, 135, 149, 154, 162, 167, 168, 171, 173, 175, 186, 188, 190–2, 210, 211, 227, 231, 239, 244, 245, 249, 256, 263, 270, 271, 284, 285, 301, 306, 320, 321, 336, 348, 350
Halictinae, 9, 65, 297, 301
Halictini, 174, 251
Halictus, 22, 75, 77, 167–9, 235, 241, 249, 251

hesperus, 134, 135, 153, 164
ligatus, 17, 69, 285, 382
lutescens, 162
Hamamelidaceae, 28
Hamelia, 156
Hedychium, 156
Helarctos malayanus, 222, 224
Heliantheae, 323
Heliconia, 156
Heliocarpus, 323
Heliotropium, 323
Hemiptera, 124, 243
Heriades, 173, 188, 190, 234, 236, 239, 250, 252
Hesperonomada, 251
Heterostylum, 240
Heterotrigona itama, 105, 133, 137, 156, 158
Hippocrateaceae, 39
Histiostoma, 246
Histiostomatidae, 245, 246
Holcopasites, 79
Homalictus, 250
Homo, 224
Homoptera, 25–7, 33, 352
Homotrigona fimbriata, 133, 135
Hopliphora, 251
Hoplitis, 166, 235, 239, 246, 269, 275, 283
Hoplocryptus, 234
Hoplognathoca, 235
Hoplomutilla, 235
Hoplostelis, 260
Hornia, 238, 239
Horstia, 246
Huartea, 235
Huntaphelenchoides, 249
Hybanthus prunifolius, 86, 147, 148
Hylaeinae, 44, 47
Hylaeus, 22, 77, 188, 232, 234, 241
Hylemya, 241
Hymenaea, 28
Hymenoptera, 233, 302, 304, 349, 386
Hyperechia, 126, 127, 241
Hypochrotaenia, 251
Hypotrigona, 15, 81, 250, 266, 354
Hyptiogaster, 234

Ichneumonidae (ichneumonid wasps), 233, 234
Ictericophyto, 241
Immanthidium repetitum, 60
Imparipes, 246
Impatiens, 40
Indicator archipelagicus, 222
Indicator indicator, 224, 225
Inquilina, 251, 254, 255
Ipomoea, 35, 114, 137

trifida, 323
Iriartea, 108
Iridaceae, 30
Iridomyrmex, 233
Isertia, 156
Ixora, 34

Jacaranda, 156
Justicia, 96, 156

Kaltenbachia, 234
Koptortosoma, 245
Krameriaceae, 31

Labiateae, 318, 321, 325
Laelapidae (laelapid mites), 244–6, 318, 321, 325
Lamiaceae (*see* Labiatae)
Langur, 225
Lantana, 156
Lanthanomelissa, 54
Lasioglossum, 22, 75, 77, 167, 195, 235, 236, 239, 241, 246, 251, 276, 303
  imitatum, 30
  malachuurum, 23
  marginatum, 382
  umbripenne, 86
  zephyrum, 23, 209, 268
Lecythidaceae, 4, 111
Leguminosae (*see also* Fabaceae) 318, 327
Leiodidae (leiodid beetles), 228
Lepidotrigona, 293
  terminata, 181
Lestrimelitta, 47, 57, 76, 81, 196, 242, 251, 254, 255, 261, 265, 267, 293, 306, 354, 355
  limao, 214, 216, 265, 266, 294, 355
Leucaena, 323, 371
Leucophora, 241
Leucospididae, 234
Leucospis, 234, 237, 333
Liliaceae, 28
Lithurge (*Lithurgus*), 166, 180, 246, 285
Locustacarus, 246
Lophotrigona canifrons, 105, 193
Lopimia dasypetala (formerly *Pavonia dasypetala*), 155, 156
Luehea seemanii, 352
Luffa, 108
Lysimachia, 31, 55, 150

Mabea occidentalis, 352
Macaca mulatta, 222
Macrochelidae, 246
Macrocheles, 246

Macrogalea, 251, 255
Macropis, 55, 150
Macrosiagnon, 238
Malpighiaceae, 31, 53, 146, 150, 322, 342
Malvaceae, 35, 318, 322, 342
Manihot esculenta, 39
Manis, 225
Mantispidae (mantispids), 243
Manuelia, 21, 22
Marantaceae, 140, 342
Megabraula, 242
Megachile, 43, 59, 78, 94, 149, 151, 154, 166, 167, 230, 234, 236, 239–41, 243, 246, 250, 252, 282, 303, 311, 319
  brevis, 121
  nana, 237, 241, 311
  rotundata, 176, 281
  zaptlana, 310
Megachilidae (megachilid bees, leafcutter bees), 6–9, 21, 22, 43, 44, 48, 63, 65, 78, 128, 166, 167, 173, 175, 179, 182, 185, 186, 188, 190, 192, 202, 203, 239, 240, 245, 249, 256, 257, 264, 269, 271, 275, 284, 306, 310, 320, 321, 336, 348, 350
Megachilinae, 150, 250
Megalopta, 96, 188, 249, 331, 333, 339
  ecuadoria, 329, 330, 339
  genalis, 329, 330, 332, 339
Meganomia, 168, 276
  binghami, 276
Melampodium, 323
Melastoma, 156
Melastomataceae, 30, 38, 152, 323
Melecta, 82, 251
Melectini, 256, 264
Meleotyphlus, 239
Meliaceae, 323
Meliplebeia, 81, 158, 387
Melipona, 28, 43, 47–9, 54, 58, 68, 80, 84, 91, 93, 103, 105, 106, 109, 120, 122, 125, 128, 132, 137, 147, 148, 158, 168, 175, 180, 81, 183, 193, 194, 199, 204, 210, 211, 215, 220, 229, 233, 246, 254, 266, 292, 296, 306, 308, 309, 319, 323, 355, 359, 363, 365, 366, 372
  beecheii, 5, 75, 198, 282, 308, 375
  compressipes, 75, 199, 211
  compressipes fasciculata, 105, 309, 375
  compressipes manaosensis, 105
  compressipes triplaridis, 68, 103, 105, 132, 137, 220
  fasciata, 68, 69, 85, 108, 130, 132, 137, 178, 211, 214, 216, 242, 266
  favosa, 211–13, 290, 294, 297, 362
  fulva, 289, 362
  fuliginosa, 28, 29, 68, 69, 86, 103, 119, 132, 137, 177, 181, 202, 208, 211, 266,
  marginata, 60, 68, 132, 182, 295, 305, 324
  quadrifasciata, 105, 181, 227, 375
  rufiventris, 294
  scutellaris, 375
  seminigra, 267, 375
  seminigra merrillae, 105
Meliponinae (meliponine bees, stingless bees), 8, 9, 10, 11, 14, 16, 20, 22, 25, 28, 36, 37, 41, 42, 47, 48, 62, 63, 74, 85, 86, 91, 102,104, 105, 109, 111, 132, 133, 137, 49, 169, 182, 200, 202, 204, 205, 210, 229, 233, 251, 265, 288, 291, 299, 300, 318, 320, 347, 348, 350, 357, 382, 385, 387
Meliponula, 76, 158
Melissodes, 235, 239, 319
  apicata, 69
  floris, 242
Melissoptila, 178
Melitoma, 189, 230, 232, 235, 236, 239, 240, 243, 246
Melitta, 77
Melittidae (melittid bees), 6, 9, 44, 49, 55, 63, 77, 128, 150, 168, 175, 179, 186, 235, 249, 263, 336, 350
Melittiphis, 246
Melittobia, 234, 311
Melitturga clavicornis, 276
Mellivora capensis, 276
Meloe, 238
Meloidae (meloid beetles), 238
Melpighamoeba, 231
Melursus, 224
Membracidae, 26
Merisus, 235
Meropidae, 124
Mesonychium, 251
Mesoplia, 232, 251, 264, 265
Mesostigmata, 244, 245
Mesotrichia, 245
Mesua, 316
Metopia, 241
Miconia, 156, 323, 324
Micrapion, 235
Microcerotermes, 243
Microdontomerus, 235
Microsphecodes, 251
Milletia, 316
Miltogrammidium, 241
Mimosa, 47, 125, 130, 323, 362
  pigra, 126
  pudica, 113, 122, 126
Mimosoideae, 39, 321, 323, 327

Monodontomerus, 237
  obscurus, 238
Monomorium, 222, 233
Moraceae, 28, 323, 324
Morgania, 251
Mouriri, 54, 156
Musa, 156
Musaceae, 342
Mutillidae (mutillid wasps), 233, 235, 237
Myriaspora, 156
Myrmecinae (myrmecine ants), 27
Myrmecophaga, 225
Myrmilla, 235
Myrmilloides, 235
Myrmosula, 235
Myrsinaceae, 28
Myrtaceae, 324, 325, 327

Nannotrigona, 76, 193, 254, 266, 267
  mellaria, 220
  testaceicornis, 281
Nasutapis, 251, 255
Nasutitermes, 364
Nemognatha, 238
Nemognathinae, 238
Neoeupetersia, 251
Neohypoaspis ampliseta, 248
Neomutilla, 235
Neoprosopis, 57, 251
Neotydeolus therapeutikos, 228, 244
Nepsera, 156
Neurospora, 39
Nogueirapis, 11, 387
  mirandula, 108
Nomada, 79, 251, 263, 264, 269
  japonica, 307, 308
  pseudops, 263
Nomadinae, 8, 256
Nomadini, 258
Nomadopsis, 79, 236
  anthidius, 276
  puellae, 276
Nomia, 22, 34, 77, 167, 230, 232, 235, 239, 241, 246, 251, 303
  melanderi, 276
Nosema, 230, 231
  apis, 233
Notocyrtus vesiculosus, 124
Notylia, 38
Nyctomelitta, 96

Ochnaceae, 39
Ochroma pyramidale, 34, 36, 156
Odontoponera, 222
Odontostelis, 260

Odyneropsis, 251
Oecophylla, 222
Oenocarpus, 323
Oenothera, 149
Olyra, 149
Onagraceae, 39, 149, 322, 325
Oncidium, 146
Opiomeloe, 239
Orchidaceae (orchids), 25, 31, 32, 37, 38, 335, 342
Orophaea katschalica, 369
Osiris, 129
Osmia, 166, 202, 234–6, 246, 307, 310
Oxaea, 57, 77, 154, 232, 252
  flavescens, 57, 77, 154, 232, 252
Oxaeidae (oxaeid bees), 6, 9, 44, 49, 52, 157, 179, 186, 232, 263, 271, 348
Oxytrigona, 27, 76, 80, 207, 208, 265

Paecilomyces, 230
Palarus, 233
Palmae (*see* Arecaceae)
Panicum, 323
Panurginae, 44
Panurginus, 78, 241
Papilionoideae, 28, 321, 325, 327
Pappognatha, 235
Paraanthidium, 235
Parabystus, 229
Paracrocisa, 251
Parafidelia, 167
  pallidula, 191
Paranthidiellum, 234
Parapartamona, 184
  zonata, 156, 158
Parasitidae, 246
Parasitus, 246
Paratetrapedia, 53, 56, 128, 129, 154, 234, 319
  calcarata, 128
Parathrincostoma, 251
Paratrigona, 82, 158, 183, 266, 297
  impunctata, 156
  subnuda, 156
Parepeolus, 251
Parevaspis, 234
Pariana, 149
Partamona, 37, 76, 103, 109, 182, 184, 195, 196, 205, 218, 221, 292, 293
  pearsoni, 75
  peckolti, 108
  pseudomusarum, 221
  vicina, 219, 221
Passiflora, 156
Passifloraceae, 321, 342

Pavonia dasypetala (see *Lopimia dasypetala*)
Peponapis, 42, 235
  pruinosa, 276
Perdita, 67, 154
  texana, 276
Pernis apivoris, 222
Pheidole, 233
Pheidologeton, 222
Philanthus, 233
Phoridae (phorid flies), 210, 242
Phytopsis, 235
Phymatidae, 123
Physocephala, 241
  testacea, 241
Pinaceae, 28
Pithecellobium, 323
Pithitis, 21, 22, 79, 232, 234
Platynopoda, 245
Plebeia, 11, 50, 254, 259, 266, 267, 296, 324
  droryana, 182
  jatiformis, 208
  remota, 153
Plega, 243
Plistophora, 230
Plodia, 243
Podapolipidae, 246
Polistomorpha, 235
Polochrum, 235
Polybiinae (polybiine wasps), 129, 364
Polygalaceae, 342
Polygonaceae, 33
Polyrachis, 222
Ponerinae (ponerine ants), 130
Pourouma, 323
Pouteria stipitata, 323
Primulaceae, 31, 55, 150
Priodontes, 225
Proplebeia, 11
Prosopis, 77, 371
Prostigmata, 244
Protium tenuifolium, 323
Protoepeolus, 262
  singularis, 260
Protoxaea, 77, 232, 252, 269
  gloriosa, 276
Proxylocopa, 179
Psaenythia, 238
Pseudagapostemon, 22, 235
Pseudaugochloropsis, 22, 66, 154
Pseudobombax, 323
Pseudococcidae, 26
Pseudodichroa, 167, 251, 262
Pseudohypocera, 242
Pseudomethoca, 235

Pseudomonas, 230
Psicidia, 323
Psithyrus, 80, 84, 226, 234, 249, 254–9, 262, 265, 312
Psychotria, 129, 156
Psyllidae, 26
Pteromalidae, 235
Ptilocleptis, 251
Ptiloglossa, 70, 77, 97, 115, 152, 246, 249, 251, 252, 269, 332, 333
  guinnae, 128, 229, 246, 276
  jonesi, 276
Ptilomelissa, 276
Ptilothrix, 235, 239
Ptilotopus, 43, 53, 182, 250
  americanus, 129
Ptilotrigona, 158, 182, 205, 218, 221, 230, 292
  lurida occidentalis, 373
Pyemotes, 246
Pyemotidae, 246
Pygmephoridae, 245, 246
Pygmephorus, 246
Pyralidae (pyralid moths), 243
Pyrobombus, 127
Pyrota, 238

Quassia, 156

Raphanus raphanistrum, 143
Rathymini, 256
Rathymus, 167, 251, 262
Rediviva, 150
  emdeorum, 55
Reduviidae (assassin bugs), 123
Rhinetula, 332, 333
Rhipiphoridae (rhiphiphorid beetles), 238
Rhipiphorus, 238
Rhizophus, 230
Rhodanthidium, 246
Rhododendron, 33
Rhyncanthera, 156
Ricinuleidae (ricinuleids), 229
Rinorea, 47
Rondaniooestrus apivoris, 241
Rubiaceae, 28, 34, 39, 129, 318, 321–3, 342
Ruizantheda, 22
Rutaceae, 28, 323

Sabicea, 342
Saccaromyces, 229
Sandemania, 156
Sapindaceae, 323, 325
Sapotaceae, 323, 325

Sapyga, 235
Sapygidae, 235
Sarcophagidae, 241
Sayapis, 173
Scaptotrigona, 29, 50, 76, 81, 82, 109, 128, 136, 185, 193, 196, 197, 244, 254, 266, 267, 270, 292–5, 306, 355
  barrocoloradensis, 132
  depilis, 291
  luteipennis, 69, 86
  mexicana, 108
  postica, 40, 213, 242
Scaura, 49, 182, 206, 266
  latitarsis, 220, 221
Schwarziana, 178, 294, 296
Scleroderma, 234
Scotocryptodes, 229
Scotocryptus, 229
Scrapter, 167, 249, 251, 262
Scrophulariaceae, 31, 55, 150
Scutacaridae, 245, 246
Scutacarus, 246
Seladonia, 134, 382
Selenarctos, 224
Senna, 37, 342
Sennertia, 246
Senotainia, 241
Serapista denticulata, 203
Sicilipes, 246
Sida, 238, 322
Sindora, 316
Siteroptes, 246
Solanaceae, 33, 152, 322, 324, 342
Solanum, 155, 156
Solenopsis invicta, 357
Sparnopolus, 241
Sphaeropthalma, 235
Sphecidae (sphecid wasps), 51
Sphecodes, 251, 261–4
Sphecodogastra, 149
Sphingidae (sphingid moths), 352
Spiroplasma, 230
Spondias, 322, 323, 342
Staphylinidae, 228
Stelis, 166, 235, 252, 260–3
Stelopolybia, 364
Stenomutilla, 235
Stenotritidae (stenotritid bees), 6, 7, 9, 44, 49, 179, 186, 227
Stenotritus, 235, 239
stingless bees (*see* Meliponinae)
Stratiomyidae, 241
Streptococcus pluton, 230, 231
Sterculiaceae, 318, 321, 342
Stromanthe, 156

Stylopidae (stylopid beetles), 238
Stylops, 238
Styracaceae, 28
Suidasiidae, 246
Svastra, 46, 235
sweat bees (*see* Halictidae)
Synaristus, 229
Synhornia, 238
Syntretus, 234

Tabebuia, 156, 196
Tachigalia versicolor, 332
Tachinidae, 241
Tamandua, 22, 225
Tanaecium nocturnum, 369
Tapinorhina, 54
Tapinotaspis, 53, 54, 251
  chalybaea, 54
Tarsipes, 225
Tarsonemidae, 247
Tarsonemus, 247
Tayra barbara (*see Eira barbara*)
Temnosoma, 252
Tenebrionidae (tenebrionid beetles), 213
Tetracera, 316
Tetragona, 11, 81, 103, 105, 308
  clavipes, 153, 374
  dorsalis, 54, 124, 132
  perangulata, 30
Tetragonisca, 193, 205, 254, 255, 266
  angustula (formerly *T. jaty*), 96, 135, 214, 215, 266
  angustula fiebrigi, 193
Tetragonula, 293, 295, 300, 303
  collina, 133, 196, 243, 387
  drescheri, 105
  laeviceps, 181, 205
  minangkabau, 86, 133, 137
Tetralonia, 149
  minuta, 276
Tetralonoidella, 252
Tetraonyx, 238, 239
Tetrapedia, 54, 56, 128, 243, 248, 319
  maura, 54
Tetrapediini, 54, 128
Thalestria, 251
Thrinchostoma, 67, 251
Thygater, 49, 114, 178, 235
Thyreus, 252, 264, 280
Tibouchina, 156
Tiliaceae, 33, 323, 342
Tillandsia, 323
Timulla, 235
Tiphiidae, 235
Tococa, 156

Tortonia, 246
Torulopsis, 230
Trachypus, 233
Trema micrantha, 323
Tremarctos, 224
Trichodes, 240
Trichotrigona, 198, 255
Tricrania, 238
Triepeolus, 46, 252, 260, 263, 269, 276
Trigona, 26, 27, 35, 37, 41, 43, 51, 59, 76, 81, 103, 107, 110, 128, 129, 154, 159, 182, 193, 218, 227, 243, 292, 355
  amalthea silvestriana (see also T. silvestriana), 108
  amazonensis, 109, 156, 162, 163
  cilipes, 54, 196, 220
  corvina, 60, 108, 156, 352
  dallatorreana, 156, 202
  eocenica, 10
  ferricauda, 96, 155, 156
  fulviventris, 39, 96, 98, 104, 108, 124, 125, 128, 131, 153, 156, 194, 225
  fulviventris guianae, 98
  fuscipennis, 39, 156
  hyalinata, 156
  hypogea, 43, 57, 58, 229
  nigerrima, 39, 105, 156, 202
  pallens, 54, 58, 59, 62, 69, 156, 220, 352
  prisca, 10
  sesquipedalis, 156
  silvestriana (see also T. amalthea silvestriana), 105, 154, 156, 281
  spinipes, 54, 109, 153, 154, 156, 182
  williana, 156
Trigonella, 293
Trigonella moorei, 133, 196, 197, 218
Trigonini (trigonine bees, formerly "trigonas"), 34, 36, 48, 68, 105, 107, 111, 125, 128, 137, 147, 150, 158, 204, 206, 215, 229, 233, 243, 319, 323, 347, 373
Trigonisca, 31, 41, 50, 81, 266
  atomaria, 242
Triplaris, 33
Trochometridium, 246
Trogoderma, 240
Tropilaelaps, 246, 248
Tupaia glis, 222
Turneraceae, 342
Tyrranidae, 124

Ulmaceae, 323, 325
Umbelliferae (Apiaceae), 26

Vaccinium, 48
Varroidae (varroid mites), 244, 247

Varroa jacobsoni, 226, 247, 248
Verbenaceae, 322, 342
Vespa tropica, 129, 222
Vespidae (vespid wasps), 123, 129, 207, 233
Vidia, 246
Vigna, 156
Viguiera, 323
Villa, 240, 241
Violaceae, 342
Virgilia divaricata, 286
Vismia, 28, 29, 202
Vitula, 243

Winteraceae, 39
Winterschmidtiidae, 247

Xanthoplyllum, 316
Xenoglossa, 235
Xeromelecta, 252
Xerospermum, 33, 34, 147
Xylocopa, 15, 20, 22, 33, 43, 47, 49, 67, 76, 79, 84, 90, 91, 93, 96, 99, 114, 116, 122, 126, 127, 128, 151, 157, 160, 165, 166, 172, 173, 175, 177, 179, 189, 203, 204, 207, 209, 211, 213, 233, 236, 239, 240, 241, 267-70, 277, 280-2, 284, 286, 287, 316, 319, 384
  aureipennis, 38
  capitata, 96, 281, 286
  combusta, 285
  fimbriata, 269
  frontalis, 239, 240, 287
  hirsutissima, 272, 273
  latipes, 285
  macrops, 241
  nigrita, 269
  nogueirai, 287
  pubescens, 281
  somalica, 21
  sulcatipes, 180, 243, 281
  tabaniformis, 96, 274
  virginica, 84
Xylocopinae (xylocopine bees, carpenter bees), 17, 18, 20, 179, 206, 244-6, 285, 301, 315, 321, 347, 383
Xylocopini (large carpenter bees), 8, 315

Zanthoxyllum, 323
Zea mays, 323
Zingiberaceae, 145, 342, 369
Zodion, 241
Zonitis, 238
Zygophyllaceae, 28